Water Activity in Foods

Fundamentals and Applications

Water Activity in Foods

Fundamentals and Applications

Gustavo V. Barbosa-Cánovas, Anthony J. Fontana, Jr.,
Shelly J. Schmidt, Theodore P. Labuza *EDITORS*

The *IFT Press* series reflects the mission of the Institute of Food Technologists—advancing the science and technology of food through the exchange of knowledge. Developed in partnership with Blackwell Publishing, *IFT Press* books serve as leading edge handbooks for industrial application and reference and as essential texts for academic programs. Crafted through rigorous peer review and meticulous research, *IFT Press* publications represent the latest, most significant resources available to food scientists and related agriculture professionals worldwide.

Gustavo V. Barbosa-Cánovas, Professor of Food Engineering, Center for Nonthermal Processing of Foods, Washington State University.

Anthony J. Fontana, Jr., Senior Research Scientist, Decagon Devices, Inc.

Shelly J. Schmidt, Professor of Food Chemistry, Department of Food Science and Human Nutrition, University of Illinois at Urbana-Champaign.

Theodore P. Labuza, Morse Alumni Distinguished Professor of Food Science and Engineering; Department of Food Science and Nutrition, University of Minnesota.

Blackwell Publishing Professional
2121 State Avenue, Ames, Iowa 50014, USA

Orders: 1-800-862-6657
Office: 1-515-292-0140
Fax: 1-515-292-3348
Web site: www.blackwellprofessional.com

Blackwell Publishing Ltd
9600 Garsington Road, Oxford OX4 2DQ, UK
Tel.: +44 (0)1865 776868

Blackwell Publishing Asia
550 Swanston Street, Carlton, Victoria 3053, Australia
Tel.: +61 (0)3 8359 1011

First edition, 2007

Library of Congress Cataloging-in-Publication Data
Water activity in foods : fundamentals and applications / edited by Gustavo Barbosa-Cánovas ... [et al.].
p. cm. — (IFT Oress series)
Includes bibliographical references and index.
ISBN-13: 978-0-8138-2408-6 (alk. paper)
ISBN-10: 0-8138-2408-7 (alk. paper)
1. Food—Water activity. 2. Food—Moisture. I. Barbosa-Cánovas, Gustavo V.

TX553.W3W358 2007
664—dc22

2007006958

The last digit is the print number: 9 8 7 6 5 4 3 2 1

Titles in the *IFT Press* series

- *Accelerating New Food Product Design and Development* (Jacqueline H.P. Beckley, Elizabeth J. Topp, M. Michele Foley, J.C. Huang and Witoon Prinyawiwatkul)
- *Biofilms in the Food Environment* (Hans P. Blaschek, Hua Wang, and Meredith E. Agle)
- *Calorimetry and Food Process Design* (Gönül Kaletunç)
- *Food Ingredients for the Global Market* (Yao-Wen Huang and Claire L. Kruger)
- *Food Irradiation Research and Technology* (Christopher H. Sommers and Xuetong Fan)
- *Food Risk and Crisis Communication* (Anthony O. Flood and Christine M. Bruhn)
- *Foodborne Pathogens in the Food Processing Environment: Sources, Detection and Control* (Sadhana Ravishankar and Vijay K. Juneja)
- *High Pressure Processing of Foods* (Christopher J. Doona, C. Patrick Dunne, and Florence E. Feeherry)
- *Hydrocolloids in Food Processing* (Thomas R. Laaman)
- *Microbiology and Technology of Fermented Foods* (Robert W. Hutkins)
- *Multivariate and Probabilistic Analyses of Sensory Science Problems* (Jean-Francois Meullenet, Rui Xiong, and Chris Findlay
- *Nondestructive Testing of Food Quality* (Joseph Irudayaraj and Christoph Reh)
- *Nonthermal Processing Technologies for Food* (Howard Q. Zhang, Gustavo V. Barbosa-Canovas, V.M. Balasubramaniam, Editors; C. Patrick Dunne, Daniel F. Farkas, James T.C. Yuan, Associate Editors)
- *Nutraceuticals, Glycemic Health and Diabetes* (Vijai K. Pasupuleti and James W. Anderson)
- *Packaging for Nonthermal Processing of Food* (J. H. Han)
- *Preharvest and Postharvest Food Safety: Contemporary Issues and Future Directions* (Ross C. Beier, Suresh D. Pillai, and Timothy D. Phillips, Editors; Richard L. Ziprin, Associate Editor)
- *Processing and Nutrition of Fats and Oils* (Ernesto M. Hernandez, Monjur Hossen, and Afaf Kamal-Eldin)
- *Regulation of Functional Foods and Nutraceuticals: A Global Perspective* (Clare M. Hasler)
- *Sensory and Consumer Research in Food Product Design and Development* (Howard R. Moskowitz, Jacqueline H. Beckley, and Anna V.A. Resurreccion)
- *Thermal Processing of Foods: Control and Automation* (K.P. Sandeep)
- *Water Activity in Foods: Fundamentals and Applications* (Gustavo V. Barbosa-Canovas, Anthony J. Fontana Jr., Shelly J. Schmidt, and Theodore P. Labuza)
- *Whey Processing, Functionality and Health Benefits* (Charles I. Onwulata and Peter J. Huth)

Contents

Dedication

Figure D1. Marcus Karel.

This book is dedicated to Marcus Karel, Emeritus Professor of Food and Chemical Engineering (Massachusetts Institute of Technology) and Emeritus State of New Jersey Professor (Rutgers University).

—Prof. Theodore Labuza (colleague, adopted science son, and confidant)

It was 1962. I had just finished my bachelor's of science degree at Massachusetts Institute of Technology, and I had an offer to go to the University of Minnesota to work on a doctorate in nutrition with Ancel Keyes. He had just received one of the first National Institutes of Health training grants in nutrition and needed a food science student because they were working on new military rations.

I had become interested in nutrition, as my undergraduate advisor was Dr. Sandy Miller (later to become the head of Food and Drug Administration's Bureau of Foods), and I did my bachelor's thesis with him. The prior year I took a course taught in part by a recently appointed assistant professor, Marcus Karel (Fig. D1). He fascinated me with his understanding of physical chemistry and kinetics and their applications to food. His work on the stability of military foods and space foods was the "chocolate" to entice me, and I took the bite to work with him, becoming his first doctoral student and getting my degree in 3 years. At the end, though, I had a draft notice in hand to go to Vietnam.

Dr. Karel helped me secure a position as an assistant professor to continue my work on reaction kinetics and stability of space foods. That allowed me to begin to work with the "giants" in the field, as that year I traveled with Marc to Aberdeen, Scotland, for an international meeting on water in foods. There, I met his colleagues from Europe, including Denise Simatos, Lou Rockland, John Hawthorne, Ron Duckworth, Felix Franks, and Grahame Gould, and was securely enthralled in being a part of this group. Several years later, in 1974, that group along with others gathered in Glasgow, Scotland, for the first ISOPOW (Fig. D2) meeting (International Symposium on the Properties of Water), setting the stage for the introduction of the concepts of water activity and eventually of glass transition and for the continuation of many meetings over the decades. We plan to meet in September 2007 in Thailand for the 10th ISOPOW. It was Marc's foresight to ensure that the meetings included invitees from other fields so that there would

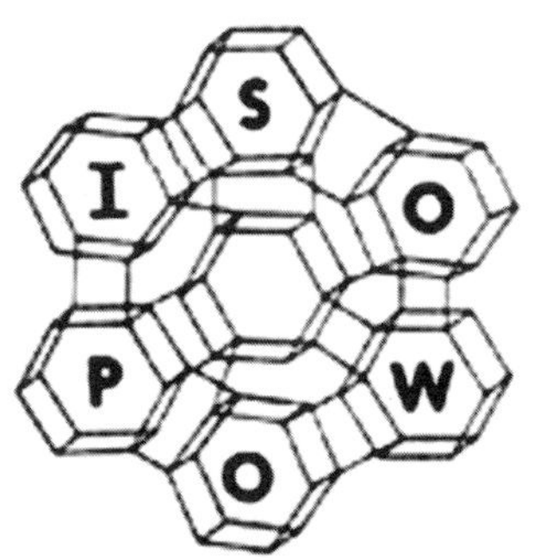

Figure D2. International Symposium on the Properties of Water (ISOPOW) logo.

Figure D3. Attendees at the third ISOPOW meeting held in Beaune, France, in 1983.

be "cross-fertilization" in both directions, a practice we continue to this day. It was at these meetings where many met Marc for the first time and added to his admirers. Unfortunately, he missed the last meeting, where we made a presentation on his impact in food science, but this book is that substitute. Figure D3 illustrates the attendance to the third ISOPOW meeting in 1983 by some of the most influential people working with water activity

Marcus Karel, to whom we dedicate this book, was born in Lwów, Poland, in 1928, before World War II. He worked in the underground in 1945–1946, helping the Jews escape across the Iron Curtain. He then studied agriculture at the Munich Technical University. Eventually, he made his way to Boston in 1951 and got a job in 1952 in the food packaging lab at the Massachusetts Institute of Technology, working for Bernie Proctor (a former IFT president) and Sam Goldblith. He certainly must have impressed them with his skills, and while working there, he finished his bachelor's of science degree at Boston University in 1955 and married Cal, his wife for over 50 years. They have four children, two of whom graduated from the Massachusetts Institute of Technology. Marc did his doctoral studies with Prof. J. T. R. Nickerson, a food microbiologist, and Marc's doctoral thesis was the first on the application of kinetics to food reactions as a function of moisture. His minor was in chemical engineering with Alan Michaels. In 1961, my senior year, I entered the picture when I took a course he taught, as mentioned earlier. He had an amazing mind, spending several days a week in the library (before the Internet), which gave him a fantastic knowledge base. One could see it when he questioned those who gave seminars, starting with a polite compliment followed by, "But let me . . .", when he would go to the board and show a better answer through a derivation. In 1969, he, Steve Tannenbaum (a fellow Massachusetts Institute of Technology faculty member in food science), and I came up with the stability map for reactions as a function of water activity, a map that still is used today (Fig. D4). It was obvious that Marc's goals included not only research but also the use and dissemination of that research through the education of students.

A challenge to the concept of water activity began in 1988 with the introduction of the concept of glass transition by two scientists at General Foods, Louise Slade and Harry Levine. The following six years were the most exciting in this field of water and stability, with much controversy. One of the original ISOPOW people, Felix Franks, wrote an editorial in *Cryoletters* saying that the concept of water activity is now dead. Again, Dr. Karel,

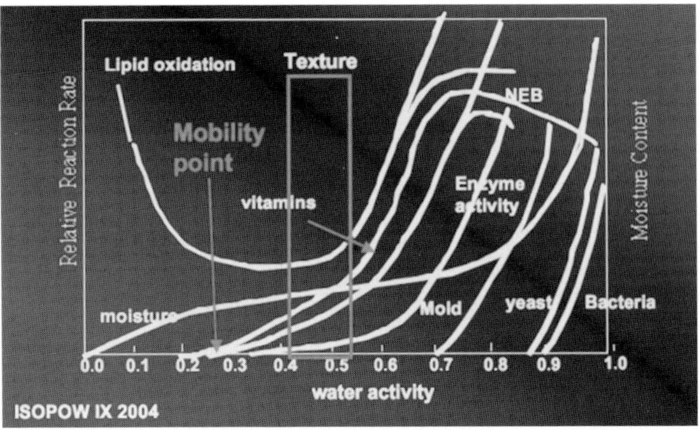

Figure D4. Stability map for reactions as a function of water activity.

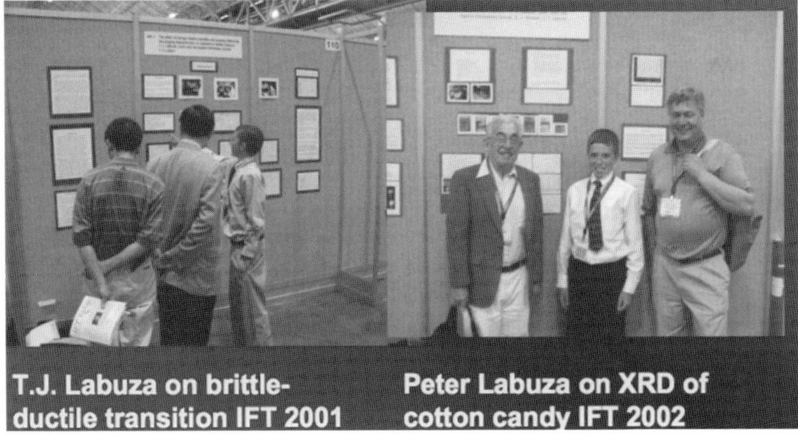

Figure D5. Poster presentations at IFT's 2001 and 2002 meetings.

along with Yrjo Roos, then at the University of Helsinki and a visiting professor at Rutgers University, came up with the idea of state diagrams, a way to bring the two concepts together, especially to explain physical states. During that time, my three children presented posters at the national Institute of Food Technologists (IFT) meeting as prime authors based on their grade school science projects (Fig. D5). They knew about Marc from me and were very worried that he would come to the meeting and ask questions. He did ask questions, and he called them his "science grandchildren," again showing his love for science and his compassion for instilling education.

My most interesting days with Marc came much later, when he and I became part of a technical advisory committee to the Pillsbury Company, headed by Dr. James Bhenke, the

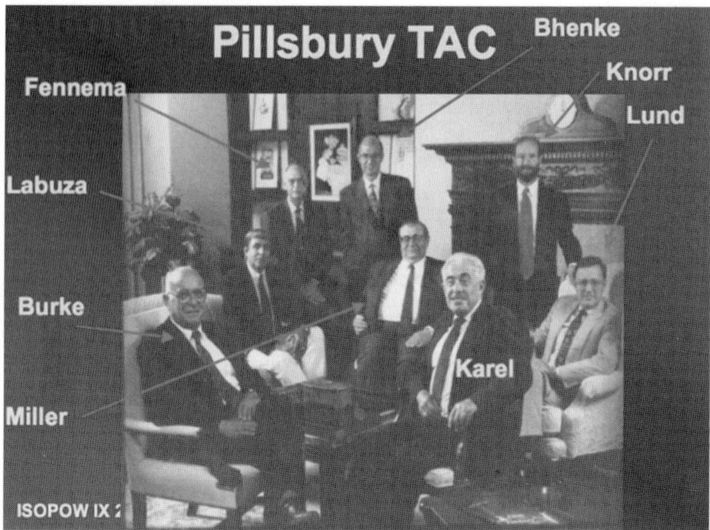

Figure D6. Technical Advisory Committee (TAC) to the Pillsbury Co.

vice president of research (Fig. D6). Marcus Karel, Owen Fennema, and Daryl Lund collaborated on one of the best food processing textbooks ever published, *Physical Principles of Food Preservation*.

This book was the outcome of an Institute of Food Technologists short course at the 2001 annual meeting. The original intent was yet another symposium book, but I suggested strongly that it should be designed as a true undergraduate textbook, as there was nothing available except books of papers from the ISOPOW meetings and the moisture sorption isotherm manual I wrote with my former student, Prof. Leonard Bell (professor of food chemistry at Auburn University and a collaborator in this book). My co-editors, Gustavo Barbosa-Cánovas, Shelly Schmidt, and Anthony Fontana, agreed heartily. We were all taught and influenced by Marcus in many different ways, and this book is dedicated to his ideals of education and the stimulation of new science. We thank you, Marc, for helping us to advance the field.

Preface

The coeditors of this book have spent a total of more than 100 years studying and researching the properties and stability of foods as related to water, and in particular, as related to two key property values: water activity (a_w) and glass transition temperature. Since 1965, there have been at least 50 symposium books devoted to these properties. One is a working manual on moisture sorption isotherms that one of the coeditors coauthored, and it describes many instruments to measure such properties. Another of the coeditors is a principal in a company that designs, fabricates, and sells many of these instruments, and a third coeditor was involved as editor of one of the many books from the ISOPOW conferences.

All of us have walked among the "giants" in this field and have learned much. We realized that many misconceptions existed out there, and this provided a good starting point on the topics of a_w and glass transition. A well-conceived and -presented short course at the Institute of Food Technologists Annual Meeting in 2001 (Anaheim, CA) served as the basis for this textbook, as well as the inspiration of Marcus Karel (see Dedication), who truly is a "giant" among the giants in the field. We have gathered the "best of the best" to bring together the physics and chemistry of water and its interactions in foods. This book complements and goes beyond the best chapter ever written at a graduate level on the subject of water, that of Owen Fennema's chapter "Water in Food" in the classic textbook *Food Chemistry*.

Our goal was to introduce basic principles and to teach applications. We have incorporated the "old" ideas on a_w from 1965 with the "new" ideas on glass transition introduced 20 years ago. This book should serve those in the fields of food science and technology, biotechnology, plant science, meat science, pharmaceuticals, chemical and food engineering, and wood and paper science, as well as assisting in the stability of medical devices. It is interesting that the foundational principle of a_w, while a principle of physical chemistry, took the combined efforts of food scientists and food microbiologists to apply and to create a new "art and science" related to the understanding of the physical and chemical stability of foods. We congratulate all those who contributed to establishing the foundation of the key topics covered in this book, and we hope we have given them due mention in the text. One cannot teach food science without teaching about a_w; the simplicity of treating a_w as a measure of free versus bound water is a misconception, which will be apparent as one learns the true meaning of the activity of water in foods. As coauthors that traveled down an arduous path, we thank each other for overcoming the hurdles along that path by always focusing on the outcome. We hope this book will help future students and current professionals understand "a sub w."

T. P. Labuza
S. J. Schmidt
A. J. Fontana Jr.
G.V. Barbosa-Cánovas

Acknowledgments

We want to express our gratitude and appreciation to Sharon Himsl, Publications Coordinator, Washington State University, for her professionalism and dedication in facilitating the steps needed to complete this challenging project. She edited and kept track of all the manuscripts, and effectively interacted with the editors, the authors, and the staff at Blackwell Publishing. Her performance was remarkable at all times, and there are not enough words to praise her fantastic job.

List of Contributors

Altunakar, Bilge (5, 9)
Center for Nonthermal Processing of Food
Department of Biological Systems Engineering
Washington State University
Pullman, Washington, USA

Alzamora, Stella M. (10, 13)
Departamento de Industrias
Facultad de Ciencias Exactas y Naturales
Universidad de Buenos Aires
Buenos Aires, Argentina

Barbosa-Cánovas, Gustavo V. (12)
Center for Nonthermal Processing of Food
Department of Biological Systems Engineering
Washington State University
Pullman, Washington, USA

Bell, Leonard N. (7)
Department of Nutrition and Food Science
Auburn University
Auburn, Alabama, USA

Buckle, Ken A. (15)
The University of New South Wales
Food Science and Technology
Sydney, Australia

Campbell, Gaylon S. (14)
Decagon Devices, Inc.
950 N.E. Nelson Court
Pullman, Washington, USA

Chirife, Jorge (1, 10)
Facultad de Ciencias Agrarias
Pontificia Universidad Católica Argentina
Buenos Aires, Argentina

Cole, Martin B. (15)
National Center for Food Safety and Technology
6502 S. Archer Road
Summit-Argo, Illinois, USA

Fontana, Anthony J., Jr. (1, 6, 14, Appendices A, B, C, D, E, F)
Decagon Devices, Inc.
950 N.E. Nelson Court
Pullman, Washington, USA

Guerrero-Beltrán, José Angel (13)
Departamento de Ingeniería Química y Alimentos
Universidad de las Américas
Puebla, México

Juliano, Pablo (12)
Center for Nonthermal Processing of Food
Department of Biological Systems Engineering
Washington State University
Pullman, Washington, USA

Labuza, Theodore P. (5, 9)
Morse Alumni Distinguished Professor of Food Science and Engineering
Department of Food Science and Nutrition
University of Minnesota
St. Paul, Minnesota, USA

Pérez, Emmy (13)
Departamento de Ingeniería Química y Alimentos
Universidad de las Américas
Puebla, México

Reid, David S. (2)
Food Science and Technology
University of California, Davis
Davis, California, USA

Richardson, Michelle (11)
Food Technologist
Advanced Processes and Packaging Team (AMSRD-CF-A)
DoD Combat Feeding Directorate
Natick, Massachusetts, USA

Roos, Yrjö H. (3)
Department of Food and Nutritional Sciences
University College Cork
Cork, Ireland

Roudaut, Gaëlle (8)
Department of Molecular Engineering of Food and Pharmaceuticals
ENSBANA: Université de Bourgogne
Bourgogne, France

Schmidt, Shelly J. (4, Appendices E and F)
Department of Food Science and Human Nutrition
University of Illinois at Urbana-Champaign
Urbana, Illinois, USA

Stewart, Cynthia M. (15)
National Center for Food Safety and Technology
6502 S. Archer Road
Summit-Argo, Illinois, USA

Taoukis, Petros S. (11)
National Technical University of Athens
School of Chemical Engineering
Division IV: Product and Process Development
Laboratory of Food Chemistry and Technology
Athens, Greece

Tapia, María S. (10)
Instituto de Ciencia y Tecnología de Alimentos
Facultad de Ciencias
Universidad Central de Venezuela
Caracas, Venezuela

Vergara-Balderas, Fidel (13)
Departamento de Ingeniería Química y Alimentos
Universidad de las Américas
Puebla, México

Welti-Chanes, Jorge (13)
Departamento de Ingeniería Química y Alimentos
Universidad de las Américas
Puebla, México

Water Activity in Foods

Fundamentals and Applications

1 Introduction: Historical Highlights of Water Activity Research

Jorge Chirife and Anthony J. Fontana, Jr.

The concept of water activity (a_w) is more than 50 years old. William James Scott showed in 1953 that microorganisms have a limiting a_w level for growth. It is now generally accepted that a_w is more closely related to the microbial, chemical, and physical properties of foods and other natural products than is total moisture content. Specific changes in color, aroma, flavor, texture, stability, and acceptability of raw and processed food products have been associated with relatively narrow a_w ranges (Rockland and Nishi 1980). Next to temperature, a_w is considered one of the most important parameters in food preservation and processing (van den Berg 1986). This chapter is not a review of the literature on a_w but rather a highlight of some early key a_w research as it relates to microbial growth, moisture sorption isotherms, prediction and measurement of a_w in foods, and, to a lesser extent, the influence of a_w on the physical and chemical stability of foods.

Australian-born microbiologist Scott (1912–1993) received his bachelor's degree from the University of Melbourne (1933) and a doctorate of science degree from the Council for Scientific and Industrial Research (CSIR) Meat Research Laboratory (1933). He then took a position as senior bacteriologist at the CSIR Division of Food Preservation and Transport from 1940–1960. In 1960, he moved to the Meat Research Laboratory, where he served as assistant chief of division until 1964 and officer-in-charge until 1972. In 1979, he became a fellow of the Australian Academy of Technological Sciences and Engineering.

Scott's early work was concerned with handling, cooling, and transport conditions that would enable chilled beef to be successfully exported to Britain. During World War II, he was concerned with the microbiology of foods supplied by Australia to the Allied Forces. After the war, he pioneered studies on the water relations of microorganisms. In 1953, Scott related the relative vapor pressure of food to the thermodynamic activity of water, using the definition $a_w = p/p_o$, where a_w is the water activity derived from the laws of equilibrium thermodynamics, p is the vapor pressure of the sample, and p_o is the vapor pressure of pure water at the same temperature and external pressure. He showed a clear correlation between the a_w of the growth medium and the rate of *Staphylococcus aureus* growth. The summary of his paper stated:

> Fourteen food-poisoning strains of *Staphylococcus aureus* have been grown in various media of known a_w at 30°C. Aerobic growth was observed at water activities between 0.999 and 0.86. The rate of growth and the yield of cells were both reduced substantially when the a_w was less than 0.94. The lower limits for growth in dried meat, dried milk, and dried soup were similar to those in liquid media. Aerobic growth proceeded at slightly lower water activities than anaerobic growth. All cells were capable of forming colonies on agar media with water activities as low as 0.92. The 14 strains proved to be homogeneous with similar water requirements.

Table 1.1 Papers by Scott and Christian.

Author	Year	Title of paper
Scott, W.J.	1953	Water relations of *Staphylococcus aureus* at 30°C
Christian, J.H.B., and Scott, W.J.	1953	Water relations of *Salmonella* at 30°C
Christian, J.H.B.	1955a	The influence of nutrition on the water relations of *Salmonella oranienburg*
Christian, J.H.B.	1955b	The water relations of growth and respiration of *Salmonella oranienburg* at 30°C
Scott, W.J.	1957	Water relations of food spoilage microorganisms

Scott's classic demonstration that it is not the water content but the a_w of a food system that governs microbial growth and toxin production was a major contribution to food microbiology. Many scientists, most notably his Australian colleague, J. H. B. Christian, expanded Scott's work. Key papers published in the 1950s by both Scott and Christian are listed in Table 1.1. These papers laid the foundation for future research into the survival and growth of microorganisms in foods at low a_w.

In the field of food science, the general acceptance and application of the concept of a minimum a_w for microbial growth began with the review by Scott published in 1957, *Water Relations of Food Spoilage Microorganisms*. Taken from the table of contents in Scott's classic review, the following are some of the aspects discussed:

III. Methods for controlling a_w:
- Equilibration with controlling solutes
- Determination of the water sorption isotherms
- Addition of solutes

IV. Water requirements for growth
- Molds
- Yeasts
- Bacteria
- General relationships

V. Factors affecting water requirements
- Nutrition, temperature, oxygen, inhibitors, adaptation

VI. Special groups
- Halophilic bacteria
- Osmophilic yeasts
- Xerophilic molds

VII. Some applications in food preservation
- Fresh foods, dried foods, concentrated foods, frozen foods, canned foods

Since the work of Scott, a_w has become one of the most important intrinsic properties used for predicting the survival and growth of microorganisms in food, due to its direct influence on product stability and quality. Thus, the minimal a_w level for growth emerged as one of the most investigated parameters for determining the water relations of microorganisms in foods. This limiting value defines, in theory, the level below which a microorganism or group of microorganisms can no longer reproduce. The limiting value will not be the same for all microorganisms, and some may be able to tolerate low a_w and still com-

Table 1.2 Selected Early Work on the Minimal Water Activity for Growth of Pathogenic and Spoilage Microorganisms.

Author	Year	Title of paper
Baird-Parker, A.C., and Freame, B.	1967	Combined effect of water activity, pH, and temperature on the growth of *Clostridium botulinum* from spore and vegetative cell inocula
Ohye, D.F., and Christian, J.H.B.	1967	Combined effects of temperature, pH, and water activity on growth and toxin production by *Clostridium botulinum* types A, B, and E
Pitt, J.I., and Christian, J.H.B.	1968	Water relations of xerophilic fungi isolated from prunes
Anand, J.C., and Brown, A.D.	1968	Growth rate patterns of the so-called osmophilic and non-osmophilic yeasts in solutions of polyethylene glycol
Ayerst, G.	1969	The effects of moisture and temperature on growth and spore germination in some fungi
Emodi, A.S., and Lechowich, R.V.	1969	Low temperature growth of type E *Clostridium botulinum* spores. II. Effects of solutes and incubation temperature
Kang, C.K., Woodburn, M., Pagenkopf, A., and Cheney, R.	1969	Growth, sporulation, and germination of *Clostridium perfringens* in media of controlled water activity
Horner, K.J., and Anagnostopoulos, G.D.	1973	Combined effects of water activity, pH, and temperature on the growth and spoilage potential of fungi
Troller, J.A.	1972	Effect of water activity on enterotoxin A production and growth of *Staphylococcus aureus*
Beuchat, L.R.	1974	Combined effects of water activity, solute, and temperature on the growth of *Vibrio parahaemolyticus*
Northolt, M.D., van Egmond, H.P., . and Paulsch, W.W	1977	Effect of water activity and temperature on aflatoxin production by *Aspergillus parasiticus*
Pitt, J.I., and Hocking, A.D.	1977	Influence of solute and hydrogen ion concentration on the water relations of some xerophilic fungi
Lotter, L.P., and Leistner, L.	1978	Minimal water activity for enterotoxin A production and growth of *Staphylococcus aureus*
Hocking, A.D., and Pitt, J.I.	1979	Water relations of some *Penicillum* species at 25°C
Briozzo, J., de Lagarde, E.A., Chirife, J., and Parada, J.L.	1986	Effect of water activity and pH on growth and toxin production by *Clostridium botulinum* type G
Tapia de Daza, MS, Villegas, Y., and Martinez, A.	1991	Minimal water activity for growth of *Listeria monocytogenes* as affected by solute and temperature

promise product safety. The understanding and control of a_w contributes to safer food storage conditions in general and forms the basis of much modern food formulation, especially for intermediate-moisture foods.

Several workers developed studies to determine the minimal a_w level for growth of bacterial pathogens, yeasts, and molds and the production of microbial toxins. Table 1.2 displays some selected papers by authors following the work of Scott and Christian. It is worth mentioning that in 1978, Troller and Christian published a book on a_w entitled *Water Activity and Food.*

In addition to the experimental determination of minimal a_w for microbial growth, researchers also were concerned with the mechanism of cell adaptation to low a_w, specifically, the intracellular composition of cells grown at reduced a_w. Some key papers discussing low a_w adaptation are listed in Table 1.3.

Two important aspects related to microbial water relations—the solute effects and the

Table 1.3 Low Water Activity Adaptation.

Author	Year	Title of paper
Christian, J.H.B., and Waltho, J.A.	1961	The sodium and potassium content of nonhalophilic bacteria in relation to salt tolerance
Christian, J.H.B., and Waltho, J.A.	1962	The water relations of staphylococci and micrococci
Christian, J.H.B., and Waltho, J.A.	1964	The composition of *Staphylococcus aureus* in relation to the water activity of the growth medium
Brown, A.D., and Simpson, J.R.	1972	Water relations of sugar-tolerant yeasts: The role of intracellular polyols
Gould, G.W., and Measures, J.C.	1977	Water relations in single cells
Brown, A.D.	1975	Microbial water relations. Effect of solute concentration on the respiratory activity of sugar-tolerant and non-tolerant yeasts
Measures, J.C.	1975	Role of amino acids in osmoregulation of non--halophilic bacteria
Chirife, J., Ferro Fontán, C., and Scorza, O.C.	1981	The intracellular water activity of bacteria in relation to the water activity of the growth medium
Anderson, C.B., and Witter, L.D.	1982	Glutamine and proline accumulation by *Staphylococcus aureus* with reduction in water activity

influence of a_w on the thermal resistance of microorganisms, specifically—were the subject of early studies by various researchers. Table 1.4 shows a compilation of some classic papers on a_w and microbial water relations.

In addition to the research on a_w and microbial control, during the 1960s and 1970s, information about a_w and its influence on the chemical, enzymatic, and physical stability of foods began to appear rapidly in the literature. Research was conducted on the influence of a_w to (1) control undesirable chemical reactions, (2) prolong the activity of enzymes, (3) understand the caking and clumping of powders, and (4) optimize the physical properties of foods such as texture and moisture migration.

Labuza (1970) presented a comprehensive review on the influence of a_w on chemical reactions in foods. Since then, extensive studies have been conducted in this area and are reviewed in Duckworth (1975), Labuza (1980), Rockland and Nishi (1980), Rockland and Stewart (1981), and Leung (1987). The a_w of a food describes the energy status of the water in that food and, hence, its availability to act as a solvent and participate in chemical or biochemical reactions (Labuza 1977). The ability of water to act as a solvent, medium, and reactant increases as a_w rises (Labuza 1975). Water activity influences nonenzymatic browning, lipid oxidation, degradation of vitamins, and other degradative reactions. The influence of a_w on the rate of nonenzymatic browning reactions, also called Maillard reactions, is described by Troller and Christian (1978b), Labuza and Saltmarch (1981), Nursten (1986), and Bell (1995). The influence of a_w on lipid oxidation has been studied extensively and reviewed by Labuza (1975), Troller and Christian (1978a), and Karel (1981, 1986).

Enzyme activity and stability are influenced significantly by a_w due to their relatively fragile nature (Blain 1962, Aker 1969, Potthast et al. 1975, Potthast 1978, Schwimmer 1980, Drapron 1985). Most enzymes and proteins must maintain conformation to remain active. Maintaining critical a_w levels to prevent or entice conformational changes in enzymes is important to food quality. Most enzymatic reactions are slowed down at water activities below 0.80, but some reactions occur even at very low a_w values.

Table 1.4 Microbial Water Relations.

Author	Year	Title of paper
Marshall, B.J., Ohye, D.F., and Christian, J.H.B.	1971	Tolerance of bacteria to high concentrations of NaCl and glycerol in the growth medium
Baird-Parker, A.C., Boothroyd, M., and Jones, E.	1970	The effect of water activity on the heat resistance of heat sensitive and heat resistant strains of salmonellae
Horner, K.J., and Anagnostopoulos, G.D.	1975	Effect of water activity on heat survival of *Staphylococcus aureus*, *Salmonella typhimurium,* and *Salmonella stefenberg*
Corry, J.E.L.	1976	The effect of sugars and polyols on the heat resistance and morphology of osmophilic yeasts
Goepfert, J.M., Iskander, I.K., and Amundson, C.H.	1970	Relation of the heat resistance of Salmonellae to the water activity of the environment
Jakobsen, M., and Murrel, W.G.	1977	The effect of water activity and a_w-controlling solute on sporulation of *Bacillus cereus T*
Christian, J.H.B.	1981	Specific solute effects on microbial water relations

Water activity affects the stability, flow, and caking and clumping of powders during storage (Peleg and Mannheim 1977, Saltmarch and Labuza 1980, Chuy and Labuza 1994, Aguilera and del Valle 1995). Controlling a_w in a powder product below critical levels maintains proper product structure, texture, flowability, density, and rehydration properties. Knowledge of the a_w of powders as a function of moisture content and temperature is essential during processing, handling, packaging, and storage to prevent the deleterious phenomenon of caking, clumping, collapse, and stickiness. Caking is dependent on a_w, time, and temperature and is related to the collapse phenomena of the powder under gravitational force (Chuy and Labuza 1994).

Water activity affects the textural properties of foods (Troller and Christian 1978a; Bourne 1987, 1992). Foods with high a_w have a texture that is described as moist, juicy, tender, and chewy. When the a_w of these products is lowered, undesirable textural attributes such as hardness, dryness, staleness, and toughness are observed. Food low in a_w normally have texture attributes described as crisp and crunchy, while at higher a_w, the texture becomes soggy. The crispness intensity and overall hedonic texture of dry snack food products are a function of a_w (Katz and Labuza 1981, Hough et al. 2001). Critical water activities are found where the product becomes unacceptable from a sensory standpoint. Glass transition theory from the study of polymer science aids in understanding textural properties and explains the changes that occur during processing and storage (Sperling 1986, Roos and Karel 1991, Roos 1993, Slade and Levine 1995). Physical structure is often altered by changes in a_w due to moisture gain, resulting in a transition from the glassy to the rubber state.

With the introduction of the concept of a_w, it is possible to describe the relationship between a_w and food moisture content, i.e., the moisture sorption isotherm. Notable among the first published papers on water sorption isotherms of foods are the works of Makower and Dehority (1943) on dehydrated vegetables, Makower (1945) on dehydrated eggs, and Gane (1950) on fruits and vegetables. It is interesting to note that these authors made references to "equilibrium relative humidity" or "relative vapor pressure" instead of "water activity."

A variety of mathematical models have been developed to describe the typical sigmoidal moisture sorption isotherm of foods. However, before the advent of computers and the availability of nonlinear regression software, it was necessary to use two-parameter models that could be transformed into a linear equation, from which the fitting parameters could be determined. Notable among two-parameter models is the BET equation (see Appendix C). Pauling (1945) applied the BET equation to water sorption by proteins to correlate the "monolayer value" with the number of polar groups. Additionally, the empirical models developed by Oswin (1946) and Henderson (1952) were extensively used for food isotherm development.

Many multicomponent foods contain ingredients that have different water activities, and during storage, moisture will be exchanged until a final equilibrium a_w is reached. Salwin and Slawson (1959) developed a simple and useful equation to predict that equilibrium a_w. Later, Labuza (1968) published a classic paper entitled "Sorption Phenomena of Foods," which reviewed the main concepts of water sorption phenomena in foods as well the most popular sorption models being used. In 1969, Rockland applied Henderson's equation to sorption data in several foods and introduced the idea of "localized isotherms." Localized isotherms divide the curve into three regions separated by intercepts that delineate major differences in the type and character of the water binding in the system. Iglesias et al. (1975, 1976) showed that a multilayer adsorption equation, originally developed by Halsey for physical adsorption on nonuniform surfaces, could be used (reasonably well) to describe the water sorption isotherms of a great variety of foods and food components. This equation became one of the most successful two-parameter models for describing the sorption behavior of foods.

Chirife and Iglesias (1978) published a review of literature on equations for fitting water sorption isotherms of foods and food products. At that time, they were able to compile 23 equations for correlating equilibrium moisture content in food systems. Compilations of moisture sorption isotherms for a large number of foods and food components were published by Wolf et al. (1973) and Iglesias and Chirife (1982).

At present, the most popular sorption isotherm model in the food area is the Guggenheim-Anderson-de Boer (GAB) equation (see Appendix C). It is an extension of the BET equation, but with an additional parameter. Van den Berg (1981) and Bizot (1983) were among the first to demonstrate that the sigmoid-shaped isotherms of food could be precisely fitted up to about 0.90 a_w using the GAB equation. Since then, several key papers have been published to further corroborate the goodness of fit of the GAB equation (Lomauro et al. 1985a, 1985b; Iglesias and Chirife 1995).

The first attempts to predict the a_w in food solutions were in confectionary products. Grover (1947) developed an empirical method to predict the a_w of confectionery solutions. He reported a relationship between the concentrations of solutions of different sugars having the same a_w. Money and Born (1951) also proposed an empirical equation for calculating a_w in solutions of sugars and sugar syrups. In contrast to Grover or Money and Born, Norrish (1966) did not propose an empirical relationship, but derived a model for predicting the a_w of nonelectrolyte solutions based on the laws of thermodynamics. This is probably the most commonly used model for predicting a_w in nonelectrolyte binary solutions, due to availability of the parameter (constant K) needed for predictions. In the 1970s and 1980s, interest in controlling a_w in intermediate moisture foods stimulated research in the prediction of the a_w of single and mixed nonelectrolyte solutions (Bone 1973, Chuang and Toledo 1976). Kaplow (1970) described the use of Raoult's law to calculate a_w in intermediate-moisture foods. Chirife et al. (1980) continued Norrish's work and reported values of the parameter K for a wide variety of nonelectrolytes (food solutes). Karel (1973) published an update on recent research and development in the field of low and intermediate moisture foods, including moisture sorption and a_w in foods.

Perhaps one of the most useful prediction equations in the food area is the Ross equation. Ross (1975) published his very simple equation for estimating the a_w of multicomponent solutions, even highly concentrated food solutes, which proved to be useful for most a_w predictions. In 1981, Ferro-Fontán and Chirife developed a refinement of the Ross equation, which allowed a better estimate of a_w of multicomponent solutions.

As knowledge about the importance of a_w increased, food scientists needed a better measurement to quantify a_w. Many early methods for the measurement of a_w of foods were adaptations of atmospheric humidity measurement techniques. There are a number of excellent reviews on a_w measurement (Smith 1971, Labuza et al. 1976, Troller and Christian 1978c, Rizvi 1986, Wiederhold 1987, Fontana and Campbell 2004). Stoloff (1978) reported the results of a collaborative study on the calibration of a_w measuring instruments and devices. Today, there are several commercially available a_w meters that allow rapid, accurate, and reproducible measurement.

In conclusion, since Scott's early work, an enormous pool of basic information about a_w and its relation to the safety, quality, and stability of foods has been generated. This increased knowledge in the understanding of a_w has led to numerous intermediate-moisture food products being developed commercially. Since the early 1960s, a large number of patents describing practical applications of a_w in foods have been issued in the United States and internationally (Bone 1987). Water activity technology also aids in the development of nutritious shelf stable food for the National Aeronautics and Space Administration (NASA) and the U.S. military in the Meals Ready-to-Eat (MRE) program. Future research and application of a_w will aid food scientists in the development of new foods that are safe, shelf-stable, easy to prepare, and highly nutritious.

References

Acker, L. 1969. Water activity and enzyme activity. *Food Technology* 23, 27–40.

Aguilera, J.M., and del Valle, J.M. 1995. Structural changes in low moisture food powders. In: *Food Preservation by Moisture Control: Fundamentals and Applications—ISOPOW Practicum II*, eds. G.V. Barbosa-Cánovas and J. Welti-Chanes, pp. 675–695. Lancaster, PA: Technomic Publishing Company.

Anand, J.C., and Brown, A.D. 1968. Growth rate patterns of the so-called osmophilic and non-osmophilic yeasts in solutions of polyethylene glycol. *Journal of General Microbiology* 52, 205–212.

Anderson, C.B., and Witter, L.D. 1982. Glutamine and proline accumulation by *Staphylococcus aureus* with reduction in water activity. *Applied And Environmental Microbiology* 43, 1501–1503.

Ayerst, G. 1969. The effect of moisture and temperature on growth and spore germination in some fungi. *Journal of Stored Products Research* 5, 127–141.

Baird-Parker, A.C, Boothroyd, M., and Jones, E. 1970. The effect of water activity on the heat resistance of heat sensitive and heat resistant strains of salmonellae. *Journal of Applied Bacteriology* 33, 515–522.

Baird-Parker, A.C., and Freame, B. 1967. Combined effect of water activity, pH and temperature on the growth of *Clostridium botulinum* from spore and vegetative cell inocula. *Journal of Applied Bacteriology* 30, 420–429.

Bell, L.N. 1995. Kinetics of non-enzymatic browning in amorphous solid systems: Distinguishing the effects of water activity and the glass transition. *Food Research International* 28, 591–597.

Beuchat, L.R. 1974. Combined effects of water activity, solute, and temperature on the growth of *Vibrio parahaemolyticus*. *Applied Microbiology* 27, 1075–1080.

Bizot, H. 1983. Using G.A.B. model to construct sorption isotherms. In: *Physical Properties of Foods,* eds. R. Jowitt, F. Escher, B. Hallström, H.F.T. Meffert, W.E.L. Spiess, and G. Vos. London: Applied Science Publishers.

Blain, J.A. 1962. Moisture levels and enzyme activity. *Recent Advances in Food Science* 2, 41–45.
Bone, D. 1973. Water activity in intermediate moisture foods. *Food Technology* 27, 71.
Bone, D.P. 1987. Practical applications of water activity and moisture relations in foods. In: *Water Activity: Theory and Applications to Food,* eds. L.B. Rockland and L.R. Beuchat, pp. 369–395. New York: Marcel Dekker Inc.
Bourne, M.C. 1987. Effects of water activity on textural properties of food. In: *Water Activity: Theory and Applications to Food,* eds. L.B. Rockland and L.R. Beuchat, pp. 75–99. New York: Marcel Dekker, Inc.
Bourne, M.C. 1992. Water activity: Food texture. In: *Encyclopedia of Food Science and Technology, Vol. 4,* ed. Y.H. Hui, pp. 2801–2815. New York: Wiley-Interscience Publication.
Briozzo, J., de Lagarde, E.A., Chirife, J., and Parada, J.L. 1986. Effect of water activity and pH on growth and toxin production by *Clostridium botulinum* type G. *Applied and Environmental Microbiology* 51, 844–848.
Brown, A.D. 1975. Microbial water relations. Effect of solute concentration on the respiratory activity of sugar-tolerant and non-tolerant yeasts. *Journal of General Microbiology* 86, 241–249.
Brown, A.D., and Simpson, J.R. 1972. Water relations of sugar-tolerant yeasts: The role of intracellular polyols. *Journal of General Microbiology* 72, 589–591.
Chirife, J., Ferro-Fontán, C., and Benmergui, E.A. 1980. The prediction of water activity in aqueous solutions in connection with intermediate moisture foods. IV. a_w prediction in aqueous non-electrolyte solutions. *Journal of Food Technology* 15, 59–70.
Chirife, J., Ferro-Fontán, C., and Scorza, O.C. 1981. The intracellular water activity of bacteria in relation to the water activity of the growth medium. *Journal of Applied Bacteriology* 50, 475–477.
Chirife, J., and Iglesias, H.A. 1978. Equations for fitting water sorption isotherms of foods: Part 1—a review. *Journal of Food Technology* 13, 159–174.
Christian, J.H.B. 1955a. The influence of nutrition on the water relations of *Salmonella oranienburg*. *Australian Journal of Biological Sciences* 8, 75–82.
Christian, J.H.B. 1955b. The water relations of growth and respiration of *Salmonella oranienburg* at 30°C. *Australian Journal of Biological Sciences* 8, 490–497.
Christian, J.H.B. 1981. Specific solute effects on microbial water relations. In: *Water Activity: Influences on Food Quality,* eds. L.B. Rockland and G.F. Stewart. New York: Academic Press.
Christian, J.H.B., and Scott, W.J. 1953. Water relations of *Salmonella* at 30°C. *Australian Journal of Biological Sciences* 6, 565–573.
Christian, J.H.B., and J.A. Waltho. 1961. The sodium and potassium content of nonhalophilic bacteria in relation to salt tolerance. *Journal of General Microbiology* 43, 354–355.
Christian, J.H.B., and Waltho, J.A. 1962. The water relations of staphylococci and micrococci. *Journal of Applied Bacteriology* 25, 369–377.
Christian, J.H.B., and Waltho, J.A. 1964. The composition of *Staphylococcus aureus* in relation to the water activity of the growth medium. *Journal of General Microbiology* 35, 205–213.
Chuang, L., and Toledo, R.T. 1976. Predicting the water activity of multicomponent systems form water sorption isotherms of individual components. *Journal of Food Science* 41, 922–927.
Chuy, L.E., and Labuza, T.P. 1994. Caking and stickiness of dairy-based food powders as related to glass transition. *Journal of Food Science* 59, 43–46.
Corry, J.E.L. 1976. The effect of sugars and polyols on the heat resistance and morphology of osmophilic yeasts. *Journal of Applied Bacteriology* 40, 269–276.
Drapron, R. 1985. Enzyme activity as a function of water activity. In: *Properties of Water in Foods in Relation to Quality and Stability,* eds. D. Simato and J.L. Multon, pp. 171–190. Dordrecht, the Netherlands: Martinus Nijhoff Publishers.
Duckworth, R. 1975. *Water Relations of Foods*. New York: Academic Press.
Emodi, A.S., and Lechowich, R.V. 1969. Low temperature growth of type E *Clostridium botulinum* spores. II. Effects of solutes and incubation temperature. *Journal of Food Science* 34, 82–87.
Ferro-Fontán, C., and Chirife, J. 1981. Technical Note: A refinement of Ross's equation for predicting the water activity of non-electrolyte mixtures. *Journal of Food Technology* 16, 219–221.
Fontana, A.J., and Campbell, C.S. 2004. Water activity. In: *Handbook of Food Analysis, Physical Characterization and Nutrient Analysis,* ed. L.M.L. Nollet, pp. 39–54. New York: Marcel Dekker, Inc.
Gane, R. 1950. Water relations of some dried fruits, vegetables and plant products. *Journal of the Science of Food and Agriculture* 1, 42.
Goepfert, J.M., Iskander, I.K., and Amundson, C.H. 1970. Relation of the heat resistance of Salmonellae to the water activity of the environment. *Applied Microbiology* 19, 429–433.

Gould, G.W., and Measures, J.C. 1977. Water relations in single cells. *Philosophical Transactions of the Royal Society, London, Series B* 278, 151–166.

Grover, D.N. 1947. The keeping properties of confectionary as influenced by its water vapor pressure. *Journal of the Society of Chemical Industry* 66, 201.

Henderson, S.M. 1952. A basic concept of equilibrium moisture. *Agricultural Engineering* 33, 29.

Hocking, A.D., and Pitt, J.I. 1979. Water relations of some *Penicillum* species at 25°C. *Transactions of the British Mycological Society* 73, 141–145.

Horner, K.J., and Anagnostopoulos, G.D. 1975. Effect of water activity on heat survival of *Staphylococcus aureus, Salmonella typhimurium* and *Salmonella stefenberg*. *Journal of Applied Bacteriology* 38, 9–17.

Horner, K.J., and Anagnostopoulos, G.D. 1973. Combined effects of water activity, pH and temperature on the growth and spoilage potential of fungi. *Journal of Applied Bacteriology* 36, 427–436.

Hough, G., del Pilar-Buera, M., Chirife, J., and Moro, O. 2001. Sensory texture of commercial biscuits as a function of water activity. *Journal of Texture Studies* 32(1): 57–74.

Iglesias, H.A., Chirife, J., and Lombardi, J.L. 1975. An equation for correlating equilibrium moisture content in foods. *Journal of Food Technology* 10, 289–297.

Iglesias, H.A., and Chirife, J. 1976. B.E.T. monolayer values in dehydrated foods and food components. *Lebensmittel Wissenschaft und Technologie* 9, 123–127.

Iglesias, H.A., and Chirife, J. 1982. *Handbook of Food Isotherms: Water Sorption Parameters for Food and Food Components,* pp. 262–319. New York: Academic Press.

Iglesias, H.A., and Chirife, J. 1995. An alternative to the Guggenheim, Anderson and De Boer model for the mathematical description of moisture sorption isotherms of foods. *Food Research International* 28, 317–321.

Jakobsen, M., and Murrel, W.G. 1977. The effect of water activity and a_w-controlling solute on sporulation of *Bacillus cereus T. Journal of Applied Bacteriology* 43, 239–245.

Kang, C.K., Woodburn, M., Pagenkopf, A., and Cheney, R. 1969. Growth, sporulation, and germination of *Clostridium perfringens* in media of controlled water activity. *Applied Microbiology* 18, 798–805.

Kaplow, M. 1970. Commercial development of intermediate moisture foods. *Food Technology* 24, 53–57.

Karel, M. 1973. Recent research and development in the field of low moisture and intermediate moisture foods. *CRC Critical Reviews in Food Technology* 3, 329–373.

Karel, M. 1986. Control of lipid oxidation in dried foods. In: *Concentration and Drying of Foods,* ed. D. MacCarthy, pp. 37–51. London: Elsevier Applied Science Publishers.

Karel, M., and Yong, S. 1981. Autoxidation-initiated reactions in foods. In: *Water Activity: Influences on Food Quality,* eds. L.B. Rockland and G.F. Stewart. New York: Academic Press.

Katz, E.E., and Labuza, T.P. 1981. Effect of water activity on the sensory crispness and mechanical deformation of snack food products. *Journal of Food Science* 46, 403–409.

Labuza, T.P. 1968. Sorption phenomena of foods. *Food Technology* 22, 262–272.

Labuza, T.P. 1970. Properties of water as related to the keeping quality of foods. Washington, DC, Proceedings of the Third International Congress of Food Science, IFT. Symposium on Physical and Chemical Properties of Foods. pp. 618–635.

Labuza, T.P. 1975. Oxidative changes in foods at low and intermediate moisture levels. In: *Water Relations of Foods,* ed. R.B. Duckworth, pp. 455–474. New York: Academic Press.

Labuza, T.P. 1977. The properties of water in relationship to water binding in foods: A review. *Journal of Food Processing and Preservation* 1, 167–190.

Labuza, T.P. 1980. Effect of water activity on reaction kinetics of food deterioration. *Food Technology* 34, 36–41, 59.

Labuza, T.P., Acott, K., Tatini, S.R., Lee, R.Y., Flink, J., and McCall, W. 1976. Water activity determination: A collaborative study of different methods. *Journal of Food Science* 41, 910–917.

Labuza, T.P., and Saltmarch, M. 1981. The nonenzymatic browning reaction as affected by water in foods. In: *Water Activity: Influences on Food Quality,* eds. L.B. Rockland and G.F. Stewart, pp. 605–650. New York: Academic Press.

Leung, H.K. 1987. Influence of water activity on chemical reactivity. In: *Water Activity: Theory and Applications to Food,* eds. L.B. Rockland and L.R. Beuchat, pp. 27–54. New York: Marcel Dekker, Inc.

Lomauro, C.J., Bakshi, A.S., and Labuza, T.P. 1985a. Evaluation of food moisture sorption isotherm equations. Part I: Fruit, vegetable and meat products. *Lebensmittel Wissenschaft und Technologie* 18, 111–117.

Lomauro, C.J., Bakshi, A.S., and Labuza, T.P. 1985b. Evaluation of food moisture sorption isotherm equations. Part II: Milk, coffee, tea, nuts, oilseeds, spices and starchy foods. *Lebensmittel Wissenschaft und Technologie* 18, 118–124.

Lotter, L.P., and Leistner, L. 1978. Minimal water activity for enterotosin: A production and growth of *Staphylococcus aureus*. *Applied and Environmental Microbiology* 36, 377—380.
Makower, B. 1945. Vapor pressure of water adsorbed on dehydrated eggs. *Industrial and Engineering Chemistry* 37, 1018–1022.
Makower, B., and Dehority, G.L. 1943. Equilibrium moisture content of dehydrated vegetables. *Industrial and Engineering Chemistry* 35, 193–197.
Marshall, B.J., Ohye, D.F., and Christian, J.H.B. 1971. Tolerance of bacteria to high concentrations of NaCl and glycerol in the growth medium. *Applied Microbiology* 21, 363–364.
Measures, J.C. 1975. Role of amino acids in osmoregulation of non-halophilic bacteria. *Nature (London)* 257, 398–400.
Money, R.W., and Born, R. 1951. Equilibrium humidity of sugar solutions. *Journal of the Science of Food and Agriculture* 2, 180.
Norrish, R.S. 1966. An equation for the activity coefficients and equilibrium relative humidities of water in confectionary syrups. *Journal of Food Technology* 1, 25–39.
Northolt, M.D., van Egmond, H.P., and Paulsch, W.E. 1977. Effect of water activity and temperature on aflatoxin production by *Aspergillus parasiticus*. *Journal of Milk and Food Technology* 39, 170–174.
Nursten, H.E. 1986. Maillard browning reactions in dried foods. In: *Concentration and Drying of Foods,* ed. D. MacCarthy, pp. 53–68. London: Elsevier Applied Science Publishers.
Ohye, D.F., and Christian, J.H.B. 1966. Combined effects of temperature pH and water activity on growth and toxin production of *Cl. botulinum* types A, B, and E. botulism, eds. T.A. Roberts and M. Ingram, *Proceedings of the International Symposium on Food Microbiology,* 1966, pp. 217–223. London: Chapman and Hall, Ltd.
Oswin, C.R. 1946. The kinetics of package life. III. The isotherm. *Journal of Chemistry and Industry (London)* 65, 419.
Pauling, L. 1945. The adsorption of water by proteins. *Journal of the American Chemical Society* 67, 655.
Peleg, M., and Mannheim, C.H. 1977. The mechanism of caking of powdered onion. *Journal of Food Processing and Preservation* 1, 3–11.
Pitt, J.I., and Christian, J.H.B. 1968. Water relations of xerophilic fungi isolated from prunes. *Applied Microbiology* 16, 1853–1859.
Pitt, J.I., and Hocking, A.D. 1977. Influence of solute and hydrogen ion concentration on the water relations of some xerophilic fungi. *Journal of General Microbiology* 101, 35–40.
Potthast, K. 1978. Influence of water activity on enzymic activity in biological systems. In: *Dry Biological Systems,* eds. J.H. Crowe and J.S. Clegg, p. 323. New York: Academic Press.
Potthast, K., Hamm, R., and Acker, L. 1975. Enzymatic reactions in low moisture foods. In: *Water Relations in Foods,* ed. R.B. Duckworth, pp. 365–377. San Diego, CA: Academic Press.
Rizvi, S.S.H. 1986. Thermodynamic properties of foods in dehydration. In: *Engineering Properties of Foods,* eds. M.A. Rao and S.S.H. Rizvi. New York: Marcel Dekker, Inc.
Rockland, L.B. 1969. Water activity and storage stability. *Food Technology* 23, 1241.
Rockland, L.B., and Nishi, S.K. 1980. Influence of water activity on food product quality and stability. *Food Technology* 34, 42–59.
Rockland, L.B., and Stewart, G.F. 1981. *Water Activity: Influences on Food Quality.* New York: Academic Press.
Roos, Y.H. 1993. Water activity and physical state effects on amorphous food stability. *Journal of Food Processing and Preservation* 16, 433–447.
Roos, Y.H., and Karel, M. 1991. Water and molecular weight effects on glass transitions in amorphous carbohydrates and carbohydrate solutions. *Journal of Food Science* 56, 1676–1681.
Ross, K.D. 1975. Estimation of water activity in intermediate moisture foods. *Food Technology* 29, 26–34.
Saltmarch, M., and Labuza, T.P. 1980. Influence of relative humidity on the physicochemical state of lactose in spray-dried sweet whey powders. *Journal of Food* Science 45, 1231–1236, 1242.
Salwin, H., and Slawson, V. 1959. Moisture transfer in combinations of dehydrated foods. *Food Technology* 8, 58–61.
Schwimmer, S. 1980. Influence of water activity on enzyme reactivity and stability. *Food Technology* 34, 64.
Scott, W.J. 1953. Water relations of *Staphylococcus aureus* at 30°C. *Australian Journal of Biological Sciences* 6, 549–564.
Scott, W.J. 1957. Water relations of food spoilage microorganisms. *Advances in Food Research* 7, 83–127.
Slade, L., and Levine, H. 1995. Glass transitions and water-food structure interactions. *Advances in Food and Nutrition Research* 38, 103–269.

Smith, P.R. 1971. The determination of equilibrium relative humidity or water activity in foods—A literature review. *BFMIRA Science and Technology Survey* No. 70.

Sperling, L.H. 1986. *Introduction to Physical Polymer Science.* pp. 224–295. New York: John Wiley & Sons.

Stoloff, L. 1978. Calibration of water activity measuring instruments and devices: Collaborative study. *Journal of the Association of Official Analytical Chemists* 61, 1166–1178.

Tapia de Daza, M.S., Villegas, Y., and Martinez, A. 1991. Minimal water activity for growth of *Listeria monocytogenes* as affected by solute and temperature. *International Journal of Food Microbiology* 14, 333–337.

Troller, J.A. 1972. Effect of water activity on enterotoxin A production and growth of *Staphylococcus aureus*. *Applied Microbiology* 24, 440–443.

Troller, J.A., and Christian, J.H.B. 1978a. Enzyme reactions and nonenzymatic browning. In: *Water Activity and Food,* eds. J.A. Troller and J.H.B. Christian, pp. 48–68. New York: Academic Press.

Troller, J.A., and Christian, J.H.B. 1978b. Lipid oxidation, changes in texture, color, and nutritional quality. In: *Water Activity and Food,* eds. J.A. Troller and J.H.B. Christian, pp. 69–85. New York: Academic Press.

Troller, J.A., and Christian, J.H.B. 1978c. Methods. In: *Water Activity and Food,* eds. J.A. Troller and J.H.B. Christian, pp. 13–47. New York: Academic Press.

van den Berg, C. 1986. Water activity. In: *Concentration and Drying of Foods,* ed. D. MacCarthy, pp. 11–36. London: Elsevier Applied Science Publishers.

van den Berg, C. 1981. Vapour sorption equilibria and other water-starch interactions; a physicochemical approach. Thesis/dissertation. Agricultural University, Wageningen, The Netherlands.

Wiederhold, P. 1987. Humidity measurements. In: *Handbook of Industrial Drying,* ed. A.S. Mujumdar. New York: Marcel Dekker.

Wolf, W., Spiess, W.E.L., and Jung, G. 1973. The water vapor sorption isotherms of foodstuffs. *Lebensmittel Wissenschaft und Technologie* 6, 94–96.

2 Water Activity: Fundamentals and Relationships

David S. Reid

Thermodynamics informs us that when a multiphase system is at equilibrium, the chemical potential μ_1 of individual components is the same in all phases. Hence, to determine chemical potential, it is sufficient to measure it in any one phase. Chemical potential is defined by the equation:

$$\mu_1(p,T,n_1,n_2,....) = (\partial G / \partial n_1)_{p,T,n_2},.... \tag{2.1}$$

where $(\partial G/\partial n_1)$ denotes the partial molar free energy.

Most foods are multiphase systems with water being an important component. This chapter will discuss how to use thermodynamics to characterize the status of water in a food system, under a variety of external conditions, including freezing temperatures, ambient temperatures, and high pressures. The most frequently used thermodynamic descriptor for water is water activity (a_w).

Activity is a thermodynamic concept, defined by Lewis and Randall (1961) as "Activity is, at a given temperature, the ratio of the fugacity, f, of a substance and its fugacity, f^o, in some state which for convenience, has been chosen as a standard state," expressed as:

$$a = \left(\frac{f}{f^o}\right)_T \tag{2.2}$$

Fugacity is a measure of the escaping tendency of a substance. It can be replaced by the vapor pressure, p, provided the vapor behaves as an ideal gas. The term *standard state* refers to the sample being characterized under some well-defined set of conditions that, by agreement, have been chosen for use as a definable reference. Note also that the reference measurement is at the same temperature as that of the sample under study, indicated by the subscript T. It is critical that one realizes this definition pertains to thermodynamics, to systems at equilibrium. As such, relationships exist between the parameter, activity (as defined above), and other defined thermodynamic properties such as free energy, enthalpy, chemical potential, osmotic pressure, and so on. Two examples of important relationships involving activity are, first,

$$\mu = \mu^o + RT \ln a \tag{2.3}$$

where μ is the chemical potential and superscript *o* indicates that quantity refers to the standard state, a is the activity, *R* is the gas constant, and T is the temperature; and second,

$$-\ln a_w = \frac{\Pi M_w}{RT\rho_w} \tag{2.4}$$

where Π is the osmotic pressure, M_W is the molar mass, and ρ is the density.

Scott (1953, 1957) sought to identify a useful measure of the "influence" of water upon the properties of a system. He realized that measures such as water content could not lead to insights relating to the energetics and the equilibrium states of the system. Accordingly, he suggested the utilization of a_W, defined, using the Lewis and Randall concept (1961), as "Water activity is, at a given temperature, the ratio of its fugacity, f_W, in a system and the fugacity, f^o_W, of pure liquid water at the same temperature."

$$a_w = \left(\frac{f_w}{f^o_w}\right)_T \tag{2.5}$$

Note here that the standard state has been chosen to be pure liquid water, a well-defined state.

He further demonstrated that the a_W of a system, as defined above, could often be correlated with other properties of the system, such as microbial stability and related characteristics. Following Scott's initial proposal, extensive literature was generated showing the wide-ranging utility of this concept. Students, particularly in the food and microbiological sciences, understand that a_W is a key parameter influencing the stability of foods and microbial systems. However, the question arises as to whether students understand the restrictions imposed by the definition of the term activity. Let us reiterate the definition but this time highlight the terms that should be discussed to better understand the consequences of the restrictions inherent in the definition: "Water activity is, at a given temperature, the ratio of the fugacity, f_W, of water in a system, and its fugacity, f^o_W, in pure liquid water at the same temperature."

It must constantly be borne in mind that this definition refers only to a system in equilibrium. Fugacity, a measure of escaping tendency, is in units of pressure and represents an effective pressure that takes into account any nonideality in the gas phase. Fortunately, to a good first approximation, fugacity can be replaced by the equilibrium vapor pressure, or by equilibrium partial vapor pressure in all systems of interest. The error associated with this approximation under normal conditions is less than 0.1%, so this substitution is routinely made in food science. Indeed, many textbooks do not mention the property known as fugacity, although it is the property required for the exact definition of activity. In the following equilibrium, partial vapor pressure will be used for all equations and in associated definitions. Hence, the working definition for a_W becomes the following: water activity is, at a given temperature, the ratio of the equilibrium partial vapor pressure of water in the system (p_W) to the equilibrium partial vapor pressure p^o_W of pure liquid water at the same temperature, which is expressed as:

$$a_w = \left(\frac{p_W}{p^o_W}\right)_T \tag{2.6}$$

Note, as indicated earlier, this definition requires that the comparison of partial pressures refers to isothermal conditions. There are some issues to consider:

1. Why is a_w a superior measure to water content?
2. Must the measuring system be isothermal?
3. Just how does one measure partial vapor pressure?

Consider carefully the primary restriction to equilibrium, and be aware of the consequences of its application. A brief example will suffice to illustrate some of the pitfalls inherent in this restriction. Assume that, at constant temperature, a mixture of water and carbon tetrachloride is shaken together and allowed to separate in a sealed, U-shaped container. After separating out, one arm of the U has water, the less dense component, as the top layer and carbon tetrachloride as a lower layer. The other arm of the U contains only carbon tetrachloride. Naturally, the water layer contains a small amount of dissolved carbon tetrachloride, and the carbon tetrachloride layer contains a small amount of water, in each case the dissolved component being at its saturation concentration. Note that the mutual saturation solubilities here are less than 10 ppm. Nevertheless, notwithstanding the huge difference in volumetric water fraction (water content) of the two layers, equilibrium thermodynamics requires that the chemical potential of water be the same in both layers, in other words, that p_w of each layer is the same. If this were not the case, water would diffuse between layers until this situation was established. This means that since the a_w in the water-rich layer is very close to 1, the same is the case for the water-poor layer, in which the water content is very small. This apparently unlikely fact is better understood by noting that in a carbon tetrachloride system, in which the water content is less than the saturation content:

$$p_{sol} = p_{sat}\ (x_{sol} / x_{sat}) \tag{2.7}$$

where x_{sol} represents the water fraction in solution in carbon tetrachloride, and x_{sat} represents the fraction of water at saturation. Hence, the a_w of this system is:

$$a_w = (x_{sol} / x_{sat})_T \tag{2.8}$$

This simple example clearly illustrates why water content is so inadequate as a predictive measure of potential stability, because the water content of the carbon tetrachloride layer is small, even though the a_w ranges from 1 down. Water content alone is in no way related to equilibrium partial vapor pressure, although a relationship exists to relative water content, with the reference being the saturation water content.

Beyond this demonstration of the inadequacy of water content per se as a measure, it is necessary to consider the consequences of the requirement for equilibrium inherent in the definition of a_w. This requirement of equilibrium is one for which it is difficult to demonstrate compliance. A steady state is not necessarily an equilibrium state. Also, an apparently steady state may still be undergoing slow change and be far from equilibrium. It is important to realize that equilibrium and kinetics are separate concepts. Unless true equilibrium can be demonstrated, the use of thermodynamic relationships to derive other quantities from the apparent a_w is risky and may be very misleading. Because equilibrium is seldom confirmed, some have advocated the use of the term *relative vapor pressure* (RVP) rather than *water activity* (a_w) to describe standard experimental data (Slade and Levine 1991). Only after clearly establishing that equilibrium has been attained should a_w be used. There is an implied onus on the author(s) of any study to clearly demonstrate that the RVP reported is indeed the equilibrium value before using the term a_w. Unless thermodynamic equilibrium has been demonstrated to exist, it is not appropriate to use thermodynamic relationships to derive other parameters. However, the uncritical use of a_w to refer to steady-state RVP is heavily entrenched in the literature, and though the purist might hope that it be eliminated, this is not practical. Rather, those who employ the results should use extreme caution and realize that the number reported as a_w in most studies is clearly an RVP; very likely it is a steady-state RVP, but seldom has it been clearly demonstrated that it is indeed an equilibrium RVP.

Determination of p_w can be challenging (Reid et al. 2005). If a direct manometric measure of vapor pressure is required, this presents some difficulty since the technique is very demanding. However, measurement of some other material property that can be correlated with p_w may be a simpler process. For example, at any given temperature, the electrical resistance and electrical capacitance of certain materials correlate well with the relative humidity (RH) = $(p_w/p^o{}_w)_T$ of the atmosphere with which they are equilibrated. Probes have been designed that measure the electrical properties of such materials exposed to a moist atmosphere, and use a correlation function between the electrical property and RH to provide an estimated RH of the atmosphere. Other materials have mechanical properties that correlate with RH, for example, the length of a human hair exposed to a given stretching force. This concept is the basis of the hair hygrometer. By measuring the change in the length of the hair, one can estimate the change in RH.

Another common method used to estimate p_w recognizes that under well-designed geometries (or with vapor circulation by gentle mechanical stirring), the partial pressures throughout the vapor space contiguous with a system should become uniform. This is the principle behind the dew point cell, which determines the temperature at which water just condenses on a cooled surface (dew forms) exposed to the vapor. Assuming p_w is the same throughout the vapor space, if the temperature of the cooled surface in contact with this vapor space when dew just forms is T_d, then p_w will equal $p^o{}_w$ at the temperature T_d, since at this temperature the vapor is saturated. This is illustrated in Figure 2.1 where for the system with partial water vapor pressure described by curve p, at temperature T_2 (point H), T_d will be T_1, as point E on the saturation curve p^o at temperature T_1 represents the same actual vapor pressure. Hence, if T_d is determined (e.g., from the change in optical appearance of a mirror as dew forms on its surface), then p_w has been found. One just has to look up $p^o{}_w$ at T_d in an appropriate table displaying the vapor pressures of water at a range of temperatures (Table 2.1). Caution must be exercised to make sure the measuring device does not significantly perturb the system. Should the surface be large enough, cooling to T_{d2} below T_d may allow a sufficient fraction of the total water content of the sample to transfer to the cooled surface such that the p_w of the sample at the now reduced water content reaches a value equal to that of $p^o{}_w$ at T_{d2}. Hence, the sample size should be large enough and the cooled surface small enough such that the amount of dew is negligible in terms of the total water content of the system. This is an example of a common potential error in systems when measuring p_w. The amount of water evaporated from the sample must be negligible, not significantly changing the water content of the sample, unless the sample water content is determined after measurement.

A careful reader will note that the term RH has been used in this discussion, and not the term *equilibrium relative humidity* (ERH). For ERH to be appropriate, the system must be at equilibrium. Clearly, in a system with temperature gradients, equilibrium cannot be

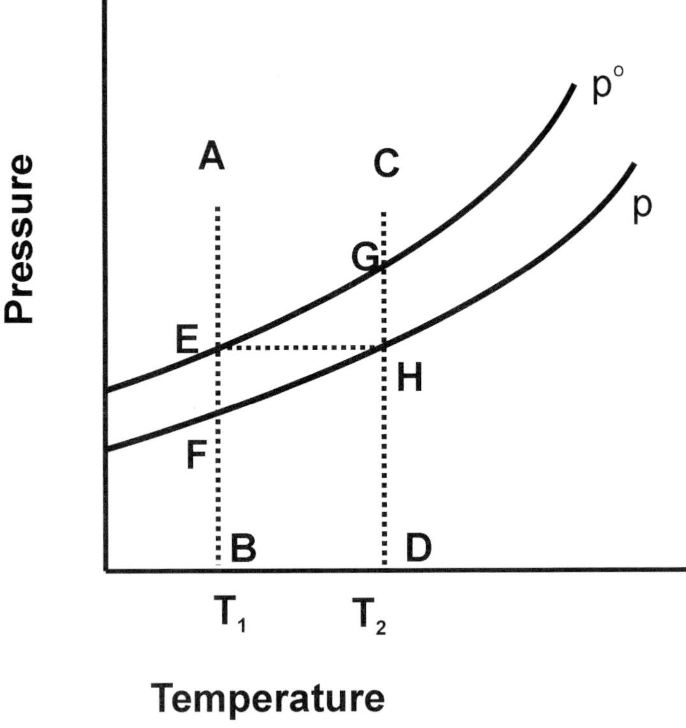

Figure 2.1. Schematic diagram of vapor pressure-temperature relationships for pure water (p^o) and an aqueous solution (p). Points E and F represent the respective vapor pressures at T_1, and points G and H the respective vapor pressures at T_2.

Table 2.1 Vapor Pressure of Pure Water at Various Temperatures.

Temperature (K)	Vapor pressure (kPa)
270	0.485
280	0.991
290	1.919
300	3.535
310	6.228
320	10.54
330	17.202
340	27.167
350	41.647
360	62.139
370	90.453
380	128.74
390	179.48
400	245.54

achieved, although a steady state is possible. Given a steady state, if a sufficiently large region around the sample is maintained isothermal, by appropriate design of the thermal enclosure, local equilibrium might be established, but clearly this must be demonstrated, not assumed. An example of such a condition is an apparatus with two temperature-controlled enclosures, with an interconnected passage. Each enclosure can be at a different temperature. Provided the vapor circulating is enabled to maintain a uniform p_W throughout the whole vapor space, each enclosure will have a steady RH characteristic of p_W at that temperature, and hence separate equilibria can be established in each enclosure.

So far, the focus has been on the vapor phase within the measuring device. Consideration must now be given to the system above which this vapor phase has been established. It is for this system (at equilibrium) that an estimate of RVP (or a_W, should it be at equilibrium) is desired. It is best to consider how ERH (or equilibrium relative vapor pressure) would be expected to change under similar conditions to those experienced by the sample, as suggested by the thermodynamic relationships. Important conditions to consider that influence sample behavior are sample concentration, system temperature, and external pressure, because each of these would influence the partial vapor pressure of water above the sample.

Concentration

The effect of concentration on p_W is best approached by first considering the effect on equilibrium relative vapor pressure. For an ideal solution, one that exhibits no solute–solvent interaction, Raoult's law states that the partial pressure of solvent is proportional to the mole fraction of solvent $(1 - x)$. Hence, any addition of solute will lower p_W (and therefore, also a_W), expressed as:

$$p_W = (1 - x) \cdot p_W^o \tag{2.9}$$

Raoult's law describes an ideal system. Many real systems with significant solute–solvent interaction exhibit lowering of equilibrium relative vapor pressure much greater than this. As a result, a_W, which for a system that obeys Raoult's law is just $(1 - x)$, is now expressed by a relation, such as $a_W = \gamma x$, where γ is known as the activity coefficient. The numerical value of γ provides a measure of nonideality.

Temperature

p^o_W is dependent on temperature, a dependence described by the Clausius-Clapeyron equation:

$$\frac{d(\ln p_W^o)}{dT} = \frac{\Delta H_{evap}}{RT^2} \tag{2.10}$$

where ΔH_{evap} is the enthalpy of evaporation of pure water. Sample values for vapor pressure of liquid water are given in Table 2.1.

p^o_W for ice is also temperature dependent, and typical values can be found in Table 2.2.

Thus, the value of p^o_W in the equation for a_W is temperature dependent. How does the value for p_W of a system change with temperature? The dependence of p_W on temperature is estimated by combining Raoult's law, or an extended form of this law, with the relationship in Equation 2.8 showing how p^o_W changes with temperature. If Raoult's law applies,

Table 2.2 Vapor Pressure of Ice at Various Temperatures.

Temperature (K)	Vapor pressure (kPa)
273	0.6105
268	0.4017
263	0.2600
258	0.1654
253	0.1035
248	0.0635
243	0.0381
238	0.0225
233	0.0129

the ratio $\left(\frac{p_w}{p_w^o}\right)_T$ will not change with temperature. Even in nonideal solutions, where Raoult's law does not hold, the change of the ratio with temperature is small.

For sorption, the appropriate relationship describing how p_W changes with temperature is:

$$\frac{\partial \ln p_w}{\partial 1/T} = \frac{\left[-\Delta H_{evap} + Q_s\right]}{R} \tag{2.11}$$

where Q_s represents the heat of sorption, defined as the total molar enthalpy change accompanying the phase change from vapor to sorbed water. The variation of $(p_W/p^o_W)_T$ (i.e., a_W) with the temperature is given by:

$$\frac{\partial \ln a_w}{\partial 1/T} = \frac{-Q_s}{R} \tag{2.12}$$

Note that since p_W is dependent upon temperature, an error in identifying the T for which the p_W has been determined can lead to an appreciable error. An error of 2K in temperature leads to a 10% error in $(p_W/p^o_W)_T$. Note also that if the temperature is lower than expected, the estimated RVP will be high. Also, in a sealed package, if a temperature gradient of 2K exists, but with p_W being the same throughout, RVP at the cool end will be 10% greater than RVP at the warm end. This is in accord with the operational concept of the dew point cell, where at the cooled surface, p_W is the same as p^o, while at the sample position, p^o is that at the sample temperature.

So far, pure liquid water at the same temperature as the sample has been used as the standard state. Recalling the definition of activity, the standard state is chosen for convenience. Note that overall pressure has not been specified. By default, the pressure of the standard state is normally assumed to be either the vapor pressure of pure liquid water or the saturation partial pressure of pure water at a total external pressure of 1 atmosphere. While these are not exactly the same, they are close enough in normal situations. The dependence of partial pressure on total external pressure is given by the equation:

$$\ln p^o_{sat} - \ln p^o_w = \left(P - p^o_w\right)V^o / RT \tag{2.13}$$

where p^o_{sat} represents the partial vapor pressure of water in saturated air at a total pressure P, p^o_W represents the vapor pressure of water at the same temperature, V^o is the molar vol-

Table 2.3 Vapor Pressure of Supercooled Water at Various Temperatures.

Temperature (K)	Vapor pressure (kPa)
273	0.6105
268	0.4217
263	0.2865
258	0.1915
253	0.1259

Table 2.4 a_w of Frozen Systems in Liquid Water Standard State.

Temperature (K)	a_w
273	1.00
268	0.95
263	0.91
258	0.86
253	0.82

ume of liquid water at temperature T, and R is the gas constant. Hence, in an external pressure range from 1 atmosphere down to the vapor pressure of water (i.e., liquid water in vacuo) the saturation partial pressure of pure liquid water varies negligibly.

Two further situations must be considered:

1. The equilibrium state of water as ice, i.e., pure solid water
2. Application of extremely high hydrostatic pressures

First, consider the frozen system. What is the a_w of a frozen food? To answer this, a standard state must first be chosen. There are two possible choices, either pure liquid water or pure solid water (ice), both which will be at the temperature of the frozen system. If pure ice is chosen as the standard state, the requirement for equilibrium means that, because ice is present in the system, the a_w based on the pure ice standard state is always 1. This may be interesting, but it is not very helpful. What if the pure liquid standard state is chosen? Technically, this state cannot exist under the above conditions, because below 273K (0°C), under atmospheric pressure, ice is the equilibrium form of water. However, undercooled liquid water can exist if care is taken to prevent the nucleation of ice, hence preventing the formation of ice. By using undercooled systems, the partial pressure above pure liquid water can be measured below 0°C. The results are shown in Table 2.3 with measured values obtained down to −15°C and extrapolated values at lower temperatures.

Using this hypothetical standard state of pure liquid water at subzero temperatures, one can determine the a_w values for frozen foods at different temperatures, as shown in Table 2.4. Note that these values are true for all frozen foods at that temperature, and represent the a_w of pure water ice (at the same temperature) using the hypothetical pure liquid water standard state. This links closely to the phenomenon of freezing point depression by solutes. Remember, as stated in the introduction, at equilibrium the chemical potential of water is the same in all phases. In a frozen system, one phase is ice, whereas the other is an aqueous solution. Both phases have the same μ_w and therefore the same p_w. For a solution, the concentration dependence of p_w can be estimated from Raoult's law (Equation 2.9), or extensions, taking into account nonideality. For any temperature, p_w for the concentration that just freezes will be p_w for ice at that temperature. The expected freezing point depression comes from taking into account this relationship. To estimate the freezing point depression for the solution, just find the temperature at which the solution p_w is equal to p_w for ice. For a system obeying Raoult's law, the relationship is expressed as:

$$\Delta T_f \cong \left(\frac{RT_{fo}^2}{\Delta H_{fus}} \right) x_s \tag{2.14}$$

where ΔT_f is the freezing point depression, T_{fo} is the freezing point of the pure solvent, ΔH_{fus} is the molar enthalpy of fusion, and x_s is the mole fraction of solute. For water, this relationship becomes:

$$\Delta T_f \cong 1.86 m_2 \tag{2.15}$$

where m_2 is the solution molality. Dissociation of ionic solutes and nonideality in nonionic solutions lead to greater depression in freezing point than would be predicted by this simple equation. Fennema (1973) lists a range of extended equations applicable to various situations. Note, however, that measurement of freezing point depression is a means of determining p_w, because p_w for ice is known as a function of temperature. From Table 2.4, by knowing T_f, the a_w at T_f can also be known, and because the temperature dependence of a_w is often small, this can be an acceptable estimate of a_w for the sample, up to room temperature.

Considering the application of high hydrostatic pressure, Equation 2.13 should apply. However, according to Equation 2.13, even pure water under high hydrostatic pressure will have a_w significantly removed from 1.0, assuming a standard state defined at 1 atmosphere. If the standard state is defined as pure water at the applied hydrostatic pressure, then the effect of solute is likely to be similar to the effect under normal pressures. Note that, at high hydrostatic pressure, V^o will not be the same as under normal pressure ranges. The utility of a_w as a system descriptor under high hydrostatic pressures has not been demonstrated and methods of measurement have not been developed.

It is appropriate at this point to summarize the effect of the various factors identified as critical to the understanding of the meaning and measurement of a_w by using a series of simple diagrams—pictures are often easier to comprehend than sets of equations on a page.

Figure 2.1 schematically represents the vapor pressure-temperature dependence of water and an aqueous system. In reality, the lines representing vapor pressure will conform to equations such as Equations 2.9 and 2.10.

Note that a_w at T_1 is given by BF/BE and a_w at T_2 is given by DH/DG. This diagram can assist in understanding the various measurements and pitfalls. Consider the situation at T_2. How is p_w at T_2 measured or estimated? Using a dew point cell, the mirror can be cooled to the unique temperature T_d at which p_w of the sample at T_2 is equal to p_w^o at T_d. In this diagram, the temperature T_d is T_1 and the saturation pressure p_w^o at T_1 is represented by point E. Both BE and DH are the same length and represent equal values of vapor pressure. Note that if the sample were cooled to T_1 rather than maintained at T_2, its vapor pressure (line p) would be represented by BF. Clearly, a source of potential error is failure to properly measure T. However, this is easily avoided with care. A more serious potential source of error is to mistake the sample temperature, which would result in using an erroneous p^o_w in the ratio. At around 25°C, both p^o_w and p_w will change approximately 5% for each 1K change in temperature, so that a 1K error in measuring T_1 or T_2 represents a 5% error in the estimated RVP.

Using this diagram, it is easy to see the precautions necessary for good measurements. For a dew point cell, both dew point and sample temperature must be precisely known. What about electrical probes? For these, an electrical characteristic of the probe has been correlated to RVP; it assumed that reported measurements represent an equilibrated sample. Such measurements are often made in a system in which the sample and probe are isothermal with one another. The actual temperature need not be known precisely, so long as the system is truly isothermal, because the temperature dependence of a_w is slight (due to similarity in the temperature dependences of p^o_w and p_w). Note, however, that even a 0.2K temperature inequality results in a 1% error.

This does not mean that such probes can only be used in an isothermal system. As long as the temperature in the region of the probe and the temperature of the sample are both precisely known, the probe reading can be used to estimate the steady state p_w maintained throughout the system, and hence, knowing p^o_w at the temperature of the sample, results in $\left(\frac{p_w}{p^o_w}\right)_T$.

Figure 2.2 illustrates a typical dew point cell configuration, with the sample being held in a controlled, constant temperature region separated from the cooled mirror surface of the dew point assembly. An optical detector provides a signal indicating the formation of dew. It is assumed that the design is such that a uniform vapor pressure is established throughout. When dew formation is just detectable, both sample temperature and mirror temperature are recorded, and used, as in Figure 2.1, to estimate sample vapor pressure and allow for calculation of the RVP at the sample temperature, which is an estimate of a_w. Figure 2.3 illustrates in a more general way the concept of how such a nonisothermal system can be used to determine a_w. In Figure 2.3, region A (the sample region) must be isothermal, as must region C (the sensor region). Region B, the connecting region or bridge, may have temperature gradients, so long as all temperatures in this region exceed T_c and T_a to prevent any possibility of condensation. The system must be designed such that, after reaching a steady state, $p_a = p_b = p_c$.

In all such measurement systems, it is critical that there is free circulation of vapor, so that the partial vapor pressure be maintained uniform throughout the system, particularly in both sample and sensor regions. As can be deduced from Figure 2.1, it is essential that the actual partial vapor pressure in the system be constant and uniform during measurement. Given the dependence of $p_w{}^o$ on temperature, it is critical that the sample region temperature be uniform throughout and, similarly, that the temperature in the vicinity of the sensor be uniform, and that both be accurately known. These systems all measure or estimate RVP after equilibration of a sample of known composition. An alternative approach to establishing the relationship between water content and a_w is illustrated in Figure 2.4. This is the isopiestic method, in which p_w is controlled rather than measured. A constant partial pressure, p_w, is established using a control solution with a known vapor pressure at the temperature of equilibration. Saturated salt solutions with excess salt (salt slurries) provide a range of known vapor pressures at convenient temperatures (Greenspan 1971). Because the slurries are saturated, the equilibrium vapor pressure is not affected by limited

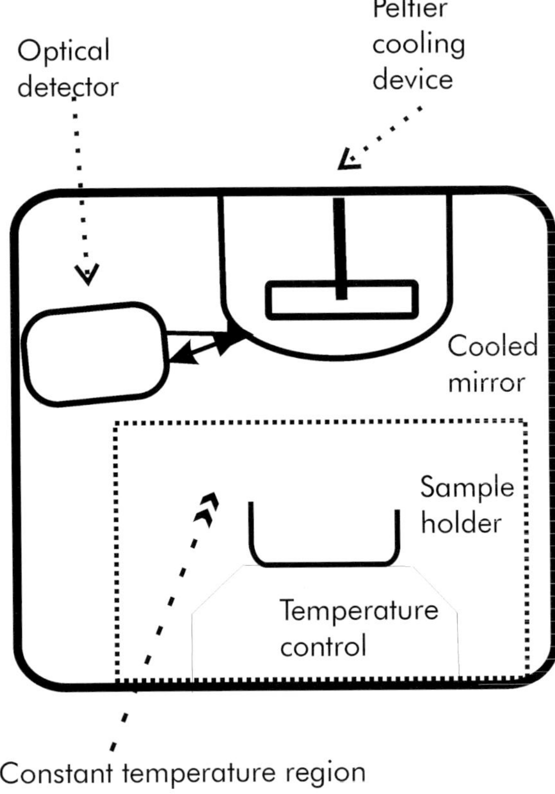

Figure 2.2. Schematic diagram of a dew point cell showing the sample holder (in a constant temperature region, and the cooled mirror outside this region. The optical detector senses the initial formation of the dew film on the cooled surface, and the temperature at that surface is immediately determined.

gain or loss of water, but it is affected by the change in saturation concentration with temperature. Hence, good temperature control is necessary. The sample equilibrates under vapor pressure established by the slurry until reaching a steady water content. The unknown attribute of the final sample to be determined is this water content. After equilibration, the difference in weight between the dry sample and equilibrated sample is the water content at that particular p_W. For an isopiestic system of this type to provide quality results, a known, constant, and stable temperature throughout the system is essential, so that the p_W can be converted into the a_W of the equilibrated sample. A variant of the method illustrated in Figure 2.5 is to control the vapor pressure within the constant temperature enclosure by connecting it to a second constant temperature enclosure containing pure water (e.g., in the relationship between A and C in Figure 2.3). The vapor pressure established throughout the whole system will be p^o_W at the temperature of the second enclosure.

Figure 2.6 shows a typical sorption isotherm, plotting water content against RVP or a_W. Sorption isotherms can be obtained using any of the measuring systems schematically illustrated in the earlier figures. Such isotherms are discussed in later chapters.

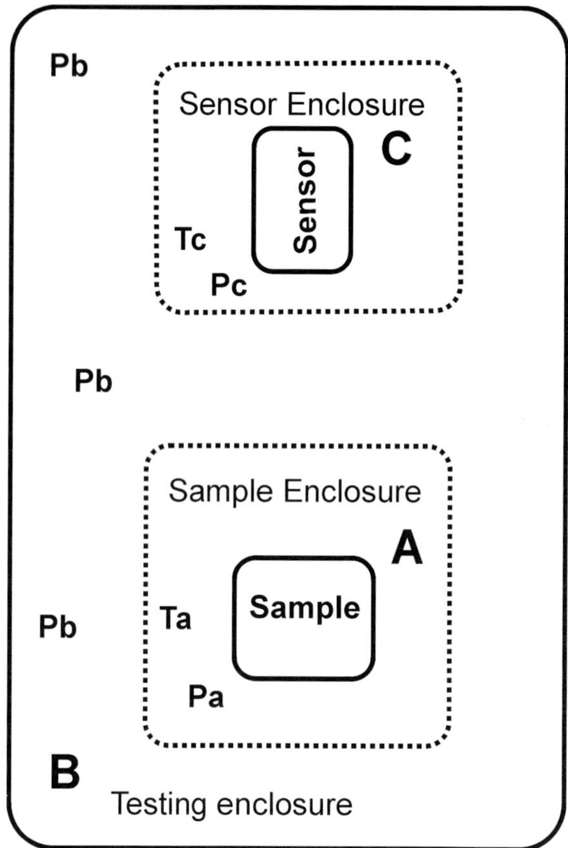

Figure 2.3. Schematic diagram of a generic system for determination of relative vapor pressure. Regions A and C are independently maintained at steady, uniform temperatures. Vapor transfer occurs through region B. The cell design ensures that the vapor pressures in each region, p_a, p_b, and p_c, are identical.

Given that a_w is a thermodynamic concept, care must be used when using the term. Absent evidence of true equilibrium, research has shown that the RVP of a food can be related in many cases to its stability, to its tendency to support microbial growth, and to the rate of change of many other quality parameters important to the acceptability of foods. This chapter has introduced the concept and identified the precautions that must be taken to obtain good-quality data, while also explaining why these precautions are necessary, as well as some of the potential errors associated with failure to meet the appropriate criteria necessary to validate a measurement.

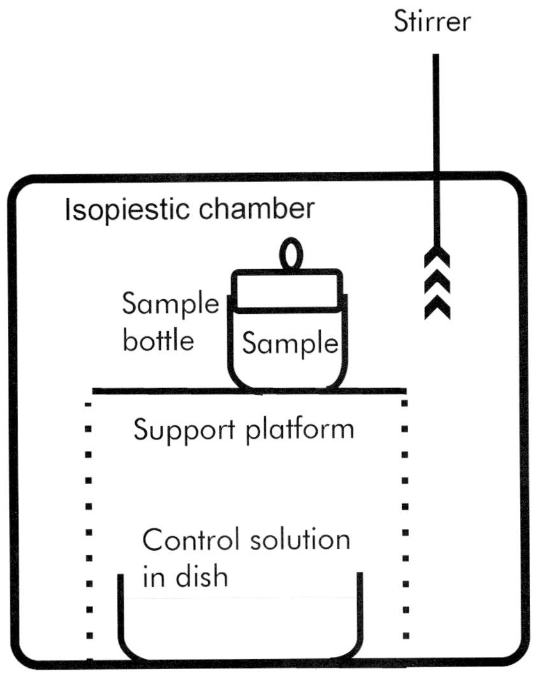

Figure 2.4. Schematic diagram of a simple isopiestic system.

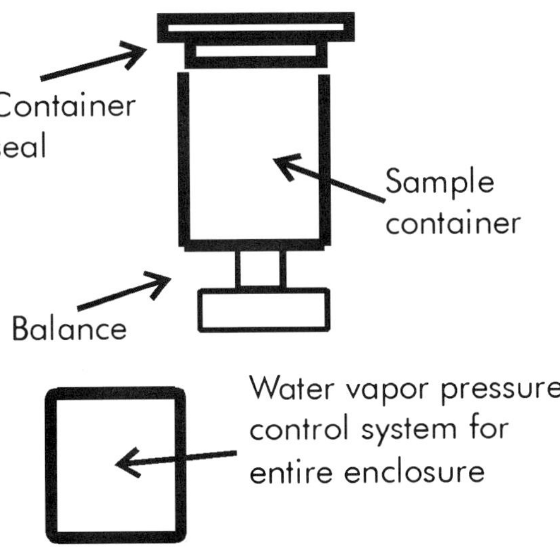

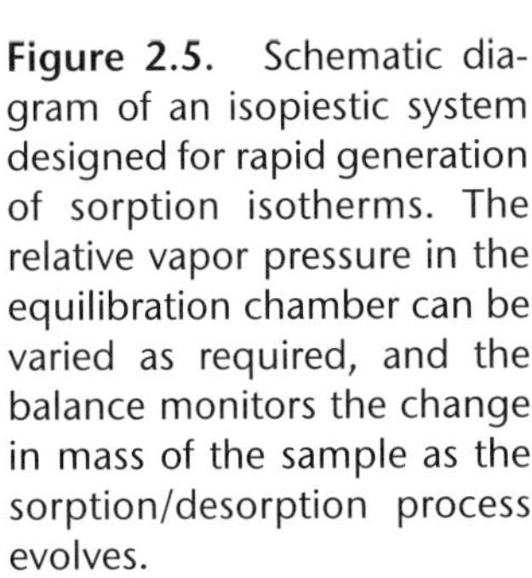

Figure 2.5. Schematic diagram of an isopiestic system designed for rapid generation of sorption isotherms. The relative vapor pressure in the equilibration chamber can be varied as required, and the balance monitors the change in mass of the sample as the sorption/desorption process evolves.

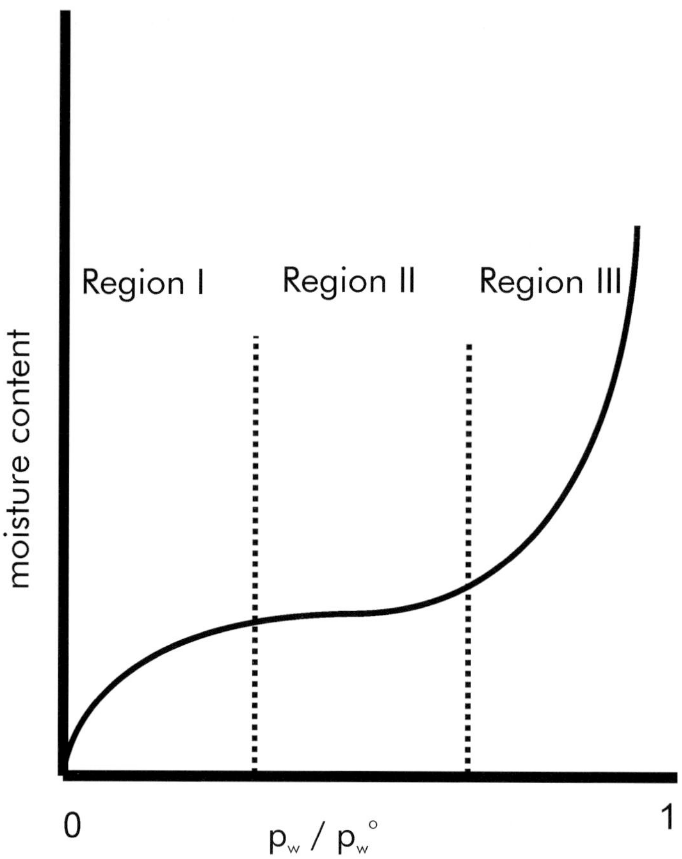

Figure 2.6. Schematic sorption isotherm. The three main regions of the isotherm are identified.

References

Fennema, O. 1973. Water and ice. In: *Low temperature Preservation of Foods and Living Matter,* eds. O. Fennema, W.D. Powrie, and E.H. Marth, pp. 1–77. New York: Marcel Dekker.

Greenspan, L. 1971. Humidity fixed points of binary saturated aqueous solutions. *Journal of Research of the National Bureau of Standards and Technology* 81A:89–96.

Lewis, G.N., and Randall, M. 1961. *Thermodynamics,* revised by K.S. Pitzer and L. Brewer, p. 242. New York: McGraw-Hill.

Reid, D.S., Fontana, A., Rahman, S., Sablami, S., and Labuza, T. 2005. Vapor pressure measurements of water. In: *Handbook of Food Analytical Chemistry*, eds. R.E. Wrolstad, T.E. Acree, E.A. Decker, M.H. Penner, D.S. Reid, S.J. Schwartz, C.F. Shoemaker, D.M. Smith, and P. Sporns, Section A2. New York: Wiley.

Scott, W.J. 1953. Water relations of *Staphylococcus aureus* at 30°C. *Australian Journal of Biology and Science* 6:549–556.

Scott, W.J. 1957. Water relations of food spoilage organisms. *Advances in Food Research* 7:83–127.

Slade, L., and Levine, H. 1991. Beyond water activity: recent advances based on an alternative approach to the assessment of food quality and safety. *Critical Reviews in Food Science and Nutrition* 30(2-3):115–360.

3 Water Activity and Glass Transition

Yrjö H. Roos

Introduction

The importance of water to all life is well recognized, as well as the role of water in controlling the growth of microorganisms in foods and other biological systems (Labuza 1968). Glass transition has also been recognized as an important parameter, affecting properties of food solids at low water contents and in frozen systems (White and Cakebread 1966, Levine and Slade 1986). Water as the main solvent of nonfat food solids is the main plasticizer of noncrystalline food components. As a plasticizer and solvent, water may dramatically affect rates of mechanical and diffusional properties of foods (Slade and Levine 1995). There are, however, fundamental differences between water activity (a_w) and glass transition as indicators of thermodynamic equilibrium and nonequilibrium phenomena of food systems (Chirife and Buera 1995, Roos 1995). *Water activity* can be defined as an equilibrium property of water in foods and other materials, whereas glass transition is a relaxation process occurring in food solids during transformation of noncrystalline solids to a more liquid-like supercooled state (Roos 2002).

The definition of a_w is based on the chemical potential of water within a food system, which at equilibrium must be the same as the chemical potential of water in the surroundings of the food. This also means that the vapor pressure of liquid water in the food and the vapor pressure of water vapor in the surroundings must be equal. Hence, a_w can be obtained as the ratio of vapor pressure of water in a food and the vapor pressure of pure water at the same temperature and pressure conditions (Labuza 1968). Therefore, the measurement of vapor pressure also gives a_w. The relationships between chemical potential, vapor pressure, and a_w indicate that a_w is a temperature-dependent property of food. The water content of food commodities are relatively constant and only the water activity of food is expected to change with temperature. The amount of water as a plasticizer corresponds to the total amount of water and, independent of a_w, any water molecules may act as plasticizers for dissolved and water miscible substances (Roos 1995).

The glass transition is clearly a property of food solids, which is affected by the extent of water plasticization of the solids. Amorphous food components plasticized by water cover mainly carbohydrates and proteins with minor hydrophilic food components and ions (Roos 1995). Hence, the fundamental distinction between a_w and glass transition is that a_w is a property of water molecules, and glass transition is a property of amorphous food components. Water activity and glass transition give extremely important information about the physicochemical properties of foods. For example, a_w is an important measure of the ability of various microorganisms to grow in a particular food (Chirife and Buera 1995, 1996). Glass transition of food controls the solid and liquid-like properties of food solids, and it affects molecular mobility within a food (Slade and Levine 1991, 1995; Roos

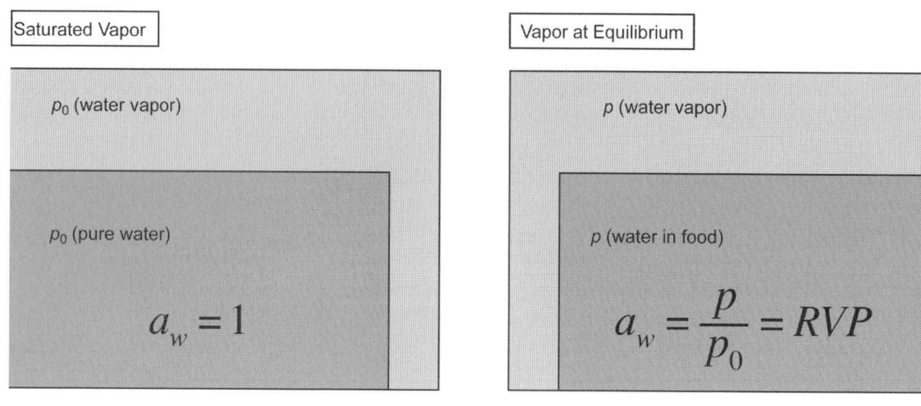

Figure 3.1 Definition of water activity and its relationship to relative vapor pressure (RVP) in a closed container.

1995; Le Meste et al. 2002; Roudaut et al. 2004). However, mobility of various substances may also be related to food microstructure and porosity (Roos 2002).

Water activity and glass transition often both contribute to food properties. This chapter will discuss the importance and relationships of these food stability parameters.

Water Activity

Water activity is, by definition, based on a thermodynamic equilibrium state of water in an equilibrium system, e.g., a dilute solution with no time-dependent properties. However, most food systems may practically exist in a steady state, but they are not in a thermodynamic equilibrium state and may undergo changes during storage (Slade and Levine 1991, 1995; Chirife and Buera 1995, 1996; Roos 1995, 2002). Long-term stability of food systems, however, can be obtained by freezing and dehydration. In both methods, food solids are concentrated either by separating water into another phase (ice) or by removing nearly all of the water (Roos 1995). Successfully processed and stored frozen and dehydrated materials are extremely stable (Chirife and Buera 1996). Stability of the materials is based on at least a partially amorphous solid (glassy) state of the solids. Although a_w can be measured for most foods, their stability at the same a_w may vary significantly, e.g., from hours to years. For example, a food powder with sucrose as a main component becomes sticky and may crystallize at 0.3 a_w within hours, whereas dairy powders with high lactose contents remain stable for years at the same a_w at typical storage temperatures.

The definition of *water activity* in terms of the vapor pressure of water in various systems is described in Figure 3.1.

Absolutely pure water always has a temperature-dependent vapor pressure, p_0, and water vapor in the headspace of a closed container with pure water at equilibrium is saturated, i.e., an equilibrium exists between the liquid and vapor phases of water. Hence, the vapor pressure of both phases is equal to p_0. The addition of solutes to water results in a decrease of the vapor pressure of water. For low solute concentrations, Raoult's law (see Equation 3.1) can be used to establish relationships between vapor pressure, p, and mole fraction of water, x. According to Raoult's law, the vapor pressure of water decreases linearly with decreasing mole fraction of water.

$$p = xp_0 \tag{3.1}$$

The vapor pressure, p, of water in food can also be obtained by measuring the headspace vapor pressure of water of the food when placed in a closed container. Because the vapor pressure of water in food is less than the vapor pressure of pure water, the vapor does not become saturated. At equilibrium, the vapor pressure of liquid water within the food and in the vapor phase is the same and a_w is given by the relative vapor pressure (RVP). It should be noted that the a_w refers to the activity of the liquid (or vapor) water inside the food and RVP refers to the vapor pressure of water vapor in the headspace of a closed container containing the food. In most cases, Raoult's law cannot be applied to complex food systems, because of the nonideality of the systems and the mole fractions of solutes and water cannot be determined.

Water Activity and Temperature

One important property of the water vapor pressure and a_w is their temperature dependence. In general, the temperature dependence of vapor pressure of water and a_w follow the Clausius-Clapeyron equation (Labuza 1968, Roos 1995). Hence, assuming that no other food properties change, the a_w applies to only one temperature, and it decreases with decreasing temperature. This temperature dependence at constant water content can be a significant factor in the storage stability of, for example, foods with high starch contents. At low temperature, the a_w may be low enough for keeping the product microbially safe. An increase in temperature may result in a fairly significant increase in a_w at a constant water content, allowing mold growth and a potential production of fungal toxins (see Figure 3.2). It should be noted that there may not be any changes in other material properties and the material may remain in a solid, glassy state at both temperatures. It should also be noted that if the a_w of a food remains constant with a change of temperature, there must also be a change of the total water content of the food or a change of water sorption properties of food solids associated with the change of temperature.

Water Activity and Time

Food solids are typically nonequilibrium materials with a few exceptions, such as crystalline food components under carefully controlled conditions. Hence, food solids are typically in a nonequilibrium state and undergo changes with time in order to attain equilibrium (Chirife and Buera 1995; Roos 1995, 2002; Jouppila et al. 1997). Relaxation times of the time-dependent changes may be extremely short or extremely long depending on the state of the solids, e.g., a liquid or an amorphous solid, respectively. However, these changes indicate the time-dependent nature of food properties, which may be observed, for example, from changes in food microstructure and a_w during storage (Chirife and Buera 1995, 1996; Roos 1995; Haque and Roos 2004).

As food properties can be time dependent, a_w may change over storage time as illustrated in Figure 3.3. Perhaps crystallization of amorphous food components causes the most dramatic of time-dependent changes in a_w (Roos 1995, Jouppila and Roos 1997,

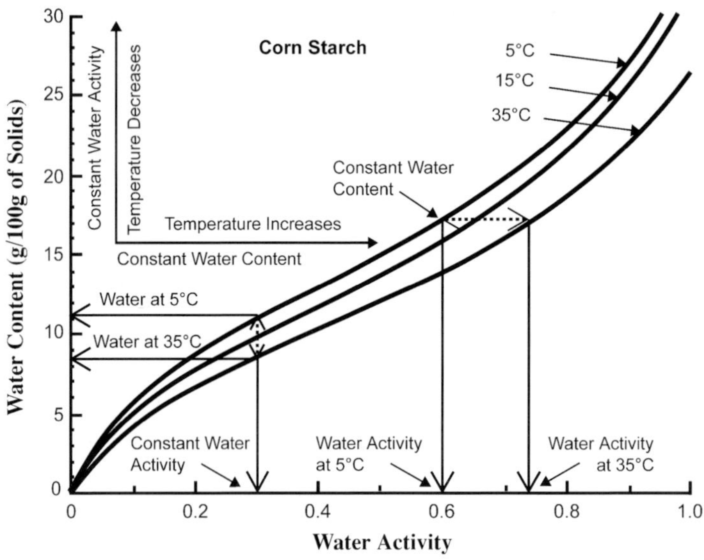

Figure 3.2 Temperature dependence of water sorption of corn starch. Changes in temperature and water content affect water activity as shown (*arrows*).

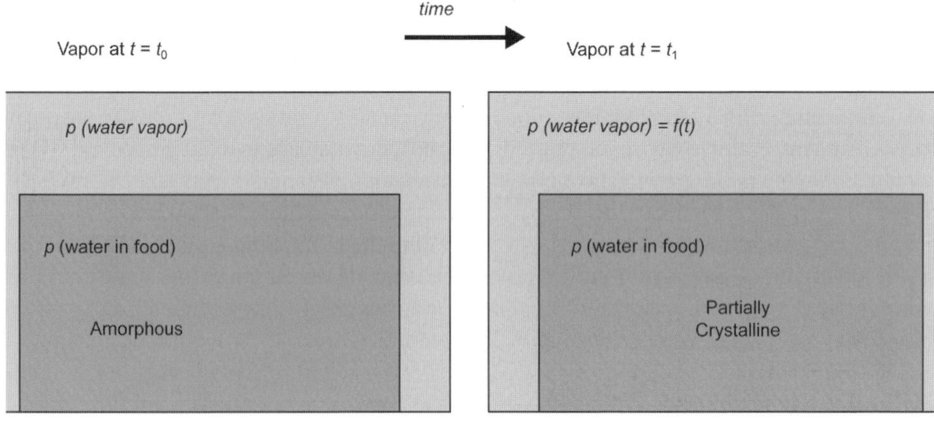

Figure 3.3 Time-dependent changes in water activity resulting from crystallization of amorphous food components.

Hartel 2001, Haque and Roos 2004). Completely amorphous food components may be very hygroscopic and contain substantial amounts of water. Water, however, causes plasticization of solids and enhances crystallization. Crystallization may occur in an anhydrous crystal form and, as a result, water associated with the amorphous solids is desorbed from the material (Jouppila et al. 1997, 1998; Haque and Roos 2004). This increases RVP and

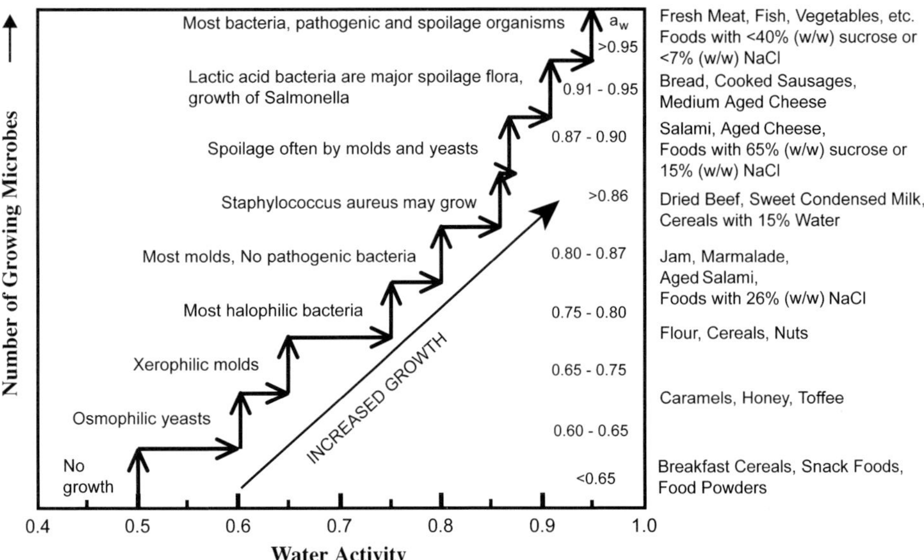

Figure 3.4 Water activity limits for growth of microorganisms in foods and examples of foods with water activities over the range of various growth limits.

apparent a_w of the system if water is not removed from the system (Katz and Labuza 1981, Roos 1995). This also means that the vapor pressure of water becomes a function of crystallization time. In addition, there can be other changes in food properties, which may affect a_w. These include, for example, starch retrogradation, chemical and biochemical changes, microstructural transformations, and other changes in food structure (Roos 1995, Bell 1996, Jouppila and Roos 1997, Haque and Roos 2004, Miao and Roos 2004).

Importance of Water Activity

Water activity is a well-established parameter for controlling the growth of microorganisms in foods (Chirife and Buera 1995, 1996). But a_w is only one factor controlling growth; in addition, there are several other factors, such as pH and temperature. However, the temperature dependence of a_w is often ignored in the determination of a_w limits for growth of microorganisms. Assuming an a_w at room temperature and limits for growth of microorganisms, there are a few important critical a_w values that apply to a number of microorganisms in various foods (see Figure 3.4). The two most important of these critical values are 0.6 a_w for the growth of any microorganisms and 0.86 aw for the lowest a_w where growth of pathogenic bacteria has been observed (Chirife and Buera 1996). These a_w limits have no relationship to the glass transition of the food solids. For example, a lactose hydrolyzed skim milk would be a free-flowing syrup-type liquid at 0.86 a_w, whereas a high-starch food system would be in the vicinity of the glass transition (see Figure 3.5) and, therefore, would be a fairly solid material, possibly a free-flowing powder. However, the a_w of these systems could be the same, allowing growth of pathogenic and other bacteria. Another important consideration is the hydrophilic and hydrophobic properties of food solids, which have an impact on a_w. In general, lipids are hydrophobic, and a_w is a prop-

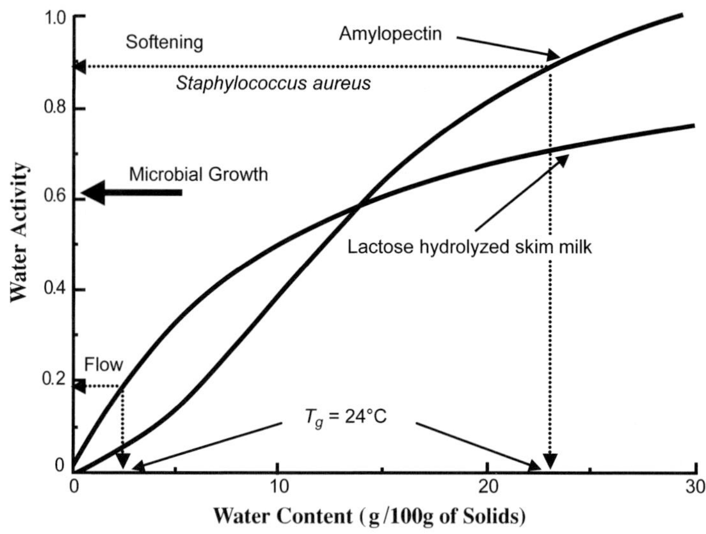

Figure 3.5 Differences in water sorption properties of amylopectin and lactose hydrolyzed skim milk (high monosaccharide, glucose plus galactose, content). Lactose hydrolyzed skim milk is in the liquid state almost over the entire water activity range, while amorphous amylopectin is a glassy material up to about 0.90 a_w at room temperature.

erty of the hydrophilic food solids. Hence, a_w of high-fat spreads can be fairly high and affected by small overall changes in the salt content, for example.

A number of attempts have been made to relate the rates of biochemical, chemical, and structural changes in foods with a_w. Rates of most changes have a relationship with the a_w of the system (Labuza et al. 1970), but such global values for rate dependence on a_w as those found for microbial growth have not been reported. It seems that a_w alone cannot be used to explain the variations in reaction rates, structural changes and flow, crystallization properties of low-moisture food components, and changes in diffusion and flavor retention.

Glass Transition

Glass transition and a_w are two completely different parameters, but they are complementary and both can be used to explain food deterioration or stability. Water activity is an extremely important parameter affecting microbial growth (Chirife and Buera 1996). Glass transition is a property of noncrystalline solids affecting mechanical properties and molecular mobility in food solids (Roos 1995, Le Meste et al. 2002). Glass transition can be defined as a transformation of a supercooled liquid to a highly viscous, solid-like glass. Glass transition occurs over a temperature range as the molecules become "frozen" and may exhibit only rotational motion and vibrations. When a glass is heated over the glass transition, molecules become mobile and gain translational mobility. Hence, glass transition is related to changes in food structure and microstructure, crystallization, rates of diffusion-controlled reactions, and, possibly, stabilization of microbial cells and spores (Roos 1995, Chirife and Buera 1996, Le Meste et al. 2002). Glass transition may occur at various tem-

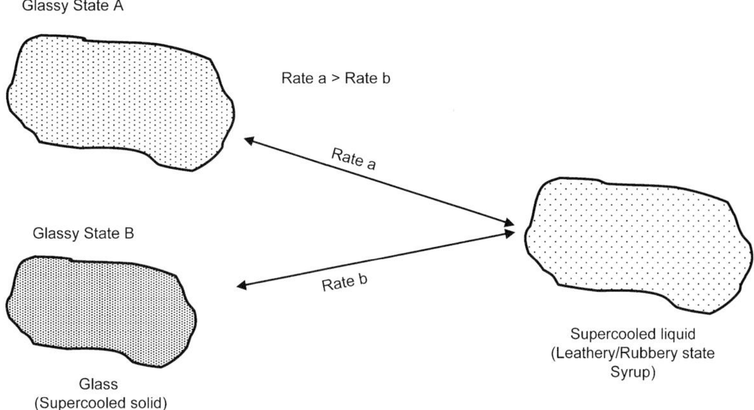

Figure 3.6 Freezing of supercooled, liquid systems into a solid glassy state. Depending on the cooling rate, various glassy structures can be formed.

peratures depending on water content and, therefore, glass transition may occur at ambient conditions at a given a_w or water content (Roos 1993a, 2002).

The amorphous state of solids in frozen and dehydrated food systems has been recognized (White and Cakebread 1966) and presumed to affect food properties (Slade and Levine 1991, Roos 1995, Le Meste et al. 2002). The glass transition is a particularly important determinant of the dehydration characteristics and stability of high carbohydrate–containing foods (Roos 2004). Polymeric food components, such as carbohydrate polymers and proteins, have high-temperature anhydrous glass transitions, and they seldom exhibit problems in dehydration or flow in a powder form. Low-molecular weight carbohydrates, particularly monosaccharides and disaccharides, have comparably low-temperature glass transitions (Roos 1993b). Foods with high monosaccharide and disaccharide contents can be extremely difficult or impossible to dehydrate, and they do not freeze completely in typical food-freezing processes (Roos 1995, 2004). Hence, a number of studies have reported glass transitions for sugars and food solids with high sugar contents.

Furthermore, glass transitions of amorphous pure sugars can be measured by a number of techniques, such as differential scanning calorimetry and dynamic mechanical analysis. Many complex food systems contain only partially amorphous components, e.g., starch and proteins, and their glass transitions cannot be measured by simple methods. Amorphous solids may also exist in dispersions, e.g., fat-containing foods and frozen systems, which are more difficult to characterize. In such systems, the glass transitions may be weak and/or interfere with other transformations, e.g., crystallization and melting of lipid components (Roos 1995, 2002).

The glassy state of a material is a nonequilibrium state, and a substance may have an infinite number of glassy states depending on how the glass has been formed (see Figure 3.6). The glass properties formed in cooling or in food processing are dependent on how molecules become “frozen” as they are immobilized. Hence, a glass may have a relatively large or low “free” volume (see Figure 3.7). Differences in glass properties can be observed from relaxations associated with the glass transition reflecting molecular arrange-

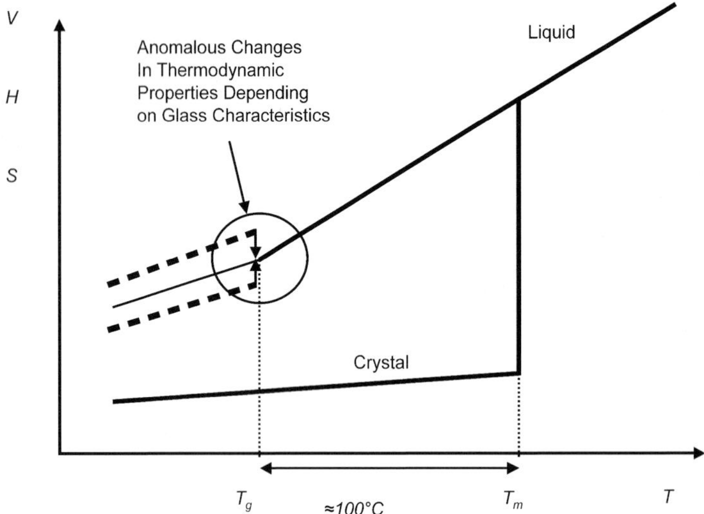

Figure 3.7 Comparison of thermodynamic properties of equilibrium solids (crystalline state) and nonequilibrium materials (glassy and supercooled liquid states). Various thermal and mechanical relaxations can be associated with the glass transition in vicinity of the glass transition temperature, T_g, depending on relative cooling and heating rates, and the state of the noncrystalline solid. T_m is the equilibrium melting temperature of the crystalline state.

ments as they gain translational mobility. A rapidly formed glass can have a high free volume and give an exotherm in a differential scanning calorimeter (DSC) scan as the extra energy is released in heating over the glass transition (relaxation to a lower free volume state) (Roos 2002). A very slowly formed "dense" glass gives an endotherm as molecules require energy when heated over the glass transition due to relaxation to a supercooled liquid state and expansion. Such relaxations indicate the nonequilibrium state of the material and its time-dependent nature. Transformations may also occur in the glassy state over storage time, which are known as physical aging (Roudaut et al. 2004).

A typical amorphous material undergoes a glass transition at 100° to 150°C below its equilibrium melting temperature (Sperling 1992, Roos 1993b). The transition of macromolecules in foods is often broad and difficult to analyze. However, the addition of water decreases the temperature of the transition and makes it more pronounced. The anhydrous transition of polymers in foods, such as starch and proteins, is very often higher than their decomposition temperature and the transition cannot be measured experimentally. The anhydrous glass transition can also be at temperatures well below room temperature, which indicates that the material cannot be solidified and dehydrated into a powder. Examples of such materials are fructose, glycerol, sorbitol, and xylitol, which are stable solids at room temperature only in the crystalline state (Roos 1993b, Talja and Roos 2001). The glass transition of most monosaccharides is in the vicinity of room temperature, and foods with high monosaccharide contents cannot be dehydrated into stable amorphous powders. The materials, however, are microbially stable but they undergo rapid physical and possibly chemical transformation, although their a_w may be near "zero."

Water Plasticization

Water is the main solvent and plasticizer (softener) of hydrophilic food solids (Slade and Levine 1991, 1995; Roos 1995). Interactions of water molecules with amorphous, hydrophilic components are observed from decreasing temperature of the glass transition with increasing water content. Such plasticization is typical of solvents in polymers, and in most biological systems water plasticization can result in significant changes of material properties (Slade and Levine 1995). As a result of water plasticization, the free volume of the material and molecular mobility increase and the glass transition is observed at a lower temperature corresponding to the increase in water content.

Water plasticization can be determined experimentally by observing the glass transition of a material at various water contents (Roos 1995, 2002). At low water contents, there is a dramatic decrease of the temperature of glass transition. The Gordon-Taylor equation, which traditionally has been applied to model solvent plasticization of synthetic polymers, has also been proved useful in modeling water plasticization of food solids (Gordon and Taylor 1952, Roos 1995). The Gordon-Taylor equation may also be obtained from a thermodynamic approach introduced by Couchman and Karasz (1978). They found that the temperature of glass transition of a binary mixture of solvent and amorphous solids was a function of the component glass transitions, their weight fractions, and the magnitude of the change in heat capacity of the components over their individual glass transitions.

The glass transition of water has been found at about −135°C (Johari et al. 1987). Addition of solutes, e.g., sugars in water, increases the glass transition toward the glass transition of the solute (Roos and Karel 1991, Roos 1995). Similarly, addition of water to an amorphous, hydrophilic glass decreases the glass transition toward that of pure water. State diagrams are often established for food components and food systems to describe their water plasticization properties as well as transitions in frozen and freeze-concentrated systems (Roos 1995, Roos et al. 1996).

Glass Transition and Water Relations

A number of low-moisture food systems have empirically established a_w limits for storage stability, crystallization of component compounds, loss of texture or crispness, and rates of deteriorative reactions. Water activity cannot be used to explain the origin of these changes, and it has been useful to combine information on water content, a_w, and glass transition data of food systems (Roos 1993a).

Hence, diagrams can be established to describe critical water content and a_w values corresponding to the extent of water plasticization (water content) depressing the glass transition to observation temperature. In such an approach, it is assumed that the increase in molecular mobility and decrease of relaxation times above the glass transition enhance structural transformations and diffusion, which accelerate many deteriorative changes.

The Guggenheim-Anderson-de Boer (GAB) water sorption model (see Equation 3.2) has proved to fit to water sorption data of most food materials (van den Berg and Bruin 1981). The model can be used to establish sorption isotherms, providing the relationship between a_w and water content (m). The Gordon-Taylor equation has proved useful in establishing relationships between the temperature of the glass transition and water content (Roos 1995, Le Meste et al. 2002). The use of these models can be advantageous in predicting glass transition temperatures at various water contents and the corresponding water activities at storage temperature (see Figure 3.8).

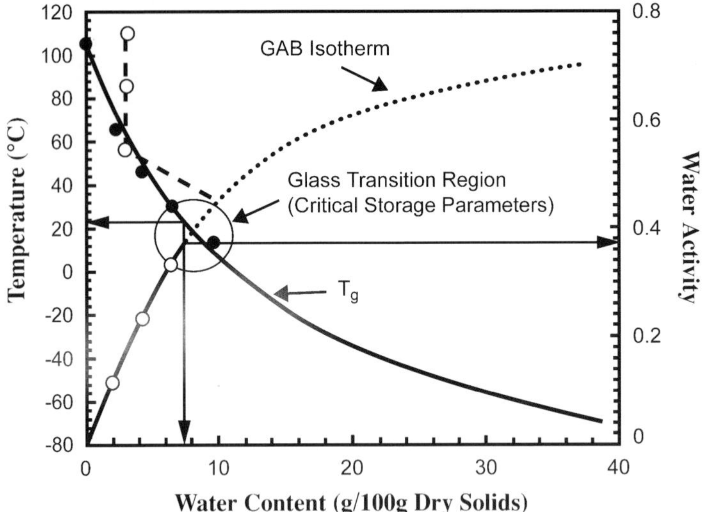

Figure 3.8 Glass transition and water activity of amorphous lactose as a function of water content. Depression of the glass transition to observation temperature because of water plasticization changes from lactose glass to a supercooled liquid. Crystallization from the supercooled liquid state occurs with a rate controlled by the extent of water plasticization. The crystals formed contain less water, which is observed from a decrease in water content as shown by the *dashed line*. Data are from Haque and Roos (2004).

$$\frac{m}{m_m} = \frac{K'Ca_w}{(1 - Ca_w)[1 + (K' - 1)a_w]} \tag{3.2}$$

where m is water content (g/100 g of solids), m_m is monolayer value, and C and K' are constants.

The critical a_w and water content values give direct information about the maximum allowed RVP for storage of amorphous or partially amorphous food systems. They also indicate that at a constant storage temperature, the glass transition may occur as a result of water sorption if the storage RVP exceeds the critical values. Hence, the water sorption may increase the water content of the material to a level, depressing the glass transition temperature (T_g) to below ambient temperature. The resultant decreases in the relaxation times of deteriorative changes can cause rapid loss of food quality (Roos 1993a). It would also be important to note that food systems with high water contents cannot be glassy at typical ambient conditions and the a_w would mainly indicate microbial stability of the product. Critical values for a number of low-molecular-weight food components cannot be defined, because their glass transitions are well below typical storage temperatures (Talja and Roos 2001). In general, the critical values increase with increasing molecular weight as shown for maltodextrins with various dextrose equivalent (DE) values in Figure 3.9.

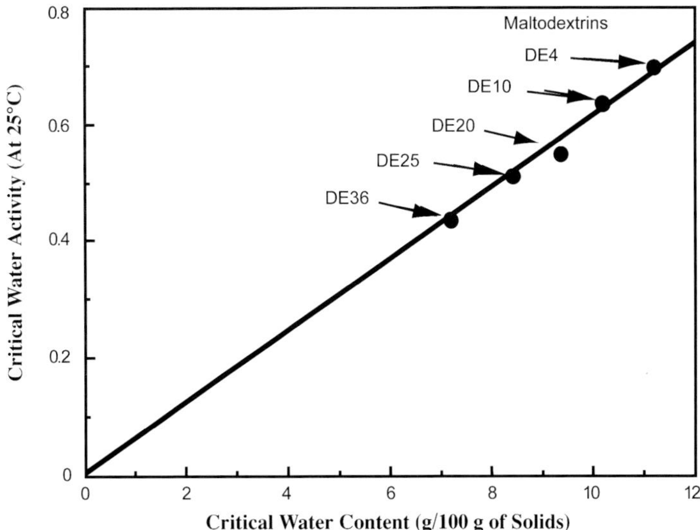

Figure 3.9 Relationship between critical water activity and critical water content of maltodextrins with various dextrose equivalent (DE) values. Data are from Roos and Karel (1991).

Glass Transition, Water Activity, and Relaxation Times

Relaxation Times

The effect of glass transition on the mechanical properties of synthetic polymers has been well established (Sperling 1992). The effect of glass transition on the mechanical properties of food systems, such as caking, stickiness, collapse of structure and crispness of low-moisture foods, has also been demonstrated (Slade and Levine 1991, Roos 1995, Roos et al. 1998, Le Meste et al. 2002). It seems that many of these phenomena are controlled by the flow properties of food solids. Increasing temperature or water content of hydrophilic food glasses may result in the glass transition. Over the glass transition, the solid-like properties of the amorphous materials disappear and they become liquid-like, supercooled materials with enhanced flow properties. This indicates a decrease in relaxation times of mechanical changes, which can be related to the viscosity of the materials (Williams et al. 1955, Sperling 1992). The viscosity of the solid, glassy materials cannot be measured. However, the decreasing moduli above the glass transition with increasing temperature or water content suggest a rapidly decreasing viscosity as a function of the temperature difference to a reference temperature, e.g., onset of the glass transition, Tg, measured as T − Tg (Williams et al. 1955; Slade and Levine 1991, 1995; Roos 1995; Le Meste et al. 2002).

There are a number of empirical equations relating the temperature of amorphous materials above the glass transition to relaxation times. The most common models are the power-law, Vogel-Tamman-Fulcher (VTF), and Williams-Landel-Ferry (WLF) models (Slade and Levine 1991, 1995; Roos 1995; Le Meste et al. 2002). According to the WLF model the relaxation times are related to the glass transition as given by Equation 3.3 (Williams et al. 1955). Hence, the relationship can be used to predict viscosity above the

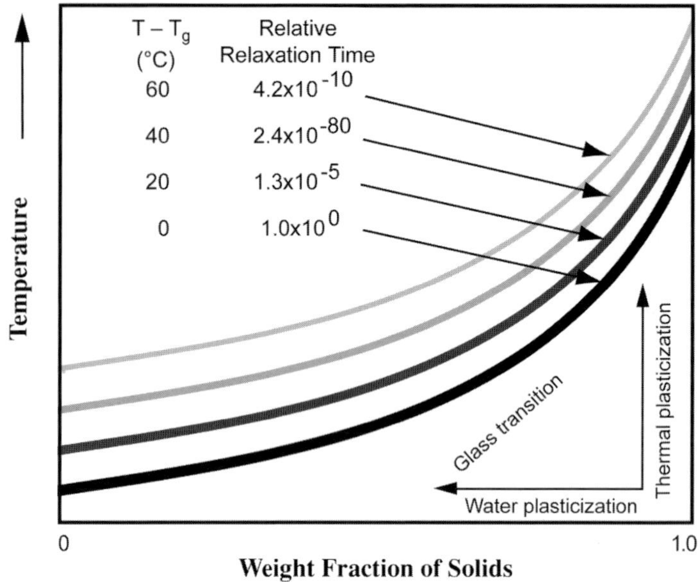

Figure 3.10 Williams-Landel-Ferry (WLF) prediction of isoviscous states of solids plasticized by water above the glass transition.

T_g, and in some cases, changes in rates of diffusion-controlled reactions (Nelson and Labuza 1994, Miao and Roos 2004). However, plasticization by an increase of temperature or water content may change relaxation times differently, and, for example, the WLF constants may vary depending on plasticization mechanism. It also seems that the changes in water content and temperature have some independent effects on reaction kinetics (Bell et al. 1998, Lievonen and Roos 2002, Miao and Roos 2004). Unfortunately, relatively few studies have measured rates of chemical reactions or viscosity as a function of water plasticization at a constant temperature. It should be noted that very small variations of water content can substantially affect a_w and glass transition (Slade and Levine 1991, Roos 1995). Therefore, around the critical a_w and water content, relaxation times can be extremely sensitive to any increase of a_w above the critical value (Roos 2002).

$$\ln a_T = \ln\frac{\tau}{\tau_s} = \ln\frac{\eta}{\eta_s} = \frac{-C_1\left(T-T_s\right)}{C_2+\left(T-T_s\right)} \tag{3.3}$$

where a_T is the ratio of relaxation times at an observation, T, and reference temperature, T_s, and η and η_s are corresponding viscosities. C_1 and C_2 are constants.

The WLF model with its universal constants, $C_1 = 17.44$ and $C_2 = 51.6$, suggests a rapid decrease in relaxation time above the glass transition, as illustrated in Figure 3.10. Figure 3.10 is a schematic state diagram showing contour lines at various temperature–water content states with the same relaxation times. Such diagrams are useful in evaluating effects of water content and temperature on amorphous, low-moisture foods, but they do not take into account a_w of the system.

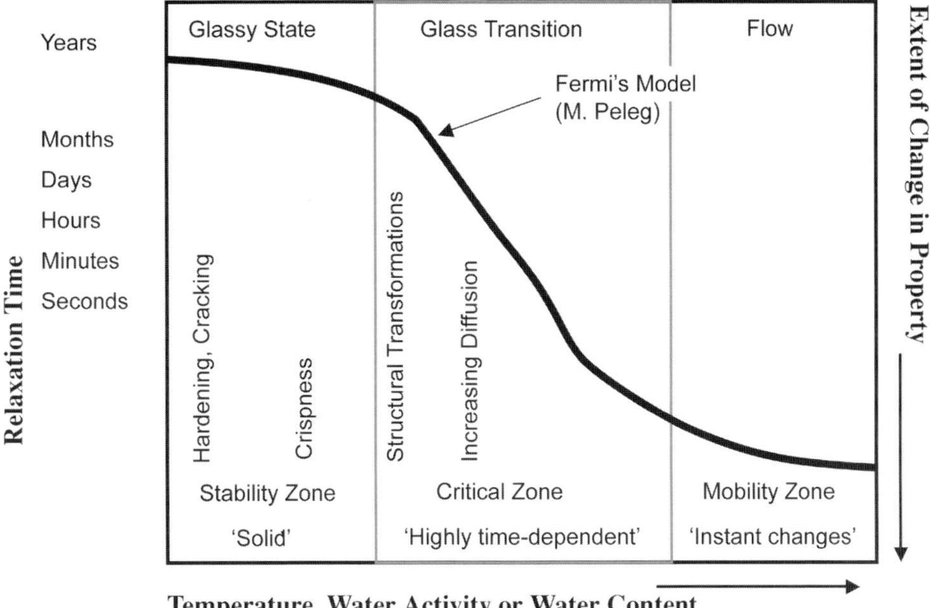

Figure 3.11 Effects of temperature, water activity, and water content on relaxation times of amorphous food solids.

Fermi's model, as suggested by Peleg (1992, 1994), can be used to describe changes in relaxation times in amorphous systems as a function of temperature, a_w, or water content (see Figure 3.11). The schematic diagram of Figure 3.11 can be used to explain stability at low water contents for systems existing in the glassy states, i.e., glass transition at low water contents occurs at high temperatures and above storage temperature. The relaxation times of structural transformations decrease rapidly at and above the glass transition within the critical zone. At and above the glass transition, changes in materials are highly time-dependent and a few degrees increase in temperature or small increases in water content may substantially decrease the relaxation times, i.e., from years to seconds. This, for example, applies to crystallization times of amorphous sugars (Makower and Dye 1956, Roos and Karel 1992, Haque and Roos 2004). Hence, the critical a_w can be defined as the a_w corresponding to water plasticization depressing the glass transition to observation (storage) temperature. The critical water content is the corresponding water content at which the glass transition occurs at the observation temperature.

Time-Dependent Changes

There are a few food properties that are directly controlled by the glass transition. These include stickiness, collapse, crispness of low-moisture foods, and crystallization of amorphous food components. These changes have characteristic relaxation times, which probably can also be related with the viscosity of the systems. Hence, the transformations have a time-dependent nature, and, depending on the $T - T_g$ conditions, changes take place at various rates. Many reactions, such as nonenzymatic browning, enzymatic changes, and, in some cases, oxidation, can be indirectly controlled by the glass transition (Roos 2002,

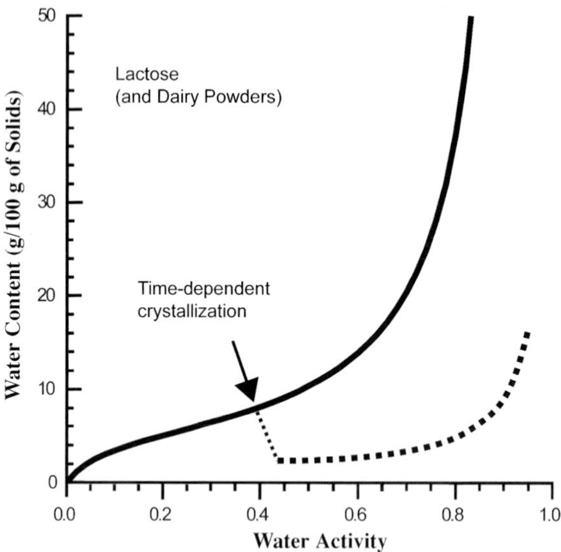

Figure 3.12 Typical water sorption isotherm of lactose in dairy systems. The dotted line indicates a decrease in sorbed water content resulting from time-dependent lactose crystallization above the critical water activity.

Roudaut et al. 2004). The important factor in such reactions is most likely the diffusion of reactants and reaction products affecting the reaction rates and thereby indirectly controlled by the glass transition. In these reactions, water often has a multiple role as at least a reaction medium, solvent, and plasticizer.

Crystallization of amorphous sugars is probably the most dramatic time-dependent change occurring in amorphous food systems above the critical a_w and corresponding water content during storage. Crystallization of amorphous lactose is typical of dairy powders at and above 40% RVP at room temperature (see Figure 3.12). Crystalline lactose may be anhydrous or monohydrate (Jouppila et al. 1997, 1998). The amount of water in amorphous lactose can be substantially higher, and crystallization may result in a dramatic increase in a_w (Katz and Labuza 1981, Chirife and Buera 1995, Roos 1995). In such materials, a_w (storage relative humidity) can be used to control glass transition and the rate of crystallization. Similarly, a_w may be used to control structural transformations affected by the glass transition.

It seems that the time-dependent properties controlled by glass transition are a result of changes in molecular mobility around the glass transition (Roudaut et al. 2004). The small water molecules seem to be mobile in solid, amorphous food systems, but as their concentration increases and the material experiences a glass transition, relaxation times decrease to values observable in short experiments. Hence, a_w can be used as a measure of glass transition at a constant temperature and related to onset of translational mobility of amorphous, hydrophilic solids in foods. However, it may be expected that some food systems are not homogeneous and such foods may exhibit local differences in glass transition behavior. Therefore, local differences in rates of deteriorative changes may occur and affect overall food stability.

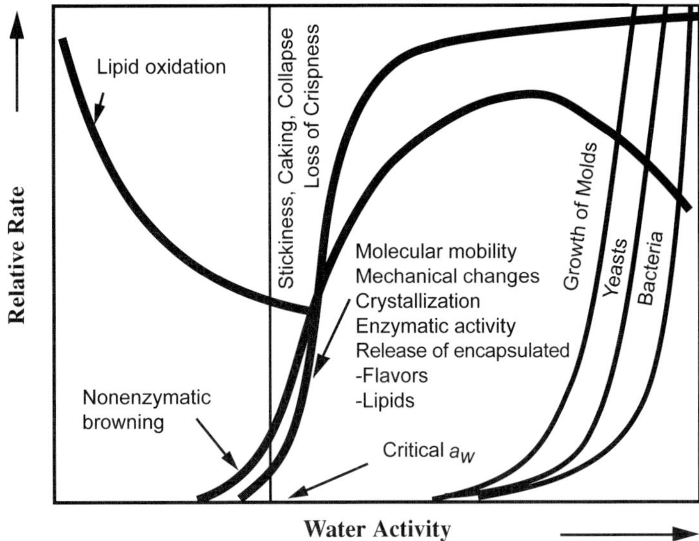

Figure 3.13 Food stability map showing effects of glass transition and water activity on rates of various deteriorative changes and microbial growth in amorphous food systems.

Overall Stability

The effects of a_w on food stability have been successfully described by using "stability maps" showing relative rates of deteriorative changes against a_w (Labuza et al. 1970). Such maps clearly show the effect of a_w on the growth of microorganisms. It is obvious that water is required for the growth of microorganisms in addition to required environmental parameters and nutrients. It has also been observed that some reactions exhibit extremely low rates at low water activities, that the rates increase above a critical a_w as diffusion becomes enhanced, and that the rates decrease at high water activities, possibly as a result of dilution (Duckworth 1981). It has been proposed that these reactions are affected by the glass transition of amorphous food solids (Slade and Levine 1991, Nelson and Labuza 1994). There seems to be some relationship between the rates of diffusion-controlled reactions and molecular mobility and the glass transition (Miao and Roos 2004). However, the rates of the reactions are also independently affected by water content (reactant concentration), pH, and temperature and possibly by other factors, such as density (collapse) and component crystallization in dehydrated systems (Bell et al. 1998, Lievonen and Roos 2002, Miao and Roos 2004).

Food stability maps can also relate glass transition–dependent changes with a_w (Roos 1995). It should, however, be noted that the rates of glass transition–dependent changes are affected by food composition and they may occur with no water present in the solids or at water activities corresponding to the critical a_w of the particular material. Hence, stability maps may be established for various foods showing critical values for a_w allowing sufficient molecular mobility to result in structural transformation, increased rates of deteriorative reactions, and component crystallization (see Figure 3.13).

Conclusions

Glass transition and a_w are independent parameters that affect food stability. Glass transition is a property of amorphous food solids and a_w is a property of water as a solvent. Glass transition and a_w in a number of foods, however, are interrelated because of water plasticization of hydrophilic, amorphous food components. Global glass transitions can be measured for food solids, but they do not necessarily explain transformations at a microstructural level in food systems. Foods are complex, heterogeneous systems with variations in local structures and possibly the glass transition, and water content may vary significantly within food microstructure. A number of structural changes in foods and crystallization of food components during storage often occur above the glass transition. Such changes are time dependent with rates controlled by the temperature difference to the glass transition. The glass transition may occur as a result of water plasticization, and critical values for a_w and water content can be defined as those corresponding to those at which the glass transition occurs at the observation temperature. More precise knowledge of the relationship between food microstructure and the heterogeneities in glass transition properties and water content is required for further understanding of the importance of glass transition and a_w in controlling food stability.

References

Bell, L.N. 1996. Kinetics of non-enzymatic browning in amorphous solid systems: Distinguishing the effects of water activity and glass transition. *Food Research International* 28:591–597.

Bell, L.N, Touma, D.E, White, K.I., and Chen, Y.H. 1998. Glycine loss and Maillard browning as related to the glass transition in a model food system. *Journal of Food Science* 63:625–628.

Chirife, J., and Buera, M.P. 1995. A critical review of some non-equilibrium situations and glass transitions on water activity values of foods in the microbiological growth range. *Journal of Food Engineering* 25:531–552.

Chirife, J., and Buera, M.P. 1996. Water activity, water glass dynamics, and the control of microbial growth in foods. *Critical Reviews in Food Science and Nutrition*. 36:465–513.

Couchman, P.R., and Karasz, F.E. 1978. A classical thermodynamic discussion of the effect of composition on glass transition temperatures. *Macromolecules* 11:117–119.

Duckworth, R.B. 1981. Solute mobility in relation to water content and water activity. In: *Water Activity: Influences on Food Quality*, eds. L.B. Rockland, G.F. Stewart, pp. 295–317. New York: Academic Press.

Gordon, M., and Taylor, J.S. 1952. Ideal copolymers and the second-order transitions of synthetic rubbers. I. Non-crystalline copolymers. *Journal of Applied Chemistry* 2:493–500.

Haque, M.d. K., and Roos, Y.H. 2004. Water plasticization and crystallization of lactose in spray-dried lactose/protein mixtures. *Journal of Food Science* 69(1):FEP23–29.

Hartel, R.W. 2001. *Crystallization in Foods*, p. 325. Gaithersburg: Aspen.

Johari, G.P., Hallbrucker, A., and Mayer E. 1987. The glass-liquid transition of hyperquenched water. *Nature* 330:552–553.

Jouppila, K., and Roos, Y.H. 1997. The physical state of amorphous corn starch and its impact on crystallization. *Carbohydrates and Polymers* 32:95–104.

Jouppila, K., Kansikas, J., and Roos, Y.H. 1997. Glass transition, water plasticization, and lactose crystallization in skim milk powder. *Journal of Dairy Science* 80:3152–3160.

Jouppila, K., Kansikas, J., and Roos, Y.H. 1998. Crystallization and X-ray diffraction of crystals formed in water-plasticized amorphous lactose. *Biotechnological Progress* 14:347–350.

Katz, E.E., and Labuza, T.P. 1981. Effect of water activity on the sensory crispness and mechanical deformation of snack food products. *Journal of Food Science* 46:403–409.

Labuza, T.P. 1968. Sorption phenomena in foods. *Food Technology* 22:263–265, 268, 270, 272.

Labuza, T.P., Tannenbaum, S.R., and Karel, M. 1970. Water content and stability of low-moisture and intermediate-moisture foods. *Food Technology* 24(5):543–544, 546–548, 550.

Le Meste, M., Champion, D., Roudaut, G., Blond, G., and Simatos, D. 2002. Glass transition and food technology: A critical appraisal. *Journal of Food Science* 67:2444–2458.

Levine, H., and Slade, L. 1986. A polymer physico-chemical approach to the study of commercial starch hydrolysis products (SHPs). *Carbohydrates and Polymers* 6:213–244.

Lievonen, S.M., and Roos, Y.H. 2002. Nonenzymatic browning in amorphous food models: Effects of glass transition and water. *Journal of Food Science* 67:2100–2106.

Makower, B., and Dye, W.B. 1956. Equilibrium moisture content and crystallization of amorphous sucrose and glucose. *Journal of Agricultural and Food Chemistry* 4:72–77.

Miao, S., and Roos, Y.H. 2004. Nonenzymatic browning kinetics of a carbohydrate-based low-moisture food system at temperatures applicable to spray drying. *Journal of Agricultural and Food Chemistry* 52:5250–5257.

Nelson, K.A., and Labuza, T.P. 1994. Water activity and food polymer science: Implications of state on Arrhenius and WLF models in predicting shelf life. *Journal of Food Engineering* 22:271–289.

Peleg, M. 1992. On the use of the WLF model in polymers and foods. *Critical Reviews in Food Science and Nutrition* 32:59–66.

Peleg, M. 1994. A model of mechanical changes in biomaterials at and around their glass transition. *Biotechnology Progress* 10:385–388.

Roos, Y. 1993a. Water activity and physical state effects on amorphous food stability. *Journal of Food Processing and Preservation* 16:433–447.

Roos, Y.H. 1993b. Melting and glass transitions of low molecular weight carbohydrates. *Carbohydrate Research* 238:39–48.

Roos, Y.H. 1995. *Phase Transitions in Foods*, p. 360. San Diego: Academic Press.

Roos, Y.H. 2002. Thermal analysis, state transitions and food quality. *Journal of Thermal Analysis and Calorimetry* 71: 197–203.

Roos, Y.H. 2004. Phase and state transitions in dehydration of biomaterials and foods. In: *Dehydration of Products of Biological Origin*, ed. A.S. Mujumdar, pp. 3–22. Enfield: Science Publishers.

Roos, Y., and Karel, M. 1991. Phase transitions of mixtures of amorphous polysaccharides and sugars. *Biotechnological Progress* 7:49–53.

Roos, Y., and Karel, M. 1992. Crystallization of amorphous lactose. *Journal of Food Science* 57:775–777.

Roos, Y.H., Karel, M., and Kokini, J.L. 1996. Glass transitions in low moisture and frozen foods: Effects on shelf life and quality. *Food Technology* 50 (11):95–108.

Roos, Y.H., Roininen, K., Jouppila, K., and Tuorila, H. 1998. Glass transition and water plasticization effects on crispness of a snack food extrudate. *International Journal of Food Properties* 1:163–180.

Roudaut, G., Simatos, D., Champion, D., Contreras-Lopez, E., and Le Meste, M. 2004. Molecular mobility around the glass transition temperature: A mini review. *Innovative Food Science and Emerging Technology* 5:127–134.

Slade, L., and Levine, H. 1991. Beyond water activity: Recent advances based on an alternative approach to the assessment of food quality and safety. *Critical Reviews in Food Science and Nutrition* 30:115–360.

Slade, L., and Levine, H. 1995. Glass transitions and water-food structure interactions. *Advances in Food Nutrition and Research* 38:103–269.

Sperling, L.H. 1992. *Introduction to Physical Polymer Science*, p. 594. New York: John Wiley.

Talja, R.A., and Roos, Y.H. 2001. Phase and state transition effects on dielectric, mechanical, and thermal properties of polyols. *Thermochimica Acta* 380:109–121.

van den Berg, C., and Bruin, S. 1981. Water activity and its estimation in food systems: Theoretical aspects. In: *Water Activity: Influences on Food Quality,* eds. L.B. Rockland, G.F. Stewart, p. 1–61. New York: Academic Press.

White, G.W., and Cakebread, S.H. 1966. The glassy state in certain sugar-containing food products. *Journal of Food Technology* 1:73–81.

Williams, M.L., Landel, R.F., and Ferry, J.D. 1955. The temperature dependence of relaxation mechanisms in amorphous polymers and other glass-forming liquids. *Journal of the American Chemical Society* 77:3701–3707.

4 Water Mobility in Foods

Shelly J. Schmidt

"If there is magic on this planet, it is contained in water"
Loran Eiseley, *The Immense Journey* (1957)

Introduction

Water is an essential molecule of life; without it, there is no life. As expressed by 1937 Nobel Laureate Albert Szent-Gyorgyi, "Water is life's mater and matrix, mother and medium." Practically all its properties are anomalous, which enables life to use it as building material for its machinery. Life is water dancing to the tune of solids (Szent-Gyorgyi 1972). Water is critically important to our daily lives—from survival and growth to temperature regulation, fire fighting, food production, transportation, hydropower, cooking, cleaning, and bathing, and recreational activities, to name only a few needs and uses. Water can also be the cause of catastrophic situations, by its lack, overabundance, or impurity (Isengard 2001a).

The wondrous nature of water has inspired countless investigations by scientists and artists alike. Much of what we know scientifically about the present state of water and aqueous systems is captured in the précis by Felix Franks[1] (2000), *Water—A Matrix of Life*. A similarly titled book, *Life's Matrix* by Philip Ball[2] (2000), is a well-written biography of water intended for the lay reader. There is, however, much we still do not know about water, so the examination of the most abundant and mysterious chemical on the face of the earth—H_2O—marches on.

With regard to foods, water is present in all food ingredients and systems, from trace amounts, as in the case of crystalline sucrose, which has an average moisture content of 0.04% wet basis, to very high amounts, as in fresh fruits and vegetables, many of which have a moisture content greater than 90% wet basis (Chirife and Fontan 1982). Water in foods is critically important because of its profound effect on the production, processing, microbial safety, and chemical and physical stability of food systems. Water's extensive involvement in food processing, stability, sensory perception, and, most important, safety has made it an essential focus of study in many fields of inquiry for numerous years. Currently, there are three main water relations in foods research avenues: water activity (a_w), molecular water mobility, and the food polymer science approach. A recent review by Schmidt (2004) critically examines and compares these three water relations in foods research avenues.

The historical development of these research avenues to study water in foods would comprise a chapter in and of itself. Briefly, however, the modern-day study of a_w in foods began taking shape when Scott and Christian (Scott 1953, 1957; Christian and Scott 1953) applied the thermodynamic concept of a_w to predict the growth of food spoilage microorganisms (van den Berg and Bruin 1981). The application of the concept of a_w to better understand the stability of food systems was a huge leap forward compared with using the moisture content of the food. However, as argued by Franks (1982, 1991), there are inherent problems with applying the concept of a_w, which is an equilibrium thermodynamic concept to food systems that are by and large nonequilibrium systems. This concern with applying the concept of a_w to nonequilibrium food systems suggested the need to develop additional means of probing water's behavior in food systems, as well as more careful use and interpretation of the a_w concept as applied to foods (Slade and Levine 1991, Fennema 1996, Schmidt 2004).

The initial use of nuclear magnetic resonance (NMR) to measure the molecular mobility of water, with an intentional focus on food systems, is difficult to trace to one researcher. However, by the time of the first meeting of the International Symposium on the Properties of Water[3] (ISOPOW) in September 1974 in Glasgow, Scotland, use of both wide-line and pulsed NMR to probe water relations in foods was well under way, as reported by a number of researchers (Duckworth 1975, conference proceedings). For the interested reader, Steinberg and Leung (1975) discuss some of the earlier uses of NMR for investigating water in foods, and Lillford and Ablett (1999) present a chronology of some of the early developments of NMR, with reference to more general food science applications.[4]

The establishment of the most recent research avenue, the food polymer science approach, is attributed to Slade and Levine (1985, 1988, 1991), who have held a copyright on the term "food polymer science" since 1988. Based on the pioneering efforts of Slade and Levine, there has been a large number of studies, review papers, book chapters, books, symposia[5], conferences, and short courses devoted to investigating, teaching, and critically evaluating applications of the glass transition concept to foods. Recently, Levine (2002a) shared his own personal chronology of key milestones in the development of polymer science concepts for understanding food stability.

Some researchers may view a_w, molecular water mobility, and the food polymer science approach as competitive approaches to solving the same problem; however, the viewpoint advocated here is that these approaches are complementary and should be used in concert to obtain a composite, multilevel (at various distance and time scales) portrait of the water and solids dynamics that govern the stability behavior of a food system. The formative principle underlying these three approaches to investigating the stability of food materials is the same "mobility"—the mobility of water, as well as the mobility of solids—from the molecular to the macroscopic distance and time scales.

Water Structure

Water is a polar compound composed of two hydrogen atoms covalently bonded to a single oxygen atom. Overall, the water molecule is electrically neutral, but the positive and negative charges are unsymmetrically distributed. The oxygen atom has a higher electron density than the two hydrogen atoms, which is often represented as a partial negative charge on the oxygen atom ($2\delta^-$) and a partial positive charge on each hydrogen atom (δ^-). This dipolar nature of individual water molecules allows them to participate in extensive hydrogen bonding between water molecules, as well as between other polar molecules, and pervasively affects the structure and behavior of water (Franks 2000, Chaplin 2007).

Types of Molecular Motions

Water molecules exhibit three types of molecular motions: vibrational, rotational, and translational. Vibrational motion is intramolecular motion (i.e., motion within the molecule) that changes the shape of the molecule (i.e., bending, stretching, and rotation of bond length or angle). Rotational motion is the in-place spinning of a whole molecule and involves a change in orientation of the molecule in three-dimensional space. Translational motion is the change in location of a whole molecule in three-dimensional space. Levitt (2001)[6] distinguishes two types of translational motion: (1) diffusion, in which the motion of the molecules is random and uncoordinated, and (2) flow, in which the motion of the molecules is directional and concerted, usually due to a driving force. The strict distinction between the terms *diffusion* and *flow* given above are often not adhered to in the literature and the term *diffusion* is commonly used for both types of translational motion. In this chapter, the distance scale of the translational motion is often added for clarity, i.e., molecular or macroscopic translational motion, and the term *diffusion* is used for both types of translational motion.

For water, a polyatomic nonlinear molecular, there are three N coordinates needed to specify the locations of the atoms (a set of x, y, and z Cartesian coordinates for each atom; where N is the total number of atoms in a molecule). This corresponds to a total of $3N$ degrees of freedom for vibrational, rotational, and translational motions. For water with $N =$ 3 atoms, there are a total of 9 degrees of freedom (3×3): 3 degrees of freedom for translational energy, 3 for rotational energy, and 3 for vibrational energy.

The types and speeds of molecular motion that can occur in water alone and water in foods are dependent on the phase of the water. In turn, the phase of the water is dependent on temperature and external pressure and, in a food system, on composition and system kinetics (i.e., changes over time) as well.

In water

To view the mobility of water molecules in water alone, we can examine the three major regions of water's phase diagram. Figure 4.1 contains computer-simulated images of the molecular-level structure of the three phases of water—solid (ice), liquid, and gas (vapor or steam)—embedded in a moderate temperature–pressure phase diagram for pure water. The "X" in Figure 4.1 indicates the location in the water phase diagram at typical room temperature (20°C) and pressure (1 atm) conditions. In the ice phase (normal, hexagonal ice), water molecules exhibit mainly vibrational motion, whereas in the liquid and gas phases, water molecules exhibit vibrational, rotational, and translational motions. In the liquid and gas phases, the individual water molecules exhibit a distribution of translational molecular speeds (units of m/sec), with the average speed being greater in the gas phase compared with the liquid phase at the same temperature and pressure. Figure 4.2 shows the Maxwell-Boltzmann distribution of water molecules (in gas phase) as a function of speed (in m/sec) at four temperatures ranging from 223K to 373K (-50 to 100°C). The Maxwell-Boltzmann distribution of speeds equation given in Atkins (1978) was used to calculate the distributions shown in Figure 4.2. Note that as temperature increases, the spread of speeds broadens and the most probable speed (peak of distribution) shifts to higher values.

Transitions between the three phases of water shown in Figure 4.1 involve enthalpy, structural, and entropy changes. For example, the phase transition from ice to liquid water (melting) occurs via the addition of heat (enthalpy), which in turn results in a solid to liquid structural change, and this in turn increases the amount of randomness or disorder (en-

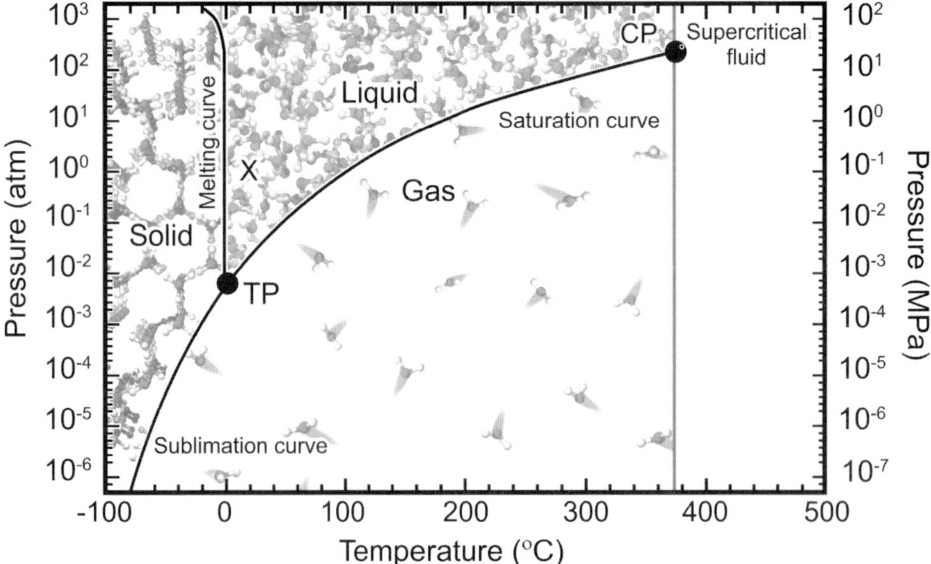

Figure 4.1 Computer-simulated images of the molecular level structure of the three phases of water—solid (ice), liquid, and gas (vapor or steam)—embedded in a moderate temperature-pressure phase diagram for pure water. The "X" marks the location on the phase diagram for water at typical room temperature (20°C) and external pressure (1 atm) conditions. TP is the triple point (273.16K and 611.657 Pa) and CP is the critical point (647.096K and 22.064 MPa) (IAPWS, 2002). The water structure computer-simulated images were generated by Christopher J. Fennell using the Object-Oriented Parallel Simulation Engine (OOPSE), which was (http://oopse.org/index.shtml) was developed by Dr. J. Daniel Gezelter and his graduate students at Notre Dame University, Department of Chemistry and Biochemistry, Notre Dame, IN.

tropy) in the system. This increase in disorder is reflected in the increase in mobility of the molecules as the system transitions from a solid with mainly vibrational motion to a liquid with vibrational, rotational, and translational motions.

In foods

Compared with water alone, the mobility of water molecules in a food system is exceedingly more complex. In general, when a component (e.g., sucrose) is added to water, the overall mobility of the water will decrease. The magnitude of the decrease depends on the number, amount, and nature of the component(s) added, as well as any processing methods used. A variety of system properties and measurement methods can be used to probe the decrease in water (and solid) mobility. However, the water and solid mobility response obtained is dependent on the property and measurement method selected to probe the system. Take, for example, a baked bread system, with 36% moisture content (wet basis) and a_w of 0.96; what types and speeds of water (and solids) motion are occurring? All three types of molecular water mobility (vibrational, rotational, and translational), each with its own distribution of molecular "speeds," are occurring throughout the bread system, as influenced by a myriad of water–component interactions, as well as by the spatial location within the bread being probed. For example, theoretically, after baking, the a_w of the crumb and crust should equil-

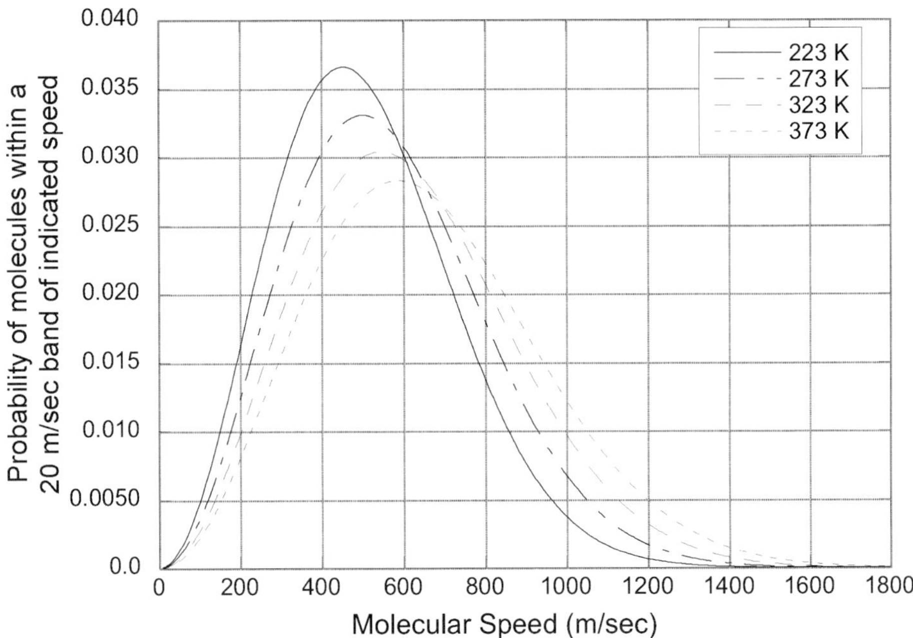

Figure 4.2 Distribution of molecular speeds (m/sec) for water molecules in the gas phase at four temperatures ranging from 223 to 373K (–50 to 100°C). The Maxwell-Boltzmann distribution of speeds equation given in Atkins (1978; Equation 24.1.13) was used to calculate the distributions.

ibrate over time; but what about the water mobility (or water mobility distribution) in each location—must they equilibrate over time, or can they remain different? These are some of the questions that need to be investigated to understand more fully the fundamental role that water mobility plays in the processing, stability, and safety of our food systems.

The three main water relations in foods research avenues

Regarding the three main water relations in foods research avenues previously mentioned, the type and distance scale of water and solids mobility that is being probed varies widely (see Figure 4.3). In the case of a_w (see Figure 4.3A), the macroscopic translational mobility (diffusion of water due to difference in chemical potential) of water molecules from inside the food to outside the food, into a small closed chamber, results in the partial vapor pressure of water (p_v), which is used in the numerator of the a_w equation ($a_w = p_v/p^o_v$, where p^o_v is the vapor pressure of pure water at the same temperature and external pressure).

In the case of molecular water mobility (see Figure 4.3B), the rotational and translational motions of the liquid phase of water are usually the types of mobility that are most connected with food stability. The most useful tools for measuring the rotational and translational motions of water in foods are NMR spectroscopy and imaging. The NMR longitudinal (T_1, seconds) and transverse (T_2, seconds) relaxation times are a measure of the molecular rotational mobility of the water, and the NMR diffusion coefficient (D, m^2/sec) is a measure of the molecular translational mobility of the water. NMR can also be used to measure the rotational and translational mobility of the solids in the food matrix, whereas

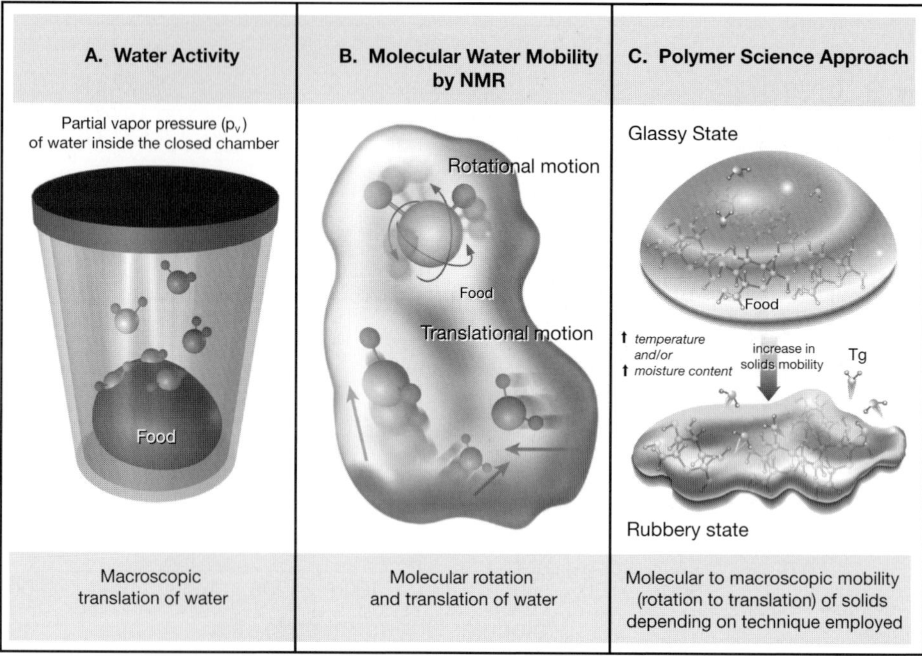

Figure 4.3 Schematic illustration of the types (rotational and translational) and distance scales (molecular to macroscopic) of water and solids mobility being probed in food materials using the three main water relations in foods research avenues: (**A**) water activity, (**B**) molecular water mobility by NMR, and (**C**) polymer science approach.

the vibrational motion of the water and solids can be measured using infrared and Raman spectroscopy (Conway 1981).

In the polymer science approach (see Figure 4.3C), the increase in mobility is measured as the material transfers from the glassy (least mobile) to the rubbery (more mobile) state as temperature and/or moisture content increase. The distance scale of mobility that is being measured depends on the specific instrumental technique used. For example, the motions resulting in a glass transition detected in the case of texture analysis are over a much larger distance scale (macroscopic) compared with the motions resulting in a glass transition detected by differential scanning calorimetry (DSC) (mesoscopic) or NMR (molecular).

Each of these approaches offers a different vantage point into the dynamics of food systems. If performed correctly, the data obtained from each technique offer its own unique view into the system being investigated. The technique(s) selected to probe a food system of interest largely depends on the specific information needed about the system and the desired end use of the data obtained. The focus of this chapter is the measurement and usefulness of the molecular mobility of water in food materials as probed by NMR techniques. For the interested reader, a recent review by Schmidt (2004) discusses all three approaches and details the merits and shortcomings of each approach.

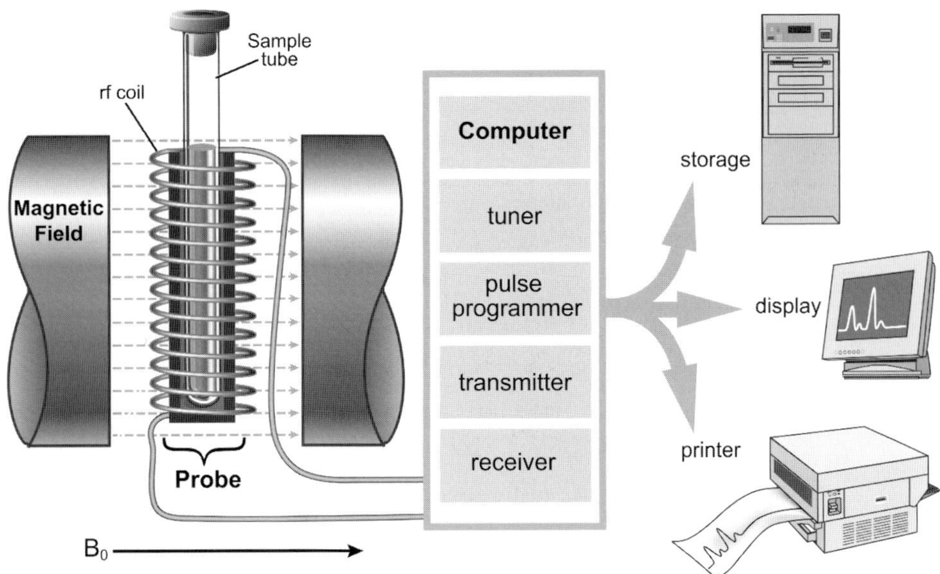

Figure 4.4 Schematic illustration of the three basic components of an NMR instrument: (1) the externally applied static magnetic field, B_0, (2) the probe, which holds the sample tube and contains the radiofrequency (RF) coil, and (3) the computer and other necessary hardware and software.

Measuring the Molecular Mobility of Water by NMR

Principles of NMR

NMR is a noninvasive, nondestructive technique that probes the system under study at the molecular level. Thus, as desired, the natural unperturbed state of the sample can be examined. NMR spectroscopy is based on the measurement of resonant radiofrequency (RF) energy adsorption by nonzero nuclear spins in the presence of an externally applied static magnetic field, B_0, when exposed to a second oscillating magnetic field (B_1). Figure 4.4 illustrates the three basic components of an NMR instrument: (1) the externally applied static magnetic field, B_0, (2) the probe, and (3) the computer and other necessary hardware and software.

The types of externally applied static magnets (B_0) used in NMR are electromagnets, permanent magnets, and superconducting magnets. However, the majority of NMR spectrometers today use cryoconducting or superconducting magnets. A superconducting coil held at the temperature of liquid helium generates the magnetic field. With superconducting magnets much higher magnetic field strengths can be achieved (Gunther 1995). The cooling of the superconducting magnet coil is done with liquid nitrogen and liquid helium, which require regular refilling. However, the increased field strengths greatly improve spectrum resolution. During an NMR experiment, it is desirable to make B_0 as homogeneous as possible so each nucleus throughout the sample volume experiences, as near as possible, the same B_0. Two instrumental features assist in creating a uniform B_0 (Macomber 1998). The first technique is to alter the shape of the B_0 field itself (within very narrow limits) by passing a very small amount of current through a complex arrange-

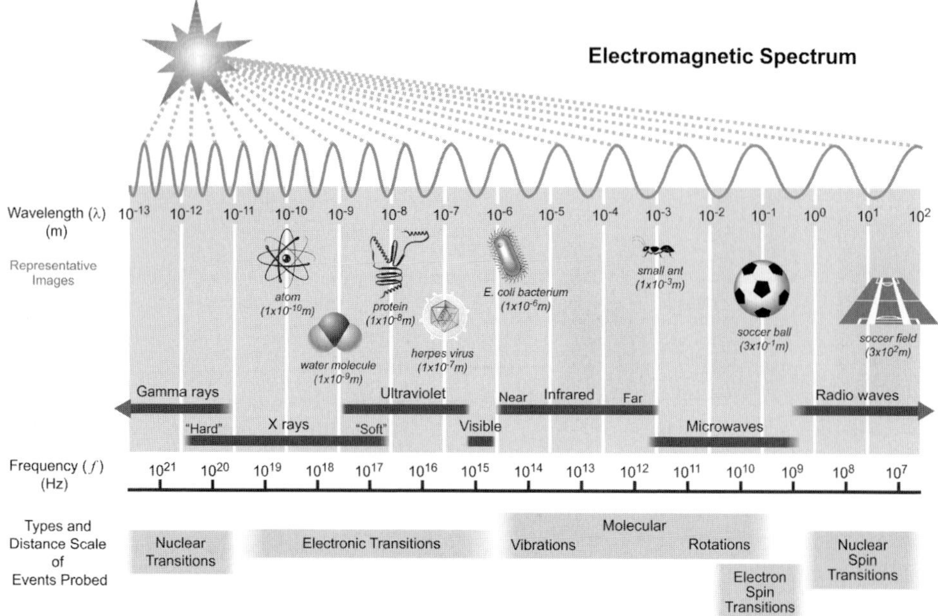

Figure 4.5 Illustration of the electromagnetic spectrum, including the types, frequencies (f), and wavelengths (λ) of radiation and the types and distance scale of events probed. The wavelength (λ) is related to the frequency of radiation (f) by $\lambda = c/f$, where c is equal to the speed of light (2.99776×10^8 m/sec).

ment of coils located around the probe cavity. This process is called *shimming* or *tuning* the magnetic field and should be done before each NMR experiment. The second technique is to spin the sample by means of an air stream that turns a small turbine attached to the top of the tube. Spinning the sample tube during an NMR experiment helps to average out small inhomogeneities in B_0.

The probe is located inside the static magnetic field (B_0) and holds the sample and houses the RF coil. The RF coil is responsible for excitation of nuclei and detection of their response to RF pulse(s). The use of radio waves to probe nuclear spin transitions is anomalous to their low-frequency, long-wavelength position in the electromagnetic spectrum (Belton 1995). The electromagnetic spectrum, including the types, frequencies (f), and wavelengths (λ) of radiation and the types and distance scale of events probed, is shown in Figure 4.5. In general, the trend is that as the frequency of the radiation decreases (and wavelength increases), the distance (or length) scale of interactions that are probed increases. For example, gamma rays (high frequency, short wavelength) operate on the subnuclear and nuclear distance scale and probe transitions within those systems, whereas infrared waves (lower frequency, longer wavelength) operate on the molecular distance scale and probe molecular vibrational (near-infrared) and some rotational (far-infrared) transitions, with additional rotational transitions probed in the microwave region. However, the trend of decreasing frequency (and increasing wave length) and increasing distance scale of interaction probed is altered starting in the microwave region of the electromagnetic spectrum where the

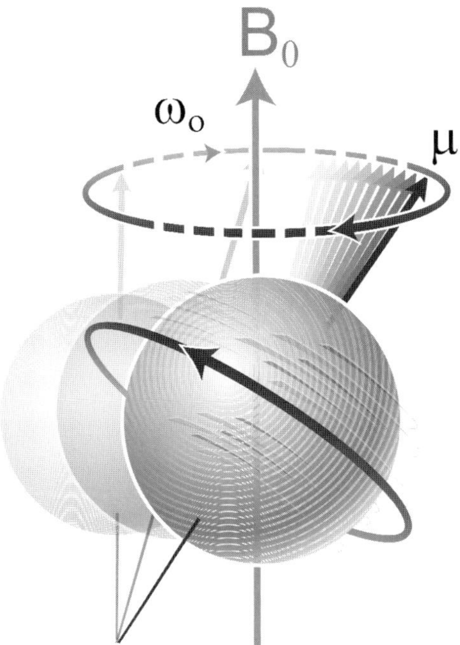

Figure 4.6 An external magnetic field (B_0) applied to a nucleus causes the nucleus to precess at a frequency (ω_0) proportional to the strength of the magnetic field (Equation 4.1). The magnetic moment (μ) is the vector quantity that is used to describe the strength and direction of the magnetic field surrounding each spinning nucleus.

magnetic resonance spectroscopies begin—first with electron spin resonance (ESR) spectroscopy and its associated electronic transitions, and next in the radio wave region with NMR spectroscopy and its associated nuclear spin transitions (Belton 1995).

The nuclei that are NMR active are those with nonzero spin values (I). Only atomic nuclei with an odd number of protons or neutrons possess a nonzero spin value, which can be thought of as being similar to the rotation of a charged nucleus. Where there is an electrical current, there is also a magnetic field. The strength and direction of the magnetic field surrounding each spinning nucleus can be described by a vector quantity known as the magnetic moment (μ), which can interact with an applied magnetic field, B_0. Thus, the first step common to all types of NMR experiments is that the sample to be analyzed is placed in a probe containing an RF coil, which is located in a strong externally applied magnetic field (B_0, in units of Tesla, or T) (see Figure 4.4). The torque exerted by B_0 on the spinning nucleus causes precession of the magnetic moment (see Figure 4.6). The frequency of precession is proportional to the strength of B_0:

$$\omega_0 = \gamma B_0 \tag{4.1}$$

where ω_0 is the angular frequency in radians per second (also called Larmor or resonance frequency) and γ is the magnetogyric ratio (or gyromagnetic ratio) ($\text{rad} \cdot T^{-1} \cdot \text{sec}^{-1}$),

Table 4.1. NMR Properties for the NMR-Active Water Nuclei.

	NMR-Active Water Nuclei			
	^{1}H	^{2}H	^{3}H	^{17}O
Properties	**Proton**	**Deuterium**	**Tritium**	**Oxygen-17**
Spin value (I)	1/2	1	1/2	5/2
Nuclear magnetic moment, μ	+2.79284	+0.85743	+2.97896	–1.89379
Gyromagnetic ratio (γ) (10^7 rad • T^{-1} • sec^{-1})	26.7510	4.1064	28.5335	–3.6264
Quadrupole moment (m^2)		2.73×10^{-31}		-2.6×10^{-30}
Relative sensitivity[a]	1.00	9.65×10^{-3}	1.21	2.91×10^{-2}
Natural abundance (%)	99.985	0.015	0	0.048
Frequency at 2.3488 T (MHz)	100.000	15.351	106.663	13.557
Number of allowed orientations (2I + 1)	2	3	2	6
Allowed values of spin, m_I	+1/2, –1/2	+1, 0, –1	+1/2, –1/2	+5/2, +3/2, +1/2, –1/2, –3/2, –5/2
Orientation vector length (in units of h/2π) $= \sqrt{I(I+1)}$	0.866	1.41	0.866	2.96
Orientation angle from B_0 (degrees) $\theta = arc\ \cos\left(m_I / \sqrt{I(I+1)}\right)$	54.7, 125.3	44.8, 90, 135.2	54.7, 125.3	32.4, 59.6, 80.3, 99.7, 120.5, 147.6

[a]Sensitivity relative to proton. To calculate the absolute sensitivity, multiply the relative sensitivity by the natural abundance.
Sources: Kemp 1986; Weast 1987.

which is a unique constant for each nucleus. The angular frequency can also be expressed in frequency units (sec^{-1} or Hz), since $\nu = \omega_0/2\pi$:

$$\nu = \frac{\gamma B_0}{2\pi} \tag{4.2}$$

Water has three stable nuclei (proton [^{1}H], deuterium [^{2}H], and oxygen-17 [^{17}O]) and one radioactive nucleus (tritium [^{3}H]) that possess nonzero spin values. The NMR properties for the NMR-active water nuclei are given in Table 4.1.

In the absence of B_0, the magnetic nuclei are randomly oriented, and all have the same energy level. However, in the presence of B_0, the nonzero spin nuclei adopt a specific number of orientations (see Table 4.1). The number of allowed orientations is dependent on the spin value and is equal to (2I + 1). For the simplest case of I = 1/2 (for ^{1}H and ^{3}H), two values of spin (m_I) are allowed, one aligned parallel to B_0 ($m_I = +1/2$) and one opposed (or antiparallel) ($m_I = -1/2$) to B_0. The values of m_I are determined by whether I is a integer or half-integer according to the follow series: I, I − 1, I− 2,..., 0,..., 2 − I, 1 − I, −I (for integer values of I) and I, I − 1, I − 2,..., 1/2, −1/2,..., 2 − I, 1 − I, −I (for half-integer values of I). Each allowed orientation has a different energy level, with the magnetic moments aligned with the applied magnetic field having slightly lower energy than those that are opposed. Figure 4.7 illustrates the energy difference, in the absence and presence of B_0, for the ^{1}H nuclei, where I = 1/2. The energy difference is given by:

$$\Delta E = h\upsilon \tag{4.3}$$

where h is Planck's constant (6.63×10^{-34} J • sec) and ν is the frequency in Hertz (sec^{-1}). For example, ΔE for ^{1}H at a magnetic field strength (B_0) of 2.35 T is calculated as (6.63

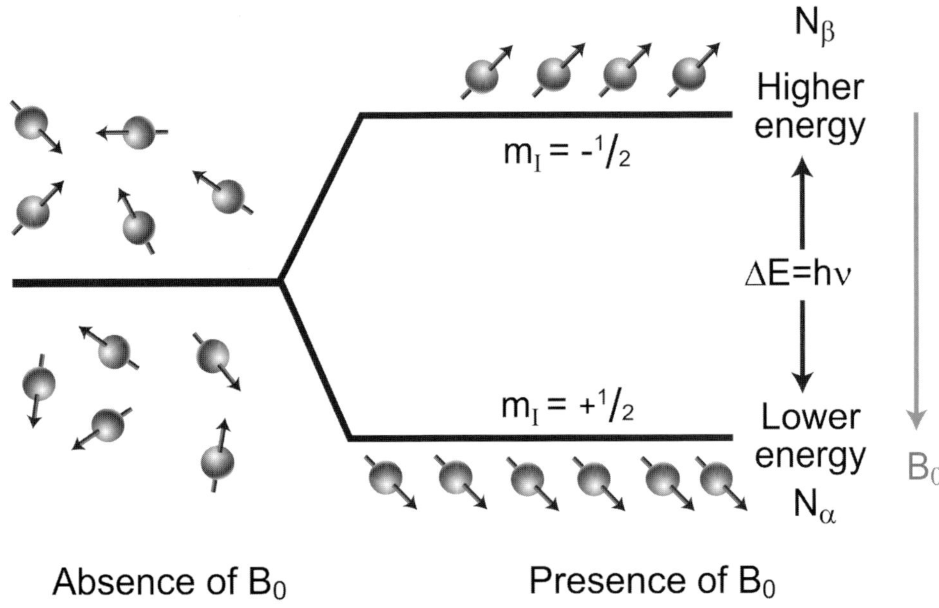

Figure 4.7 Schematic illustration of magnetic nuclei, for the case of I = 1/2 (e.g., ^{1}H), in the absence and presence of B_0. In the presence of B_0, the nuclei aligned with B_0 (N_α with $m_I = +1/2$) are of lower energy and those opposed to B_0 (N_β with $m_I = -1/2$) are of higher energy.

$\times$ 10^{-34} J • sec) (100 $\times$ 10^6 Hz), which equals 6.63 $\times$ 10^{-26} J, a very small but important energy difference.

The number of nuclei in the aligned and opposed positions is determined by the Boltzmann distribution (see Equation 4.4):

$$\frac{N_\alpha}{N_\beta} = e^{\Delta E / kT} \tag{4.4}$$

where N_α and N_β are the numbers of nuclei in the lower energy (α) and higher energy (β) positions, respectively, ΔE is the energy difference between the states and is equal to $h\nu$, k is the Boltzmann constant (1.38 $\times$ 10^{-23} J • K^{-1}), and T is temperature (K). Using Equation 4.4, the ratio of lower energy nuclei (N_α) to higher energy nuclei (N_β) at 20°C (293K) is approximately 1.0000164, a very small but important difference, because this slight excess of lower energy nuclei (N_α) results in the creation of sample magnetization and undergoes the observable NMR transition resulting in an NMR spectrum. Thus, the NMR spectrum is a measure of the energy required to cause a transition between energy levels and depends on the strength of B_0 (Belton 1995). Typical NMR magnetic field strengths (0.2 to 21 T, which corresponds to ^{1}H resonance frequencies of 8.5 to 900 MHz, respectively) result in large differences in resonance frequency (see Table 4.1 for frequencies of ^{1}H and ^{17}O at B_0 = 2.3488 T), making it possible to independently observe each NMR-active nucleus.

What about the water nuclei with spin values other than 1/2 (^{2}H and ^{17}O)? In addition

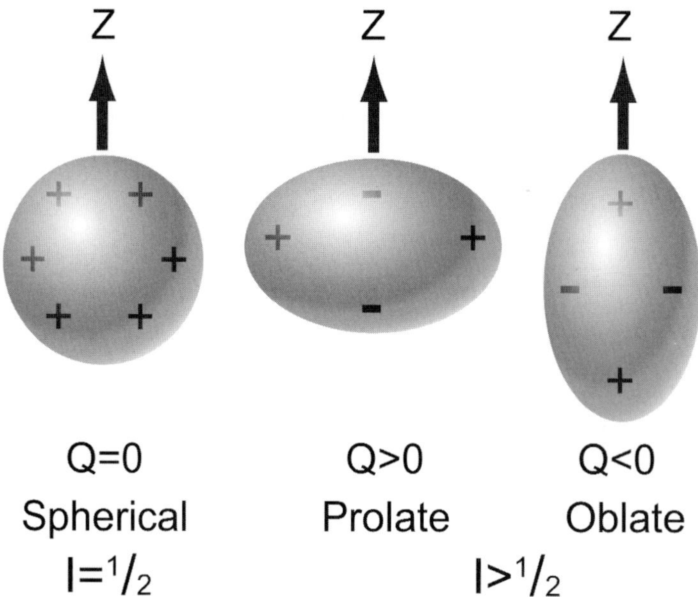

Figure 4.8 Schematic illustration of the nonspherical distribution of nuclear charge that results in an electrical quadrupole moment, Q, compared to the spherical distribution where Q = 0. By convention, the value of Q is taken to be positive if the ellipsoid is prolate (Q > 0) and negative if it is oblate (Q < 0).

to the interaction of nonzero spin nuclei with B_0, the nuclei with I > 1/2 (called quadrupolar nuclei) possess an electrical quadrupole moment that allows them to interact with electric fields produced by neighboring nuclei and electrons. The electrical quadrupole moment results from the nonspherical distribution of nuclear charge (Gunther 1995). The nuclear electric quadrupole moment parameter (Q) describes the effective shape of the ellipsoid of nuclear charge distribution. The quadrupole moments for ^2H and ^{17}O are given in Table 4.1. A nonzero quadrupole moment Q indicates that the charge distribution is not spherically symmetric. Figure 4.8 is a schematic illustration of the nonspherical distribution of nuclear charge that results in an electrical quadrupole moment (Q > 0 or < 0) compared with the spherical charge distribution with no quadrupole moment (Q = 0). By convention, the value of Q is taken to be positive if the ellipsoid is prolate and negative if it is oblate. The quadrupole moment allows ^2H and ^{17}O nuclei to interact with the electric fields produced by neighboring electrons and nuclei. This interaction is often very strong and dominates the NMR relaxation behavior of the quadrupolar nuclei, which results in the shortening of relaxation times and broadening of spectral line widths.

A Basic NMR Experiment

In a basic pulsed NMR experiment (for I = 1/2), after a sample is placed in the applied magnetic field (B_0), the nuclear spins are distributed between the aligned and opposed positions, according to the Boltzmann distribution (see Equation 4.4) (see Figure 4.7). This small excess population of spins in the lower energy position (aligned with B_0), which can be represented as a collection of spins distributed randomly about the precessional cone aligned with the *z*-axis, results in the bulk magnetization vector (M_z) along the *z*-axis (solid

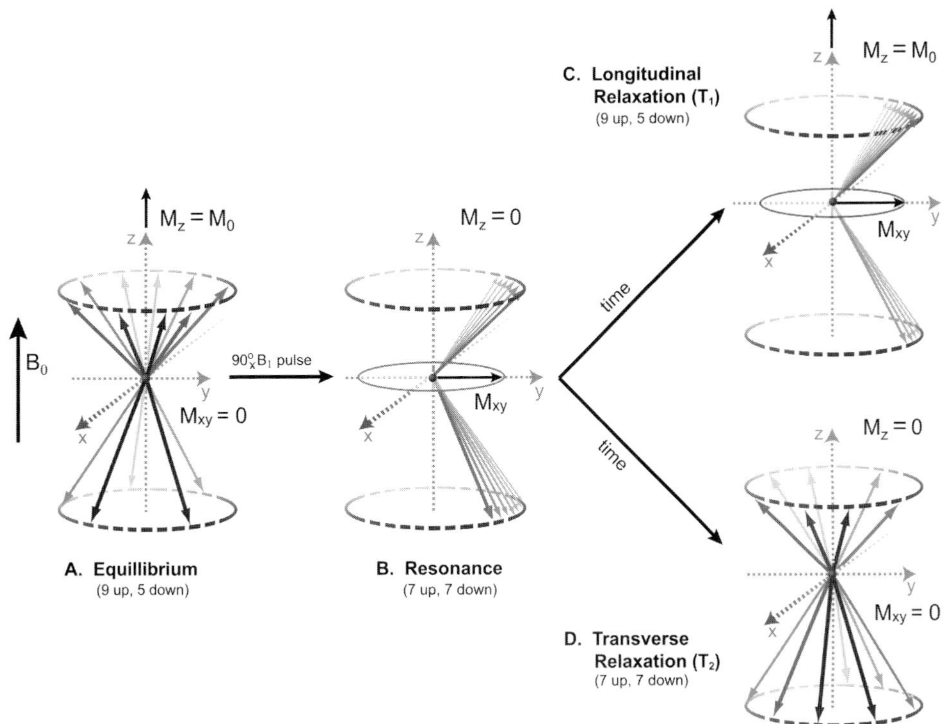

Figure 4.9 Schematic illustration of the changes in spin states experienced by I = 1/2 nuclei during a basic NMR experiment. **(A)** In the presence of B_0, the nuclear spins are distributed in the aligned (lower energy) and opposed (higher energy) positions, according to the Boltzmann distribution (Equation 4.4). The spins are precessing randomly (i.e., lack phase coherence, $M_{xy} = 0$). Net magnetization along the z-axis (M_z) is at a maximum (solid, black arrow) and is equal to the equilibrium magnetization (M_0). **(B)** Immediately following application of a $90°_x$ B_1 pulse, the populations of aligned and opposed positions are equalized, and the spins precess in-phase (i.e., exhibit phase coherence) at the Larmor frequency. The sample magnetization (M_{xy}) is at a maximum value, but $M_z = 0$. **(C)** The longitudinal relaxation process restores the equilibrium distribution of spins to the aligned and opposed positions ($M_z = M_0$). **(D)** The transverse relaxation process is a loss of phase coherence of the spins ($M_{xy} = 0$). It is important to note that in an actual NMR experiment, the longitudinal and transverse relaxation processes actually occur simultaneously.

black arrow above the *z*-axis) (see Figure 4.9A) (Claridge 1999). The magnitude of M_z at equilibrium is denoted as M_0. At this point in the experiment (at equilibrium in the presence of B_0), there is nothing to cause the spins in the transverse direction (*x-y* plane) to adopt a preferred orientation; thus, the spins are precessing randomly (i.e., they lack phase coherence) about the precessional cone and consequently there is no net magnetization in the transverse plane ($M_{xy} = 0$).

The next step in the experiment is to apply a RF pulse of energy to the specific nuclei being probed in the sample. The pulse is supplied to the nuclei by an oscillating magnetic field (B_1) at right angles to B_0, through the coils of the NMR probe, with a frequency equal to the Larmor frequency. As is the case with B_0, the RF electromagnetic field imposes a torque on the bulk magnetization vector in a direction that is perpendicular to the direction of the B_1 field, which rotates the vector from the *z*-axis toward the *x-y* plane. For example, applying the RF field along the *x*-axis will drive the vector toward the *y*-axis (see Figure 4.9B). The angle through which the vector rotates (θ; the pulse or flip, or tip angle; measured from the *z*-axis), which depends on the duration (t_p, the pulse width) and strength (or amplitude) of the B_1 pulse, determines the magnitude of M_{xy} and is given in degrees by Claridge (1999) as:

$$\theta = 360 \gamma\!\!\!- B_1 t_p \tag{4.5}$$

where $\gamma\!\!\!-$ is the gyromagnetic ratio of the nuclei being probed (γ) divided by 2π. A hard pulse consists of a short burst (usually a few microseconds) of high amplitude, while a soft pulse consists of a longer burst of lower amplitude. Two important hard pulses are $\theta = 90°$ (or $2/\pi$) and $180°$ (π). A 90° pulse angle focuses the entire sample magnetization in the *x-y* plane, and M_{xy} is at a maximum value (see Figure 4.9B), whereas a pulse angle of 180° inverts the net magnetization in the $-z$-axis direction ($-M_z$ is at a maximum), and M_{xy} is equal to 0.

When the 90° B_1 pulse is applied (see Figure 4.9; moving from A to B) two events occur: (1) the populations of aligned and opposed spin positions are equalized with a resultant loss in the net *z*-magnetization ($M_z = 0$; recall that the *z*-magnetization arises because of the population difference between the spin states) and (2) the magnetic moments begin to precess in phase with one another (called phase coherence) resulting in net magnetization in the *x-y* plane (M_{xy} is at a maximum).

Immediately afterward, the 90° B_1 pulse is turned off and the two events caused by the 90° pulse begin to relax, i.e., the system over time returns to its equilibrium condition. The relaxation process contains both a longitudinal component (along the *z*-axis; spins returning to their equilibrium lower energy, aligned position; see Figure 4.9C) and a transverse component (in the *x-y* plane; loss of phase coherence; see Figure 4.9D). The objective of separating the simultaneously occurring longitudinal and transverse components of the relaxation (see Figures 4.9C and 4.9D) is to be able to view their individual contributions to the relaxation process (see Figure 4.9A). In Figure 4.9C, longitudinal relaxation is complete (discussed later, time $> 5T_1$)—the excess spins have returned to their equilibrium lower energy, aligned position ($M_z = M_0$)—but transverse relaxation remains (spins are still in phase; M_{xy} is at a maximum). In Figure 4.9D, transverse relaxation is complete—phase coherence is lost ($M_{xy} = 0$)—but longitudinal relaxation remains ($M_z = 0$; excess spins remain in the higher energy, opposed position).

The relaxation process (illustrated schematically by the dephasing arrows and decaying spiral in Figure 4.10A) produces an electromotive force or voltage as a function of time. Detection of the loss of energy (through the process of relaxation) occurs via the same coil used for excitation. The coil is located such that signal detection occurs only in the *x*-y plane (see Figure 4.10A). The magnitude or intensity of the transverse relaxing signal, as a function of time [$M_y(t)$], is called the free induction decay (FID) and is shown schematically in Figure 4.10B. The FID can be analyzed directly or can be Fourier transformed (FT) from the time-domain into the frequency domain (in units either in ppm or Hz), yielding an NMR spectrum (see Figure 4.10C). The two most common peak shapes (also called lineshapes) observed in an NMR spectrum are the Lorentzian and Gaussian lineshapes. An FID that exhibits exponential decay upon FT produces a natural Lorentzian lineshape,

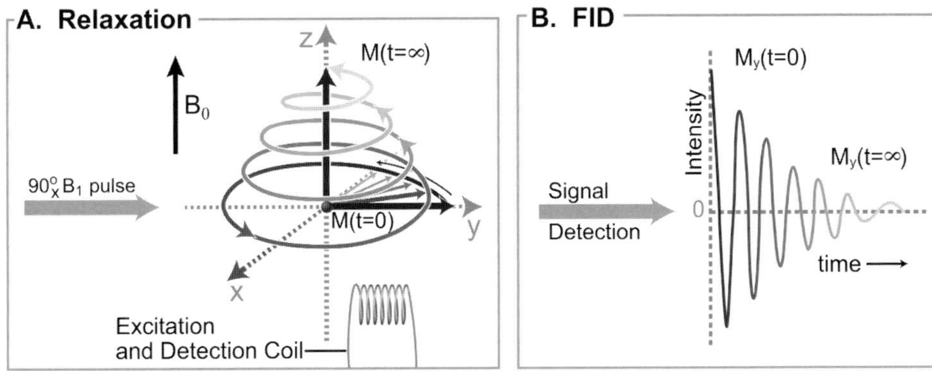

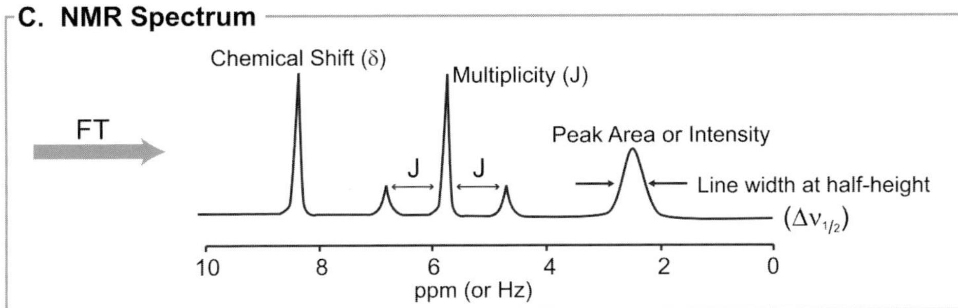

Figure 4.10 (**A**) Following a B_1 pulse, the magnetization [M(t = 0)] relaxes overtime [M(t = ∞)] as shown schematically by the dephasing arrows and decaying spiral. At any instant, the magnetization has a longitudinal component, Mz, aligned with B_0, along the z-axis, and a transverse component, My, in the plane perpendicular to B_0. M(t = ∞) is used to indicate complete relaxation back to equilibrium conditions (t = ∞ is any time > $5T_1$). (**B**) The relaxation of the transverse signal [My(t)] as a function of time (t) is called the free induction decay (FID). The FID can itself be analyzed or it can be transformed from the time domain to the frequency domain to yield an NMR spectrum. (**C**) The NMR spectrum contains a wealth of chemical and dynamic information, including (1) chemical shift, δ, (2) peak area or intensity (measured as peak height), (3) line width at half-height ($\Delta\nu_{1/2}$), and (4) multiplicity (called spin-spin or J coupling).

whereas the Gaussian lineshape is only a first approximation to the peak shape. In general, a Lorentzian lineshape possesses broader wings, which indicates that the nuclei possess higher molecular mobility, compared with a Gaussian lineshape, which indicates that the nuclei are less mobile and more solid-like.

As detailed in a number of NMR textbooks (e.g., Gunther 1995, Claridge 1999) a great deal of chemical information (e.g., molecular structure and dynamics, and chemical identification) can be extracted from an NMR spectrum (see Figure 4.10C), since each resonance peak provides fingerprint-like information. These include:

1. position (called chemical shift, δ), which provides general immediate chemical environment information
2. peak area (or intensity as measured by peak height), which is proportional to the number of spins of the specific nucleus being probed

3. line width at half-height ($\Delta\nu_{1/2}$), which contains information about the molecular mobility, interactions, and field homogeneity
4. multiplicity (called spin-spin or J coupling), which contains information about bonding patterns

NMR relaxation time measurements

NMR relaxation time measurements are powerful tools for investigating the molecular dynamics of water in food systems. The two most often measured relaxation times are longitudinal and transverse relaxation and are discussed in detail below. In addition, there exists a third NMR relaxation time, rotating-frame relaxation ($T_{1\rho}$), which contains the characteristics of both longitudinal and transverse relaxation. To obtain $T_{1\rho}$, the equilibrium spin magnetization is first subjected to a 90° pulse, which rotates the equilibrium spin magnetization along the *y*-axis. A spin-locking frequency field (amplitude $\omega_1 = \gamma B_1$) is immediately applied for a time, τ, along the *y*-axis. The spin-locking field is then turned off and the resultant FID is recorded. A plot of the signal amplitude as a function of the spin-locking time, τ, exhibits an exponential decay with time constant $T_{1\rho}$. In general, $T_{1\rho}$ is useful for characterizing molecular motions in solids (Bakhmutov 2004). A discussion of the potential usefulness of $T_{1\rho}$ to probe the molecular dynamics of water is given by Hills (1998).

Longitudinal relaxation

The decay of the longitudinal component of magnetization is called longitudinal relaxation or spin-lattice relaxation and is characterized by a time constant T_1 (sec) or a rate constant R_1 (sec^{-1}, which equals $1/T_1$). T_1 characterizes the rate at which the *z*-vector component of magnetization (M_z) returns to its equilibrium value, M_0. The longitudinal relaxation process restores the equilibrium distribution of spins to the aligned (lower energy) and opposed (higher energy) positions (see Figure 4.9C). Longitudinal relaxation occurs because of the existence of magnetic fields fluctuating at the correct frequency, which are able to induce transitions between the opposed and aligned positions of the spins in the applied magnetic field. If these fluctuations are associated with a lattice, then exchange of energy can occur between the spin system (the nuclei being probed) and the lattice (the molecular assembly in which the spins are embedded). There are many physical processes that result in locally fluctuating magnetic fields. The most important such interaction in liquids is the dipole–dipole interaction. Additional relaxation mechanisms are mentioned in the next section.

Since the excitation/detection coil is located in the *x*-*y* plane and the longitudinal component relaxes along the *z*-axis, T_1 cannot be measured directly from an NMR spectrum, but must be obtained using a pulse sequence. The most commonly used pulse sequence to measure T_1 is an inversion recovery pulse sequence (Kemp 1986; Eads 1998) and is illustrated in Figure 4.11. Other commonly used pulse sequences for measuring T_1 are given in Ernst et al. (1987).

The inversion recovery pulse sequence (see Figure 4.11) consists of two pulses, a 180° pulse followed by a wait time τ followed by a 90° detection pulse. This pulse sequence is repeated for several τ values from short to long. As the name of the pulse sequence implies, the first pulse *inverts* the spin population and then the wait time τ allows the spins to *recover* toward their equilibrium position. The 90° pulse is used to detect the amount of magnetization recovered as a function of τ, because the longitudinal (or *z*) component cannot be measured directly because the detection coil is in the *x*-*y* plane.

As mentioned above, the inversion recovery pulse sequence is repeated with a range of

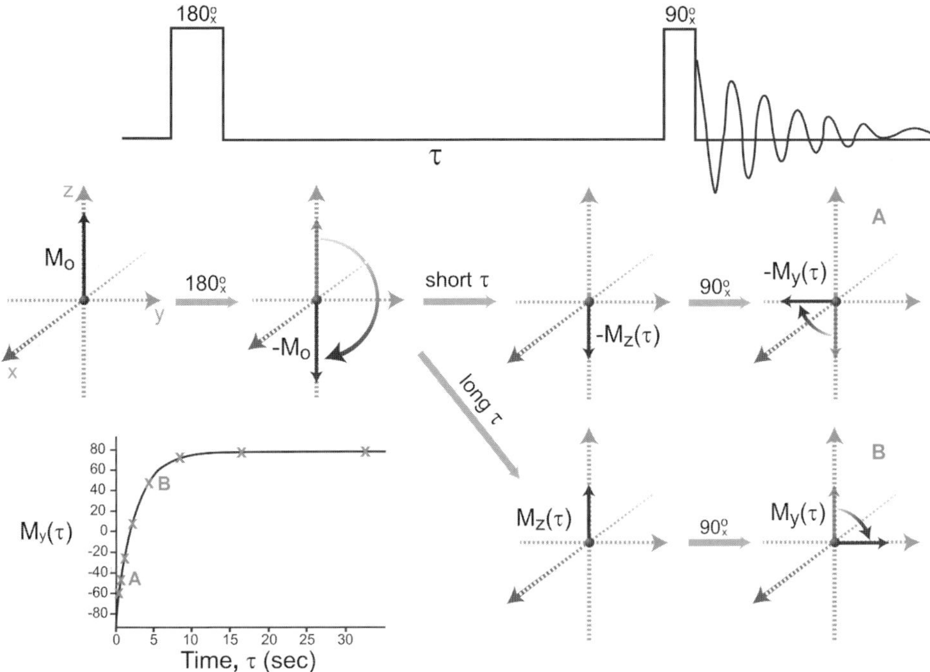

Figure 4.11. Schematic diagram of the inversion-recovery pulse sequence used to measure longitudinal relaxation (or spin-lattice relaxation, T_1) for a short and a long τ value, the rotating frame vector model, and the plot of the peak intensity [$M_y(\tau)$] as a function of the waiting time τ between pulses. The 180°_x pulse *inverts* the spin population and then the wait time τ allows the spins to *recover* toward their equilibrium position. The 90°_x pulse is used to detect the amount of magnetization recovered as a function of τ, because the longitudinal (or z) component cannot be measured directly because the detection coil is in the *x-y* plane. A short τ value results in a negative peak as shown by rotating frame vector model A and data point A on the plot, whereas a long t value results in a positive peak as shown by rotating frame vector model B and data point B on the plot.

τ values to obtain the recovery plot (M_y as a function of τ) shown in Figure 4.11. It is assumed that longitudinal relaxation is a first-order process and thus T_1 can be obtained by fitting the recovery data to the exponential expression:

$$M_z(\tau) = M_0(1 - 2e^{\frac{-\tau}{T_1}}) \tag{4.6}$$

Where M_z is the recovery of the magnetization upon relaxation as a function of τ measured as $M_y(\tau)$ using a 90°_x pulse, and M_0 is the maximum net magnetization (same as the equilibrium magnetization) obtained for values of $\tau > 5T_1$. T_1 can also be obtained by taking minus the slope of the plot of $\ln[(M_0 - M_z(\tau))/M_0]$ versus τ.

T_1 is a measure of both molecular rotational mobility and molecular interactions. The relation between T_1 and molecular mobility (as measured by the molecular correlation time, τ_c) exhibits a minimum when τ_c equals $1/\omega_0$ (where ω_0 is the frequency of precession defined previously, see Equation 4.1), which is discussed in detail later in this chapter (see Figure 4.15). One additional point regarding T_1 is that the value obtained depends on the frequency of the spectrometer employed (called T_1 frequency dispersion). In general, increasing the spectrometer frequency decreases the observed longitudinal relaxation rate (Hills 1998). It is possible to measure the T_1 frequency dispersion using spectrometers of varying magnetic field strength or, more easily, using a field-cycling spectrometer. Noack and others (1997) and Belton and Wang (2001) provide recent reviews of applications of field-cycling NMR.

Transverse relaxation

The decay of the transverse component of magnetization is called transverse relaxation or spin-spin relaxation and is characterized by a time constant T_2 (sec) or a rate constant R_2 (sec^{-1}, which equals $1/T_2$). The transverse relaxation process is a loss of phase coherence (dephasing) in the spins after an excitation pulse (B_1) is over, which is due to differences in the magnetic field experienced by individual magnetic moments and to the exchange of energy between identical nuclei (see Figure 4.9D). These processes result in a loss of magnetization in the *x-y* plane. Magnetic field differences arise from two distinct sources: static magnetic field inhomogeneity (imperfections in B_0, minimized by shimming and sample tube spinning) and local magnetic fields arising from intramolecular and intermolecular interactions in a sample (true or natural transverse relaxation). The total relaxation time constant, designated as $T_{2(\Delta B_0)}$, is a combination of both sources:

$$\frac{1}{T_2^*} = \frac{1}{T_2} + \frac{1}{T_{2(\Delta B_0)}} \tag{4.7}$$

where T_2 refers to the contribution from the true relaxation process and $T_{2(\Delta B0)}$ to that from the field inhomogeneity (ΔB_0). $T_{2(\Delta B0)}$ equals $1/\pi\gamma\Delta B_0$ (on a frequency scale), where $\gamma\Delta B_0$ is the spread in frequencies caused by field inhomogeneity (Eads 1998).

For single exponential relaxation (i.e., Lorentzian lineshape), T^*_2 can be obtained from the line width at half-height ($\Delta\nu_{1/2}$) of an NMR spectrum resonance (peak):

$$T_2^* = \frac{1}{\pi\Delta\nu_{1/2}} \tag{4.8}$$

The NMR T_2^* value obtained depends on the characteristics of the sample being probed. For a low-viscosity liquid sample, such as water, the relaxation time is long (i.e., a long FID) but Fourier transforms (FT) into a sharp peak with a small $\Delta\nu_{1/2}$ and thus a long T_2 value. Whereas, a viscous liquid sample exhibits fast relaxation (i.e., a short FID), a broader FT peak with a larger $\Delta\nu_{1/2}$ and thus a short T_2 value (see Figure 4.16 for further details).

Measurement of a true T_2 can be obtained using a spin-echo pulse sequence, such as the Carr-Purcell-Meiboom-Gill (CPMG) sequence, which minimizes the loss of phase coherence caused by static magnetic field inhomogeneities (Kemp 1986). The CPMG pulse sequence consists of an initial 90°_x pulse along the *x*-axis followed by a train of 180°_y pulses along the *y*-axis (see Figure 4.12). After the initial 90°_x pulse, the inhomogeneities of the static magnetic field cause the contributing spins to fan out (or diphase). The 180°_y pulse inverts the dephasing spins, causing the spins to refocus (or form an echo) after time τ. The train of the 180° pulses along the *y*-axis is then applied, which continually inverts the dephasing spins giving rise to a series of echoes at times 2τ, 4τ, 6τ, etc., until the echo inten-

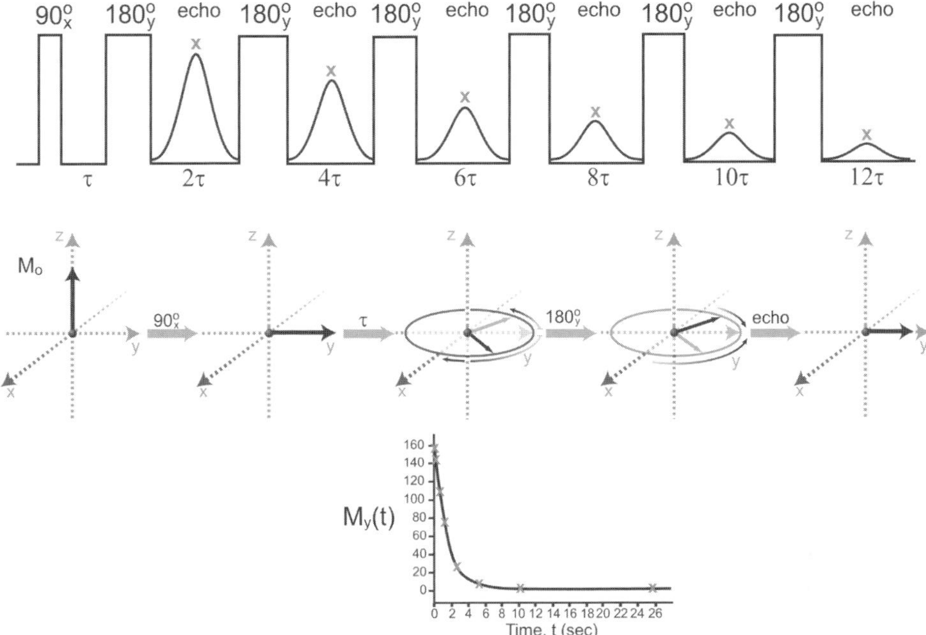

Figure 4.12 Schematic diagram of the Carr-Purcell-Meiboom-Gill (CPMG) spin-echo pulse sequence, the rotating frame vector model, and the plot of the peak intensity [My(t)] as a function of time, t. After the initial 90°_x pulse, which results in a maximum value of M_{xy}, the inhomogeneity of the static magnetic field causes the contributing spins to fan out (or diphase). The 180°_x pulse inverts the dephasing spins causing the spins to refocus (or form an echo) after time τ. The train of 180°_x pulses along the *y*-axis is then applied, which continually inverts the dephasing spins giving rise to a series of echoes at times 2τ, 4τ, 6τ, etc., until the echo intensity dies away (via true transverse relaxation processes).

sity dies away (via true transverse relaxation processes). The concept of inverting the dephasing spins with a 180° pulse to form an echo can be likened to athletes at different running speeds spreading out after starting a race; if after 5 minutes the runners would all turn around and start running back to where they started, they would all arrive at the starting line at the same time (Kemp 1986). One major advantage of the CPMG pulse sequence is that phase shifted 180° pulses (applied along the *y*-axis as opposed to the *x*-axis) are used in order to minimize the effects of imperfections in the 180° pulse (i.e., slight inaccuracies in the pulse angle). Additional details regarding the CPMG pulse sequence, as well as other spin-echo pulse sequences, are given by Kemp (1986), Derome (1987), Claridge (1999), and Levitt (2001).

It is important to point out that the spins being refocused by the 180°_y pulse are those that were dephased by static magnetic field inhomogeneities. The loss of phase coherence due to true transverse relaxation processes is also occurring; however, these spins are not refocused by the 180°_y pulse because, basically, there is no phase memory associated with the true transverse process to be undone (Claridge 1999). This means that at the time of the echo, the intensity of the observed magnetization will have decayed according to the true T_2 time constant, apart from the influence of static magnetic field inhomogeneities.

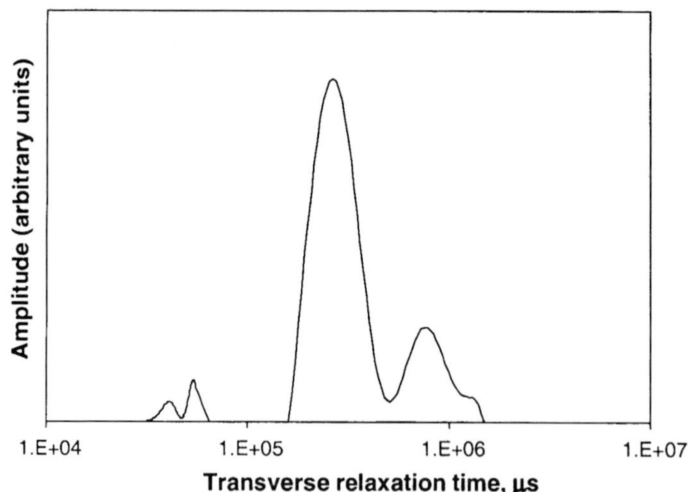

Figure 4.13 The distribution of proton transverse relaxation times for fresh egg yolk at 23°C obtained by the inverse Laplace transformation of the CPMG signal envelope (reprinted from Laghi, L., Cremonini, M.A., Placucci, G., Sykora, S., Wright, K., and Hills, B. 2005. A proton NMR relaxation study of hen egg quality. *Magnetic Resonance Imaging* 23:501-510, with permission from Elsevier).

Thus, spin-echo pulse sequences are designed to distinguish between the nonrandom dephasing of the spins caused by static field inhomogeneities and the random dephasing of the spins caused by true transverse relaxation processes.

It is assumed that transverse relaxation is a first-order process and thus T_2 can be obtained by fitting the echo decay data to the exponential expression:

$$M_{y(t)} = e^{\frac{-t}{T_2}} = e^{\frac{-2\tau n}{T_2}} \tag{4.9}$$

where $M_{y(t)}$ is magnetization at decay time t and for the CPMG pulse sequence t equals $2\tau n$, where τ is the delay time between the 90° and 180° pulse, and n is an integer. An alternative data analysis method for obtaining transverse relaxation times in systems that are not adequately described by a single average T_2 value (e.g., in complex food systems) is to present the data as a continuous distribution of relaxation times by deconvolution of single pulse FID or CPMG echo decay envelopes with an inverse Laplace transform as used in the CONTIN or WINDXP software packages (Hills 1998, Tang and others 2000, Laghi and others 2005). Figure 4.13 is an example of such a distribution for proton transverse relaxation times for fresh egg yolk at 23°C obtained by the inverse Laplace transformation of the CPMG signal envelope (Laghi and others 2005). Even though it was beyond the scope of the paper to specifically assign the five peaks to the various lipoprotein fractions, the authors hypothesized that the dominant peak at approximately 12 ms was attributed to water protons.

Despite the attention to detail paid in determining T_2, using the various pulse sequences discussed above, problems still arise such as homonuclear couplings that are not refocused

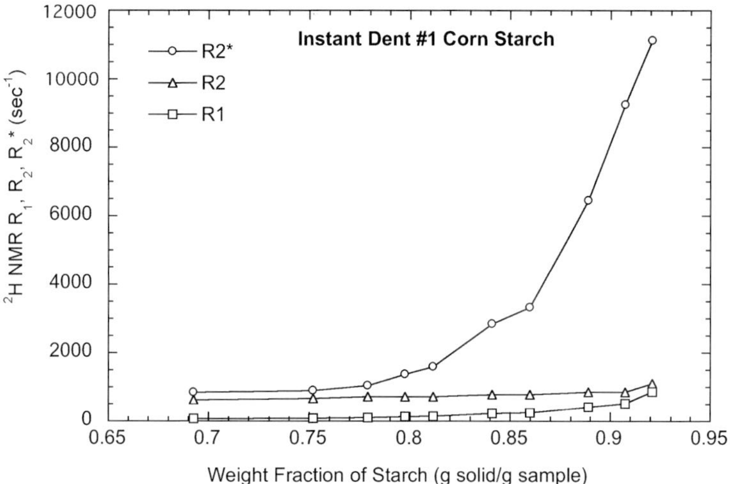

Figure 4.14 ^2H NMR R_1, R_2, and R^*_2 values for instant (pregelatinized) dent #1 corn starch samples containing D_2O as a function of weight fraction of starch solids, illustrating the divergence in the ^2H NMR R_1, R_2, and R^*_2 values as starch concentration increases (plotted using data from Kou et al. 1999).

by the spin-echo (Claridge 1999). Regardless of the difficulties encountered with measuring T_2, it is important to continue exploring the relationship between T_2 (and T^*_2) and various food stability parameters, such as water activity, glass transition temperature (T_g), mold germination times, etc. In general, at high moisture contents, T^*_2 is more similar to T_2, than at lower moisture contents (i.e., increasing solids content), where T^*_2 becomes much smaller than T_2 (and correspondingly, $R^*_2 >> R_2$). In contrast to T_1, T^*_2 and T_2 exhibit a continual decrease as a function of τ_c until the rigid lattice limit is reached, where little to no change in T_2 occurs as a function of temperature or increasing solids concentration (further details are given in Figure 4.15).

Relaxation time comparison

The values for T_1, T_2, and T^*_2 vary in magnitude for the three water nuclei of interest, ^1H, ^2H, and ^{17}O. For example, at 14.1 T and 20°C, the ^1H NMR T_1 relaxation rates obtained using the inversion-recovery pulse sequence for water alone are on the order of 2.7 seconds, whereas the ^2H and ^{17}O NMR T_1 relaxation times for water alone are on the order of 0.5 and 0.0068 second, respectively. Within each nucleus, the three relaxation time values for water when measured carefully are nearly equal (i.e., $T_1 = T_2 = T^*_2$), but will diverge as solids are added to the water. Figure 4.14 illustrates the divergence in ^2H NMR R_1, R_2, and R^*_2 values (recall $R_{1,}$ $R_{2,\text{ or}}$ $R^*_2 = 1/T_{1,}$ $T_{2,\text{ or}}$ T^*_2) as instant (pregelatinized) dent #1 corn starch concentration increases from 0.69 to 0.92 g solids/g sample (Kou and others 1999). As observed in Figure 4.14, ^2H NMR R^*_2 values become much larger than the ^2H NMR R_1 and R_2 values at starch concentrations above approximately 0.78 g solids/g sample. The large increase in ^2H NMR R^*_2 was attributed at high solids to chemical shift anisotropy produced by the electrons that shield the nucleus from the applied magnetic

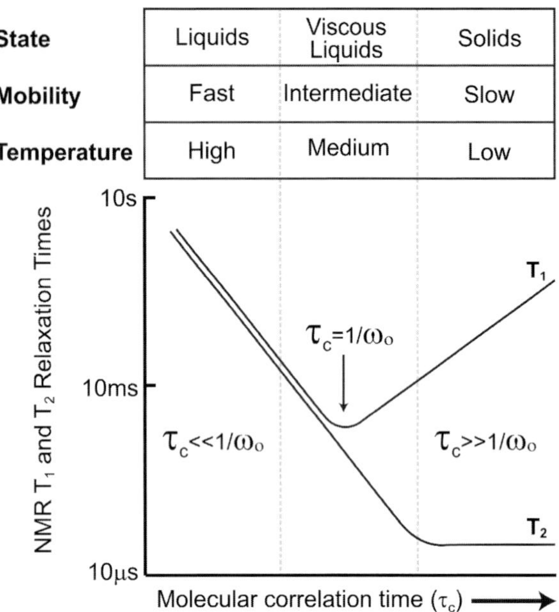

Figure 4.15 A schematic illustration of the dependence of the NMR T_1 and T_2 relaxation times on the molecular rotational correlation time, τ_c, characterizing molecular mobility in a single-component system. Both slow and fast motions are effective for T_2 relaxation, but only fast motions near ω_0 are effective in T_1 relaxation. Thus, T_1 goes through a minimum, whereas T_2 continues to decrease as the correlation time becomes longer until the rigid-lattice limit is reached.

field, which causes excessive line broadening in the ^2H NMR spectrum. The difference between R_2 and R_1 values ($R_2 - R_1$) ranges from a low of 244 to a high of 614 sec^{-1} over the starch concentrations shown in Figure 4.14. It is important to note that not only are these relaxation parameters sensitive to concentration, they are also influenced by a number of other factors including, magnetic field strength, nucleus being probed, temperature, pH, the presence of additional relaxation pathways, and, in some cases, instrumental errors and experimental parameters. For additional details regarding the effects of instrumental errors and experimental parameters on T_1 and T_2 measurements and calculations, see Chapter 3 in Bakhmutov (2004). Additionally, the molecular origins of water ^1H relaxation in solutions, gels, rubbers, and glasses are thoroughly discussed by Hills (1998).

Connecting NMR relaxation times and molecular mobility

NMR spin relaxation is not a spontaneous process. Rather, it requires stimulation by a suitable fluctuating field to induce an appropriate spin transition to reestablish equilibrium magnetization. There are four main mechanisms for obtaining relaxation: (1) dipole–dipole (most significant relaxation mechanism for $I = 1/2$ nuclei), (2) chemical shift anisotropy, (3) spin rotation, and (4) quadrupolar (most significant relaxation mechanism for $I > 1/2$ nuclei) (Claridge 1999).

Both T_1 and T_2 relaxation times are coupled to molecular mobility, but the details of

Table 4.2. Some τ_c Values for the Three NMR-Active Water Nuclei in H_2O or D_2O at the Specified Temperatures Obtained or Calculated From the Literature.

Nuclei and Sample	NMR Correlation Time τ_c	Temperature	References
^{1}H H_2O	2.0×10^{-12} sec	30°C	Krynicki (1966)
	2.6×10^{-12} sec	25°C	Smith and Powles (1966)
^{2}H 85% D_2O and 15% H_2O	2.8×10^{-12} to 1.6×10^{-12} sec	30°C	Calculated based on ^{2}H quadrupole coupling constant value (= 213.2 KHz) in Harris and Mann (1978), T_1 value (= 0.53) in Woessner (1964), and the order parameters, η ranging from 0 to 1, respectively
^{17}O H_2O	2.7×10^{-12} sec	25°C	Glasel (1966)
	3.1×10^{-12} sec	27°C	Calculated based on values in Halle et al. (1981) [R (H_2O) = 131.0 sec^{-1} and ^{17}O quadrupole coupling constant = 6.67 MHz]
^{17}O D_2O	4.0×10^{-12} sec	27°C	Calculated based on values in Halle et al. 1981 [R (D_2O) = 167.4 sec^{-1} and ^{17}O quadrupole coupling constant = 6.67 MHz]

Sources: Adapted from Table 4.6 in Eisenberg and Kauzmann 1969.
Other sources: Krynicki 1966; Smith and Powles 1966; Harris and Mann 1978; Woessner 1964; Glasel 1966; Halle et al. 1981.

their relationship are different and vary depending on the nucleus being probed (Campbell and Dwek 1984, Belton 1995, Eads 1998). In the simplest case of relaxation in a single proton pool (e.g., pure water), each water proton experiences a randomly fluctuating local magnetic field, due to transient dipolar interactions with other water protons as water molecules rotate and translate (via Brownian motion). Fluctuating fields at an appropriate frequency (Larmor frequency, ω_0) are able to induce transitions between opposed and aligned spin states, resulting in longitudinal relaxation. Hills (1998) and Claridge (1999) give details on the relationship between the rate of relaxation and the amplitude and frequency of fluctuating fields.

A schematic illustration of the dependence of the NMR T_1 and T_2 relaxation times as a function of the molecular rotational correlation time (τ_c) is shown in Figure 4.15. The usual definition of τ_c is the average time required for a molecule to rotate through an angle of one radian about any axis (Claridge 1999). Thus, rapidly tumbling molecules possess shorter correlation times (see left side of Figure 4.15), while molecules that tumble more slowly possess large correlation times (see right side of Figure 4.15). The dependence of T_1 on mobility shows a minimum, when $\tau_c = 1/\omega_0$ (or the product $\omega_0\tau_c = 1$). The location of the minimum and, correspondingly, the point at which the slow motion regimen is encountered, is magnetic field strength dependent since ω_0 itself is field dependent (Claridge 1999). When $\tau_c << 1/\omega_0$, the system is in the extreme narrowing limit and when $\tau_c >> 1/\omega_0$, the system is in a slow motion regime, and T_1 is again large. Thus, as can be seen from Figure 4.15, it is possible to obtain T_1 values of equal magnitude for both liquid and solid domains, as well as for high and low temperatures. Examples of τ_c values from the literature obtained for H_2O and deuterium oxide (D_2O) samples located in the extreme narrowing limit for all three NMR-active water nuclei are given in Table 4.2.

Figure 4.16 illustrates the molecular motions and associated T_2 relaxation curve (FID) and FT peak shapes for the three major domains in foods—liquid, viscous liquid, and solid (crystalline and glassy). ^{1}H T_2 relaxation time values typically observed in these domains,

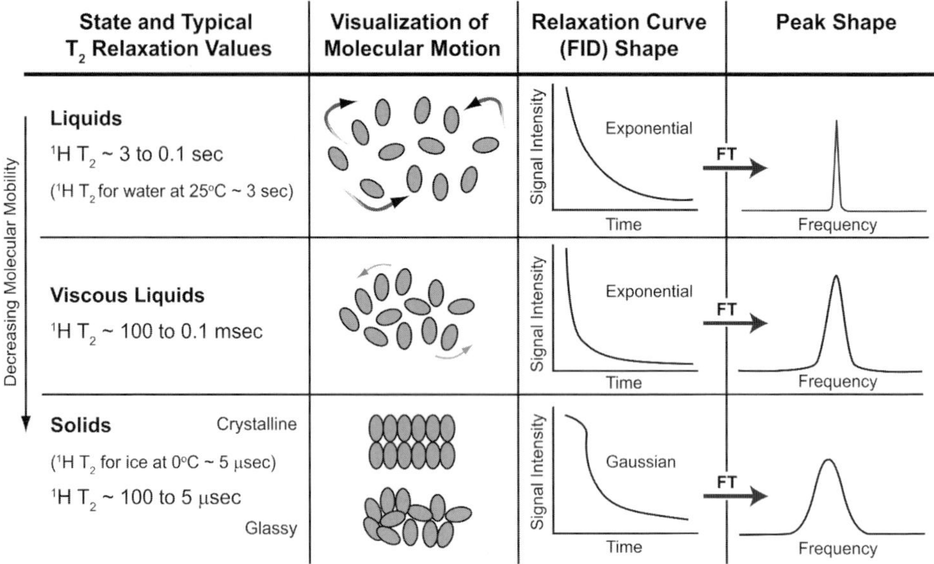

Figure 4.16 A schematic illustration of the molecular motions and associated T_2 relaxation curve (FID) shape and Fourier transform (FT) peak shapes for the three major domains in foods—liquid, viscous liquid, and solid (crystalline and glassy). Typical ^{1}H T_2 NMR relaxation time values observed in these domains, and values specific for water in liquid and crystalline domains, are listed.

as well as ^{1}H T_2 values specific for water in liquid and crystalline solid domains, are also given in Figure 4.16. The difference in T_2 relaxation behavior between liquids and solids is very dramatic and is the basis for using NMR for determining water content (Schmidt 1991) and solid fat content (Gribnau 1992). The dependence of T_2 relaxation on molecular correlation time is also illustrated in Figure 4.15. In the extreme narrowing limit, $T_2 = T_1$, whereas in the slow motion regime, T_1 becomes much longer than T_2. Until very long τ_c values are encountered, the shorter the T_2 value obtained, the less mobility is in the system. However, at very long values of τ_c, T_2 approaches a very small and constant value. At this point, the system is said to be in the rigid-lattice limit where further increases in solid concentrations or decreases in temperature do not result in a further decrease in T_2.

It is important to note that the relationship between relaxation times and molecular correlation times shown in Figure 4.15 is dependent on the strength of the externally applied magnetic field (B_0). In general, in the case of T_1, the T_1 minimum is shifted toward shorter correlation times for high magnetic field strengths. T_2 is usually less affected by magnetic field strength, but it does exhibit a small shift toward shorter correlation times for lower magnetic field strengths in the intermediate mobility region. For further discussion on the nature of the relationship between NMR T_1 and T_2 relaxation rates and molecular correlation times, as well as specific details on the influence of magnetic field strength, interested readers are encouraged to see Campbell and Dwek (1984), Hills (1998), Macomber (1998), and Mirau (2003).

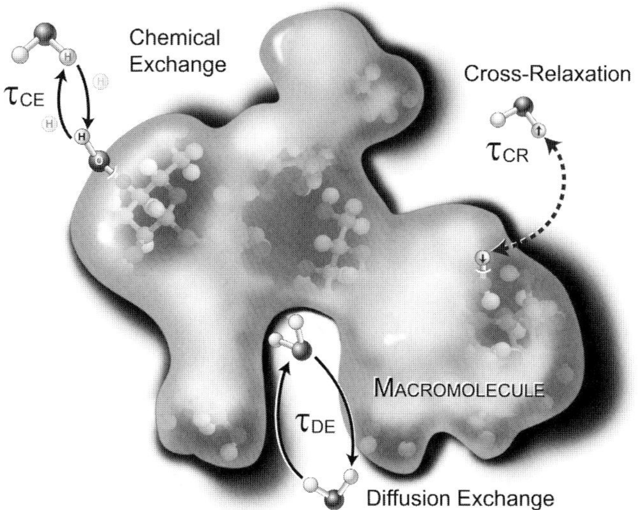

Figure 4.17 A schematic illustration of the additional pathways that can contribute to relaxation characterized by correlation times: (1) chemical exchange (with correlation time, τ_{CE})—physical exchange between water and exchangeable solute protons, such as hydroxyl, amine, and carboxyl groups within homogeneous regions; (2) diffusion exchange (with correlation time, τ_{DE})—physical movement of whole molecules between spatially separate regions (i.e., the exchange of whole water molecules between a bound environment on the solid phase and bulk phase water molecules); and (3) cross-relaxation (with correlation time, τ_{CR})—through-space transfer of *z*-magnetization between spin states having different T_1 values. The dotted lines indicate a physical exchange of atoms or molecules, whereas the dotted lines indicate a through-space exchange of magnetization.

NMR relaxation time measurement factors and considerations

As described in the previous section, NMR relaxation time measurements (T_1 and T_2) can provide valuable information for investigating the molecular dynamics of water in food systems. However, a number of factors can seriously complicate the analysis and quantitative interpretation of the relaxation behavior of water, such as the presence of additional relaxation pathways, the water nuclei chosen, and the complex nature of most food systems.

In addition to the four main relaxation pathways that were previously mentioned in this section, there exists three additional pathways (sometimes referred to as magnetization transfer mechanisms) that can contribute to relaxation, the presence and magnitude of which depend on the relaxation time being measured (T_1 or T_2) and the water nuclei chosen (^1H, ^2H, or ^{17}O). The additional pathways that can contribute to relaxation are illustrated in Figure 4.17 and include (1) chemical exchange—physical exchange of protons between water and exchangeable solute protons, such as hydroxyl, amine, and carboxyl groups within homogeneous regions (also called proton exchange) (affects both T_1 and T_2); (2) diffusion exchange—physical movement of whole molecules between spatially separate regions, for example, the exchange of whole water molecules between a bound environment on the solid phase (i.e., water molecules bound to specific binding sites on a protein macromolecule) and bulk phase water molecules (affects both T_1 and T_2); and (3) cross-relaxation—through-space transfer of z-magnetization between spin states having different T_1 values, such as between water molecule protons near the surface of the biopolymer and the protons on the biopolymer (also called magnetization exchange) (affects T_1).

The contribution of each of the three additional pathways listed above to the total relaxation rate is dependent on the type of sample being probed (e.g., liquid or solid or both liquid and solid). The magnitude of the contribution of the cross-relaxation pathway (listed as pathway 3 above) to the total relaxation is still somewhat controversial. For the interested reader, these additional relaxation pathways, as well as their contribution to the total relaxation rate, are discussed further by Belton (1990), Bryant et al. (1995), Bryant (1996), Hills (1998), Eads (1999), and Bakhmutov (2004).

Before proceeding, it is important to mention the consequences that chemical exchange can have on the magnitude of the observed transverse relaxation rate as measured by the CPMG pulse sequence (Hills 1998). In general, as the pulse frequency, $1/\tau$, in the CPMG pulse sequence (see Figure 4.12) is increased the observed transverse relaxation rate exhibits a strong dispersion, characteristic of chemical exchange. When the pulsing frequency is much greater than the proton exchange rate, the dephasing caused by the protons exchanging between the two sites differing in resonance frequency is completely refocused by the train of 180° pulses of the CPMG pulse sequence, so proton exchange does not contribute any additional dephasing effects. However, when the pulsing frequency is much lower than the proton exchange rate, the frequency variation experienced by the protons as they exchange between the two sites, differing in resonance frequency, causes rapid dephasing of the transverse magnetization and the observed transverse relaxation rate increases to a maximum. The midpoint of the dispersion occurs when the pulsing and proton exchange rates are equal. An example of the dependence of the observed transverse relaxation rate on the CPMG pulse frequency for a 6% glucose solution (pH = 5.8) at 100 MHz at 283K is shown in Figure 4.18. A detailed explanation of the transverse relaxation rate dispersion, as well as its quantitative analysis and usefulness for probing molecule behavior (i.e., changes in the conformation and mobility of biopolymers in solution during gelatination, gelation, denaturation, and retrogradation) is given by Hills (1998).

One additional magnetization transfer mechanism that can occur in samples containing paramagnetic species, such as iron, copper, manganese, and dissolved oxygen, is paramagnetic relaxation—which is the interaction of nuclear spins with unpaired electrons (affects T_1 and T_2). Paramagnetic species can be intentionally added to the sample (e.g., paramagnetic metal ions added as NMR shift reagents) to simplify complicated NMR spectra (Bakhmutov 2004); some paramagnetic species may be naturally present in the sample (e.g., the presence of dissolved oxygen in an aqueous sample or dietary iron in a food sample), or unknowingly present in the sample as a contaminate (e.g., using a previously used NMR tube containing residual paramagnetic metal ions). To avoid any undesirable paramagnetic relaxation effects from dissolved oxygen, samples can be degassed prior to T_1 and/or T_2 measurements. As described by Claridge (1999), there are two general approaches to degassing a sample. The first approach is to slowly bubble an inert gas, such as oxygen-free nitrogen or argon, through the solution for a few minutes to displace the dissolved oxygen. The second approach is to freeze the sample with liquid nitrogen or dry ice and subsequently place the frozen solid sample under vacuum for 5 to 10 minutes. This procedure, referred to as the "freeze-pump-thaw" technique, is typically repeated twice.

The water nucleus chosen also has an impact on the resultant relaxation data. Proton (^1H) relaxation is affected by chemical exchange, diffusion exchange, and cross-relaxation;

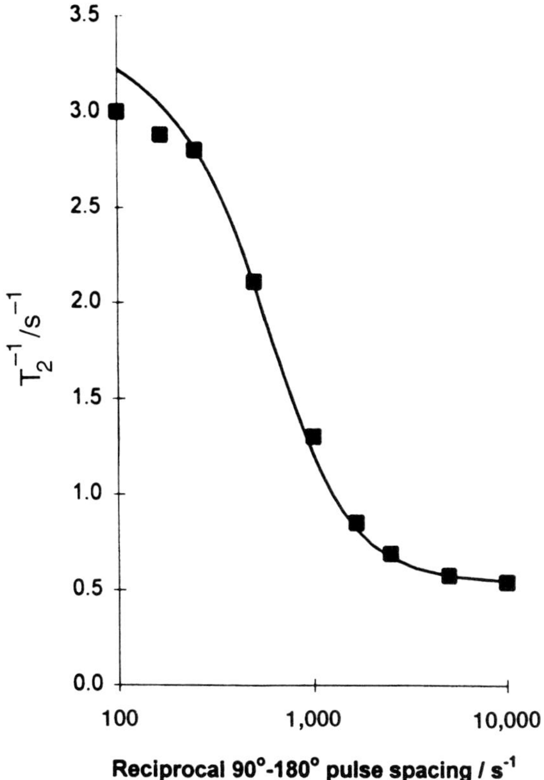

Figure 4.18 Dependence of the observed transverse relaxation rate (T_2^{-1}, sec^{-1}) on the CPMG pulse frequency (labeled as Reciprocal 90° to 180° pulse spacing, sec^{-1}) for a 6% glucose solution (pH 5.8) at 100 MHz at 283K. The line is a fit of the data points to the proton exchange theory given and explained in detail in Chapter 12 and Appendix A of Hills (1998) (reprinted with permission).

^2H by chemical exchange and diffusion exchange; and ^{17}O by proton-exchange (scalar spin-spin coupling between ^1H and ^{17}O nuclei; affects T_2 but not T_1) and diffusion exchange (Glasel 1972; Halle and others 1981; Halle and Karlstrom 1983a, 1983b). As illustrated in Figure 4.19 for H_2O and D_2O and native corn starch in both H_2O and D_2O, the effects of ^{17}O proton-exchange broadening can be eliminated by proton decoupling (Richardson 1989, Schmidt and Lai 1991). Adjusting the pH of the system to acid or basic conditions can also control the effects of ^{17}O proton-exchange broadening; however, adjusting the pH is an invasive action and will affect the nature of the system, which is usually undesirable. Deuterium and ^{17}O are quadrupolar nuclei ($I > 1/2$) and are not affected by magnetization transfer, because quadrupolar relaxation is already very efficient, and these nuclei intrinsically possess weak dipolar interactions (Belton 1990).

Another factor regarding the choice of water nuclei to be probed is the natural abundance of individual nuclei (see Table 4.1). ^1H is highly abundant, where as both ^2H and ^{17}O are much less abundant, which means their signal-to-noise ratio is much smaller.

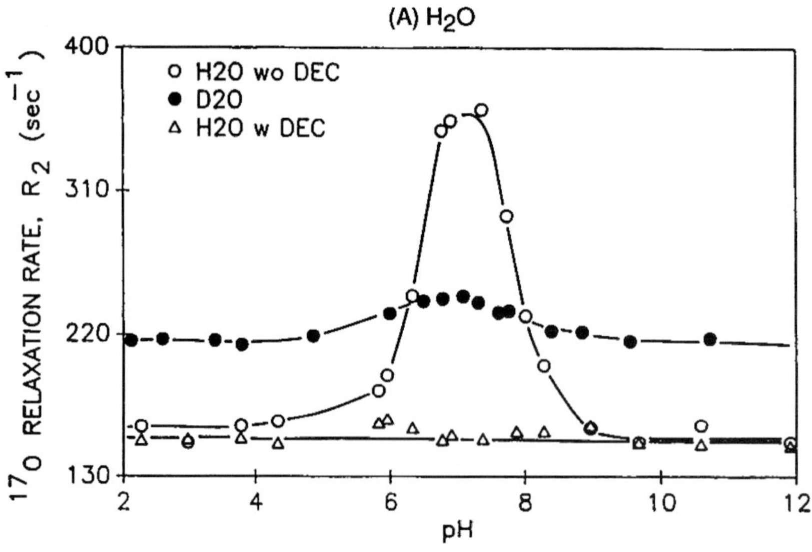

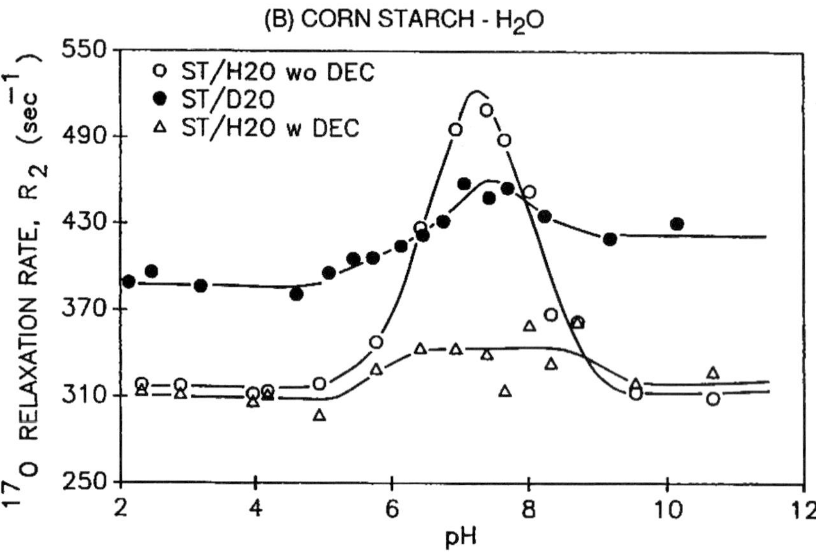

Figure 4.19 ^{17}O NMR transverse relaxation rate (R_2^*) in (**A**) H_2O without (wo) and with (w) ^{1}H decoupling and in D_2O and (**B**) corn starch and H_2O (ST/H_2O wo and w ^{1}H decoupling and ST/D_2O) and in corn starch and D_2O as a function of pH at 20°C (reprinted from Schmidt, S.J., and Lai, H.M. 1991. Use of NMR and MRI to study water relations in foods. In: *Water Relationships in Foods,* eds. H. Levine and L. Slade, pp. 599–613; originally in Richardson, S.J. 1989. Contribution of proton exchange to the ^{17}O NMR transverse relaxation rate in water and starch-water systems. *Cereal Chemistry* 66:244–246.; with kind permission of Springer Science and Business Media).

Thus, the number of scans (n) needed to obtain an adequate signal-to-noise ratio is greater for ^2H and ^{17}O compared with that for ^1H. However, because both ^2H and ^{17}O are quadrupolar nuclei, their relaxation times are much faster than that for ^1H. So the total time (called acquisition time) needed to obtain relaxation time data is dependent on the nuclei and sample being probed, where the acquisition time equals slightly greater than 5 $\times$ T_1 (for the nuclei and sample being measured) $\times$ n (the number of scans needed to obtain an adequate signal-to-noise ratio). A waiting period of slightly greater than 5 $\times$ T_1 is used to allow the nuclei in the sample to relax back to equilibrium before receiving the next pulse if n is greater than 1, because $5T_1$ results in 99.33% magnetization recovery. For example, for water alone, using the NMR T_1 values previously given in the text (with B_0 = 14.1 T and temperature = 20°C), the minimum acquisition time for ^1H is 13.5 seconds (5 $\times$ 2.7 seconds $\times$ 1 = 13.5 seconds), for ^2H, 80 seconds (5 $\times$ 0.5 second $\times$ 32), and for ^{17}O, 2.176 seconds (5 $\times$ 0.0068 second $\times$ 64). In this example, the ^2H measurement yields the longest minimum acquisition time; however, the acquisition time for ^2H in D_2O could be shortened to 2.5 seconds (5 $\times$ 0.5 second $\times$ 1), because the number of scans required for a good signal-to-noise ratio decreases from 32 to 1, since the abundance of ^2H increases from approximately 0.015% in H_2O to 99.9% in D_2O. This comparison also assumes that the ^2H NMR T_1 relaxation rate is similar in D_2O compared with water, which appears to be a good assumption based on the value for ^2H NMR T_1 of 0.53 second given in Table 4.2 for a 85% D_2O and 15% H_2O mixture. In a food sample with intermediate moisture contents, the number of scans needed for all three NMR-active nuclei increases greatly.

To decrease the acquisition time (and increase the signal strength) when doing ^2H NMR, as mentioned above, the sample can be prepared using D_2O instead of H_2O and in the case of a solid sample, such as starch, the sample can even be exchanged with D_2O (e.g., in order to replace the exchangeable protons with deuterons before NMR measurement). Samples can also be prepared with ^{17}O-enriched water, but compared with D_2O at \$3.25 per 10 ml (99.9% D_2O), ^{17}O-enriched water is very costly at \$1,650 per 10 ml (10% ^{17}O-enriched H_2O). In a study done by Richardson et al. (1987), ^{17}O-enriched water was used to increase the ^{17}O signal strength in high-solid dent corn starch samples. However, the researchers reported not being able to go beyond 83% solids concentration (g solids/100 g total) because the line widths for greater concentrations were larger than the bandwidth of the 250-MHz spectrometer used in the study. Thus, for high-solid, low-moisture content systems, ^2H NMR may be the nucleus of choice, both from an economic point of view and from the point of view of being able to obtain a spectrum at very high solid concentrations without the interpretation problems associated with ^1H NMR.

One additional complication associated with ^1H NMR that needs to be mentioned is the possibility of radiation damping. The effects of radiation damping may be observed in samples that produce an intense amount of signal (even after minimizing the signal amplification, i.e., gain), such as samples containing a large amount of water. The signal is strong enough to induce an electromagnetic force in the RF coil that significantly acts back on the sample, disturbing the actual amplitude and phase of the RF pulse. There are a number of erroneous and/or distorting effects of this feedback on the NMR spectrum, such as peak broadening, peak asymmetry, and phase shifting. The effects of radiation damping can be observed, for example, in the inversion recovery experiment for high water content samples. The best way to get rid of radiation damping is to significantly reduce the amount of sample being used, for example, by employing a specially designed NMR tube insert that holds only a small amount of sample. Another way to minimize the effects of radiation damping, in the case of the inversion-recovery pulse sequence, is to instead use the saturation-recovery pulse sequence (Bakhmutov 2004). In addition to decreasing the effects of radiation damping, the saturation-recovery pulse sequence decreases the acquisition time. Because the initial 180° pulse of the inversion-recovery pulse sequence (see Figure 4.11) is replaced by a 90° saturation pulse, the beginning position of the spins in the saturation-recovery pulse sequence is not the equilibrium state (with the associated need to wait >5T_1 before repeating the pulse sequence) but rather the saturated state, which can be induced soon after the second 90° detection pulse (the second pulse in both the inversion- and saturation-recovery pulse sequences). In general, radiation damping becomes greater as magnetic field strength (B_0) increases and effects high-power pulses more than low-power pulses.

Sample complexity is an additional factor that needs to be taken into account when measuring and interpreting NMR relaxation data to probe water mobility in foods. Compared with the type of chemically pure samples most often introduced into the NMR (i.e., a single, homogeneous compound), food samples can be exceedingly complex, including systems that are heterogeneous, multicomponent, and multiphase. The effects of spatial (or structural), compositional, and dynamical complexity of foods on relaxation time measurements are discussed in detail by Eads (1999). Because of the aforementioned complications, quantitative interpretation of T_1 and T_2 relaxation time measurements for water in foods often requires the use of nucleus- and system-specific, sophisticated, and detailed analysis and modeling (Belton 1995). For interested readers, Hills (1998) presents a comprehensive overview of water proton relaxation, ranging from that in dilute (solutions and gels) to more concentrated (rubbers and glasses) model food systems.

NMR Pulse Sequences for Measuring Water Diffusion

Up to this point we have introduced NMR methods to measure the rotational mobility of water. What is still needed is a method to measure the translational motion (also called diffusion) of water in foods. There are a number of pulse sequences that can be used to measure translational water motion (Hills 1998, Sorland and Aksnes 2002, Weingartner and Holz 2002, Metais and Mariette 2003, Avram and Cohen 2005); however, the most basic is the pulse-field gradient spin-echo (PGSE) pulse sequence (also known as the pulse-field gradient [PFG] pulse sequence) introduced by Stejskal and Tanner (1965).

The principle underlying the measurement of diffusion using the PGSE pulse sequence is that when a molecule (e.g., a water molecule) moves to a new location where the magnetic field is different from that in the original location, its resonance frequency changes. In PGSE NMR, a relatively small (compared with B_0) linear magnetic field difference (called a magnetic field gradient, *g*) is imposed on the sample. The change in frequency experienced by a moving molecule results in a more rapid loss of transverse magnetization, which can be measured using a spin-echo experiment, such as the CPMG experiment described previously to measure T_2. The PGSE pulse sequence is shown in Figure 4.20. The pulse sequence is repeated with different gradient strength values (*g*, in Gauss/cm), yielding a series of spectra with all other parameters held constant. The resultant data can be analyzed graphically according to Equation 4.10, yielding the simulated attenuation curve displayed in Figure 4.20 (Eads 1998):

$$\ln\left[S(2\tau)/S(2\tau_{g=0})\right] = -\gamma^2 g^2 \delta^2 D(\Delta - \delta/3) \qquad (4.10)$$

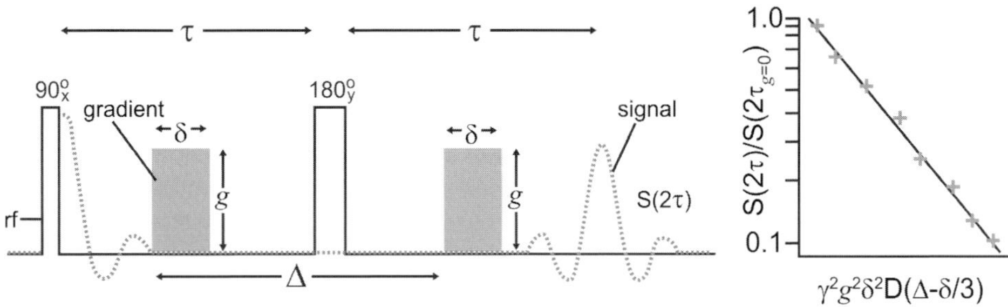

Figure 4.20 The principle of pulsed field gradient NMR (PFGNMR) measurement of diffusion coefficients in liquids. The pulse sequence is on the left, and a simulated attenuation curve on the right. The diffusion coefficient (*D*, m^2/sec) is proportional to the slope of the line.

where $S(2\tau)$and $S(2\tau_g{=}0)$ are the amplitudes of the resultant NMR signal at 2τ in the presence and absence of g, respectively, γ is the gyromagnetic ratio, D is the translation diffusion constant (in cm^2/sec), and Δ is the diffusion time. Thus, a plot of the natural logarithm of the attenuation ($\ln[S(2\tau)/S(2\tau_g{=}0)]$) versus g_2 is predicted to be a straight line with the slope proportional to D. As mentioned by Price (1997), it is also possible to perform the PGSE NMR experiment varying δ or Δ instead of g. The diffusion coefficient of water in water, as measured by PGSE NMR, is given as $2.3 \times 10^{-9}\ m^2 \bullet s^{-1}$ at 25°C and atmospheric pressure (Mills 1973, Weingartner 1982, Holz and Weingartner 1991, Holz et al. 2000).

The spin echo obtained in the Stejskal and Tanner spin echo pulse sequence is attenuated not only by diffusion but also by transverse relaxation. As discussed by Price (1997) and Hills (1998), the transverse relaxation attenuation effect can be normalized in systems exhibiting single-exponential relaxation by measuring the ratio of the NMR signals in the presence and absence of g (see ratio on left side of Equation 4.10). However, the involvement of the transverse relaxation places an upper limit on the maximum diffusion time possible in the Stejskal and Tanner pulse sequence; in heterogeneous systems with short T_2 values, this can be a problem, because the diffusion distance needs to be larger than the scale of the structural heterogeneity being probed. An important variant of the PGSE pulse sequence that extends the upper limit of the maximum diffusion time is the pulsed-field gradient stimulated echo (PGSTE) pulse sequence (Tanner 1970). The PGSTE pulse sequence is attenuated mainly by longitudinal relaxation (rather than by transverse relaxation, as in the PGSE experiment), which is usually much slower than the transverse relaxation for systems containing water plus solute(s), such as foods. Thus, the PGSTE pulse sequence is useful for measuring diffusion in viscous phases and for determination of droplet size distribution and pore size distribution (Eads 1998, Hills 1998).

An important consideration when performing PGSE NMR diffusion measurements is whether the diffusing species is free or restricted in its diffusion. If the attenuation plot exhibits deviations from linearity, it is possible the molecule is diffusing within a restricted geometry (i.e., exhibiting restricted diffusion) (Price 1997). It is also important to note that the diffusion coefficient measured with the PGSE NMR method is an average D for the entire sample. Diffusion-ordered NMR spectroscopy (DOSY), a relatively recent adaptation of pulsed-field gradient spin-echo pulse sequence, can be used to separate the NMR signals of different components of a mixture based on their diffusion characteristics (Morris 2002b). The DOSY pulse sequence was recently used to attempt to measure the change in the distribution of water diffusion coefficients in gelatinized dent corn starch–water systems during retrogradation (Olson 1999). Preliminary two-dimensional DOSY spectra revealed a distribution of water diffusion coefficients in retrograding starch gels, which increased as a function of both increasing starch concentration and time. It is important to note that both the NMR diffusion pulse sequence and experimental parameters selected for use may have an effect on the diffusion coefficient obtained, especially for systems exhibiting exchange and nuclear Overhauser effect (NOE) (Avram and Cohen 2005). For interested readers, a number of comprehensive reviews exist on the topic of diffusion by Sun and Schmidt (1995), Price (1996, 1997, 1998a), Morris (2002a, 2002b), and Weingartner and Holz (2002).

NMR Imaging for Water Mobility Mapping

Magnetic resonance imaging (MRI) techniques allow one to gain spatially dependent information about the sample of interest; for example, instead of obtaining spatially averaged T_1, T_2, and diffusion coefficient values, one can obtain water mobility information as a function of location within the sample (T_1, T_2, and diffusion coefficient images or maps). The basic principles for obtaining an image are simply described in Figure 4.21, using as a sample two spatially separated tubes filled to different levels with water.

The first step in obtaining a magnetic resonance image is to cause every position in the sample to have a different resonance frequency. This is achieved by applying, in addition to the static magnetic field (B_0), smaller, linear magnetic field gradients to the sample in combinations of *x, y,* and *z* directions (the number of gradients corresponds to the number of dimensions to be resolved). This first step is called the frequency-encoding step (see Figure 4.21A). The second step is to detect the NMR signal, which is achieved by carrying out a pulse sequence (such as a single pulse or spin-echo experiment) and at the same time applying the magnetic field gradient(s) (see Figure 4.21B). The third step is to transform the NMR frequency signal back to position (called image reconstruction). The result is a one-dimensional (called a projection), two-dimensional (called a slice), or three-dimensional (viewed as a series of slices or as a solid surface) image or map of the original sample. The NMR spectra shown in Figure 4.21C are without (left) and with (right) the application of one magnetic field gradient along the *x* direction. Without the gradient, all the water protons have the same Larmor frequency (neglecting the small susceptibility and slight inhomogeneity in B_0) and a familiar spatially averaged spectrum is obtained. While application of the gradient results in the water protons having spatially dependent Larmor frequencies and a spatially dependent spectrum is obtained (spatial displacements in the direction of the *x*-gradient become frequency displacements in the NMR spectrum). The spectrum obtained with the gradient is a one-dimensional spatially resolved "image" or profile (also called a projection) of the proton density of the two tubes filled to different water levels. A slice (two-dimensional image) of a sample can be obtained by using a specially shaped ("soft") 90° or 180° pulse and by applying additional magnetic gradients. A thorough description of MRI theory and applications can be found in McCarthy (1994), Hills (1998), and Price (1998b).

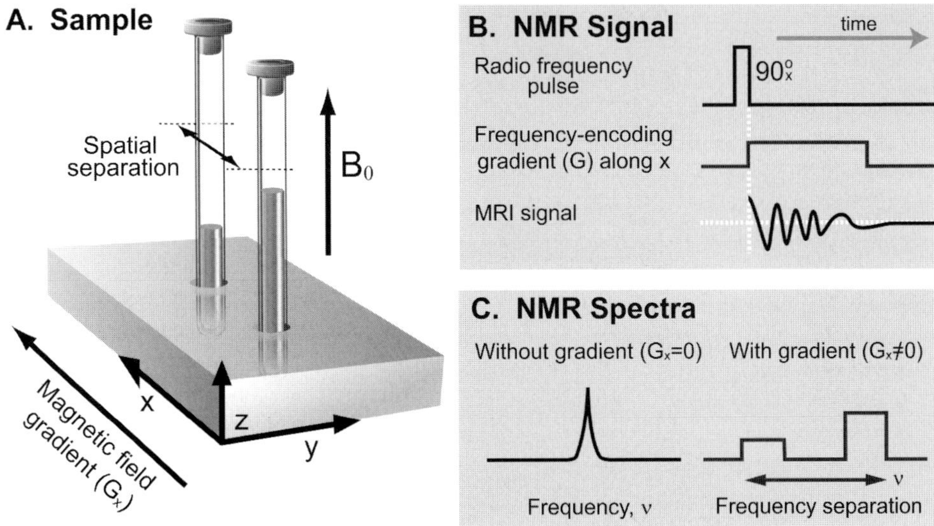

Figure 4.21 (**A**) The first step in obtaining a magnetic resonance image is the application of a small, linear magnetic field gradient to the sample (in this case, two tubes of water filled to different levels) to cause every position in the sample to have a different resonance frequency. This is called the frequency-encoding step. (**B**) The second step is to detect the NMR signal, which is achieved by carrying out a pulse sequence at the same time the magnetic field gradient is applied. (**C**) The third step is to transform the NMR frequency signal back to position. The NMR spectrum on the left is without application of the gradient ($G_x = 0$), while the spectrum on the right is with application of the gradient ($G_x \neq 0$). Without the gradient, all the water protons have the same Larmor frequency and a familiar spatially averaged spectrum is obtained, whereas application of the gradient results in the water protons having spatially dependent Larmor frequencies and a spatially dependent spectrum is obtained. The spectrum obtained with one gradient is a one-dimensional spatially resolved "image" or profile (also called a projection) of the proton density of the two tubes filled to different water levels.

Usefulness of the Molecular Water Mobility Approach for Food Materials

General Advantages and Disadvantages of NMR

NMR is an incredibly versatile tool that has been used for a wide array of applications, including determination of molecular structure, monitoring of molecular dynamics, chemical analysis, and imaging. NMR spectroscopy and imaging have found broad application in the food science and food processing areas (O'Brien 1992; Belton et al. 1993, 1995, 1999; Colquhoun and Goodfellow 1994; McCarthy 1994; McCarthy et al. 1994; Duce and Hall 1995; Webb et al. 1995, 2001; Gil et al. 1996; Sun and Schmidt 1995; Schmidt et al. 1996; Hills, 1995, 1998, 1999; Eads 1999; Rutledge 2001).

There are a number of significant advantages to using NMR to examine food systems and processes, including (1) the noninvasive and nondestructive nature of the NMR technique, (2) the limitless types of food samples that can be probed, (3) the ease of sample preparation, and (4) the numerous classes of NMR measurements and pulse sequences that can be used to either embrace or overcome sample complexity (Eads 1999), including the ability of NMR to quantify spatial and temporal variation within a sample. In addition, advances in NMR instrumentation, techniques, and applications are developing at an incredibly rapid rate, as discussed by Grant and Harris (2002), and can now be applied to food systems and processes to further advance the study of water relations in foods.

There are limitless types of food samples that can be probed using NMR because food materials are transparent to the RF electromagnetic radiation required in an NMR experiment. Thus, NMR can be used to probe virtually any type of food sample, from liquids such as beverages, oils, and broths to semisolids such as butter, cheese, mayonnaise, and bread to solids such as flour, powdered drink mixes, and potato chips. Intact, "as is" food materials can be placed directly into NMR tubes or holders, in the case of NMR imaging, for analysis. Thus, little to no sample preparation is needed, beyond shaping the sample (when necessary) to fit into the NMR probe. Sample size is limited only by the size of the NMR probe. The volume of space available for a sample in an NMR probe currently ranges from around 0.1 cm^3 to 2.5×10^5 cm^3, depending on the type of instrument (i.e., high-resolution spectrometer to an imaging spectrometer).

Despite these extraordinary advantages to food science and technology from the use of NMR, there are three general disadvantages: (1) poor sensitivity, (2) high equipment costs, and (3) the frequent need for highly trained/experienced personnel for data collection and interpretation. The relatively low sensitivity of NMR, compared with other spectroscopic techniques, such as infrared or ultraviolet spectroscopy, arises from the small differences between spin energy levels and, thus, the small population differences exploited in an NMR technique. However, with the advent of higher external magnetic field strengths, the sensitivity of NMR has improved to nanomolar levels (Eads 1999). In addition, NMR can be coupled with other techniques (referred to as hyphenated NMR), such as liquid chromatography–NMR, to take advantage of the benefits of each technique, while overcoming individual disadvantages (Spraul et al. 2001; Duarte et al. 2003).

The initial investment in NMR equipment can be rather expensive, with the magnitude of the cost depending on the equipment required and the type(s) of experiments to be performed. For example, a 400-MHz high-resolution spectrometer, for running a variety of advanced-level experiments, costs approximately $400,000 for a well-equipped liquids system to $500,000 for an instrument with solids capabilities. On the other hand, a 20-MHz low-resolution tabletop NMR, for running relatively routine relaxometry analyses, costs $50,000, plus an additional $10,000 for a gradient accessory for enabling diffusion measurements. Another important cost consideration, not reflected in the above prices, is the operational costs associated with using and maintaining the instruments, such as personnel costs, refilling of liquid nitrogen and helium for high-resolution instruments, and use of consumables.

Measurements obtained using NMR techniques can be made for virtually any food or ingredient. However, depending on the nature of these measurements, the training needed by a person obtaining such measurements varies widely. For example, in the case of a routine online analysis, such as using NMR to measure sample moisture content, quality assurance personnel can be trained to obtain such measurements; however, calibration and upkeep of an instrument by more highly trained personnel may still be required. In contrast, in the case of experimental research, such as using NMR to probe water dynamics during processing, design of experiments and collection and interpretation of resultant re-

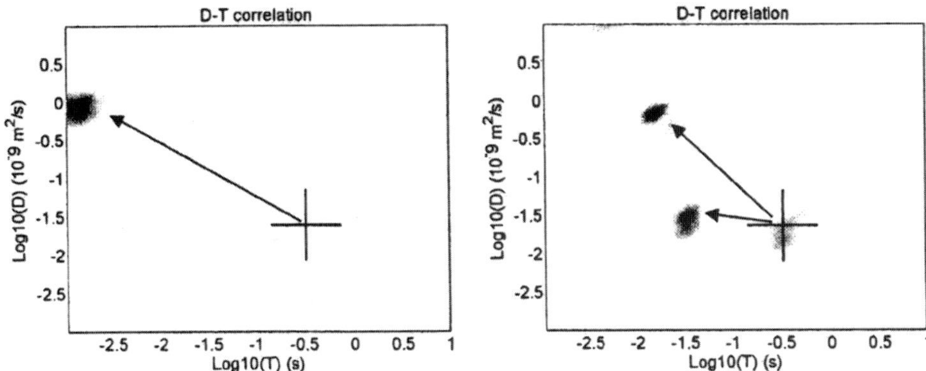

Figure 4.22 Diffusion (D) versus transverse relaxation time (T_2) distribution maps of water for mozzarella (*left*) and Gouda (*right*) cheese samples, measured at 40°C and aged for 74 days. The black cross on each plot corresponds to the position of the peaks at 1 day (reprinted from Godefroy, S., and Callaghan, P.T. 2003. 2D relaxation/diffusion correlations in porous media. *Magnetic Resonance Imaging* 21:381–383, with permission from John Wiley & Sons, Inc.).

laxation or diffusion data require highly trained personnel with experience in carrying out sophisticated and detailed data analysis.

In addition to the three general advantages/disadvantages of NMR discussed above, there are also specific advantages/disadvantages associated with particular magnetic resonance techniques and experiments. For a thorough discussion of the pros and cons associated with various NMR techniques for investigating intact food materials see Eads (1999).

Examples of Using NMR to Probe Water Relations in Foods

Because of the vast amount of information already published on the use of NMR to probe water relations in foods, it is not feasible to present a comprehensive survey of the topic in this chapter. Rather, the approach used here is to illustrate the usefulness of NMR in studying water relations in foods by presenting selected papers from recent literature organized broadly into three main topic areas: (1) characterization of water dynamics and distribution, (2) correlation of NMR water mobility parameters to food stability and quality, and (3) determination of water properties in food materials as affected by various processes (i.e., drying, freezing, and cooking).

Characterization of water dynamics and distribution

A number of NMR techniques have been used to determine water dynamics and distribution in food systems. A recent example is the research by Godefroy and Callaghan (2003). They used two-dimensional spin-relaxation and diffusion experiments to correlate molecular dynamics and interactions in cheese and microemulsions. Figure 4.22 shows the diffusion versus transverse relaxation time distribution maps of water for mozzarella (left) and Gouda (right) cheese samples, as measured at 40°C and aged for 74 days. The black cross on each plot corresponds to the position of the peaks at an age of 1 day. The authors reported that as the cheeses aged, an exchange in intensity was observed for the peaks at low diffusion coefficient with a peak at higher D and very short T_2 for the mozzarella sample and two different peaks at shorter T_2 with a faster diffusion for the Gouda sample. The authors propose that one explanation for this phenomenon could be the migration of water molecules from pools between fat globules to adsorption sites on proteins. A similar but much more extensive study by Hills et al. (2004) explored the potential of using two-dimensional T_1–T_2 correlation spectroscopy for quality control in three types of food samples, egg (white and yolk), cellular tissue (fruit and vegetable), and hydrocolloids (creams and baked products).

Another example is the thorough description of the molecular origins of water proton relaxation for solutions, gels, rubbers, and glasses given by Hills (1998). Of current interest is the dynamics of water in the glassy state. One of the first studies to comprehensively measure the mobility of water near the glass transition using NMR was by Ablett et al. (1993), who measured the glass transition temperature with DSC and molecular mobility as a function of temperature with ^{1}H NMR (T_1, T_2, and D). They reported that water had high rotational and translational mobility, even below the DSC-measured glass transition temperature of the malto-oligomer systems they had studied. A number of subsequent studies (e.g., Hills and Pardoe 1995; Tromp et al. 1997; Roudaut et al. 1999; Kou et al. 1999, 2000; Hills et al. 2001; Le Meste et al. 2002) have also shown that water retains a high degree of rotational and transitional mobility in the glassy state, relative to the glass transition temperature of the solid components in a variety of materials. The studies on sugar-water glasses (i.e., sucrose and xylose) above and below T_g by Hills et al. (2001) have shown that the water and sugar dynamics are decoupled from one another in the glassy state. The emerging composite picture of sugar and water in a sugar–water system below its glass transition temperature (see Figure 4.23) is that the sugar molecules in the glassy system are irrotationally frozen in a rigid three-dimensional amorphous matrix or lattice, exhibiting mainly vibrational motions and experiencing a gradual relaxation (i.e., physical aging) toward equilibrium. The water molecules are able to diffuse in the interstitial spaces formed by the amorphous sugar matrix, exhibiting vibrational, rotational, and translational motions, albeit at rates slower than in bulk water. Physical aging of the amorphous sugar–water matrix in the glassy state results in an overall decrease in the DSC T_g overtime (Wungtanagorn and Schmidt 2001).

Other recent examples of research using NMR techniques to study the dynamics and/or distribution of water in model or real food systems includes Gottwald et al. (2005), who investigated water diffusion, relaxation, and chemical exchange using NMR techniques in casein gels; Tang et al. (2000), who studied the microscopic distribution and dynamics state of water within native maize (A-type), potato (B-type), and pea (C-type) starch granules, using NMR relaxometry and diffusometry; Cornillon and Salim (2000), who characterized water mobility and distribution in low- and intermediate-moisture food systems such as cereals, chocolate chip cookies, soft caramel candies, and corn starch–water model systems; Choi and Kerr (2003), who determined the moisture content and molecular mobility of wheat starch suspensions at a_w values between 0 and 0.93 using ^{1}H NMR T_2 values obtained using both single-pulse and CPMG pulse sequence experiments; Pitombo and Lima (2003), who determined the ^{1}H NMR transverse relaxation times using the CPMG pulse sequence and MARAN WinDXP software in Pintado fish fillets (at −70° and 60°C) and freeze-dried fillets (reconstituted to moisture contents from 1% to 32% at 10°, 25°, and 40°C); Wang et al. (2004), who investigated the dynamics of water in white bread and starch gels as affected by water and gluten content; and Hermansson et al. (2006), who

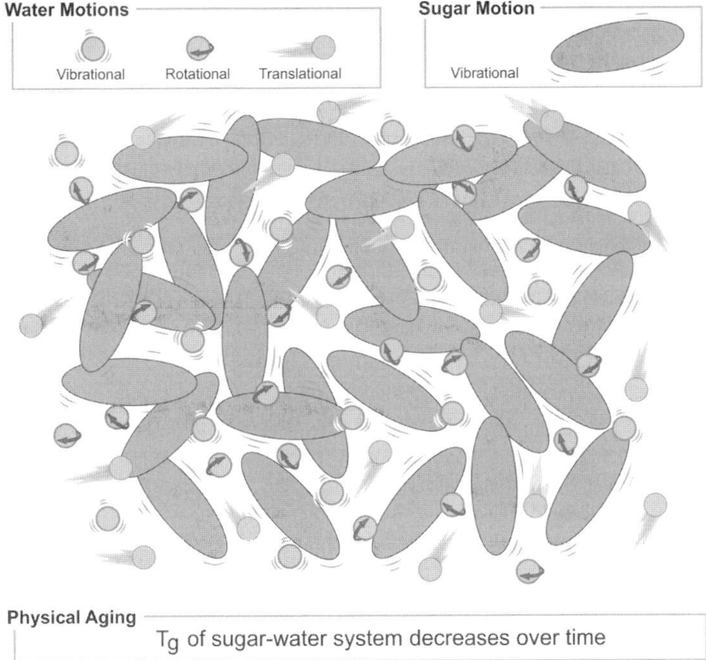

Figure 4.23 Schematic illustration of the emerging composite picture of a sugar and water in an amorphous sugar–water system below its glass transition temperature (based mainly on NMR mobility and DSC T_g data). The sugar and water dynamics are decoupled in the glassy state. The sugar molecules in the glassy system are irrotationally frozen in a rigid three-dimensional amorphous matrix or lattice, exhibiting mainly vibrational motions and experiencing a gradual relaxation (i.e., physical aging) toward equilibrium, resulting in a decrease in the glass transition temperature over time. The water molecules are able to diffuse in the interstitial spaces formed by the amorphous sugar matrix, exhibiting vibrational, rotational, and translational motions, albeit at rates slower than in bulk water.

combined NMR diffusometry with microscopy, image analysis, and mathematical modeling to investigate the relationship between microstructure (e.g., in gels and emulsions) and mass transport of solvents and solutes.

NMR measures the mobility of water in foods, but it also can be used to measure the mobility of solids in food systems. Of specific relevance here is the usefulness of NMR to measure the glass transition temperature of low-moisture or frozen systems. The NMR method involves measuring the change in molecular mobility (rotational and translational mobility) experienced by nuclei associated with the solid component (e.g., ^1H and ^{13}C). The temperature associated with an increase in solid component mobility is assigned as T_g. Additional information on the various NMR methods used to measure the T_g of food materials can be found in Ruan and Chen (1998), Ruan et al. (1998, 1999), van den Dries et al. (1998, 2000), Grattard et al. (2002), Kumagai et al. (2002), Sherwin et al. (2002), and Farhat (2004).

In addition to characterizing the distribution and dynamics of water and solids in foods, NMR is used to determine water content. In the case of determining water content, a low-resolution NMR (small B_0) instrument can be used. In general, as mentioned previously, the lower the magnetic field strength, the less costly is the NMR instrument. These instruments can be used for quality assurance measurements of moisture content and online or near-online monitoring. NMR is classified as an indirect method for measuring moisture content. In indirect methods, the moisture is not removed from the material, but rather parameters of the wet solid that depend on the quantity of water or number of constituent hydrogen atoms present are measured. The main advantage of indirect methods is the speed at which they can be carried out. ^{1}H NMR spectroscopy determines the number of water nuclei in a substance (NMR signal strength). This can then be used to determine the amount of water in the sample. A calibration curve must be developed correlating the NMR signal strength to the moisture content as measured by a direct method, such as air or vacuum oven. The calibration curve obtained is product specific. Additional information regarding the use of NMR for measuring product moisture can be found in Schmidt (1991), Isengard (1995, 2001b), Ruan and Chen (1998), Rutledge (2001), and Chinachoti and Krygsman (2005).

Correlation of NMR water mobility parameters to food stability and quality
Understanding and predicting the stability (chemical, biochemical, microbial, and textural) and quality of food materials are of utmost importance in providing consumers worldwide with a safe, nutritious, and high-quality food supply (Schmidt 1999, Lee et al. 2005). Recent examples of research involving NMR techniques that attempt to probe the stability and/or quality of food systems are summarized in Table 4.3.

The ability of the measured NMR parameters to correlate successfully with the targeted attribute (or attributes) varies widely in these examples (see Table 4.3). For example, Bell et al. (2002), who investigated the relationship between water mobility as measured by the ^{17}O NMR transverse relaxation rate and chemical stability in glassy and rubbery polyvinylpyrrolidone (PVP) systems, reported that water mobility was not related to T_g and that there was no link found between water mobility and reaction kinetics data (half-lives) for four chemical reactions studied (degradation of aspartame, loss of thiamin and glycine, and stability of invertase). However, one important point to consider is that the mobility measured in this study was ^{17}O NMR rotational mobility, whereas it is hypothesized that translational mobility (diffusion) may be a better NMR measure to relate to the reaction data, because it is expected that diffusion-limited reaction kinetics would be better correlated to the translational rather than the rotational mobility of water. In the Pham et al. (1999) study, which investigated the role of water mobility, as measured by ^{2}H NMR CPMG T_2, on mold spore germination, it was reported that T_2 correlated well with spore germination time. For the interested reader, a recent review by Chinachoti and Vittadini (2006) further explores water stress in bacteria and molds as probed by NMR water mobility measurements.

Determination of water properties in food materials as affected by various processes
Both NMR spectroscopy and imaging have been used to investigate the effects of various food processes on water properties, including dynamics, mass transfer, component interactions, and distribution. NMR imaging has been an especially useful tool in this regard because of its capacity to measure spatial and temporal changes during processing. McCarthy (1994) and Hills (1998) give a number of examples for the qualitative and quantitative use of MRI in investigating the behavior and/or location of water in food materials

Table 4.3. Summary of Recent Examples of Research Involving NMR Techniques That Attempt to Probe the Stability and/or Quality of Food Systems.

Stability or Quality Attribute	NMR Details	Reference
Chemical Stability—degradation of aspartame, loss of thiamin and glycine, and stability of invertase	^{17}O line width at half-height at 54.219 MHz	Bell et al. (2002)
Chemical Stability—oxidation rate of encapsulated flaxseed oil	^{1}H NMR at 23 MHz; FID was fit to obtain the fraction of mobile and immobile protons	Grattard et al. (2002)
Textural Stability—prediction of caking behaviors in powdered foods	^{1}H NMR T_2 at 20 MHz as a function of temperature	Chung et al. (2003)
Textural Stability—retrogradation in low corn starch concentration gels using DSC, rheology, and NMR	^{17}O T_1 by infrared and proton decoupled T_2 by line width at half-height; ^{1}H NMR T_1; ^{1}H PGSTE diffusion coefficient; ^{1}H cross-relaxation spectroscopy	Lewen et al. (2003)
Textural Stability—mealiness in apples	MRI and ^{1}H NMR T_2	Barreiro et al. (2002)
Textural Stability—correlation between the retrogradation enthalpy and water mobility in different rice starches	^{17}O proton decoupled NMR R_2 (line width at half height) at 27.13 MHz	Lin et al. (2001)
Microbial Stability—investigates the relationship between NMR relaxation rates, water activity, electrical conductivity, and bacterial survival in porous media	NMR relaxation rates	Hills et al. (1996a); Hills et al. (1997)
Microbial Stability—effect of physicochemical and molecular mobility parameters on *Staphylococcus aureus* growth	^{17}O proton decoupled NMR R_2, correlation time of the anisotropic slow relaxing component, and population of bound water	Vittadini and Chinachoti (2003)
Microbial Stability—mold conidia germination in starch, sucrose, and starch: sucrose systems	^{2}H R_1 and R_2 and ^{1}H PGSE NMR diffusion coefficient	Kou et al. (1999)
Microbial Stability—mold spore germination in nutrient media with varying a_w and varying amounts of nonnutritive and nontoxic carbohydrates (L-sorbose and cellulose)	^{2}H NMR T_2 at 46.7 MHz	Pham et al. (1999)
Quality—internal hen egg quality (oiled and nonoiled) fresh and at 7 days	^{1}H NMR T_1 dispersions (NMRD curves), T_2 CPMG with varying τ values at 23.4 and 300.15 MHz; T_2 with inverse Laplace transformation of the CPMG signal envelope	Laghi et al. (2005)
Quality—rehydration properties of foodstuffs	^{1}H NMR CPMG T_2	Weerts et al. (2005)
Quality—protein hydration and aging of dry-salted Gouda and mozzarella style cheeses	Magnetic field dependence (0.01 to 20 MHz) of ^{1}H T_1 values (NMRD curves)	Godefroy et al. (2003)
Quality—predicting sensory attributes of potato samples	^{1}H NMR T_2 CPMG at 23.2 MHz	Povlsen et al. (2003)
Quality—rehydration and solubilization of milk powders	^{1}H NMR CPMG T_2 at 10 MHz	Davenel et al. (2002)

Sources: Bell et al. 2002; Grattard et al. 2002; Chung et al. 2003; Lewen et al. 2003; Barreiro et al. 2002; Hills et al. 1996a, 1997; Vittadini and Chinachoti 2003; Kou et al. 1999; Pham et al. 1999; Laghi et al. 2005; Weerts et al. 2005; Godefroy et al. 2003; Povlsen et al. 2003; Davenel et al. 2002.

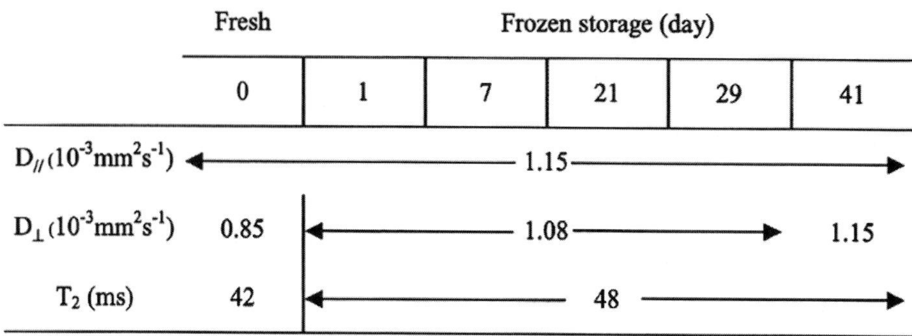

Figure 4.24 The ^{1}H NMR parameters for water in unfrozen (fresh) and frozen (–20°C), and stored (1 to 41 days), whole Rainbow Trout muscle obtained using a 4.7-T magnet. Thawing and fresh trout storage were at 4°C for 24 hours before analysis at 20°C [additional experimental details are given in Foucat et al. (2001)]. $D_{//}$ and $D_{\perp}$ are the axially and radially measured apparent diffusion coefficient, respectively (reprinted from Renou, J.P., Faoucat, L., and Bonny, J.M. 2003. Magnetic resonance imaging studies of water interactions in meat. Food Chemistry 82:35–39, with permission from Elsevier).

during processing, which includes drying, rehydration, freezing, freeze-drying, thawing, frying, and cooking; they suggest as well the use of MRI to explore these effects in newer processing technologies, such as ultrasonic and high-pressure treatments, pulsed electric fields, vacuum (*sous-vide*), and microwave heating. In addition, MRI has been used to measure and map other material characteristics during processing, such as rheology (McCarthy 1994, Wichchukit et al. 2005), temperature (Kantt et al. 1998, Nott and Hall 2005), and pH (Evans and Hall 2005).

Drying is a fundamental processing technology for many low-moisture shelf-stable ingredients and food products. Various MRI techniques have been used to study moisture migration and loss during drying. For example, Ziegler et al. (2003) used MRI techniques to obtain moisture profiles during drying of starch-molded confectionery gels. Conventional spin-echo imaging was used to monitor the water within the gel component, because the relaxation times were relatively long. Rapid formation of a very dry "skin" on the gel component was reported, which resulted in case hardening. Drying was effectively diffusion controlled after the first 30 minutes. Moisture profiles within the porous bed of molding starch were imaged using single-point ramped imaging with T_1 enhancement (SPRITE), because the total proton density was low and T_2^* was short. Other recent examples of using MRI to quantify the behavior of water during the drying process include moisture loss from harvested rice seeds (Ishida et al. 2004) and assessment of water diffusivity in gelatine gels (Ruiz-Cabrera et al. 2005a, 2005b).

Freezing is another common technology used by the food industry to extend the shelf-life of foods that has been investigated using NMR spectroscopy and imaging techniques. Research by Renou and others (Foucat et al. 2001, Renou et al. 2003) used MRI to investigate the effects of freezing on whole Rainbow trout muscle. Figure 4.24 summarizes the data extracted from the MRI experiments (where $D_{//}$ is the axially measured apparent diffusion coefficient, $D_{\perp}$ is the radially measured apparent diffusion coefficient, and T_2 is the transverse

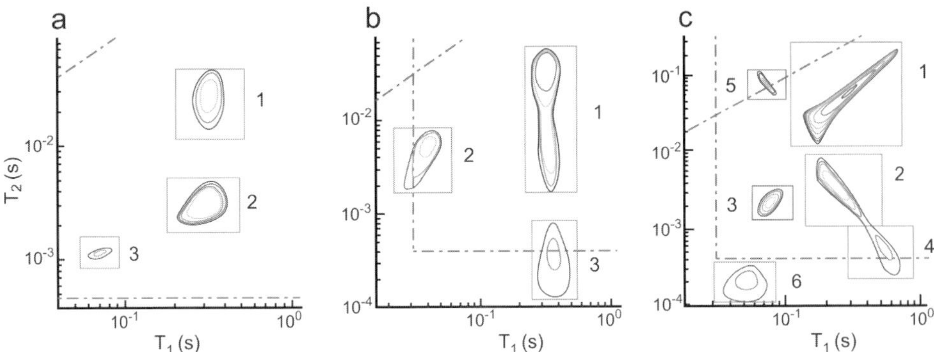

Figure 4.25 T_1–T_2 correlation spectra acquired at 23.4 MHz with a CMPG 90° to 180° pulse sequence of 200 μsec at 298K of a water-saturated packed bed of native waxy maize starch (**a**); after pressure treatment of the sample in (**a**) for 15 min at 500 MPa (**b**); and after microwave treatment –20 seconds on high setting [900 W] (**c**). The diagonal dashed line corresponds to $T_1 = T_2$. The vertical and horizontal lines show the shortest T_1 and T_2, respectively, that can be measured with the set of experimental; NMR acquisition parameters (adapted from Hills, B., Costa, A., Marigheto, N., and Wright, K. 2005. T_1–T_2 NMR correlation studies of high- pressure-processed starch and potato tissue. *Applied Magnetic Resonance* 28:13–27, with permission from Springer-Verlag.).

relaxation time) on fresh and frozen and thawed trout. In comparison with fresh trout, neither the frozen storage duration (from 1 to 41 days) nor the freeze-thaw process seemed to have an effect on $D_{//}$. On the other hand, both the $D_{\perp}$ and T_2 values exhibited a difference between the fresh and frozen and thawed trout. The longer T_2 values observed in the frozen and thawed trout were attributed to protein denaturation, which results in redistribution of the water within the muscle due to partial extrusion of water from the intracellular compartment. The variations observed in the MRI parameters as a function of freezing and/or storage duration were reported to be consistent with histological observations and can be used to differentiate fresh from frozen fish. Other food materials that have recently been studied during freezing or in the frozen state using NMR techniques include partially baked bread (Lucas et al. 2005a, 2005b), beef (Lee et al. 2002), orange juice (Lee et al. 2002), dough (Lee et al. 2002, Lucas et al. 2005c), and corn (Borompichaichartkul et al. 2005).

Hills and others have used NMR techniques to study the effects of high-pressure processing on strawberries (Marigheto et al. 2004) and starch and potato tissue (Hills et al. 2005). T_1–T_2 NMR correlation spectroscopy revealed that B-type starch (potato) was more resistant to pressure treatment (500 MPa for 15 minutes) than was A-type starch (waxy maize). In addition, the authors reported that high-pressure–induced waxy maize gels were found to be radically different compared with the correspondingly thermally (microwave)-induced gels. A comparison of the T_1–T_2 correlation spectra of waxy maize starch before treatment (native) and after high-pressure and microwave treatments is shown in Figure 4.25. For native waxy maize starch (see Figure 4.25a), peak 1 was assigned to extragranular water, peak 2 to water inside the granules, and peak 3 to mobile starch CH protons. For the high-pressure–treated waxy maize starch (see Figure 4.25b), peak 1 was greatly elongated to encompass shorter T_2 values, which is consistent with high-pressure swelling and gelatinization of the granule bed; peak 2 was shifted to much shorter T_1 values, indicative of a major dynamical and/or microstructural change inside the granule; and peak 3 shifted to longer T_1 values. For the microwave-treated waxy maize starch (see Figure 4.25c), peaks 1 through 3 were approximately similar to those of the native waxy maize starch, although the peak shapes are greatly distorted; in addition, three new peaks (4 through 6) emerged. However, the identity of these peaks has yet to be assigned.

NMR spectroscopy and imaging have also been applied to study the effects of cooking on various food materials, including recent research on changes during the cooking of meat (Micklander et al. 2002), white salted noodles (Lai and Hwang 2004), rice kernels (Mohoric et al. 2004), pasta (McCarthy et al. 2002, Irie et al. 2004), and starch/gum/sugar solutions (Gonera and Cornillon 2002). Micklander and others (2002) obtained low-field NMR data during the cooking of meat, resulting in the identification of five important temperatures (42°, 46°, 57°, 66°, and 76°C) at which major changes in meat structure occurred. In addition, the authors reported the formation of a new water population above approximately 40°C. They hypothesize that the new population of water is probably expelled water from the myofibrillar lattice or perhaps the formation of a porous myosin gel.

Relationship between NMR Relaxation Rates and a_w

As illustrated in Figure 4.3, both NMR and a_w are used to measure the mobility of water in food systems, although NMR reflects molecular-level rotational and translational mobility and a_w reflects macroscopic-level translational mobility. Research papers that were found containing plots of NMR relaxation times or rates as a function of a_w are listed in Table 4.4. It is interesting to note that in the cases listed in Table 4.4, where the T_1 values were plotted as a function of a_w, a curve with a minimum was observed (see Figure 4.26), corresponding to the T_1 minimum previously illustrated in Figure 4.15. In cases where the T_2 values were plotted as a function of a_w, a continuously increasing curve was observed (see Figure 4.27). Lastly, in cases where the R_2 values (recall $R_2 = 1/T_2$) were plotted as a function of a_w, a linear relationship was observed (see Figure 4.28).

An empirically derived relationship between NMR relaxation rates and a_w was proposed by Hills and others (Hills et al. 1996a, 1999; Hills 1998, 1999). The model describes observed "average" values for NMR (T_{av}^{-1}) and a_w (a_{av}) parameters as a weighted average of values over all water states:

$$T_{av}^{-1} = \sum_i p_i T_i^{-1} \tag{4.11}$$

$$a_{av} = \sum_i p_i a_i \tag{4.12}$$

where T_i^{-1} and a_i are the intrinsic relaxation rate and a_w of the *i*th state of water in a system, and p_i is the fractional population of that state. According to Hills' multistate theory of water relations (Hills et al. 1996a, Hills 1999), it is usually sufficient to consider three states of water in biopolymer systems: (1) "structural" or strongly interacting water hydrogen bonded inside the cavities or grooves of polysaccharides and globular proteins; sometimes the structural water state is referred to as "bound" water, however, it must be kept in mind that this water has exchange lifetimes ranging from nanoseconds to milliseconds, depending on the nature of the water-biopolymer interaction, thus it is not permanently bound to the biopoly-

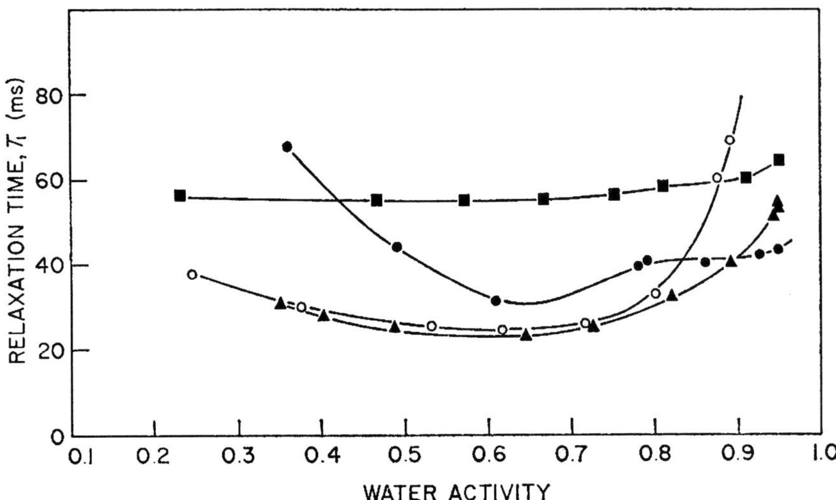

Figure 4.26 Plot of ^{1}H NMR T_1 as a function of water activity for casein (●), pectin (▲), corn starch (■), and sodium alginate (○) macromolecules at 22°C (reprinted from Steinberg, M.P., and Leung, H. 1975. Some applications of wide-line and pulsed N.M.R. in investigations of water in foods. In: *Water Relations of Foods,* ed. R.B. Duckworth, pp. 233–248. New York: Academic Press, with permission from Elsevier).

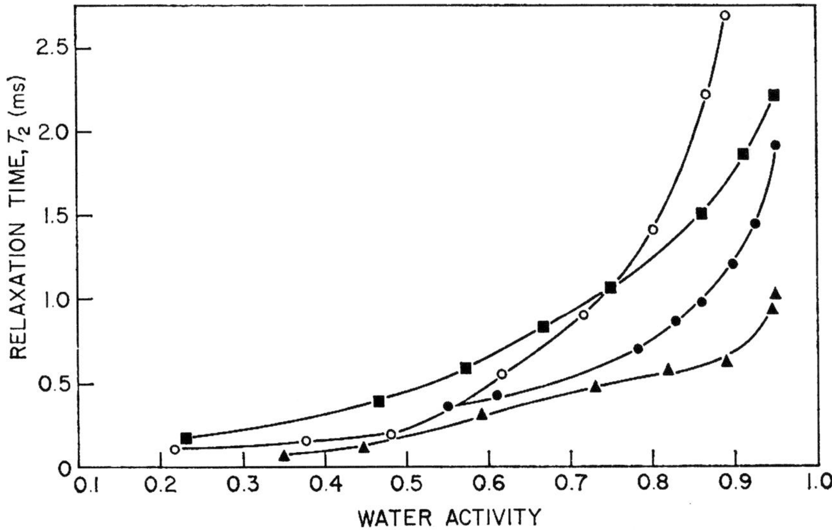

Figure 4.27 Plot of ^{1}H NMR T_2 as a function of water activity for casein (●), pectin (▲), corn starch (■), and sodium alginate (○) macromolecules at 22°C (reprinted from Steinberg, M.P., and Leung, H. 1975. Some applications of wide-line and pulsed N.M.R. in investigations of water in foods. In: *Water Relations of Foods,* ed. R.B. Duckworth, pp. 233–248. New York: Academic Press, with permission from Elsevier).

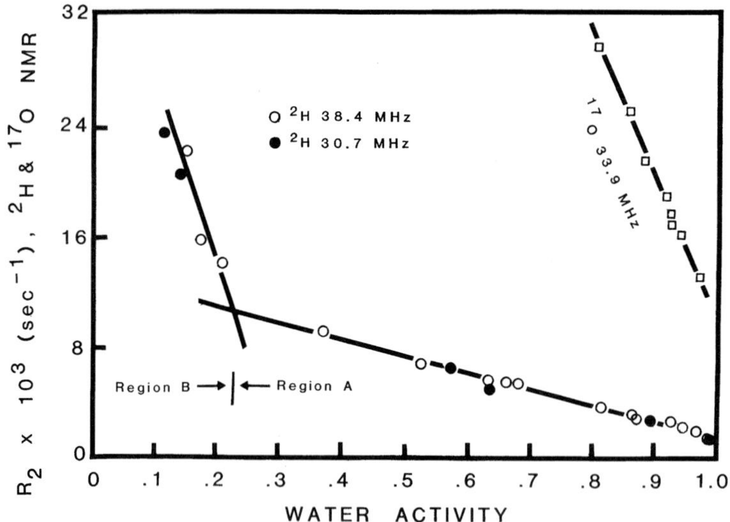

Figure 4.28 Plot of ^2H (at two frequencies) and ^{17}O NMR R_2 as a function of water activity of native corn starch in D_2O or enriched with $H_2{}^{17}O$, respectively at 20°C (reprinted from Richardson, S.J., Baianu, I.C., and Steinberg, M.P. 1987. Mobility of water in starch powders by nuclear magnetic resonance. *Starch* 39:198–203, with permission from John Wiley & Sons, Inc.)

Table 4.4. Research Papers That Were Found Containing Plots of NMR Relaxation Times or Rates as a Function of a_w.

Source	NMR Nucleus	Relaxation Parameters Plotted as a Function of a_w	Food Materials
Steinberg and Leung (1975) and Leung et al. (1976)	^1H	T_1 and T_2	Casein, pectin, corn starch, and sodium alginate
Richardson et al. (1987)	^2H and ^{17}O	R_2	Native corn starch
Curme et al. (1990)	^2H	R_2	Casein and casein-NaCl (95:5 ratio)
Monteiro Marques et al. (1991)	^1H	T_1 and T_2	Grated carrot samples

Sources: Steinberg and Leung 1975; Leung et al. 1976; Richardson et al. 1987; Curme et al. 1990; Monteiro Marques et al. 1991.

mer; (2) "multilayer" or "surface" or "hydration" water, which is water that is perturbed by the presence of the biopolymer; multilayer water extends for a number of layers from the surface of the biopolymer and is very mobile, having exchange lifetimes on the order of nanoseconds or less; and (3) "bulk" or "free" water, which is water that is undisturbed by the presence of the biopolymer and has mobility similar to that of pure water.

Equation 4.11, which was originally postulated by Zimmerman and Brittin (1957), assumes fast exchange between all water states (i) and neglects the complexities of proton exchange and cross-relaxation. Equation 4.12 is consistent with the Ergodic theorem of statistical thermodynamics, which states that at equilibrium, a time-averaged property of an individual water molecule, as it diffuses between different states in a system, is equal to a time-independent ensemble-averaged property (Equation 4.12 being an ensemble-averaged expression) but ignores configurational entropy effects (Hills 1998, 1999; Hills et al. 1999). Next, Equations 4.11 and 4.12 were simplified to the case of two states of water in fast exchange (e.g., bulk water [subscript b] exchanging with hydration water [subscript h (also called multilayer water)]), yielding the following two equations, respectively:

$$T_{av}^{-1} = T_b^{-1} - \frac{m_o\left(T_b^{-1} - T_h^{-1}\right)}{W} \tag{4.13}$$

$$a_{av} = a_b - \frac{m_o\left(a_b - a_h\right)}{W} \tag{4.14}$$

where m_o is the weight of hydration water per unit weight of dry biopolymer, and W is water content, defined as the weight of water per unit weight of dry solid. It should be noted that Equation 4.13 is only valid when water relaxation is single-exponential (requires rapid diffusive exchange between all water states), and Equation 4.14 is only valid for $W > m_o$, because W values less than m_o correspond to the removal of hydration water, a region of an isotherm where the BET isotherm equation is applicable (Hills et al. 1996b).

Based on the correspondence between Equations 4.13 and 4.14, Hills combined the two equations by setting the a_w of bulk water (a_b) equal to 1 in Equation 4.14, rearranging Equation 4.14 to solve for m_o/W and then substituting the results of this term into Equation 4.13, yielding Equation 4.15, which linearly correlates NMR relaxation rates to a_w:

$$T_{av}^{-1} = T_b^{-1} + C(1 - a_{av}) \tag{4.15}$$

where the constant C is equal to $(T_h^{-1} - T_b^{-1})/(1 - a_h)$. This combined equation is valid within the limitations given for the individual equations (two states of water in exchange, water relaxation is single-exponential, and $W > m_o$). Breaks in linearity (i.e., changes in slope) between otherwise linear regions indicate that one of the states of water has been removed or a new one has been introduced.

The linear relationship between ^{1}H NMR transverse relaxation rate (which is equivalent to T_{av}^{-1}) and ($1-a_{av}$) is shown in Figure 4.29 for pregelled potato starch (Hills et al. 1999). The change in slope at about 0.90 a_w corresponds to the bulk water break (i.e., the removal of bulk water) in a corresponding adsorption isotherm. Equation 4.15 has been successfully applied to beds of Sephadex microspheres and silica particles, using both ^{1}H NMR (Hills et al. 1996b) and ^{17}O NMR, the latter of which is free from the complication of proton exchange when measured using proton decoupling (Hills and Manning 1998).

Figure 4.30 shows the application of Equation 4.15 to the ^{2}H NMR R_2^* and a_w data of Kou et al. (1999) for instant (pregelatinized) dent #1 corn starch. Figure 4.30 also illustrates the linear dependence between NMR relaxation rates and a_w predicted by Equation 4.15. The break, which occurs at a a_w of approximately 0.90 (with a corresponding moisture content of approximately 21% wet basis), indicates the removal of bulk water and is somewhat lower than the a_w associated with the DSC measured T_g at the experimental temperature (20°C) of 0.936 (with a corresponding moisture content of approximately 22% wet basis). Regarding the comparison of the above a_w and moisture content values, it should be noted that the NMR associated values were for saturated salt solutions made with D_2O, whereas the T_g associated values were for saturated salt solutions made with

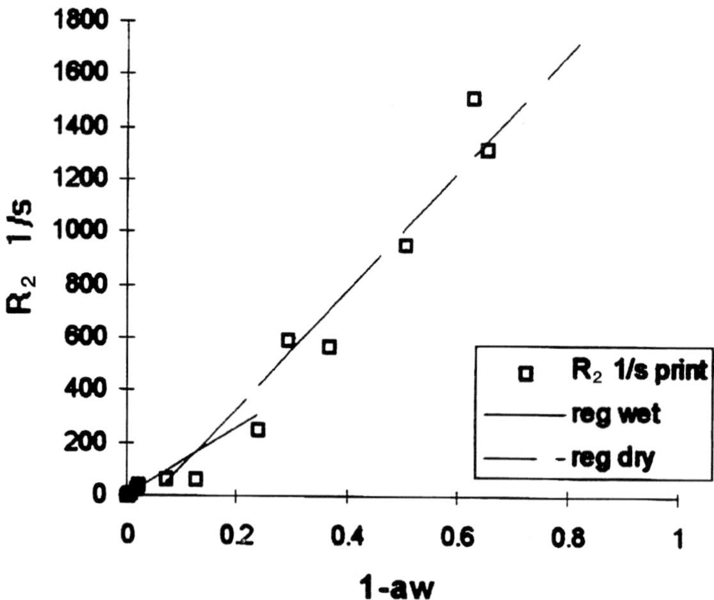

Figure 4.29 Linear relationship between ^{1}H NMR transverse relaxation rate (recall $R_2 = 1/T_2$) and $(1 - a_w)$ for pregelled potato starch (reprinted from Hills, B.P., Manning, C.E., and Godward, J. 1999. A multistate theory of water relations in biopolymer systems. In: *Advances in Magnetic Resonance in Food Science,* eds. P.S. Belton, B.P. Hill, and G.A. Webb, pp. 45–62. Cambridge, UK: Royal Society of Chemistry, with permission from Woodhead Publishing Limited).

H_2O. The a_w values were on average only slightly higher (0.003 a_w in the a_w range of interest) for the D_2O versus the H_2O saturated salt solutions, whereas the moisture content was on average 1.9% (wet basis) higher for the D_2O starch samples compared with the H_2O starch samples (comparison for the moisture content range of interest).

The ^{2}H NMR R_2^* relaxation rate data of Richardson et al. (1987) for native (ungelatinized) corn starch shown in Figure 4.28 was also plotted as a function of $(1 - a_w)$ and is shown in Figure 4.31. Three linear regions and two breaks were observed. The first two linear regions (starting at high a_w values) and related break correspond to the application of Equation 4.15, as previously discussed for Figures 4.29 and 30. This first break, in the case of native (ungelatinized) corn starch, occurs at a_w of approximately 0.97 and moisture content of approximately 24% (wet basis). This break corresponds to the removal of the bulk water in the system, which was not noted in the original paper. The a_w value for the first break in Richardson et al. data (1987) of 0.97 (see Figure 4.31) is somewhat higher than the value estimated for the break in Kou et al. data (1999) of 0.90 (see Figure 4.30). However, the Kou et al. data refers to instant (pregelatinized) dent #1 corn starch, which would be expected to adsorb more bulk water than the native (ungelatinized) corn starch used in the Richardson et al. study.

The second break in Figure 4.31 occurs at an a_w of 0.25 (the break in the original paper was reported at approximately 0.23 a_w). The moisture content of the second break is ap-

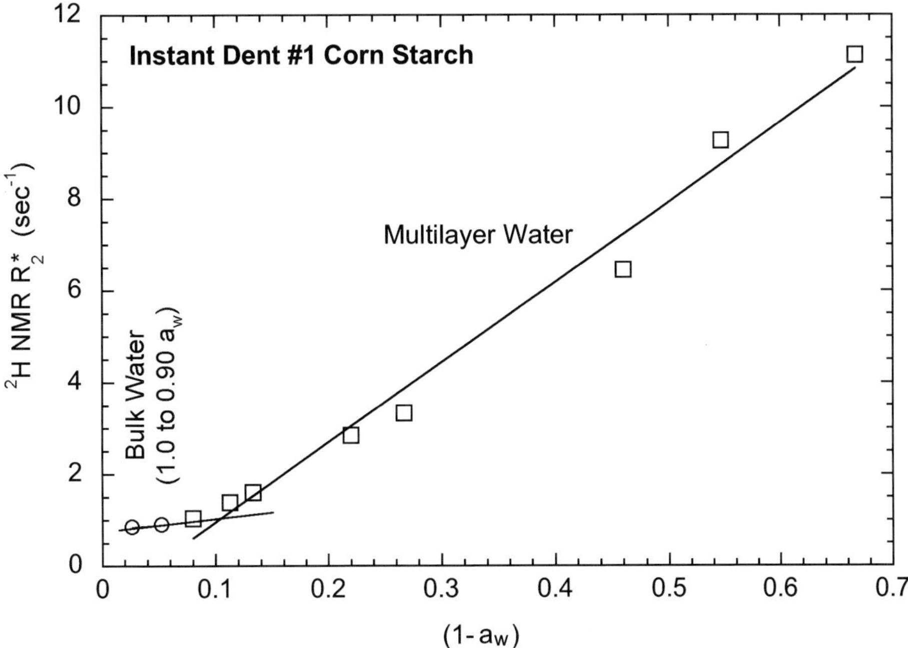

Figure 4.30 Linear relationship between ^{2}H NMR R_2^* relaxation rate and $(1 - a_w)$ for the instant (pregelatinized) dent #1 corn starch data of Kou et al. (1999). Linear regions and break are discussed in the text.

proximately 8.4% (wet basis), which is about 1.5% higher than the calculated BET monolayer value of 6.89% (wet basis) given in the Richardson et al. (1987) paper. This second break corresponds to the removal of the hydration or multilayer water in the system. It is important to note that the observation of the second break and the third linear region (structural water) in Figure 4.31 is new and not explained by Equation 4.15, because this third linear region occurs where $W < m_o$, violating one of the assumptions underlying Equation 4.15. However, as observed in Figure 4.31, there remains a linear relationship between R_2 and a_w in this low-a_w region, indicating, perhaps, that there is more than one type of structural water state present, all of which are in fast exchange, which is supported by the research of Tang et al. (2000), discussed next.

Figure 4.32 shows a schematic illustration of the three states of water—structural, multilayer, and bulk—in native (ungelatinized) corn starch (A-type starch) and corresponds to the three linear regions in Figure 4.31. Figure 4.32 is a composite picture of the water in a native corn starch–water system based on NMR research results, especially those of Tang et al. (2000). The general distribution and dynamics of the three water types are:

1. Structural water is intragranular water that is associated with the amorphous growth ring regions and the semicrystalline lamellae regions, which are in fast exchange. Intragranular water has slowed rotational and translational motion compared with bulk

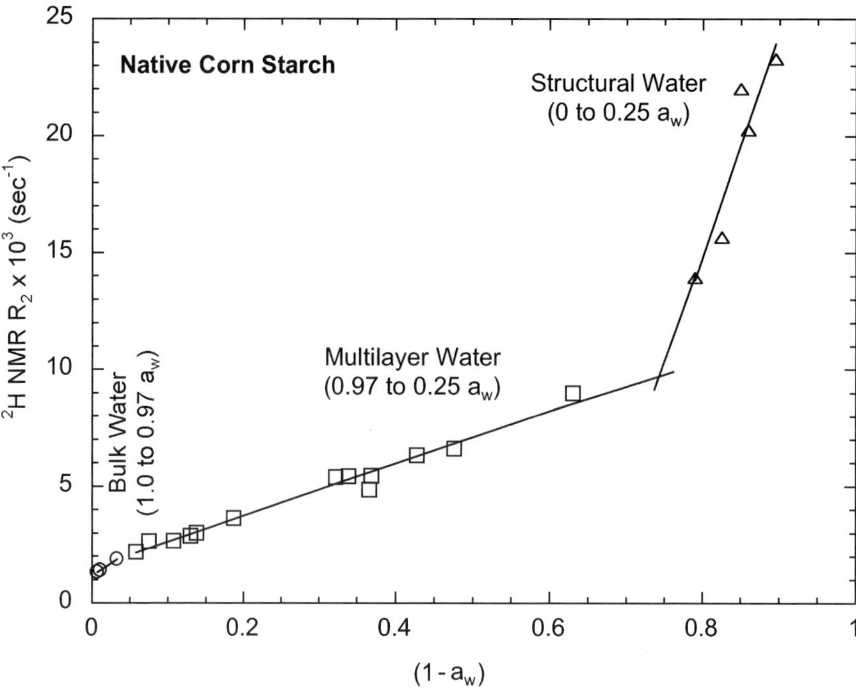

Figure 4.31 Linear relationship between ^{2}H NMR R_2^* relaxation rate and (1 – a_w) for the native (ungelatinized) corn starch data of Richardson et al. (1987). Linear regions and breaks are discussed in the text.

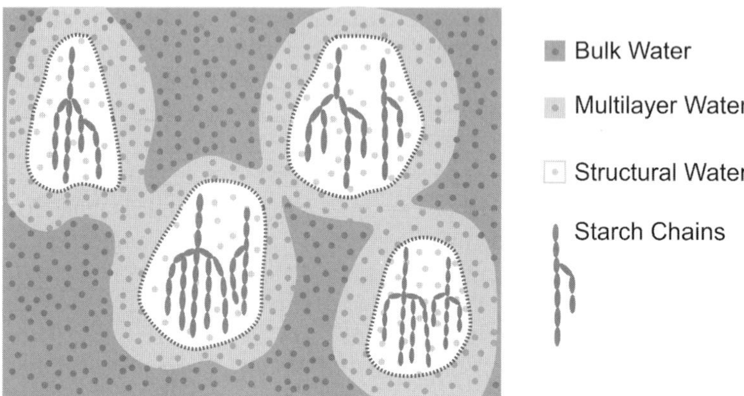

Figure 4.32 Schematic illustration of the three states of water—structural, multilayer, and bulk—in native (ungelatinized) starch, corresponding to the three liner regions in Figure 4.31. Attributes of each water state are given in the text.

water, but it is much more mobile than the solid starch chains. However, structural water does not freeze even at very low temperatures.

2. Multilayer water is extragranular, surface-adsorbed water. This water has slower rotational and translational motion compared with bulk water, but its motion is still quite fast, and it is freezable.
3. Bulk water is also extragranular, but it is minimally perturbed by the presence of starch granules. Bulk water is similar to free water, exhibiting fast rotational and translational motion, and it is freezable.

Hills et al. (1999) clearly expressed the point that there is no implied fundamental physical relationship between a_w, an equilibrium thermodynamic quantity, and NMR relaxation, a nonequilibrium kinetic event in this theory. Rather, all that is suggested here is that the changing states of water in a biopolymer system affect both a_w and NMR relaxation similarly.

Future Needs and Directions

NMR spectroscopy and imaging have been proved to be exceedingly powerful tools for noninvasive and nondestructive probing of the molecular nature of water in food materials. The numerous advantages of using NMR to study water in foods, as well as the bountiful and continuous adaptations of NMR techniques for studying water in food problems, make it an indispensable tool for future inquiries. At the same time, it must be recognized that the increasing complexity of NMR instrumentation and experiments, as well as its relatively high cost, limits its usefulness in and by the food industry. What is needed, in addition to the continued development of highly technical research uses of NMR, is the development of lower cost, more direct, and more immediately beneficial applications of NMR for use in and by the food industry, which is similar in concept, for example, to the use of NMR for measuring moisture and solid fat contents. Quality assurance problem areas may be a good target for development and application of these "food industry–friendly" NMR and MRI systems and techniques.

In addition, continued efforts are needed to further explore and exploit the interrelationship among the three main water relations in food research avenues presented at the beginning of this chapter: a_w, molecular water mobility, and the food polymer science approach. Each of these avenues has its advantages and drawbacks (details can be found in Schmidt 2004), yet together they can provide a composite, multilevel (at various distance and time scales) portrait of the water and solids dynamics that govern the stability behavior of food systems. Figure 4.33 illustrates a vision toward this integrative approach by way of superimposed general ranges for water mobility by NMR and T_g by DSC (compared with the temperature of the composite isotherm, room temperature, RT) techniques on a composite food sorption isotherm partitioned into the traditional three water-type regions or zones. The division of the isotherm into three regions is based on Labuza's original stability map (Labuza et al. 1970), where Region I ranges from 0 to 0.25 a_w, Region II from 0.25 to 0.80 a_w, and Region III from 0.80 to 1.00 a_w. Different reproductions of the map have subsequently used slightly different a_w partitioning values (±0.05). The a_w partitions used here for the Region I to II transition was 0.25 a_w, but 0.75 a_w was used for the transition from Region II to III, because there is a noticeable increase in the moisture content of the composite food sorption isotherm just after 0.75 a_w. Additional details regarding the development of the composite food sorption isotherm can be found in Schmidt (2004). The extended composite food isotherm shown in Figure 4.33 is based on a conceptual figure presented by Slade and Levine (2004), which combines the food stability map of Labuza et al. (1970) and general T_g ranges compared with experimental temperature conditions (T) (see Figure 4.34). It is important to note the conceptual difference between plotting the isotherm for a single material or compatible blend (see Figure 4.34) and the extended composite food isotherm (see Figure 4.33), which is constructed using individual a_w and moisture content pairs for a variety of actual food materials (many of which are multicomponent) ranging from low to high a_w.

Even though Figure 4.33 is rather generic, it is intended to stimulate the integration of the viewpoints offered by different analytical approaches, including the approaches discussed in this chapter, as well as others such as electron spin resonance and optical luminescence. Because food systems vary widely in their composition, processing, and history, what is needed is more research that obtains and evaluates actual measured values (e.g., a_w, NMR, and T_g values) for specific food ingredients and systems and their stability (chemical, biochemical, microbial, and textural) and quality attributes.

An example of such a plot is shown in Figure 4.35, where the ^{1}H PGSE NMR diffusion coefficient of water (D, m^2/sec), the DSC T_g (midpoint, °C), and the mold conidia germination time (days; >30 days indicates no mold conidia germination was observed even after 30 days) are plotted as a function of a_w on the isotherm (done at 20°C) for instant (pregelatinized) dent #1 corn starch (data from Kou et al. 1999). Figure 4.35 integrates the water (via D) and solid (via T_g) dynamics, as well as the microbial stability (as probed by mold conidia germination), of the instant (pregelatinized) dent #1 corn starch system. In the case of instant (pregelatinized) dent #1 corn starch system, both the diffusion coefficient of water and the mold germination time exhibit a significant change above the critical a_w value, where the critical a_w is defined as the a_w and the T_g value equals the temperature at which the isotherm was obtained, in this case 20°C. The critical a_w value given by Kou et al. (1999) was 0.936 using the Smith (1947) isotherm equation (using the Guggenheim-Anderson-De Boer [GAB] equation yields a slightly lower critical a_w value of 0.92, as discussed in Schmidt 2004). Above this critical a_w value (or, as discussed above, the relative vapor pressure), there is a step change increase in the diffusion coefficient, and mold conidia germination begins.

It is important to point out the ongoing contributions of Roos (e.g., in 1993, 1995, 1998, and 2003) in producing a number of extremely useful multiparameter diagrams, which are rooted in the interconnectedness of the various stability-related measurements. For example, Roos (1993) simultaneously plotted the glass transition temperature and a_w data as a function of moisture content for a single food material and subsequently extracted (and defined) the critical a_w and moisture content parameters that correspond to depressing the T_g to ambient temperature. This T_g–water sorption plot is useful as a "map" for the selection of storage conditions for low- and intermediate-moisture foods.

The ultimate goal of pooling and mining the mobility and stability data from an array of different perspectives, as discussed and demonstrated above, is to assist in understanding the stability behavior of the complex food systems that we have come to rely on, as well as to assist in developing foods that are safe, nutritious, and high in quality for the peoples of the world to enjoy.

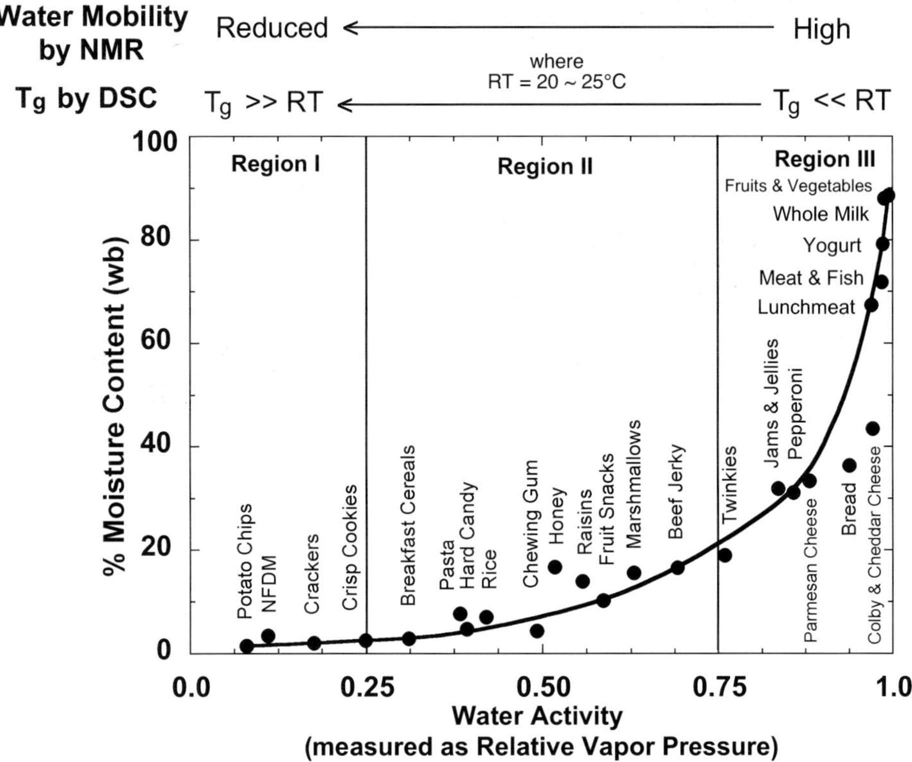

Figure 4.33 Illustrates the interrelationship among water activity, molecular water mobility by NMR, and T_g by DSC. The base of this illustration is the composite (or universal) food sorption isotherm at 20° to 25°C (the a_w of some food items were measured at 20°C, others at 25°C) partitioned into the traditional three water type regions or zones (based on Labuza et al. 1970), upon which are superimposed ranges for general water mobility by NMR and T_g by DSC (compared with the temperature of the composite isotherm, room temperature, RT). Water activity is included as one of the *x*-axis labels for consistency, however, as advocated by Slade and Levine (1991) and Fennema (1996) the term *relative vapor pressure* (RVP, the measured term) is a more accurate term than a_w (the theoretical term). A few additional comments regarding the figure are (1) Tg values are only obtainable for food systems that contain amorphous component(s); (2) there is an important difference between plotting the isotherm for a single material or compatible blend (several water activity and moisture content values for one material) and plotting single water activity and moisture content pairs for several food materials, which is how the composite food isotherm was constructed. Thus, water mobility by NMR and T_g by DSC are illustrated as continuous parameters since specific demarcations are food material specific. For example, in the case of water mobility by NMR, the difference in location of the high a_w breaks in Figure 4.30 [at 0.89 a_w] versus 4.31 [at 0.97]. In order to give more specific changes (add additional demarcations) for both DSC by T_g and water mobility by NMR, T_g and NMR values would need to be obtained for each food material for a series of different water activity and moisture content combinations (i.e., the traditional isotherm); and (3) the curve is to guide the eye—it is not a fitted line.

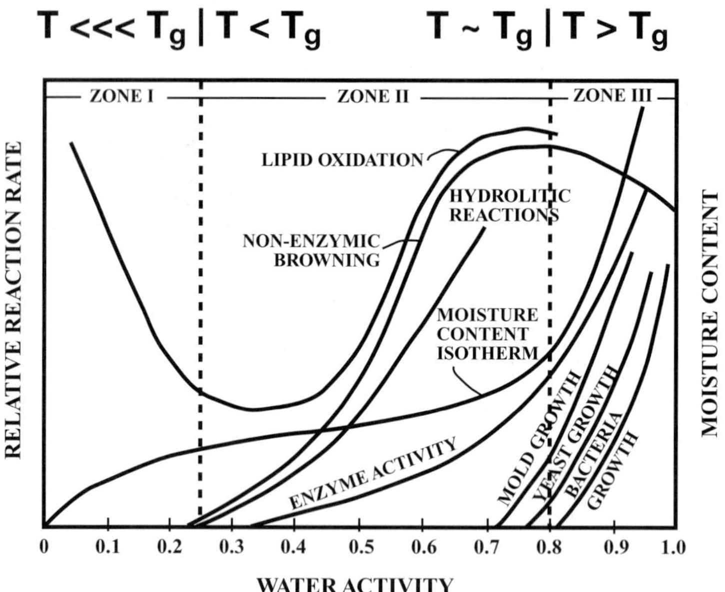

Figure 4.34 Conceptual figure regarding how to interpret the three sorption zones presented by Slade and Levine (2004), which combined the food stability map of Labuza (Labuza et al. 1970) and T_g values compared with experimental temperature conditions (T).

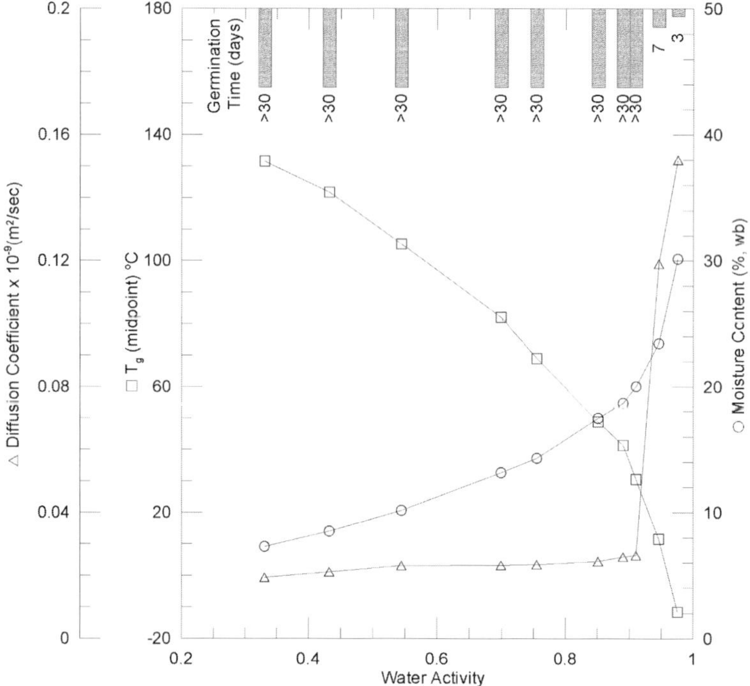

Figure 4.35 The diffusion coefficient of water (D, m^2/sec) obtained by ^{1}H PGSE NMR, the T_g (midpoint, °C) obtained by DSC, and the germination time of mold conidia (days; >30 days indicates no germination was observed even after 30 days), each as a function of a_w, are plotted on the isotherm (done at 20°C) for instant (pregelatinized) dent #1 corn starch. Data are from Kou et al. (1999).

Endnotes

[1]Felix Franks is editor of a seven-volume series on water and aqueous solutions, *Water—A Comprehensive Treatise*, published by Plenum Press, New York. The title and publication date of each volume are (1) *The Physics and Physical Chemistry of Water* (1972); (2) *Water in Crystalline Hydrates; Aqueous Solutions of Simple Nonelectrolytes* (1973); (3) *Aqueous Solutions of Simple Electrolytes* (1973); (4) *Aqueous Solutions of Amphiphiles and Macromolecules* (1975); (5) *Water in Disperse Systems* (1975); (6) *Recent Advances* (1979); and (7) *Water and Aqueous Solutions at Subzero Temperatures* (1982).

[2]Interestingly, Philip Ball, writer and editor of *Nature* magazine, in the preface of the book *Life's Matrix,* acknowledges a number of scientists and colleagues in the writing of this book, including Felix Franks.

[3]International Symposium on the Properties of Water (ISOPOW) is a nonprofit scientific organization. Its activities aim at progressing the understanding of the properties of water in food, pharmaceutical, and biological materials, for the improvement of raw materials, food products and processes, agrofood, or related industries. The first symposium was organized in Glasgow, Scotland, in 1974 under the pioneering initiative of the late Professor Ron B. Duckworth and Dr. Lou Rockland. To date, eight additional ISOPOW meetings have been held, and two practicums. Following are previous symposia and practicums, as well as associated publications (based on information at http://www.u-bourgogne.fr/ISOPOW/Articles_of_Association.htm).

Meeting Type and No.	Location and Year	Meeting Title and Associated Publication
ISOPOW 1	Glasgow (UK), 1974	**Water Relations of Foods**, 1975, R.B. Duckworth, Ed., Academic Press.
ISOPOW 2	Osaka (Japan), 1978	**Water Activity: Influences on Food Quality**, 1981, L.B. Rockland and G.F. Stewart, Eds., Academic Press.
ISOPOW 3	Beaune (France), 1983	**Properties of Water in Foods**, 1985, D. Simatos and J.L. Multon, Eds., Martinus Nijhoff Publishers.
ISOPOW 4	Banff (Canada), 1987	Not Published, Meeting Chair - Dr. Marc Le Maguer
Practicum I	Penang (Malaysia), 1987	**Food Preservation by Moisture Control**, 1988, C.C. Seow, Ed., Elsevier Applied Science.
ISOPOW 5	Peniscola (Spain), 1992	**Water in Foods**, 1994, P. Fito, A. Mulet, and B. MacKenna, Eds., Elsevier Applied Science, Papers were also published in the *Journal of Food Engineering*, 1994, 22:1-532.
Practicum II	Puebla (Mexico), 1994	**Food Preservation by Moisture Control: Fundamental and Applications**, 1995, G.V. Barbosa-Cánovas and J. Welti-Chanes, Eds., Technomics Pub. Co.
ISOPOW 6	Santa Rosa (USA), 1996	**The Properties of Water in Foods**, 1998, D.S. Reid, Ed., Blackie Academic and Professional.
ISOPOW 7	Helsinki (Finland), 1998	**Water Management in the Design and Distribution of Quality Foods**, 1999, Y.H. Roos, J.B. Leslie and P.J. Lillford, Eds., Technomics Pub. Co.
ISOPOW 8	Zichron Yaakov (Israel), 2000	**Water Science for Food, Health, Agriculture and Environment**, 2001, Z. Berk , P. Lillford, J. Leslie and S. Mizrahi, Eds., Technomics Pub. Co.
ISOPOW 9	Mar del Plata (Argentina), 2004	**Water Properties Related to the Technology and Stability of Food, Pharmaceutical and Biological Materials**, 2006, M. del Pilar Buera, J. Welti-Chanes, P.J. Lillford, and H.R. Corti, Eds., Taylor & Francis, CRC Press.

In addition to the ISOPOW symposia and practicums, the first International Workshop on Water in Foods was held in 2000. Following are the workshops that have been held, as well as associated publications.

Workshop No.	Location and Year	Workshop Title and Associated Publication
1	Ispra (Italy), 2000	**Water Determination in Food - A Challenge for the Analysis**. Proceedings published in a special issue of *Food Control* 2001, 12(7).
2	Reims (France), 2002	**Water Structure, Water Activity, Water Determination in Food**, Proceedings published in a special issue of *Food Chemistry* 2003, 82(1).
3	Lausanne (Switzerland), 2004	**Water Functionality in Food, Water and Food Structure, Water Determination in Food, Water Activity in Food**, Proceedings published in a special issue of *Food Chemistry* 2006, 96(3).
4	Brussels (Belgium), 2006	**International Workshop on Water in Foods**

[4]To further advance the applications of NMR in food science, an international conference, the "1st International Conference on Applications of Magnetic Resonance in Food Science," was held in 1992 in Surrey, UK. Since that time the conference has been held biennially. Listed below are previous conferences, as well as associated publications. The 8th International Conference on the Applications of Magnetic Resonance in Food Science is currently planned for July 2006 in Nottingham, UK.

Conference No.	Location and Year	Conference Title and Associated Publication
1	Surrey (UK), 1992	**Annual Reports on NMR Spectroscopy**, special issue, 1995, G.A. Webb, P.S. Belton, and M.J. McCarthy, Eds., vol. 31.
2	Aveiro (Portugal), 1994	**Magnetic Resonance in Food Science: The Developing Scene**, 1995, P.S. Belton, I. Delgadillo, A.M. Gil, G.A. Webb, Eds., Royal Society of Chemistry, UK. A Conference report was published by Eads, T.M. in *Trends in Food Science and Technology*, 1994, 5:368-371.
3	Nantes (France), 1996	**Magnetic Resonance in Chemistry**, special issue, 1997, M.L. Martin and G.J. Martin, Eds., vol. 35.
4	Norwich (UK), 1998	**Advances in Magnetic Resonance in Food Science**, 1999, P.S. Belton, B.P. Hills, G.A. Webb, Eds., Royal Society of Chemistry, UK.
5	Aveiro (Portugal), 2000	**Magnetic Resonance in Food Science: A View to the Future**, 2001, G.A. Webb, P.S. Belton, A.M. Gil, I. Delgadillo, Eds., Royal Society of Chemistry, UK.
6	Paris (France) 2002	**Magnetic Resonance in Food Science: Latest Developments**, 2003, P.S. Belton, A.M. Gil, G.A. Webb, Eds., Royal Society of Chemistry, UK.
7	Copenhagen (Denmark), 2004	**Magnetic Resonance in Food Science: The Multivariate Challenge**, 2005, S.B. Engelsen, P.S. Belton, H.J. Jakobsen, Eds., Royal Society of Chemistry, UK.
8	Nottingham (UK), 2006	**Applications of Magnetic Resonance in Food Science**

[5]Of special interest is the Amorph symposium series. The original objective of the first symposium was "to discuss and define the relevant problems, rank them in some order of importance and suggest effective experimental, theoretical and computational approaches for their study" (Levine 2002b). Following are the Amorph symposium titles and their associated publications.

Symposium No.	Location and Year	Symposium Title and Associated Publication
Amorph 1	Girton College, Cambridge (UK), 1995	**Chemistry and Application Technology of Amorphous Carbohydrates** Summary report published as Appendix I in Amorph 2 publication
Amorph 2	Churchhill College, Cambridge (UK), 2001	**The Amorphous State - A Critical Review,** Publication: Amorphous Food and Pharmaceutical Systems, 2002 (H. Levine, Ed.), Royal Society of Chemistry, Cambridge, UK.
Amorph 3	Churchhill College, Cambridge (UK), 2006	**Molecular Basis of Stability in Pharmaceutical and Food Glasses**

[6]According to Levitt (2001), the term used to describe the macroscopic translational mobility of water to establish the equilibrium partial vapor pressure in Figure 4.3A is flow.

Acknowledgments

I am grateful for the cohort of colleagues, students, and friends all over the world who share my excitement for investigating edible things using NMR spectroscopy and imaging. A special thanks is extended to Dr. Peter S. Belton, School of Chemical Sciences and Pharmacy, University of East Anglia, Norwich, United Kingdom, for reviewing and editing a draft copy of this chapter and to Dr. Harry Levine and Dr. Louise Slade for their guidance regarding the introduction, as well as the future needs and directions sections. I would like to thank Christopher J. Fennell and Dr. J. Daniel Gezelter, Department of Chemistry and Biochemistry, Notre Dame University, Notre Dame, IN, for generating and allowing me to use the water structure computer-simulated images embedded in Figure 4.1; Dr. Arthur R. Schmidt, Department of Civil and Environmental Engineering, University of Illinois at Urbana-Champaign, for calculating the Maxwell-Boltzmann distribution of speeds for water molecules used in the construction of Figure 4.2 and for assistance in constructing Figure 4.35; and Dr. Paul Molitor, VOICE NMR Lab, University of Illinois at Urbana-Champaign, for assistance with collecting the relaxation rate values at 14.1 T reported in the text. The artwork assistance of Janet Sinn-Hanlon and Carl Burton from the Visualization, Media, and Imaging Laboratory, Beckman Institute for Advanced Science and Technology, is also gratefully acknowledged. I appreciate and could not do what I do without the love and support of my wonderful family, Art, Robbie, and Annie Schmidt, as well as our many friends, especially Mevanee and Phill Parmer and the Byers family.

Bibliography

Ablett, S., Darke, A.H., Izzard, M.J., and Lillford, P.J. 1993. Studies of the glass transition in malto-oligomers. In: *The Glassy State in Foods*, eds. J.M.V. Blanshard and P.J. Lillford, pp. 189–206. Loughborough, Leicestershire: Nottingham University Press.

Atkins, P.W. 1978. *Physical Chemistry*. San Francisco, CA: W.H. Freeman and Company. p. 1022.

Avram, L., and Cohen, Y. 2005. Diffusion measurements for molecular capsules: Pulse sequences effects on water signal decay. *Journal of American Chemical Society* 127:5714–5719.

Bakhmutov, V.I. 2004. *Practical NMR Relaxation for Chemists*. West Sussex, England: John Wiley & Sons, Ltd. p. 202.

Ball, P. 2000. *Life's Matrix: A Biography of Water*. New York: Farrar, Straus and Giroux. p. 417.

Barreiro, P., Moya, A., Correa, E., Ruiz-Altisent, M., Fernandez-Valle, M., Peirs, A., Wright, K.M., and Hills, B.P. 2002. Prospects for the rapid detection of mealiness in apples by nondestructive NMR relaxometry. *Applied Magnetic Resonance* 22:387–400.

Bell, L.N., Bell, H.M., and Glass, T.E. 2002. Water mobility in glassy and rubbery solids as determined by oxygen-17 nuclear magnetic resonance: Impact on chemical stability. *Lebensittel-Wissenschaft und-Technologi* 35(2):108–113.

Belton, P.S. 1990. Can nuclear magnetic resonance give useful information about the state of water in foodstuffs? *Comments on Agricultureal and Food Chemistry* 2(3):179–209.

Belton, P.S. 1995. NMR in context. In: *Annual Reports on NMR Spectroscopy*, eds. G.A. Webb, P.S. Belton, and M.J. McCarthy, vol. 31, pp. 1–18. New York: Academic Press.

Belton, P.S., Colquhoun, I.J., and Hills, B.P. 1993. Applications of NMR to food science. In: *Annual Reports on NMR Spectroscopy*, ed. G.A. Webb, vol. 26, pp. 1–45. New York: Academic Press.

Belton, P.S., Delgadillo, I., Gil, A.M., and Webb, G. A., Eds. 1995. *Magnetic Resonance in Food Science*. Cambridge, UK: Royal Society of Chemistry.

Belton, P.S., Hills, B.P., and Webb, G.A., Eds. 1999. *Advances in Magnetic Resonance in Food Science*. Cambridge, UK: Royal Society of Chemistry.

Belton, P.S., and Wang, Y. 2001. Fast field cycling NMR - Applications to foods. In: *Magnetic Resonance in Food Science: A View to the Future*, eds. G.A. Webb, P.S. Belton, A.M. Gil, and I. Delgadillo, pp. 145–156. Cambridge, UK: Royal Society of Chemistry.

Borompichaichartkul, C., Moran, G., Srzednicki, G., and Price, W.S. 2005. Nuclear magnetic resonance (NMR) and magnetic resonance imaging (MRI) studies of corn at subzero temperatures. *Journal of Food Engineering* 69:199–205.

Bryant, R.G. 1996. The dynamics of water-protein interactions. *Annual Review of Biophysics and Biomolecular Structure* 25:29–53.

Bryant, R.G., Hilton, D.P., Jiao, X., and Zhou, D. 1995. Proton magnetic resonance and relaxation in dynamically heterogeneous systems. In: *Magnetic Resonance in Food Science*, eds. P.S. Belton, I. Delgadillo, A.M. Gil., and G.A. Webb, pp. 219–229. Cambridge: The Royal Society of Chemistry.

Campbell, I.D., and Dwek, R.A. 1984. *Biological Spectroscopy*. Menlo Park, CA: The Benjamin/Cummings Publishing Company, Inc.

Chaplin, M. 2007. Water structure and science, May 13, 2007. Available at www.lsbu.ac.uk/water. Accessed May 27, 2007.

Chinachoti, P., and Krygsman, P.H. 2005. Applications of low-resolution NMR for simultaneous moisture and oil determination in food (oilseeds). In: *Handbook of Food Analytical Chemistry: Water, Proteins, Enzymes, Lipids, and Carbohydrates*, eds. R.E. Wrolstad, T.E. Acree, E.A. Decker, M.H. Penner, D.S. Reid, S.J. Schwartz, C.F. Shoemaker, D.M. Smith, and P. Sporns, p.768. Hoboken, NJ: John Wiley & Sons, Inc.

Chinachoti, P., and Vittadini, E. 2006. Water stress of bacteria and molds from an NMR water mobility standpoint. In: *Water Properties of Foods, Pharmaceutical, and Biological Materials*, eds. M. del Pilar Buera, J. Welti-Chanes, P.J. Lillford, and H.R. Corti, p. 747. Boca Raton, FL: CRC Press, Taylor & Francis Group.

Chirife, J., and Fontan, C.F. 1982. Water activity of fresh foods. *Journal of Food Science* 47:661–663.

Choi, S.G., and Kerr, W.L. 2003. 1H NMR studies of molecular mobility in wheat starch. *Food Research International* 36:341–348.

Christian, J.H.B., and Scott, W.J. 1953. Water relations of Salmonellae at 30°C. *Australian Journal of Biological Science* 6:565–573.

Chung, M.S., Ruan, R., Chen, P., Kim., J.H., Ahn, T.H., and Baik, C.K. 2003. Predicting caking behaviors in powdered foods using a low-field nuclear magnetic resonance (NMR) technique. *Lebensittel-Wissenschaft und-Technologie* 36:751–761.

Claridge, T.D.W. 1999. *High-Resolution NMR Techniques in Organic Chemistry.* New York: Pergamon, imprint of Elsevier Science. p. 382.

Colquhoun, I.J., and Goodfellow, B.J. 1994. Nuclear magnetic resonance spectroscopy. In: *Spectroscopic Techniques for Food Analysis*, ed. R.H. Wilson, pp. 87–145. New York: VCH Publishers, Inc.

Conway, B.E. 1981. *Ionic Hydration in Chemistry and Biophysics.* New York: Elsevier Scientific Publishing Company. p. 774.

Cornillon, P., and Salim, L.C. 2000. Characterization of water mobility and distribution in low- and intermediate-moisture food systems. *Magnetic Resonance Imaging* 18:335–341.

Curme, A.G., Schmidt, S.J., and Steinberg, M.P. 1990. Mobility and activity of water in casein model systems as determined by 2H NMR and sorption isotherms. *Journal of Food Science* 55(2):430–433.

Davenel, A., Schuck, P., Mariette, F., and Brule, G. 2002. NMR relaxometry as a non-invasive tool to characterize milk powders. *Lait* 82:465–473.

Derome, A.E. 1987. *Modern NMR Techniques for Chemistry Research*. Oxford, UK: Pergamon Press. p. 280.

Duarte, I.F., Godejohann, M., Braumann, U., Spraul, M., and Gil, A.M. 2003. Application of NMR spectroscopy and LC-NMR/MS to the identification of carbohydrates in beer. *Journal of Food Chemistry* 51: 4847–4852.

Duce, S.L., and Hall, L.D. 1995. Visualization of the hydration of food by nuclear magnetic resonance imaging. *Journal of Food Engineering* 26:251–257.

Duckworth, R.B. 1975. *Water Relations of Foods.* New York: Academic Press. p. 716.

Eads, T.M. 1998. Magnetic resonance. In: *Food Analysis*, 2nd Edition, ed. S.S. Nielsen, pp. 455–481. Gaithersburg, MD: Aspen Publishers, Inc.

Eads, T.M. 1999. Principles for nuclear magnetic resonance analysis of intact food materials. In: *Spectral Methods in Food Analysis*, ed. M.M. Mossoba, pp. 1–88. New York: Marcel Dekker, Inc.

Eisenberg, D., and Kauzmann, W. 1969. *The Structure and Properties of Water*. New York: Oxford University Press. p. 296.

Ernst, R.R., Bodenhausen, G., and Wokaun, A. 1987. *Principles of Nuclear Magnetic Resonance in One and Two Dimensions*. New York: Oxford University Press, Oxford Science Publications.

Evans, S., and Hall, L. 2005. Measurement of pH in food systems by magnetic resonance imaging. *Canadian Journal of Chemical Engineering* 83(1):73–77.

Farhat, I.A. 2004. Measuring and modelling the glass transition temperature. In: *Understanding and Measuring the Shelf-life of Foods*, ed. R. Steele, pp. 218–232. Cambridge, England: Woodhead Publishing Limited.

Fennema, O. 1996. Water and ice. In: *Food Chemistry*, ed. O. Fennema. New York: Marcel Dekker.

Foucat, L., Taylor, R.G., Labas, R., and Renou, J.P. 2001. Characterization of frozen fish by NMR imaging and histology. *American Laboratory* 33(16):38–43.

Franks, F. 1982. Water activity as a measure of biological viability and quality control. *Cereal Foods World* 27(9):403–407.

Franks, F. 1991. Water activity: A credible measure of food safety and quality? *Trends in Food Science and Technology* 2(3):68–72.

Franks, F. 2000. *Water: A Matrix of Life*. Cambridge, UK: Royal Society of Chemistry, p. 225.

Gil, A.M., Belton, P.S., and Hills, B.P. 1996. Applications of NMR to food science. In: *Annual Reports on NMR Spectroscopy*, ed. G.A. Webb, vol. 32, pp. 1–43. New York: Academic Press.

Glasel, J.A. 1966. A study of water in biological systems by ^{17}O magnetic resonance spectroscopy, I. Preliminary studies and xenon hydrates. *Physics* 55:479–485.

Glasel, J.A. 1972. Nuclear magnetic resonance studies on water and ice. In: *Water—A Comprehensive Treatise. The Physics and Physical Chemistry of Water*, ed. F. Franks, vol. 1, pp. 215–254. New York: Plenum Press.

Godefroy, S., and Callaghan, P.T. 2003. 2D relaxation/diffusion correlations in porous media. *Magnetic Resonance Imaging* 21:381–383.

Godefroy, S., Korb, J.P., Creamer, L.K., Watkinson, P.J., and Callaghan, P.T. 2003. Probing protein hydration and aging of food materials by the magnetic field dependence pf proton spin-lattice relaxation times. *Journal of Colloid and Interface Science* 267:337–342.

Gonera, A., and Cornillon, P. 2002. Gelatinization of starch/gum/sugar systems studied by using DSC, NMR, and CSLM. *Starch* 54:508–516.

Gottwald, A., Creamer, L.K., Hubbard, P.L., and Callaghan, P.T. 2005. Diffusion, relaxation, and chemical exchange in casein gels: A nuclear magnetic resonance study. *The Journal of Chemical Physics* 122:034506, 1–10.

Grant, D.M., and Harris, R.K. 2002. *Encyclopedia of Nuclear Magnetic Resonance*, Vol. 9. *Advances in NMR*. New York: John Wiley.

Grattard, N., Salaun, F., Champion, D., Roudaut, G., and Le Meste, M. 2002. Influence of physical state and molecular mobility of freeze-dried maltodextrin matrices on the oxidation rate of encapsulated lipids. *Journal of Food Science* 67(8):3002–3010.

Gribnau, M.C.M. 1992. Determination of solid/liquid ratios of fats and oils by low-resolution pulsed NMR. *Trends in Food Science and Technology* 3:186–190.

Gunther, H. 1995. *NMR Spectroscopy: Basic Principles, Concepts, and Applications in Chemistry*, 2nd ed. New York: John Wiley and Sons, Ltd. p. 581.

Halle, B., Andersson, T., Forsen, S., and Lindman, B. 1981. Protein hydration from water oxygen-17 magnetic relaxation. *Journal of the American Chemical Society* 103:500–508.

Halle, B., and Karlstrom, G. 1983a. Prototropic charge migration in water. Part 1. Rate constants in light and heavy water and in salt solutions from oxygen-17 spin relaxation. *Journal of the Chemical Society, Faraday Transactions 2* 79:1031–1046.

Halle, B., and Karlstrom, G. 1983b. Prototropic charge migration in water. Part 2. Interpretation of nuclear magnetic resonance and conductivity data in terms of model mechanisms. *Journal of the Chemical Society, Faraday Transactions 2* 79:1047–1073.

Harris, R.K., and Mann, B.E. 1978. *NMR and the Periodic Table*. New York: Academic Press. p. 459.

Hermansson, A.M., Loren, N., and Nyden, M. 2006. The effect of microstructure on solvent and solute diffusion on the micro- and nanolength scales. In: *Water Properties of Foods, Pharmaceutical, and Biological Materials*, eds. M. del Pilar Buera, J. Welti-Chanes, P.J. Lillford, and H.R. Corti, p. 747. Boca Raton, FL: CRC Press, Taylor & Francis Group.

Hills, B. 1995. Food processing: An MRI perspective. *Trends in Food Science and Technology* 6, 111–117.

Hills, B. 1998. *Magnetic Resonance Imaging in Food Science*. New York: John Wiley & Sons, Inc.

Hills, B., Benaira, S., Marigheto, N., and Wright, K. 2004. T1-T2 correlation analysis of complex foods. *Applied Magnetic Resonance* 26:543–560.

Hills, B., Costa, A., Marigheto, N., and Wright, K. 2005. T1-T2 NMR correlation studies of high-pressure-processed starch and potato tissue. *Applied Magnetic Resonance* 28(1-2):13–27.

Hills, B.P. 1999. NMR studies of water mobility in foods. *In: Water Management in the Design and Distribution of Quality Foods*, ISOPOW 7, eds. Y.H. Roos, R.B. Leslie, and P.J. Lillford. Lancaster, PA: Technomic Publishing Co. Inc.

Hills, B.P., and Manning, C.E. 1998. NMR oxygen-17 studies of water dynamics in heterogeneous gel and particulate systems. *Journal of Molecular Liquids* 75, 61–76.

Hills, B.P., Manning, C.E., and Godward, J. 1999. A multistate theory of water relations in biopolymer systems. *In: Advances in Magnetic Resonance in Food Science*, eds. P.S. Belton, B.P. Hills, and G.A. Webb, pp. 45–62. Cambridge, UK: Royal Society of Chemistry.

Hills, B.P., Manning, C.E., and Ridge, Y. 1996b. New theory of water activity in heterogeneous systems. *Journal of the Chemical Society, Faraday Transactions* 92(6):979–983.

Hills, B.P., Manning, C.E., Ridge, Y., and Brocklehurst, T. 1996a. NMR water relaxation, water activity, and bacterial survival in porous media. *Journal of the Science of Food and Agriculture* 71(2):185–194.

Hills, B.P., Manning, C.E., Ridge, Y., and Brocklehurst, T. 1997. Water availability and the survival of *Salmonella typhimurium* in porous media. *International Journal of Food Microbiology* 36:187–198.

Hills, B.P., and Pardoe, K. 1995. Proton and deuterium NMR studies of the glass transition in a 10% water-maltose solution. *Journal of Molecular Liquids* 62:229–237.

Hills, B.P., Wang, Y.L., and Tang, H.R. 2001. Molecular dynamics in concentrated sugar solutions and glasses: An NMR field cycling study. *Molecular Physics* 99(19):1679–1687.

Holz, M., Heil, S.R., and Sacco, A. 2000. Temperature-dependent self-diffusion coefficients of water and six selected molecular liquids for calibration in accurate 1H NMR PFG measurements. *Journal of Physical Chemistry Chemical Physics* 2:470–4742.

Holz, M., and Weingartner, H. 1991. Calibration in accurate spin-echo self-diffusion measurements using 1H and less-common nuclei. *Journal of Magnetic Resonance* 92:115–125.

IAPWS—The International Association for the Properties of Water and Steam. 2002. Guidelines for the use of fundamental physical constants and basic constants of water. Revision of September 2001 guidelines, Gaitherburg, MA. Available at http://www.iapws.org. Accessed March 15, 2007.

Irie, K., Horigane, A.K., Naito, S., Motoi, H., and Yoshida, M. 2004. Moisture distribution and texture of various types of cooked spaghetti. *Cereal Chemistry* 81(3):350–355.

Isengard, H.D. 1995. Rapid water determination in foodstuffs. *Trends in Food Science and Technology* 6: 155–162.

Isengard, H.D. 2001a. Water—A very common and yet a particular substance. *Food Control* 12:393.

Isengard, H.D. 2001b. Water content, one of the most important properties of food. *Food Control* 12:395–400.

Isengard, H.D. 2006. Water—A simple substance? *Food Chemistry* 96(3):345.

Ishida, N., Naito, S., and Kano, H. 2004. Loss of moisture from harvested rice seeds on MRI. *Magnetic Resonance Imaging* 22:871–875.

Kantt, C.A., Schmidt, S.J., Sizer, C., Palaniappan, S., and Litchfield, J.B. 1998. Temperature mapping of particles during aseptic processing with magnetic resonance imaging. *Journal of Food Science* 63(2):305–311.

Kemp, W. 1986. *NMR in Chemistry: A Multinuclear Introduction*. London, UK: MacMillan Education, Ltd.

Kou, Y., Dickerson, L.C., and Chinachoti, P. 2000. Mobility characterization of waxy corn starch using wide-line ^{1}H Nuclear Magnetic Resonance. *Journal of Agricultural and Food Chemistry* 48:5489–5495.

Kou, Y., Molitor, P.F., and Schmidt, S.J. 1999. Mobility and stability characterizing of model food systems using NMR, DSC, and conidia germination techniques. *Journal of Food Science* 64(6):950–959.

Krynicki, K. 1966. Proton spin-lattice relaxation in pure water between 0 °C and 100 °C. *Physica* 32:167–178.

Kumagai, H., MacNaughtan, W., Farhat, I.A., and Mitchell, J.R. 2002. The influence of carrageenan on molecular mobility in low moisture amorphous sugars. *Carbohydrate Polymers* 48:341–349.

Labuza, T.P., Tannenbaum, S.R., and Karel, M. 1970. Water content and stability of low moisture and intermediate moisture foods. *Food Technology* 24:543–550.

Laghi, L., Cremonini, M.A., Placucci, G., Sykora, S., Wright, K., and Hills, B. 2005. A proton NMR relaxation study of hen egg quality. *Magnetic Resonance Imaging* 23:501–510.

Lai, H.M., and Hwang, S.C. 2004. Water status of cooked white salted noodles evaluated by MRI. *Food Research Interna*tional 37:957–966.

Lee, S., Cornillon, P., and Kim, Y.R. 2002. Spatial investigation of the non-frozen water distribution in frozen foods using NMR SPRITE. *Journal of Food Science* 67(6):2251–2255.

Lee, Y., Lee, S., and Schmidt, S.J. 2005. Probing the sensory properties of food materials with nuclear magnetic resonance spectroscopy and imaging. *In: Handbook of Modern Magnetic Resonance,* ed. G. Webb. London: Kluwer Academic Publishers.

Le Meste, M., Champion, D., Roudaut, G., Blond, G., and Simatos, D. 2002. Glass transition and food technology: A critical appraisal. *Journal of Food Science* 67(7):2444–2458.

Leung, H.K., Steinberg, M.P., Wei, L.S., and Nelson, A.I. 1976. Water binding of macromolecules determined by pulsed NMR. *Journal of Food Science* 41:297–300.

Levine, H. 2002a. Introduction—Progress in amorphous food and pharmaceutical systems. In: *Amorphous Food and Pharmaceutical Systems*, ed. H. Levine. Cambridge, UK: Royal Society of Chemistry.

Levine, H. 2002b. Appendix I: Summary report of the discussion symposium on chemistry and application technology of amorphous carbohydrates. In: *Amorphous Food and Pharmaceutical Systems*, ed. H. Levine. Cambridge, UK: Royal Society of Chemistry.

Levitt, M.H. 2001. *Spin Dynamics: Basics of Nuclear Magnetic Resonance*. Chichester, England: John Wiley & Sons, LTD. p. 686.

Lewen, K.S., Paeschke, T., Reid, J., Molitor, P., and Schmidt, S.J. 2003. Analysis of the retrogradation of low starch concentration gels using differential scanning calorimetry, rheology, and nuclear magnetic resonance spectroscopy. *Journal of Agricultural and Food Chemistry* 51:2348–2358.

Lillford, P.J., and Ablett, S. 1999. From solid-liquid ratios to real time tomography—The development of NMR in food applications. In: *Advances in Magnetic Resonance in Food Science*, eds. P.S. Belton, B.P. Hills, and G.A. Webb. Cambridge, UK: Royal Society of Chemistry.

Lin, Y., Yeh, A.I., and Lii, C. 2001. Correlation between starch retrogradation and water mobility as determined by differential scanning calorimetery (DSC) and nuclear magnetic resonance (NMR). *Cereal Chemistry* 78(6):647–653.

Lucas, T., Grenier, A., Quellec, S., Le Bail, A., and Davenel, A. 2005c. MRI quantification of ice gradients in dough during freezing or thawing processes. *Journal of Food Engineering* 71:98–108.

Lucas, T., Le Ray, D., and Davenel, A. 2005a. Chilling and Freezing of part-baked bread. Part I: An MRI signal analysis. *Journal of Food Engineering* 70:139–149.

Lucas, T., Le Ray, D., and Davenel, A. 2005b. Chilling and Freezing of part-baked bread. Part II: Experimental assessment of water phase changes and structural collapse. *Journal of Food Engineering* 70:151–164.

Macomber, R.S. 1998. *A Complete Introduction to Modern NMR Spectroscopy*. New York: John Wiley and Sons, Ltd. p. 382.

Marigheto, N., Vial, A., Wright, K., and Hills, B. 2004. A combined NMR and microstructural study of the effects of high-pressure processing on strawberries. *Applied Magnetic Resonance* (26):521–531.

McCarthy, K.L., Gonzalez, J.J., and McCarthy, M.J. 2002. Changes in moisture distribution in lasagna pasta post cooking. *Journal of Food Science* 67(5):1785–1789.

McCarthy, M.J. 1994. *Magnetic Resonance Imaging in Foods*. New York: Chapman and Hall.

McCarthy, M.J., Lasseux, D., and Maneval, J.E. 1994. NMR imaging in the study of diffusion of water in foods. *Journal of Food Engineering* 22:211–224.

Metais, A., and Mariette, F. 2003. Determination of water self-diffusion coefficients in complex food products by low field 1H PFG-NMR: Comparison between the standard spin-echo sequence and the T_1-weighted spin-echo sequence. *Journal of Magnetic Resonance* 165:265–275.

Micklander, E., Peshlov, B., Purlow, P.P., and Engelsen, S.B. 2002. NMR-cooking: monitoring the changes in meat during cooking by low-filed 1H NMR. *Trends in Food Science and Technology* 13:341–346.

Mills, R. 1973. Self-diffusion in normal and heavy water in the range 1-45 degrees. *Journal of Physical Chemistry* 77:685–688.

Mirau, P.A. 2003. Nuclear magnetic resonance spectroscopy. In: *Comprehensive Desk Reference of Polymer Characterization and Analysis*, ed. R.F. Brady, Jr. Oxford, UK: Oxford University Press.

Mohoric, A., Vergeldt, F., Gerkema, E., de Jager, A., van Duynhoven, J., van Dalen, G., and As, H.V. 2004. Magnetic resonance imaging of single rice kernels during cooking. *Journal of Magnetic Resonance* 171:157–162.

Monteiro Marques, J.P., Rutledge, D.N., and Ducauze, C.J. 1991. Low resolution pulse nuclear magnetic resonance study of carrots equilibrated at various water activities and temperatures. *Sciences Des Aliments* 11:513–525.

Morris, G.A. 2002a. Special Issue: NMR and diffusion. *Magnetic Resonance in Chemistry* 40(13):S1–S152.

Morris, G.A. 2002b. Diffusion-ordered spectroscopy. In: *Encyclopedia of NMR*, eds. M. Grant and R.K. Harris, Vol. 9D, pp. 35–44. New York: Wiley.

Noack, F., Becker, S., and Struppe, J. 1997. Applications of field-cycling NMR. In: *Annual Reports on NMR Spectroscopy*, vol. 33, ed. G.A. Webb, pp. 1–35. New York: Academic Press.

Nott, K.R., and Hall, L.D. 2005. Heating of foods studied by magnetic resonance imaging. *Canadian Journal of Chemical Engineering* 83(1):78–82.

O'Brien, J. 1992. Special Issue: Applications of nuclear magnetic resonance techniques in food research. *Trends in Food Science and Technology* 3(8/9):177–249.

Olson, B.F. 1999. Analysis of the diffusion coefficients in gelatinizated starch-water systems over time using diffusion-ordered spectroscopy. Master's thesis, University of Illinois, Urbana-Champaign, IL.

Pham, X., Vittadini, E., Levin, R.E., and Chinachoti, P. 1999. Role of water mobility on mold spore germination. *Journal of Agricultural and Food Chemistry* 47:4976–4983.

Pitombo, R.N.M., and Lima, G.A.M.R. 2003. Nuclear magnetic resonance and water activity in measuring the water mobility in Pintado (*Pseudoplatystoma corruscans*) fish. *Journal of Food Engineering* 58:59–66.

Povlsen, V.T., Rinnan, A., van den Berg, F., Andersen, H.J., and Thybo, A.K. 2003. Direct decomposition of NMR relaxation profiles and prediction of sensory attributes of potato samples. *Lebensittel-Wissenschaft und-Technologie* 36:423–432.

Price, W.S. 1996. Gradient NMR. In: *Annual Reports on NMR Spectroscopy,* ed. G.A. Webb, vol. 32, pp. 52–142. New York: Academic Press.

Price, W.S. 1997. Pulsed-field gradient nuclear magnetic resonance as a tool for studying translational diffusion: Part 1. Basic theory. *Concepts in Magnetic Resonance* 9(5):299–336.

Price, W.S. 1998a. Pulsed-field gradient nuclear magnetic resonance as a tool for studying translational diffusion: Part II. Experimental aspects. Basic theory. *Concepts in Magnetic Resonance* 10(4):197–237.

Price, W.S. 1998b. NMR imaging. In: *Annual Reports on NMR Spectroscopy*, ed. G.A. Webb, vol. 34, pp. 140–216. New York: Academic Press.

Renou, J.P., Faoucat, L., and Bonny, J.M. 2003. Magnetic resonance imaging studies of water interactions in meat. *Food Chemistry* 82:35–39.

Richardson, S.J. 1989. Contribution of proton exchange to the ^{17}O NMR transverse relaxation rate in water and starch-water systems. *Cereal Chemistry* 66(3):244–246.

Richardson, S.J., Baianu, I.C., and Steinberg, M.P. 1987. Mobility of water in starch powders by nuclear magnetic resonance. *Starch* 39(6):198–203.

Roos, Y.H. 1993. Water activity and physical state effects on amorphous food stability. *Journal of Food Processing and Preservation* 16(6):433–447.

Roos, Y.H. 1995. *Phase Transitions in Foods*. New York: Academic Press

Roos, Y.H. 1998. Role of water in phase-transition phenomena in foods. In: *Phase/State Transitions in Foods*, eds. M.A. Rao and R.W. Hartel, pp. 57–93. New York: Marcel Dekker, Inc.

Roos, Y.H. 2003. Thermal analysis, state transitions and food quality. *Journal of Thermal Analysis and Calorimetry* 71, 197–203.

Roudaut, G., Maglione, M., van Dusschoten, D., and Le Meste, M. 1999. Molecular mobility in glassy bread: A multispectroscopy approach. *Cereal Chemistry* 76(1):70–77.

Ruan, R.R., and Chen, P.L. 1998. *Water in Foods and Biological Materials: A Nuclear Magnetic Resonance Approach*. Lancaster, PA: Technomic Publishing Company, Inc.

Ruan, R.R., Long, Z., Song, A., and Chen, P.L. 1998. Determination of the glass transition temperature of food polymers using low field NMR. *Lebensittel-Wissenschaft und-Technologie* 31:516–521.

Ruan, R., Long, Z., Chen, P., Huang, V., Almaer, S., and Taub, I. 1999. Pulsed NMR study of the glass transition in maltodextrin. *Journal of Food Science* 64(1):6–9.

Ruiz-Cabrera, M.A., Foucat, L., Bonny, J.M., Renou, J.P., and Daudin, J.D. 2005a. Assessment of water diffusivity in gelatine gel from moisture profiles. I. Non-destructive measurement of 1D moisture profiles during drying from 2D nuclear magnetic resonance images. *Journal of Food Engineering* 68:209–219.

Ruiz-Cabrera, M.A., Foucat, L., Bonny, J.M., Renou, J.P., and Daudin, J.D. 2005b. Assessment of water diffusivity in gelatine gel from moisture profiles. II. Data processing adapted to material shrinkage. *Journal of Food Engineering* 68:221–231.

Rutledge, D.N. 2001. Characterization of water in agro-food products by time-domain-NMR. *Food Control* 12:437–445.

Schmidt, S.J. 1991. Determination of moisture content by pulsed nuclear magnetic resonance spectroscopy. In: *Water Relationships in Foods*, eds. H. Levine and L. Slade, pp. 599–613. New York: Plenum Press.

Schmidt, S.J. 1999. Probing the physical and sensory properties of food systems using NMR spectroscopy. In: *Advances in Magnetic Resonance in Food Science*, eds. P.S. Belton, B.P. Hills, and G.A. Webb. Cambridge, UK: The Royal Society of Chemistry.

Schmidt, S.J. 2004. Water and solids mobility in foods. *In: Advances in Food and Nutrition Research*, Vol. 48, pp. 1–101. London, UK: Academic Press.

Schmidt, S.J., and Lai, H.M. 1991. Use of NMR and MRI to study water relations in foods. In: *Water Relationships in Foods*, eds. H. Levine and L. Slade, pp. 599–613. New York: Plenum Press.

Schmidt, S.J., Sun, X., and Litchfield, J.B. 1996. Applications of magnetic resonance imaging in food science. *Critical Reviews in Food Science and Nutrition* 36(4):357–385.

Scott, W.J. 1953. Water relations of *Staphylococcus aureus* at 30°C. *Australian Journal of Biological Science* 6:549–564.

Scott, W.J. 1957. Water relations of food spoilage microorganisms. *Advances in Food Research* 7:83–127.

Sherwin, C.P., Labuza, T.P., McCormick, A., and Chen, B. 2002. Cross-polarization/magic angle spinning NMR to study glucose mobility in a model intermediate-moisture foods system. *Journal of Agricultural and Food Chemistry* 50(26):7677–7683.

Slade, L., and Levine, H. 1985. Intermediate moisture systems; concentrated and supersaturated solutions; pastes and dispersions; water as plasticizer; the mystique of "bound" water; thermodynamics versus kinetics (No. 24). Presented at Faraday Division, Royal Society of Chemistry Discussion Conference—Water Activity: A Credible Measure of Technological Performance and Physiological Viability? Cambridge, July 1–3.

Slade, L., and Levine, H. 1988. Structural stability of intermediate moisture foods—A new understanding? In. *Food Structure—Its Creation and Evaluation*, eds. J.M.V. Blanshard and J.R. Mitchell, pp. 115–147. London: Butterworths.

Slade, L., and Levine, H. 1991. Beyond water activity: Recent advances based on an alternative approach to the assessment of food quality and safety. *Critical Reviews in Food Science and Nutrition* 30(2–3), 115–360.

Slade, L., and Levine, H. 2004. The food polymer science approach to understanding glass transitions in foods. Presented at the 42nd Fred W. Tanner Award Lecture, Chicago IFT Section, Illinois, May 10, 2004.

Smith, D.W.G., and Powles, J.G. 1966. Proton spin-lattice relaxation in liquid water and liquid ammonia. *Molecular Physics* 10:451–463.

Smith, S.E. 1947. The sorption of water by high polymers. *Journal of the American Chemical Society* 69:646–651.

Sorland, G.H., and Aksnes, D. 2002. Artifacts and pitfalls in diffusion measurements by NMR. *Magnetic Resonance in Chemistry* 40:S139–S146.

Spraul, M.S., Braumann, U., Godejohann, M., and Hofmann, M. 2001. Hyphenated methods in NMR. In: *Magnetic Resonance in Food Science: A View to the Future*, eds. G.A. Webb, P.S. Belton, A.M. Gil, and I. Delgadillo, pp. 54–66. Cambridge, UK: Royal Society of Chemistry.

Steinberg, M.P., and Leung, H. 1975. Some applications of wide-line and pulsed N.M.R. in investigations of water in foods. In: *Water Relations of Foods*, ed. R.B. Duckworth, pp. 233–248. New York: Academic Press.

Stejskal, E.O., and Tanner, J.E. 1965. Spin diffusion measurements: Spin echoes in the presence of a time-dependent field gradient. *Journal of Chemical Physics* 42:288–292.

Sun, X., and Schmidt, S.J. 1995. Probing water relations in foods using magnetic resonance techniques. In: *Annual Reports on NMR Spectroscopy*, eds. G.A. Webb, P.S. Belton, and M.J. McCarthy, pp. 239–273. New York: Academic Press.

Szent-Gyorgyi, A. 1972. Water, the hub of life. In: *The Living State with Observations on Cancer.* New York: Academic Press. p. 86.

Tang, H.R., Godward, J., and Hills, B. 2000. The distribution of water in native starch granules—A multinuclear NMR study. *Carbohydrate Polymers* 43, 375–387.

Tanner, J.E. 1970. Use of the stimulated echo in NMR diffusion studies. *Journal of Chemical Physics* 52:2523–2526.

Tromp, R.H., Parker, R., and Ring, S.G. 1997. Water diffusion in glasses of carbohydrates. *Carbohydrate Research* 303:199–205.

van den Berg, C., and Bruin, S. 1981. Water activity and its estimation in food systems: Theoretical aspects. In: *Water Activity: Influences on Food Quality*, eds. L.B. Rockland and G.F. Stewart. New York: Academic Press.

van den Dries, I.J., Besseling, N.A.M., van Dusschoten, D., Hemminga, M.A., and van der Linden, E. 2000. Relation between a transition in molecular mobility and collapse phenomenon in glucose-water systems. *Journal of Physical Chemistry B* 104:9260–9266.

van den Dries, I.J., van Dusschoten, D., and Hemminga, M.A. 1998. Mobility in maltose-water glasses studied with ^{1}H NMR. *Journal of Physical Chemistry B* 102:10483–10489.

Vittadini, E., and Chinachoti, P. 2003. Effect pf physico-chemical and molecular mobility parameters on *Staphylococcus aureus* growth. *International Journal of Food Science and Technology* 38:841–847.

Wang, X., Choi, S.G., and Kerr, W.L. 2004. Water dynamics in white bread and starch gels as affected by water and gluten content. *Lebensittel-Wissenschaft und-Technologie*. 37:377–384.

Weast, R.C., ed. 1987. *The Handbook of Physics and Chemistry. Nuclear Spins, Moments, and Magnetic Resonance Frequencies.* Boca Raton, FL: CRC Press. pp. E78–E82.

Webb, G.A., Belton, P.S., Gil, A.M., and Delgadillo, I., eds. 2001. *Magnetic Resonance in Food Science: A View to the Future*. Cambridge, UK: Royal Society of Chemistry.

Webb, G.A., Belton, P.S., and McCarthy, M.J., eds. 1995. *Annual Reports on NMR Spectroscopy*. New York: Academic Press.

Weerts, A.H., Martin, D.R., Lian, G., and Melrose, J.R. 2005. Modelling the hydration of foodstuffs. *Simulation Modelling Practice and Theory* 13:119–128.

Weingartner, H. 1982. Self-diffusion in liquid water. A reassessment. *Zeitschrift für Physikalische Chemie* 132:129–149.

Weingartner, H., and Holz, M. 2002. NMR studies of self-diffusion in liquids. *Annual Reports on the Progress* of *Chemistry Section* C 98:121–155.

Wichchukit, S., McCarthy, M.J., and McCarthy, K.L. 2005. Flow behavior of milk chocolate melt and the application to coating flow. *Journal of Food Science* 70(3): E165–E171.

Woessner, D.E. 1964. Molecular reorientation in liquids. Deuteron quadrupole relaxation in liquid deuterium oxide and perdeuterobenzene. *Journal of Chemical Physics* 40(8):2341–2348.

Wungtanagorn, R., and Schmidt, S.J. 2001. Thermodynamic properties and kinetics of the physical aging of amorphous glucose, fructose, and their mixture. *Journal of Thermal Analysis and Calorimetry* 65:9–35.

Ziegler, G.R., MacMillan, B., and Balcom, B.J. 2003. Moisture migration in starch molding operations as observed by magnetic resonance imaging. *Food Research International* 36:331–340.

Zimmerman, J.R., and Brittin, W.E. 1957. Nuclear magnetic resonance studies in multiple phase systems: Lifetime of a water molecule in an absorbing phase on silica gel. *Journal of Physical Chemistry* 61:1328–1333.

5 Water Activity Prediction and Moisture Sorption Isotherms

Theodore P. Labuza and Bilge Altunakar

The role of water in food stability can be described to a significant extent by the potential of water to contribute to both physical and chemical deteriorative reactions through its ability to dissolve reactants, mobilize them, and participate in the reactions as well. Water activity (a_w), which is the measure of the state of water in foods, is a successful concept commonly used in correlation with food safety and quality. Water activity is a unique factor in food stability that enables the development of generalized limits within ranges where certain types of deteriorative reactions are dominant (Scott 1957). One of the main preservation methods to ensure food safety against microbial and chemical deterioration is controlling the a_w in food, which can extend shelf-life and create convenience with new food products. Therefore, several food preservation techniques rely on lowering the a_w so as to reduce the rates of microbial growth and chemical reactions, as described in the global stability map (Fig. 5.1) developed by Labuza et al. (1970) and Labuza (1970). Although general in nature, the stability map indicates that for reactions requiring an aqueous phase, such as nonenzymatic browning and enzyme reactions, there is a lower limit, usually at a_w between 0.2 and 0.3, below which the reactivity is 0; above that, the reaction rate increases until reaching a maximum at an a_w essentially between 0.6 and 0.8 and then decreases again, reaching 0 at a_w of 1.0 (Sherwin and Labuza 2006). Lipid oxidation, on the other hand, shows a minimum in the 0.2 to 0.35 a_w range and increases in rate on both sides, i.e., an increase or a decrease in a_w (Labuza 1971). The stability map also shows that molds can generally grow at lower a_w limits than yeast; such limits are lower than that for bacteria. Each of these organism types shows a minimum at both the high and low ends and a maximum in-between for growth rate as a function of a_w. Although this phenomenon has been attributed to "free" versus "bound" water (e.g., Troller and Christian 1978), the control of growth rate is much more complex, being ruled by ionic balance, the amino acid pool each organism uses, and porin genes that manipulate water transport across the membrane. What is known is that no microbes can grow at ≤ 0.6 a_w. However, although they cannot grow, they can survive at low a_w and, if added to a suitable medium, they may resuscitate and begin to grow again. Given this, it is important to know the relationship between the moisture content and a_w of a food. This relationship is called the moisture sorption isotherm, which is superimposed on the stability map (see Figure 5.1).

The relationship derived between the moisture content of food components and the relative water vapor pressure (p/p_0) of the atmosphere in equilibrium with the material can be used to compute the corresponding equilibrium relative humidity [ERH = $100(p/p_0) = 100\ a_w$]. Thus, under isothermal conditions, this equilibrium relationship is represented by the moisture sorption isotherm shown in Figure 5.1.

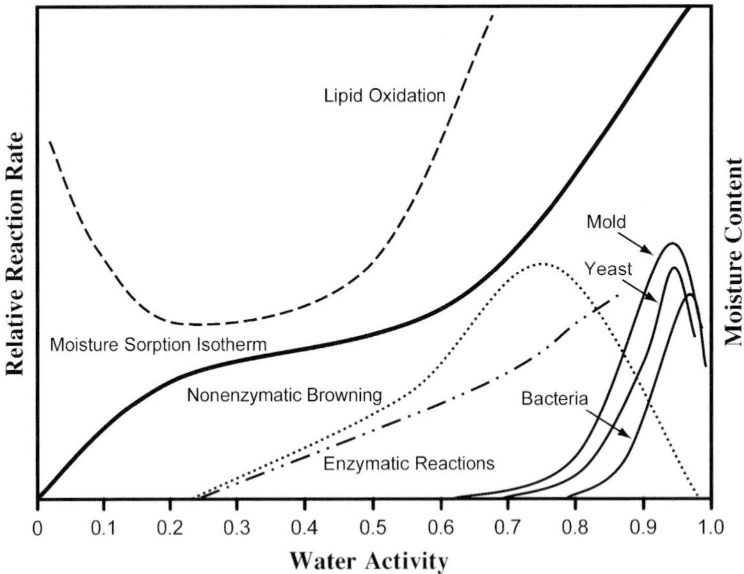

Figure 5.1 Food stability map as a function of water activity (adapted from Labuza, T.P., Tannenbaum, S.R., and Karel, M. 1970. Water content and stability of low moisture and intermediate moisture foods. *Journal of Food Technology* 24:543–550).

Water Activity Prediction

As discussed previously, knowledge of a_w is a major principle that can be used to estimate the limiting reaction(s) a food would support at that a_w. As shown above, a very dry food ($a_w < 0.2$) would most likely be degraded by lipid oxidation, whereas tissue foods such as meat with $a_w > 0.98$ would be spoiled by bacterial growth. As seen in Figure 5.1, the region between ~0.35 to 0.45 a_w is where physical state changes occur, such as loss of crispness, as was originally proposed by Katz and Labuza (1981). Physical changes were added later to the stability map by Roos and Karel (1991) and Labuza and Labuza (2004), in terms of reconciling another new concept, the glass transition (T_g) temperature of a system. The latter paper showed that T_g corresponded to an a_w of ~0.33, at which point recrystallization of amorphous sugars began to occur, an important physical change that leads to stickiness and caking of powders and hardening of soft cookies. Further practical examples can be found in Labuza et al. (2005). These changes include:

1. The uptake of water by amorphous sugars leads to stickiness, caking, and collapse of the structure (Labuza and Labuza 2004).
2. The loss in crispiness of products like potato chips above a_w of 0.45, which became soggy (Payne and Labuza 2005a, 2004b).
3. The hardening of soft products like raisins as they dessicate below a_w of 0.45 (e.g., when mixed with cereal).

These three phenomena are better understood by combining concepts of a_w with the concept of glass transition, as done by Labuza et al. (2005).

Although the a_w concept should theoretically be used for systems at equilibrium, this does not preclude its use in understanding what may happen in a nonequilibrium system. Thus, for a multidomain food system, e.g., a crisp outer layer at a_w of 0.3 encasing a soft pasty filling at a_w of 0.7, water will flow from the higher a_w zone to the lower zone driven by the thermodynamic energy state difference between the two layers. This energy is the chemical potential (J/mol) in each domain where:

$$\mu = \mu_0 + RT \ln a_w = J / mol \tag{5.1}$$

In this case, μ is the chemical potential, μ_o is the value at $a_w = 1$ for temperature T (K), and *R* is the gas constant. As can be seen, the a_w is an exponential function of the free energy. Thus, the crisp encasing material could become soggy while the center could become tough or leathery.

Thermodynamics

According to the first law of thermodynamics, for a system undergoing change, the driving force is the difference in the free energy of the substance undergoing state change between the two states as noted previously. Note that although the term is called *free energy,* one should not look at energy as free or bound; it is only a name. On a molar basis, the chemical potential μ, which is the free energy per mole of the substance under consideration (*G/n*), comes from thermodynamics where:

$$\mu = \mu_0 + RT \ln \frac{f}{p_0} \tag{5.2}$$

In this equation, μ_0 is the chemical potential (J/mol) at some standard state, *R* is the gas constant (8.314 J/mol K), T is the temperature in Kelvin, and *f* is the fugacity of the particular substance in the system divided by p_o, the true vapor pressure of the pure substance at that temperature. When the chemical potential of a substance is the same in two different states, the substance will be in equilibrium between the two states; thus, they will have the same fugacity ratio. Fugacity is a very misunderstood principle but can be defined as the escaping tendency of a particular component within a system (Glasstone 1946). The fugacity ratio (f/p_0) under normal temperature and pressure conditions can be replaced by the activity of the substance "a." When water is being considered, the activity of water is the equilibrium water pressure of water above the system (p) at some temperature T, divided by the vapor pressure of pure water at temperature T. Therefore,

$$\mu = \mu_0 + RT \ln \frac{f}{p_0} = \mu_0 + RT \ln a \tag{5.3}$$

Note that *f* represents the escaping tendency of the material.

Thus, to review, the chemical potential of a substance μ_v in the vapor state under typical storage conditions for foods is equal to:

$$\mu = \mu_0 + RT \ln \frac{f}{p_0} = \mu_0 + RT \ln \frac{p}{p_0} = \mu_0 + RT \ln a_w = \mu_v \tag{5.4}$$

Note that p/p_0 is termed the relative vapor pressure. From engineering, the term *percent relative humidity* (% RH) is simply 100 times the ratio of the mass of water in a unit volume of air in equilibrium at some temperature with the food divided by the mass of water that would be in the air if the food was replaced by pure water. This comes from the gas law in Equation 5.5:

$$pV = nRT \tag{5.5}$$

Because n = moles = (mass/molecular weight), then

$$pV = \frac{m}{MW}RT \tag{5.6}$$

Rearranging Equation 5.6, then

$$m = \frac{pV(MW)}{RT} \tag{5.7}$$

Going back to the definition of % RH, then % RH = $100(m/m_0)$, and canceling out V, MW, and RT (total volume and temperature are the same), then

$$\%RH = 100\frac{p}{p_0} \tag{5.8}$$

Thus, in Equation 5.4, given that we are looking at the properties of water at a given temperature, then the activity of water is:

$$a_w = e^{\frac{\mu-\mu_0}{RT}} = \frac{\%RH}{100} \tag{5.9}$$

This explains that if there are two products with the same a_w, the water inside the products will have the same chemical potential μ and thus there will be no change in weight for each product at equilibrium. Chapter 6 discusses how this is used to create instruments to measure a_w.

In Equations 5.4 and 5.9, "a" was replaced with "a_w" where a_w is the activity of water at equilibrium in the system. The last term, μ_v, refers to the fact that the water in the vapor phase in equilibrium with the water in the complex food both have the same chemical potential, i.e., $\mu_v = \mu_L$ where μ_L is the chemical potential of the water in the liquid phase. Assuming the vapor phase behaves as an ideal gas and the escaping tendencies of the liquids are independent of pressure, then Raoult's law can be applied to the liquid water in the food phase to independently predict the activity of the water. As a final note, one can see in this definition that we do not introduce any notation of "free" versus "bound" water, i.e., a_w of 0.8 does not mean that 20% of water is bound and 80% is free. This wrong concept was introduced early on in the application of a_w to microbial growth, but as can be seen in the derivation, a_w refers to an exponential function of the chemical potential of water compared with the standard state divided by RT.

Raoult's law

For ideal solutions, the relationship between the molar concentration of solutes and the relative vapor pressure (p/p_0) can be evaluated by Raoult's law. At constant pressure and temperature, the a_w of the substance is equal to the mole fraction of water in the solution where mole fraction X_{water} is ($X = n_{water}/(n_{water}+n_{solute})$. Because nothing behaves as a true ideal system, the actual a_w is:

$$a_w = \gamma_s X_{water} = \frac{n_{water}}{n_{water} + n_{solute}} \tag{5.10}$$

where γ_s is the activity coefficient, X_{water} is the mole fraction of water, and n is the moles of solute and water in terms of colligative units. Raoult's law is fairly good at predicting the a_w of multicomponent solutions of solutes used in foods of low solute concentration (Sloan and Labuza 1976). If the activity coefficient can be determined, Equation 5.10 is fairly good for predicting a_w. The term "colligative" as used above means that the lowering of a_w is based on the number of particles of substance X in the solution with water. The theory is based on the principle that these particles reduce the escaping tendency (fugacity) of the water. In addition, the principle also states that substance X does not interact with the water, which is obviously not true for any of the solutes (salts, sugars, proteins, etc.) commonly added to foods. And finally it states that particle X is approximately the same size as the water molecule, again violated by most of the food solutes. Thus, there is need for a correction factor, i.e., the activity coefficient.

Activity coefficients of many solutes are reported by the National Research Council (1926). Electrolytes decrease the a_w more than expected on a molecular weight basis because they can dissociate. For example, 1 mole of NaCl can theoretically produce a total of 2 moles of the solutes Na^+ and Cl^-, if fully dissociated (i.e., 2 moles of colligative entities). Dipole–dipole interactions with water resulting from the change in these electrolyte solutes may also decrease or increase the activity coefficient γ, as discussed extensively by Fontan et al. (1979). Leung (1986) showed that electrolyte solutions deviate from ideality even at low concentrations. Table 5.1 gives the a_w values for NaCl solutions and Table 5.2 for glycerol, both common humectants.

Table 5.1 Moisture Sorption Data for NaCl

a_w	m (Adsorption)	m (Desorption)
0.11	0.13	16.97
0.33	0.12	15.92
0.44	0.12	16.12
0.52	0.12	17.19
0.68	0.13	15.93
0.75	6.44	330.85
0.85	391.46	437.58
0.88	473.15	545.68
0.91	600.31	752.99
0.97	1037.31	2770.15

Source: Adapted from Bell, L.N. and Labuza, T.P. 2000. *Moisture Sorption: Practical Aspects of Isotherms Measurements and Use*. St. Paul, MN: AACC International Publishing.

Table 5.2 Moisture Sorption Data for Glycerol (g water/100g glycerol)

a_w	m (Adsorption)	m(Desorption)
0.10	5	5
0.20	10	10
0.30	12	12
0.40	18	18
0.50	27	27
0.60	41	27
0.70	64	41
0.80	108	108
0.90	215	215
0.95	625	625

Source: Adapted from Bell, L.N. and Labuza, T.P. 2000. *Moisture Sorption: Practical Aspects of Isotherms Measurements and Use*. St. Paul, MN: AACC International Publishing.

Example 5.1: Prediction of a_w Lowering With Raoult's Law

If 20 g glycerol (MW = 92.1) is added to 100 g meat emulsion at 50% moisture (wet basis), the amount of water in 100 g meat is 50 g, where 20 g glycerol is theoretically dissolved in 50 g water. Assuming the activity coefficient is equal to 1 for this solution, i.e., assume it is ideal so that $\gamma = 1$, the a_w is calculated by Raoult's law as:

$$a_w = \gamma \frac{n_{H_2O}}{n_{H_2O} + n_{solute}} = 1 \frac{\left[\frac{50}{18}\right]}{\left[\frac{50}{18}\right] + \left[\frac{20}{92.1}\right]} = 0.928$$

In this calculation, we assume that glycerol does not interact with the food components and that all the water is available to the glycerol. When the isotherm data (i.e., g water/g glycerol as a function of a_w) (see Table 5.2) are plotted and the deviation from ideality is determined directly, we see that the actual a_w of this solution is equal to 0.915.

Thus, the glycol reduces the a_w more than that predicted by Raoult's law, and from this the activity coefficient is calculated as $\gamma = \frac{0.918}{0.928} = 0.989$ which is lower than 1, indicating deviation from ideality.

Deviations from ideality in a solution

There are several factors contributing to deviations from ideality in a solution. These factors are related to specific molecular properties of the species in the solution including solute size, intermolecular forces, and solvation effects, which increase with concentration. Table 5.3 shows this quite well for sucrose solutions. As concentration increases, the activity coefficient decreases, depressing a_w significantly.

For salts, given that they can dissociate to some degree when dissolved, the number of kinetic units (i.e., colligative particles) increases so that the a_w is further decreased, as shown in Example 5.2.

Table 5.3 Non-Ideality of Sucrose Solution

Mole fraction of water	Water Activity	Mole fraction of Sucrose	Sucrose Activity	Sucrose Molality
0.994	0.994	0.006	0.006	0.34
0.986	0.993	0.013	0.014	0.77
0.982	0.979	0.018	0.017	1.00
0.976	0.970	0.024	0.030	1.35
0.967	0.962	0.034	0.049	1.92
0.960	0.948	0.044	0.072	2.56
0.944	0.930	0.056	0.104	3.30
0.932	0.904	0.068	0.139	4.03
0.910	0.876	0.090	0.219	5.51
0.891	0.814	0.109	0.305	6.79

Source: Adapted from Labuza, T.P. 1984. *Moisture Sorption: Practical Aspects of Isotherms Measurements and Use*. St. Paul, MN: AACC International Publishing.

Example 5.2: Adding Salt to Water

Predict the lowering of a_w for 1 mole NaCl (MW = 58.45) added to 1000 g water. Note that in physical chemistry, we use molality, moles of solute added to 1000 g solvent at a given temperature. Assuming $\gamma = 1$, the a_w for a 1 molal solution of NaCl is:

$$a_w = \gamma \frac{n_{H_2O}}{n_{H_2O} + n_{solute}} = 1 \frac{\left[\frac{1000}{18}\right]}{\left[\frac{1000}{18}\right] + \left[\frac{58.45}{58.45}\right] 2} = 0.965$$

Note that the number 2 in the denominator denotes twice as many particles (kinetic units), assuming complete dissociation of NaCl. The activity coefficient (γ) was assumed as 1 for an ideal solution. Considering the actual measured a_w value of 1 molal NaCl is 0.969, it can be calculated that sodium chloride at this concentration does not lower the a_w as much as predicted because $\gamma = \frac{0.969}{0.965} = 1.004$, which is still close to 1. From thermodynamics, it can be shown that the activity coefficient can be predicted from Equation 5.11, which indicates that γ is a function of concentration:

$$\ln \gamma = K_s X_s^2 \tag{5.11}$$

where X_s is the mole fraction of solute, K_s is a constant for each solute, and γ is the activity coefficient. Table 5.4 lists K values for some typical solutes added to foods. Going back to Example 5.1 for glycerol and using the K value of −1 from Table 5.4 where $(1 - X_N)^2 = 0.077$, we obtain a_w activity coefficient of $\gamma_{glycerol} = e^{-0.077} = 0.93$ because

$$a_w = \gamma X_w \text{ and } X_s = 1 - X_w \tag{5.12}$$

then

$$\gamma = \frac{a_w}{X_w} \tag{5.13}$$

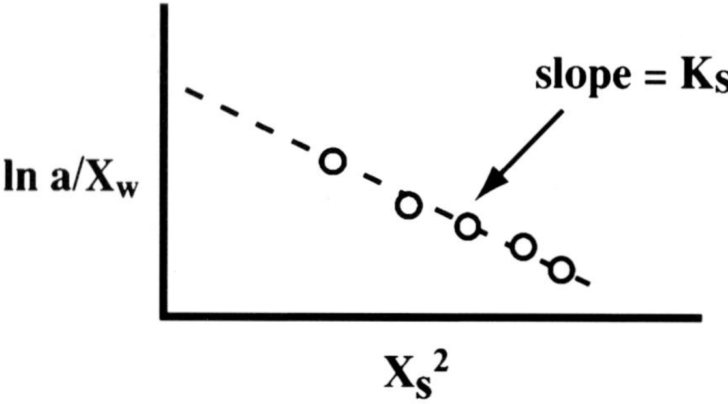

Figure 5.2 Graphic expression of *K* value derivation for Norrish equation, X_s, X_w, and a denoted mole fraction of solids, mole fraction of water, and water activity, respectively.

and taking

$$\ln \gamma = \ln \left[\frac{a_w}{X_w} \right] \tag{5.14}$$

and then substituting in Equation 5.11, we obtain

$$\ln \gamma = \ln \left[\frac{a_w}{X_w} \right] = K_s X_s^2 = K_s \left[1 - X_w \right]^2 \tag{5.15}$$

Thus, plotting $\ln \left[\frac{a_w}{X_w} \right]$ versus $\left[X_s \right]^2$ from measurements of a_w at different solute concentrations, as done by Robinson and Stokes (1956), gives a straight line with a slope of K_s. From this, γ can be predicted for any solute concentration.

Other factors that lower water activity

A second effect that depresses a_w is the capillary effect. The change in degree of intermolecular hydrogen bonding between water molecules due to surface curvature leads to a difference in the vapor pressure of water above a curved liquid meniscus versus that of an infinite plane of pure water. In the concave curved surface of a meniscus, there are nearer neighbors, i.e., more water molecules that interact with each other. This depresses the escaping tendency compared with a planar surface. Considering that foods contain a large number of pores (capillaries) with water, the result is a lowering of a_w. The Kelvin equation is employed to predict this lowering:

$$a_w = e^{\frac{-2\gamma_s (\cos\theta) V_L}{rRT}} = e^{\frac{-\Delta P V_L}{RT}} \tag{5.16}$$

where γ_s is the surface tension of the liquid in a pore, θ is the wetting angle with the wall surface, V_L is the molar volume of liquid (cm^3/mol), r is the capillary radius, *R* is the gas constant (8.314×10^7 ergs/K mole), T is the temperature in Kelvin (K), and ΔP is the suction pressure. The pore size ranges from 10 to 300 μm in most foods depending on the type

Table 5.4 Norrish Constants *K* for Some Compounds.

Compound	***K***	**Reference**
PEG* 600	–56	Chirife and Fontan 1980a
PEG* 400	–26.6 ± 0.8	Chirife and Fontan 1980a
Lactose	–10.2	Miracco et al. 1981
DE 42	–5.31	Norrish 1966
Tartaric acid	–4.68 ± 0.5	Chirife and Fontan 1980b
Alanine	–2.52 ± 0.37	Chirife et al. 1980
Mannose	–2.28 ± 0.22	Leiras et al. 1990
Glucose	–2.25 ± 0.02	Chirife et al. 1980
	–2.11 ± 0.11	Chirife et al. 1982
Galactose	–2.24 ± 0.07	Leiras et al. 1990
Fructose	–2.15 ± 0.08	Chirife et al. 1982
Malic acid	–1.82 ± 0.13	Chirife and Fontan 1980b
Glycerol	–1.16 ± 0.01	Chirife et al. 1980
Propylene glycol	–1.00	Chirife et al. 1980
Mannitol	–1.91 ± 0.27	Chirife et al. 1980
Glycine	+0.87 ± 0.11	Chirife et al. 1980
Lactic acid	+1.59 ± 0.20	Chirife and Fontan 1980b

* Polyethylene glycol

Source: Adapted from Bell, L.N. and Labuza, T.P. 2000. *Moisture Sorption: Practical Aspects of Isotherms Measurements and Use*. St. Paul, MN: AACC International Publishing.

Table 5.5 Moisture Sorption Data for Glucose.

a_w	m (Adsorption)	m (Desorption)
0.11	0.08	2.98
0.33	0.07	3.66
0.44	0.27	3.56
0.52	0.70	10.23
0.68	9.83	11.86
0.75	10.06	12.00
0.88	9.99	12.73
0.91	90.85	117.83
0.97	185.87	384.05

Source: Adapted from Bell, L.N. and Labuza, T.P. 2000. *Moisture Sorption: Practical Aspects of Isotherms Measurements and Use*. St. Paul, MN: AACC International Publishing.

of food and its processing history (Farkas and Singh 1990). Assuming ideal conditions with complete wetting $\theta = 0$ and $\cos \theta = 1$) and pure water ($\gamma_s = 72.3$ dyne/cm) in the pores, the Kelvin equation predicts a_w in the range of 0.989 to 0.999, indicating a very small lowering of a_w by the capillarity effect. However, for very small pore sizes (0.01 to 0.001 µm), the a_w can be lowered above the meniscus to values of 0.28. Note that the capillary effect can occur over the whole range of the moisture sorption isotherm. Thus, importantly, on adsorption of water, small pores will fill first while larger ones will fill at higher water activities. One can look at this capillary effect as suction force in Equation 5.17:

$$\Delta P = \Delta \rho g h \tag{5.17}$$

where ΔP is the suction force, $\Delta \rho$ is the density difference between water and air, and h is the height the water is pulled up in the capillary from a flat plane of water. In water binding studies of meats, for example, one applies an external pressure and some water is expressed. This has been called "free" water, which is an improper term, and really means the water being pushed out of the capillaries down to the equivalent a_w.

As a final side note, one can have an a_w of greater than 1. This occurs for very small droplets of water, in which case the curvature is opposite that of a liquid meniscus in a capillary. Because of this curvature the number of nearest neighbors is decreased so the escaping tendency increases. A droplet of pure water at a size of 10^{-7} cm will have an a_w of 2.95. The Kelvin equation still applies but the negative sign becomes positive. This fact alone indicates that the concept of a_w being a measure of "free" water, as used in some texts, e.g., at a_w of 0.8, 80% of the water is free, makes no sense; one cannot have 295% of the water as free.

Surface interaction and water

The third major factor affecting a_w is that water interacts directly with other chemical groups of molecules through dipole–dipole forces, ionic bonds, van der Waals forces, and hydrogen bonding. These water molecules require that extra energy beyond the heat of vaporization (ΔH_{vap}) be available to transfer a molecule from the liquid state into the vapor state. However, note even at equilibrium, this so-called "bound water" has a lifetime at the interacting site of usually less than 10^{-11} seconds unless it is strongly bound, as in the water of hydration of crystals. Associated with this binding is the so-called monolayer moisture content (m_o), which theoretically assumes that each hydrophilic group has a water molecule associated with it. Even though this has been associated with the formation of a continuous liquid like phase, in reality there are little pools of water in very small capillaries and some water on H-bonding sites such as on carbohydrates. One cannot think of this as a true aqueous phase, but it is important, starting with the work of Duckworth in 1963, that molecules other than water clearly begin to have mobility at this m_o point and in fact can also enter into chemical reactions. Thus, as will be pointed out elsewhere, a dry food just below its monolayer moisture will limit most deteriorative reactions except for lipid oxidation (Labuza et al. 1970). The monolayer concept and determination of monolayer moisture content using mathematical models are discussed in the following sections.

As stated earlier, all three major factors (colligative solution effects, capillary effects, and surface interaction) occur over the whole moisture range in biological systems and yield the characteristic sigmoid-shaped moisture sorption isotherm. Foods with different moisture contents will have different water activities depending on the interactions between the water and the food solids and its ensuing matrix, and thus each food will have its own unique moisture sorption isotherm. Although done in the early work on a_w of foods, it is not correct to segment the characteristic isotherm shape by these three factors. Moisture sorption isotherms will be discussed in detail in the following sections. It should be noted that at very high water activities, for example, 0.98 for strawberries, the total water content is limited because of the inherent structure. Thus strawberries have a swelling limit of about 95% water or a water-to-solids ratio of 19:1. Its water content cannot be increased unless we grind it up and destroy the native structure. Then, of course, we will have strawberry juice that can be infinitely diluted to approach an a_w of 1.0. Table 5.6 below shows the a_w of many tissue foods.

Another critical issue is how we define moisture content. This is needed since terms are thrown around with some inconsistency and water content can be measured by many techniques, such as by drying in an oven at different temperatures, the use of a vacuum, or by

Table 5.6 Food Tissue a_W Values

Fruits	a_w	**Vegetables**	a_w
Apple juice	0.986	Artichokes	0.976–0.987
Apples	0.976–0.988	Asparagus	0.992–0.994
Apricots	0.977–0.987	Avocadoes	0.989
Bananas	0.964–0.987	Beans, green	0.984–0.996
Bilberries	0.989	Beans, lima	0.994
Blackberries	0.986–0.989	Broccoli	0.990
Blueberries	0.982	Brussels sprouts	0.991
Cherries	0.959–0.986	Cabbages	0.992
Cherry juice	0.986	Carrots	0.983–0.993
Cranberries	0.989	Cauliflower	0.984–0.990
Currants	0.990	Celeriac	0.990
Dates	0.974	Celery	0.987–0.994
Dewberries	0.985	Celery leaves	0.992–0.997
Figs	0.974	Corn, sweet	0.994
Gooseberries	0.989	Cucumbers	0.985–0.992
Grape juice	0.983	Eggplant	0.987–0.993
Grapefruit	0.980–0.985	Endive	0.995
Grapes	0.963–0.986	Green onions	0.992–0.996
Lemons	0.982–0.989	Leeks	0.991–0.976
Limes	0.980	Lettuce	0.996
Mangoes	0.986	Mushrooms	0.995–0.969
Melons	0.970–0.991	Onions	0.974–0.990
Nectarines	0.94	Parsnips	0.988
Orange juice	0.988	Peas, green	0.982–0.990
Oranges	0.90	Peppers	0.982–0.997
Papayas	0.990	Potatoes	0.982–0.988
Pears	0.979–0.989	Potatoes, sweet	0.985
Persimmons	0.976	Pumpkins	0.984–0.992
Plums	0.969–0.982	Radishes	0.980–0.990
Quinces	0.961–0.979	Rhubarb	0.989
Raspberries	0.984–0.994	Rutabagas	0.988
Raspberry juice	0.988	Radishes, small	0.992–0.996
Sour cherries	0.971–0.983	Spinach	0.988–0.996
Strawberries	0.986–0.997	Squash	0.994–0.996
Strawberry juice	0.991	Tomato pulp	0.993
Sweet cherries	0.975	Tomatoes	0.991–0.998
Tangerines	0.987	Turnips	0.986
Watermelons	0.992		
		Other products	a_w
		Beef	0.980–0.990
		Lamb	0.990
		Pork	0.990
		Fish, cod	0.990–0.994
		Cream, 40% fat	0.979
		Milk	0.994–0.9

Source: Adapted from Labuza, T.P. 1984. *Moisture Sorption: Practical Aspects of Isotherm Measurement and Use*. St. Paul, MN: AACC International Publishing.

the Karl Fischer technique. Drying methods introduce error, as some water such as in crystal hydrates may not come out, and may be counted as solids. This occurs in measuring the moisture content of milk powders where the lactose is present as a crystalline hydrate. In addition, other volatiles may be lost. Although a more complicated technique, the Karl

Fischer method usually can measure all the water. Thus, for a low-moisture food such as dry milk powder, the Karl Fischer value can be almost twice as much as that determined by the oven method.

In terms of units, moisture content can be expressed as a percentage, i.e., as *X*% or *x* g of water/100 g of food. In physical chemistry and engineering it is important to express moisture content as grams of water/100 g solids (m) where:

$$m = \left[\frac{x\ g\ water}{100 - x}\right]100 \tag{5.18}$$

This water content can be properly called the water-holding capacity (WHC) of the food, if indeed it was brought to equilibrium at a given % RH. Of concern is the term *water binding capacity* (WBC), which is a nonequilibrium value determined after some external stress is applied to the food, e.g., squeezing a piece of meat at some pressure and then stating that the water left is the bound water or WBC. Obviously using a different measure will give a different value. This is useful when comparing the effects of different ingredients on a food but it is not the equilibrium moisture content.

To sum this up graphically, a_w can be plotted versus m (g H_2O/100 g solids) as shown in Figure 5.3. We see that most tissue foods have a_w values close to 1 as was seen in Table 5.6, and it is not until moisture is either removed by some means (air drying, baking, roasting, or extruding) or by adding solutes such as in making jams, we begin to reduce a_w significantly below 1. For dry and intermediate moisture we usually only use an a_w range of 0 to 0.9 and reverse the axis to m versus a_w, as in Figure 5.4.

In general, we obtain the sigmoid shape shown with moisture in the range shown. In fact, we can estimate the a_w of a food from its texture. If a food is hard or brittle, its a_w is generally less than 0.4, with leathery foods ranging from 0.5 to 0.7, soft gummy foods from 0.7 to 0.9, and liquids or crisp wet foods at a_w higher than 0.9.

Finally, another term used for some food ingredients is *humectant*. This term is applied to ingredients easily soluble in water, such as sugars, sugar alcohols, and polyols, that are nonelectrolytes due to the lack of the dissociation step of the solute when introduced into a solvent. Depression of a_w by addition of humectants is a common method of preservation to create intermediate-moisture foods (Labuza 1978, Taoukis et al. 1988). They reduce a_w because of their colligative effects (Lindsay 1985, Barbosa-Cánovas and Vega-Mercado 1996). Prediction of a_w using Raoult's law was discussed earlier. For solutions deviating from ideality, several other mathematical solutions are available to predict a_w, which are discussed next.

Models for Predicting Water Activity

Norrish equation

Norrish (1966) derived a model for predicting the a_w of nonelectrolyte solutions so as to correct for the nonideality of nonelectrolyte solutions containing both single and multiple solutes. The model basically relates the partial excess free energy of mixing to the concentration of the species in the mixture. One should note that the equation can also include electrolytes, such as NaCl.

For a single solute, we can use the solution for the activity coefficient as a function of solute concentration (see Equation 5.12) and derive the following:

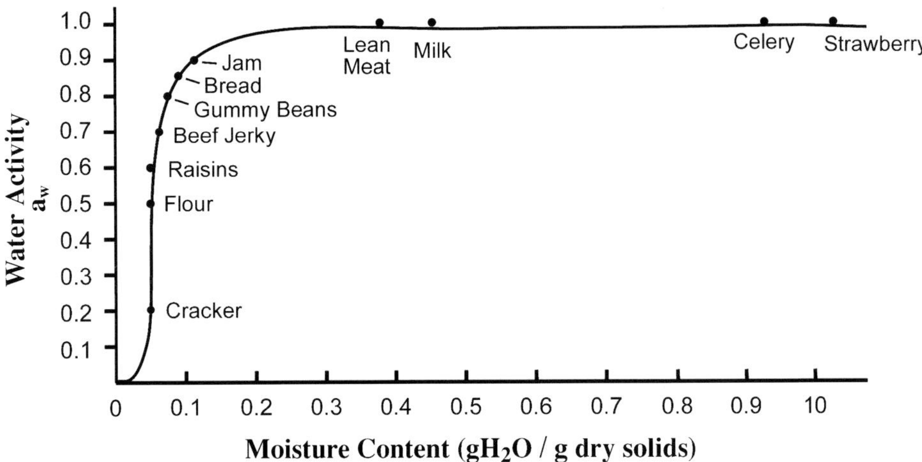

Figure 5.3 General moisture sorption curve for the whole moisture range of food (adapted from Labuza, T.P. 1984. *Moisture Sorption: Practical Aspects of Isotherm Measurement and Use.* St. Paul, MN: AACC International Publishing).

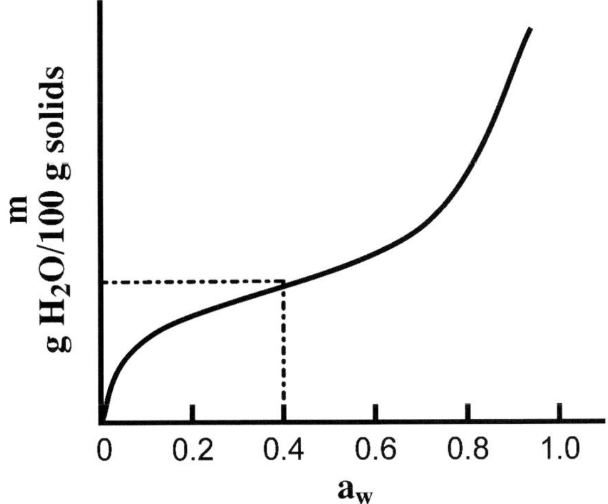

Figure 5.4 Moisture sorption isotherm for a typical food material.

$$
\begin{aligned}
a_w &= \gamma X_{H_2O} \\
\ln a_w &= \ln X_{H_2O} + \ln \gamma \\
\ln a_w &= \ln[X_{H_2O}] + K[X_s]^2
\end{aligned}
\tag{5.19}
$$

For multiple solutes, Norrish showed that

$$
\ln a_w = \ln X_{H_2O} + (K_2^{1/2} X_2 + K_3^{1/2} X_3 + \ldots)^2 \tag{5.20}
$$

where X_{H_2O} is the mole fraction of water (moles water/total moles)and X_i is the mole fraction of each solute added. In a generalized form, for an infinite number of solutes, the equation can be expressed as:

$$\ln a_w = \ln X_{H_2O} + \frac{\Sigma K_i (X_i)^2}{\Sigma (X_i)^2}(1 - X_w)^2 \quad (5.21)$$

Using the K values, as previously derived, we can determine the effect of solute addition on a_w lowering as given in the following example.

Example 5.3: Prediction of Water Activity With Norrish Equation
Water activity lowering of 20 g 20:50 glycerol to water solution can be estimated by using the Norrish equation. First, the number of moles for both solute and the water is calculated.

$n_{solute} = 20/92.1 = 0.217$ and $n_{H_2O} = 50/18.015 = 2.776$

Then, the mole fraction of water is calculated to determine the mole fraction of solute (glycerol):

$$X_{H_2O} = \frac{2.786}{2.786 + 0.217} = 0.928$$

As before, the mole fraction of solute is $x_{solute} = (1 - X_{H_2O}) = 0.0865$.
From Table 5.4, the constant K is -1.16. Thus, the a_w of the using Equation 5.19 is:

$\ln a_w = \ln [X_{H_2O}] + K[X_s]^2$
$\ln a_w = \ln [0.928] - 1.16\ (0.072)^2 = -0.07472 - 0.006013$
$a_w = 0.922$

This value of a_w is higher than values from the isotherm data (0.89 from Table 5.2) for glycerol (Bell and Labuza 2000) and less than that calculated from Raoult's law (0.928).

Grover equation
The Grover equation (1947) is an empirical approach used to estimate a_w. The model is based on a polynomial expression derived for confectionery ingredients; it assigns a sucrose equivalency conversion factor for a_w lowering to each ingredient in a confectionary product as:

$$a_w = 1.04 - 0.1E_s^\infty + 0.0045(E_s^\infty)^2 \quad (5.22)$$

Thus,

$$E_s^\circ = \Sigma \frac{E_i}{m_i} \quad (5.23)$$

where E_i is a Grover value for ingredient i (see Table 5.7) and m_i is the ingredient for moisture content in grams of water per grams of the ingredient, i being for each ingredient. The

Table 5.7 Grover Constants Ei for Some Compounds.

Compound	E_i
Water and fat	0.0
Ethanol	0.8
Starch (DE < 50)	0.8
Gums	0.8
Starch	0.8
Lactose	1.0
Sucrose	1.0
Glucose, fructose	1.3
Protein	1.3
Egg white	1.4
Sorbitol	2.0
Acids	2.5
Glycerol	4.0

Source: Adapted from Grover, D.W. 1947. The keeping properties of confectionary as influenced by its water vapor pressure. *J. Soc. Chem. Ind.* 66:201–205.

E_i values are the sucrose equivalent values for different ingredients such as lactose (1.0), invert sugar (1.3), corn syrup 45DE (0.8), and gelatin (1.3) converted by setting the E_i value of sucrose to 1, then comparing the relative water binding of other ingredients. The maximum value for $E_s\gamma$ is limited to a value of 10. Obviously this empirical approach will not be very accurate because of the assumptions.

Compared with other mathematical models derived for nonelectrolyte solutions, the Grover equation does not apply at very low moisture contents (Karel 1975), so to use this model, more information is required about the composition of the system. Recalling the same example of glycerol addition to 100 g food at 50% moisture, the following example illustrates the prediction of a_w lowering with the Grover equation.

Example 5.4: Prediction of Water Activity With Grover Equation
We will calculate the a_w of 100 g partially dried meat emulsion at 50% moisture (50 g water) when 20 g glycerol is added to it. The solids composition will be assumed to be 25 g protein, 1 g carbohydrate, and 34 g fat.

Table 5.8 Composition of Meat (100 g) at 50% Moisture Calculated With Grover Equation.

Component	g	m_i	E_i	E_i/m_i
Moisture	50	0.0	—	—
Protein	25	2.0	1.3	0.65
Starch	1	50.0	0.8	0.016
Fat	24	0.0	0.0	0.0

$\Sigma E_i/m_i = Es° = 0.666$

Therefore, the initial a_w of the product is $a_w = 1.04 - 0.0666 + 0.0045\ (0.666)^2 = 0.99$, which is reasonable as it should be just below 1. Note that based on the measurement of the freezing point depression, the a_w value of fresh meat ranges from 0.99 to 0.998 (Chirife and Fontan 1982), so the a_w of this partially dried product seems reasonable.

Considering E_i for glycerol is 4 (see Table 5.7), adding the 20 g glycerol (m_i of glycerol = 50/20 = 2.5) the new value of $\Sigma E_s°$ becomes:

$\Sigma E_s° = 1.6 + 0.666 = 2.267$

Using Equation 5.22, we then find that

$a_w = 1.04 - (0.1)2.267 + 0.0045[2.267]^2 = 0.84$

As seen, this is considerably lower than that predicted by previous equations, calling into question this predictive equation.

Money and Born equation
The Money and Born equation (1951), another very empirical equation using sucrose as the basic humectant, is generally used to determine a_w of sugar confections jams, fondant creams, and boiled sweets expressed as:

$$a_w = \frac{1}{1 + 0.27N} \tag{5.24}$$

where N is the moles of sugar/100 g water. The equation is formally based on Raoult's law but can be applied over a wide range of concentrations. Let us consider the same example using the Money and Born equation to predict a_w lowering.

Example 5.5: Prediction of Water Activity With Money and Born Equation
The a_w of a solution when 20 g glycerol is added to 100 g partially dried meat product at 50% moisture (50 g water) is now estimated with the Money and Born equation:

$$a_w = \frac{1}{1 + 0.27(40/92.1)} = 0.895$$

which is higher than the a_w calculated with the Grover equation but lower than a_w calculated with Raoult's law and the Norrish equation. Thus, we have the dilemma of determining which is the right equation to choose. Experience dictates that equations based on formal physical chemical concepts, as long as the limitations are considered, are the correct choice compared with empirical equations.

Electrolyte solutions
Ionizable substances like salts, acids, bases, and polyelectrolytes partly dissociate into ions when dissolved in water. Electrolytes are substances that exist in the form of ions when in solution, forming electrolyte solutions. The effect of an electrolyte on a_w is a function of all the ionic species released when they dissociate in solution rather than as a single chemical species or compound (Stokes 1979). This is the main difference compared with nonelectrolyte compounds. As a consequence of dissociation, the ionic species generally are reactive because of their electric charge, and the charge is generally shielded to a certain extent by the presence of ions of opposite charge, called counterions. This implies that the activity coefficient may be greatly diminished if the concentration of counterions is high. In order to simplify some thermodynamic calculations when working with the activity co-

efficients of solvents, the osmotic coefficient, Φ, was defined as:

$$\Phi = -1000 \ln \frac{a_s}{M_s \Sigma u m_s} \tag{5.25}$$

If the equation is rearranged in the case of water as the solvent,

$$a_s = e^{-0.018 \Phi u m_s} \tag{5.26}$$

where Φ is the osmotic coefficient, a_S is the solvent activity, i.e., a_W, $m_S = 0.018$ is the moles of solute per kg of water, and u is the number of ions per species. The osmotic coefficient approaches zero at infinite dilution of all solutes (Stokes 1979), which recognizes the fact that Raoult's law becomes exact at an infinite dilution condition. Several empirical models were proposed to predict a_W of electrolyte solutions considering the contribution of multiple ionic species dissolved in the solution. We will briefly recall some basic facts about dissociation and its consequences before discussing the proposed empirical models.

The Debye-Huckel theory is one of the commonly used theories derived to predict the ion activity coefficient of electrolyte species. Several factors affect the ion activity coefficient, however, the main factor resulting in low ionic strength is explained by the limiting law of this theory. An ion is normally surrounded by more counterions (opposite charge) than co-ions (same charge) in a solution, thereby to some extent shielding the charge of the ion. The attraction between counterions is stronger when the ions are closer to each other, and consequently when the concentration is higher the shielding is stronger. Therefore, the total ionic strength, I, is defined as:

$$I = \frac{1}{2} \sum_i m_i z_i^2 \tag{5.27}$$

where m denotes molality and z the valence of the ions. So, the theory suggests ion activity coefficient for ionic strength of less than 0.01 m as:

$$\log \gamma_{\pm} = -|z_+ z_-| A\sqrt{I} \tag{5.28}$$

where A is an empirical constant ($A = 0.509$ for water at 25°C).

Note that for nonideal systems, as discussed earlier, the deviation from ideality is compensated by the activity coefficient, which may markedly deviate from unity. At high concentrations of solutes the activity coefficient is generally greater than unity; however, for ionizable substances, in the case of electrolyte solutions, the activity coefficient is generally smaller than unity. The higher the total ionic strength is, the smaller the activity coefficient where opposite charges result in decreasing the activity coefficient.

Pitzer equation

Pitzer (1973) recognized the effect of short-range forces in binary attractions on ionic strength, which is an effect neglected by Debye-Huckel theory. As stated earlier the Debye-Huckel model approximates the thermodynamic properties by considering the columbic forces, omitting the effect of short range forces (Pitzer 1973), where this limiting theory is generally accepted for ionic strength of less than 0.01 M (Leung 1986). The Pitzer equation enables estimation of osmotic coefficient for aqueous electrolytes with an ionic strength of up to 6 m where the osmotic coefficient of a single electrolyte is:

$$\Phi = 1\left|Z_z Z_x\right| F + 2m\left[\frac{u_m u_z}{u_{total}}\right] B_{mx} + 2m^2\left[\frac{u_m u_z}{u_{total}}\right]^{3/2} C_{mx} \tag{5.29}$$

$$F = -A\left[\frac{I^{1/2}}{1+I^{1/2}B}\right] \text{ and } I = 1/2\Sigma(m_i Z_i^2) \tag{5.30}$$

where u_m and u_x are the number of ions for species M and X, respectively, and where the total number of ions for species M and X is:

$$u_{total} = u_m + u_x \tag{5.31}$$

Z_i is the charge on species M and X and B_{mx} is the second virial coefficient (second coefficient of virial expansion function relating original function to points of application) calculated from B_{mx0} and B_{mx1}, which represents the short-range interaction of a single electrolyte:

$$B_{mx} = B_{mx}(0) + B_{mx}(1)\exp(-\alpha I^{0.5}) \tag{5.32}$$

C_{mx} is the third virial coefficient, and I denotes the ionic strength. A and B are Debye-Huckel coefficients where A is 0.392 and B is 1.2 for water at 25°C. Pitzer constants for selected electrolytes are given in Table 5.9. The a_w can be evaluated by rearranging Equation 5.32 as:

$$a_w = \exp(-0.1802\Phi\Sigma_i m_i) \tag{5.33}$$

The following example illustrates the use of Pitzer model for a 1:1 type (solution of NaCl) solution.

Example 5.6: Pitzer equation

For an electrolyte solution of NaCl, 2.32 molal at 25°C, the Pitzer's B_{mx} (0), B_{mx} (1), and C_{mx} are 0.0765, 0.2664, and 0.00127, respectively, where the charges Z_{Na} and Z_{Cl} and respective number of ions u_{Na} and u_{Cl} are equal to 1: $m_{Na} = m_{Cl} = 2.31$ and the ionic strength I is 2.31 where F = −0.211 and B_{NaCl} = 0.089; thus, the osmotic coefficient is determined from Equation 5.29 as:

$\Phi = 1.001$

The a_w is determined from Equation 5.33 as:

$a_w = \exp(-0.01802\times(2.31+2.31))$
$a_w = 0.92$

The experimental a_w of this mixture is 0.92 (Teng and Seow 1981).

Bromley equation

The Bromley (1973) equation is a slightly different version of the Pitzer equation, which approximates the osmotic coefficient by estimating the second virial coefficient, B_{mx}, as the sum of individual B values of ions in the solution. Thus, the osmotic coefficient is expressed as:

$$\Phi = 1 + 2.303\left[T_1 + (0.06 + 0.6B)T_2 + 0.5BI\right] \tag{5.34}$$

$$T_1 = A\left|Z_z Z_x\right| \frac{\left[1 + 2(1 + I^{0.5})\ln(1 + I^{0.5}) - (1 + I^{0.5})^2\right]}{I(1 + I^{0.5})} \tag{5.35}$$

$$T_2 = \frac{(1 + 2aI)}{a(1 + aI)^2} - \frac{\ln(1 + aI)}{a^2 I} \text{ and } a = \frac{1.5}{\left|Z_m Z_x\right|} \tag{5.36}$$

where T_1 and T_2 are used to simplify Equation 5.34, A is equal to 0.511 at 25°C, and B values are reported by Bromley (1973) for different salt solutions as an estimation from Bromley's individual B and δ values for the ions as:

$$B = B_m + B_x + \delta_m \delta_x \tag{5.37}$$

Some examples of the Bromley constant for selected electrolytes are given in Table 5.8. Compared with Pitzer's equation the Bromley model is less accurate; however, Bromley's equation is still quite effective for thermodynamic calculations of electrolyte solutions (Pitzer and Mayorga 1973). To illustrate the difference, let us use Bromley's model for the previous NaCl solution example.

Example 5.7: Bromley equation

The Bromley B parameter is equal to 0.0574 for NaCl. The ionic strength and the charge product $|Z_m Z_x|$ are the same as the previous example. Replacing these values in Bromley's equation gives:

$T1 = -0.0615$
$T2 = -0.0226$
$a = 1.5$

The osmotic coefficient is then evaluated and a_w from Equation 5.33 as:

$\Phi = 1.0062$
$a_w = 0.919$

In the case of multicomponent systems, where different types of solutes are mixed in the solution, prediction of a_w becomes rather difficult due to both solute–solute and solute–solvent interactions, which are not considered in Raoult's law. Different levels of a_w could be obtained in these cases. There are several models proposed for multicomponent solutions, most of which neglect the effect of solute–solute interactions. The individ-

Table 5.9 Pitzer and Bromley Constants for Some Electrolytes.

Electrolytes	*B*(0)[a]	*B*(1)	*C*	*B*[b]
NaCl	0.0765	0.2664	0.00127	0.0574
KCl	0.0483	0.2122	–0.0008	0.0240
HCl	0.1775	0.2945	0.00080	0.1433
KOH	0.1298	0.3200	0.0041	0.1131
KH_2PO_4	–0.0678	–0.1042	–0.1124	—
NaOH	0.0864	0.2530	0.0044	0.0747
NaH_2PO_4	–0.533	0.0396	0.00795	–0.0460

[a] *B*(0), *B*(1), and *C* are Pitzer constants.
[b] *B* is the Bromley constant.
Sources: Adapted from Bromley, L.A. 1973. Thermodynamic properties of strong electrolytes in aqueous solutions. *AICHE Journal* 19(2):313–320; and from Pitzer, K.S., and Mayorga, G. 1973. Thermodynamics of electrolytes. 2. Activity and osmotic coefficients for strong electrolytes with one or both ions univalent. *Journal of Physical Chemistry* 77:2300–2308.

ual binary a_w data are assumed available at the same total ionic strength of the solution to predict the a_w of the mixture (Teng and Seow 1981). For these mixed solutions, the Ross equation is generally accepted as the best model to predict a_w.

Ross equation
The most useful a_w prediction equation has been derived by Ross (1975), who provided a reasonable estimation for multicomponent solutions over the intermediate and high a_w ranges. The equation assumes that each solute behaves independently and dissolves in all of the water in system where solute–solute interaction cancels in the mixture on average. The equation is:

$$a_w = \Pi a_i \tag{5.38}$$

$$a_w = a_1 a_2 a_3 \ldots a_n \tag{5.39}$$

where a_i is the a_w of each component *i* assuming all the solute is dissolved in all the water. Chirife et al. (1980) proposed a correction to the Ross equation and the possibility of using this equation for a_w determination for mixtures of salts, sugars, and polyols, expressed as:

$$a_w = \Pi a_{SiT}^{\ -\frac{n_i}{n_T}} \tag{5.40}$$

where n_i is the moles of solute *i*, n_T is the total moles of solutes, and a_{SiT} is the a_w at total molality. The Ross model gives fairly good estimations for a_w down to about 0.6 and the relative error is generally less than 10%. However, it has been shown that the model may give significant deviations at reduced water activities in solutions of mixed strong electrolytes, but this is not that common in foods. The following food product example illustrates use of the Ross equation.

Example 5.8: Prediction of Water Activity With Ross Equation
Calculate a_w for the same product used in the previous examples with the Ross equation in which 20 g glycerol is added to 100 g partially dried meat product at 50% moisture (50 g water/50 g solids).

The initial a_w of the product will be chosen as 0.99 as established earlier, and the a_w of the added glycerol determined from the desorption isotherm at a moisture of 250 g H_2O/100 g glycerol is 0.906. Therefore the final a_w is:

$$a_w = a_{initial} \times a_{glycerol}$$
$$a_w = 0.99 \times 0.918 = 0.908$$

This is lower than the value from the theoretical Raoult's law and Norrish equations but higher than the value from the empirical Grover equation. Industrial experience shows that the Ross equation is the best one to use when comparing calculated values to actual measured a_w values. In fact, if the isotherms of individual ingredients are known, the Ross equation can be used to predict these contributions. For example, if 100 g texturized carbohydrate is added to the product in the above example, and if the carbohydrate's a_w is 0.95 at 50 g H_2O/100 g, the final a_w would be:

$$a_{wf} = a_{food} \times a_{glycerol} \times a_{CHO}$$
$$a_w = 0.99 \times 0.918 \times 0.95 = 0.863$$

This brings the finished product into the IMF range with a moisture content of 50 g water divided by [50 g meat solids + 20 g glycerol + 100 g CHO]; or 50/170 = 29.4 g H_2O/100 g solid.

Given that a finished food product may undergo further processing, such as extrusion, and include multiple ingredients where solute–solute interaction, solute–water interaction, and capillary action effects are not easy to quantify for complex mixtures such as foods. Thus, measurement of moisture sorption isotherms in the specific product may be required. The next section discusses these isotherms in general.

Moisture Sorption Isotherms

Understanding the water relationships in a food system requires determination of a_w levels corresponding to the range of water content to which the food may be subject. As noted earlier, there are many critical water activities corresponding to minimum and maximum values for both physical and chemical reactions. These are summarized below:

1. a_w (0.2 to 0.3 range) in which the monolayer moisture (m_o) represents the optimal moisture content region where dehydrated foods have a maximum shelf-life. Above the monolayer, chemical reactions that require a water phase begin, while above and below the region, the rate of oxidation of lipids increases compromising shelf-life.
2. a_w (0.35 to 0.45 range) where physical state changes begin, such as loss of crispiness, stickiness of powders and hard candies, followed by the recrystallization of amorphous state sugars causing irreversible caking. These physical state changes are also controlled by the T_g, which can more accurately define the critical moisture content where such changes begin (Labuza et al. 2004).

   Below this a_w region or the T_g, a material is glassy or brittle/hard, whereas above it, materials show a rubbery texture, from leathery to soft to sticky. Knowing the isotherm slope and the package permeability to moisture, one could predict the time to loss of crispiness for packaged potato chips stored at different external % RH (see Chapter 9).

3. a_w (0.4 to 0.5 range) where soft materials (e.g., raisins) become hard as they dry out; this is also related to T_g.

   For multidomain foods with regions of different a_w, it is the difference in the a_w of the two domains that drives the water transfer from the higher region to the lower. A Holy Grail in the processed food arena is to find the means to retard such transfer and to maintain separate crisp versus soft domains. The chocolate coating on the inside of the cone, separating the cone (a_w ~0.2) from the ice cream (a_w = 0.85) of a frozen novelty stored at −15°C is one such example where an edible barrier is used to help the cone remain crisp.
4. a_w (0.6 critical point) at which there is the potential for growth of microbes if the moisture content goes higher. Generally, in the 0.6 to 0.75 range, mold growth would dominate.
5. a_w (0.6 to 0.8 range) in which, for intermediate-moisture foods, the rates of chemical reactions that require an aqueous phase, and which deteriorate foods, reach a maximum and then fall at higher moisture content.

   This fall in rate is due to the fact that at these high a_w levels, a small increase in a_w means a large gain in moisture. This essentially dilutes the concentration of the reactants in the adsorbed aqueous phase, and because rate of reaction is a function of concentration, the rate falls. One such reaction is the Maillard reaction (nonenzymatic browning [NEB]) between reducing sugars and protein. Formulating a food to an a_w above the 0.6 to 0.8 range to lower chemical reaction rates, such as NEB, or to make the food softer, could result in the growth of bacteria spoiling the food, as seen in the next criterion.
6. a_w (0.85 critical point) at which bacterial pathogens and spoilage bacteria begin to grow. One of the changes in the U.S. Food and Drug Administration (FDA) 2005 Food Code amends the definition of potentially hazardous foods (PHF) as foods having a finished equilibrium pH greater than 4.6 and an a_w greater than 0.85 and can thus support the growth of pathogens. In addition, for 21CFR 113, the regulation for canning (heat sterilization) of foods, the FDA also requires that if the a_w of a canned food is greater than 0.85 and the pH is greater than 4.6, the process must deliver a minimum of a 12-log cycle kill for botulinum.

Water sorption isotherms illustrate the steady-state amount of water held (i.e., water-holding capacity) by the food solids as a function of a_w or storage % RH at constant temperature (Labuza 1968). Water vapor sorption by foods depends on many factors, including chemical composition, physical-chemical state of ingredients, and physical structure. It has been noted previously that this includes the pore structure, and as pointed out, deteriorative reactions including microbial growth and physical states follow fairly well-derived patterns with respect to a_w. Thus, the moisture sorption isotherm is an extremely valuable tool for food scientists because it can be used to predict which reactions will decrease stability at a given moisture; it allows for ingredient selection to change the a_w, to increase the stability and can be used to predict moisture gain or loss in a package with known moisture permeability (Bell and Labuza 2000).

The sorption isotherms of most food materials are nonlinear and generally sigmoid shaped, as noted earlier, but do differ based on the chemical composition and physicochemical state of a food's constituents. Brunauer et al. (1945) classified sorption isotherms as being related to five general types on the basis of the van der Waals adsorption of nonpolar gases adsorbing on various nonporous solid substrates (see Figure 5.5). Note, how-

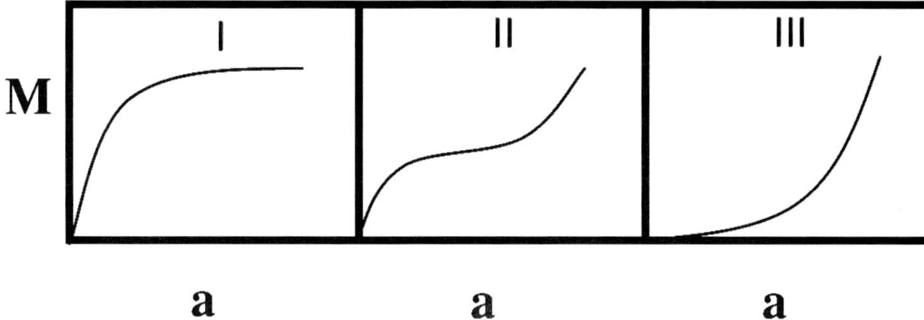

Figure 5.5 Three types of sorption isotherms showing adsorbent content (M) versus activity (a) of adsorbent (adapted from Brunauer, S. 1945. *The Adsorption of Gases and Vapors*. Princeton, NJ: Princeton University Press).

ever, that with foods we are dealing with a polar gas (water) and mostly polar adsorbents like carbohydrate and proteins. Despite this, the applicability of the BET model to water sorption in food is remarkable and will lead to the ability to understand and control many stability problems. According to this, classification types I, II, and III are the most commonly observed moisture sorption isotherms for food materials.

The shape of the moisture sorption isotherms are characterized by the states of the system constituents. A pure solid can exist in two basic states, as a crystal or as an amorphous solid, what physicists call soft condensed matter. A crystalline state of a material is characterized by a regular three-dimensional arrangement of the molecules with different crystal structures based on differences in orientation, including polymorphs and hydrates. The type of crystal determines the solubility, which in turn affects the shape of the moisture sorption isotherm. The amorphous state is a thermodynamically unstable state compared with the crystalline state and is lacking in molecular order. The amorphous state of material can consist of either a rubbery or a glassy state and the transition temperature at which the glassy state is converted to a rubbery state, and vice versa, is known as the glass transition temperature (see Figure 5.6). For a given moisture content the transition can be achieved by either increasing the temperature or adding low-molecular-weight substances called plasticizers (e.g., water). Different properties of these states can also influence sorption isotherms. The relation between glass transition and a_w will be discussed in detail in the following chapters; however, some examples are useful here.

1. The desired state of a sugar snap cookie or a potato chip is crisp and brittle when bitten into. If held under abuse conditions such as high humidity, it can be transferred into a soft soggy state. This transition from state A to state B in Figure 5.6 is the glass transition from a glassy to a rubbery state. Interestingly, this transfer occurs above the monolayer moisture and generally in the range of 0.4 to 0.45 (Payne and Labuza 2005).
2. Cotton candy is formed by heating crystalline sugar above the melting temperature and then spinning it out into the air to cool to a glassy state. In fact, it looks like glass wool insulation if below the glass transition line. When held above a given temperature–humidity combination (e.g., 33% RH and 23 °C, which is the T_g for room temperature), it picks up moisture and begins to flow and collapse under the force of gravity as man-

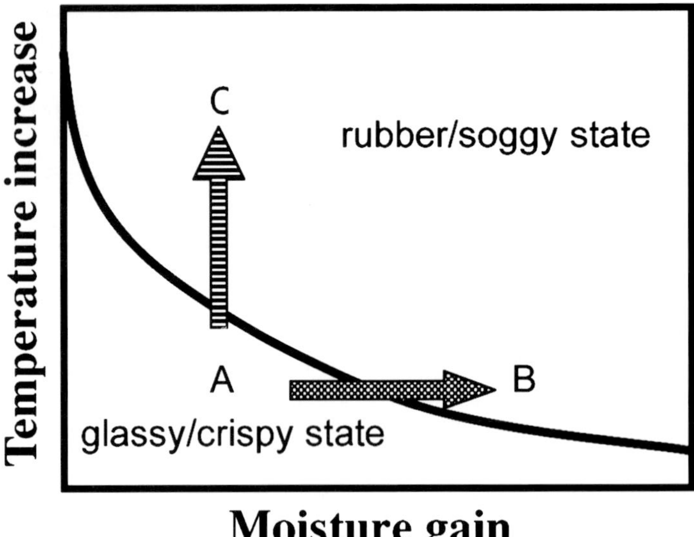

Figure 5.6 General glass transition phase diagram showing glass transition temperature versus moisture content and the effect of a temperature or moisture increase (adapted from Bell, L.N., and Labuza, T.P. 2000. *Moisture Sorption: Practical Aspects of Isotherms Measurements and Use*. St. Paul, MN: AACC International Publishing).

ifestation of the rubbery state. The higher the humidity, the faster is the activity, so as Labuza and Labuza (2004) found, it takes 3 days at 33% RH, 1 day at 44% RH, and 1 hour at 75% RH. In this example, the sugar molecules are close enough together that they can nucleate and grow to form crystals, wherein the sticky mass that forms the rubber is infiltrated by granules as the sucrose crystallizes (Labuza and Labuza 2004).

The type I isotherm is typical of anticaking agents as they can hold large amounts of water at low water activities. This type of ingredient absorbs water via chemisorption onto specific sites, so the excess binding energy Q_s is very large. Once all the binding sites are occupied (including narrow capillaries), there is little increase in moisture content at increasing humidity since the pores are all filled and there is no swelling or solutes present that could dissolve. For most processed foods, e.g., ready-to-eat cereals, where the a_w is below 0.95, the general moisture sorption isotherm is represented by Figure 5.3, which is a sigmoid-shaped curve, type II isotherm. The resultant shape is caused by the additive effects of Raoult's law, capillary effects, and surface water interactions. Two bending regions are noted in this type of isotherm, one around an a_w of 0.2 to 0.4 and another at 0.6 to 0.7, the results of changes in magnitude from the separate physical and chemical effects, i.e., the build-up of multilayers and filling of small pores in the lower region, followed by swelling, the filling of large pores, and solute dissolution in the upper region. And as noted earlier, dividing this into three regions of highly bound, capillary, and "free" water is just incorrect.

Food systems composed mainly of crystalline components such as sugars and salt are represented by a type III isotherm. Moisture gain is very low up to the point where the crystals begin to dissolve in the absorbed water at the surface of the crystal. This point is referred to as the *deliquescent point.* The type III isotherm, for pure crystalline sugar, shows very little moisture gain until the a_w goes above 0.8. At 0.81 a_w the solution at the surface is saturated in sugar and below this a_w water is hydrogen-bonded to the −OH groups that stick out on the surface of the crystal. Because this is a surface effect, grinding the sugar into smaller particles, possibly resulting in loss of crystal properties, will increase the moisture content at the low water activities by the amount of surface increase. Thus, for true crystalline materials, sieve size is an important factor for water content as a function of a_w.

A large number of sorption isotherms for ingredients and commodities have been determined and are available in the literature (Iglesias and Chirife 1982). Many prediction models for moisture sorption isotherms have been proposed and are discussed in the following section.

Isotherm Prediction Models

As stated earlier, it is essential to have a good understanding of moisture sorption isotherms for establishing critical moisture contents of biological materials as well as to predict potential changes in food stability. To predict moisture sorption behavior, mathematical models with two or more parameters have been used and more than 270 isotherm equations have been proposed to fit the moisture sorption isotherms of biological materials (Van der Berg and Bruin 1981). These models are classified as theoretical, semi-empirical, or empirical.

Theoretical (Kinetic Based) Models

Kinetic models based on a monolayer or multilayer sorption and a condensed film are grouped as theoretical models. The constants of the kinetic models, in contrast with those of the empirical models, are a material's physical properties. Therefore, the determination of a kinetic model requires establishment of the appropriate sorption mechanism and verification of the model's parameter magnitude by an independent physical test (Peleg 1992).

As noted, earlier studies of the rates of chemical reactions in foods have shown that for most dry foods, there is a moisture content below which the rates of quality loss are negligible. This moisture content corresponds to the monolayer value (m_o), at a_w around 0.2 to 0.3, generally used as the starting point of theoretical models (Salwin 1959). At this a_w, the adsorbed water content becomes significant and changes the overall characteristics of the food material, since chemical species can dissolve. In addition with an incremental increase in a_w, there is increased mobility and reactivity within the system (Sherwin and Labuza 2006). In general, the rate of quality loss increases above an a_w of 0.3 for most chemical reactions, because at this a_w level the amount of water absorbed on the surfaces and in capillaries is enough to affect the overall dielectric properties so that water behaves as a solvent. Therefore, chemical species can dissolve, become mobile, and react. Furthermore, the higher the a_w, the faster is the reaction rate due to the greater solubility and increased mobility. However, at some higher water activities, the chemical species present have completely dissolved and the reaction rate decreases as a_w increases, because water dilutes the reacting species. Shelf-life studies and reaction rate determinations are based on the linear region between this maximum point and the monolayer value. Some typical monolayer values are given in Table 5.10.

Table 5.10 Typical Monolayer Values.

Food	m_0 (g water/100 g solids)
Kidney beans	4.2
Cocoa	3.9
Ground and roasted coffee	3.5
Chicken	5.2
Egg	6.8
Fish meal	3.5
Gelatin	8.7
Meat	5.0
NFDM	5.7
Potato	6.0
RTE cereals	5.2
Whole milk powder	2.2

Source: Adapted from Bell, L.N., and Labuza, T.P. 2000. *Moisture Sorption: Practical Aspects of Isotherms Measurements and Use.* St. Paul, MN: AACC International Publishing.

Langmuir isotherm

The Langmuir isotherm was developed by Irving Langmuir in 1916 to describe the dependence of the surface coverage of an adsorbed gas on the pressure of the gas above the surface at a fixed temperature. It is one of the simplest of kinetic models, still providing useful insight into the pressure dependence of the extent of surface adsorption for biological materials. The model gives fairly good predictions within an a_w range of 0 to 0.3 relying on a number of assumptions. The model assumes that adsorption is limited to the monolayer coverage while all surface sites are equivalent, which can accommodate at most one adsorbed atom. Also, the ability of a molecule to adsorb water molecules at a given site is assumed independent of the occupation of neighboring sites. In theory, if a gas is in contact with a solid, the equilibrium will be established between the molecules in the gas phase and the corresponding adsorbed species (molecules or atoms) bound to the surface of the solid. The adsorption process between vapor molecules in the gas phase (*A*) is represented in Figure 5.7, where *B* is the total number of available vacant sites on the surface and *AB* denotes the occupied surface sites. As seen, there are some pockets of water caused by condensation in very small diameter capillaries. The Langmuir isotherm ignores this, which leads to error.

The effect of temperature on moisture sorption isotherms will be discussed in detail; however, it would be useful to define heat of sorption to understand Langmuir equation derivations. Heat of sorption basically determines the effect of temperature on a_w. The Clausius-Clapeyron equation from thermodynamics is used to compute the heat of sorption by measuring the temperature dependence of a_w. Knowing the heat of sorption and a_w at a given temperature, a_w at any other temperature can be calculated. Thus, if all the surface sites are assumed to have the same constant excess energy of binding Q_s, the residence time τ on the surface can be calculated as:

$$\tau = \tau_0 e^{\frac{Q_s}{RT}} \tag{5.41}$$

The Langmuir equation can either be thermodynamically derived or based on kinetics of adsorption. For both cases, the fraction of surface sites occupied is represented by the sur-

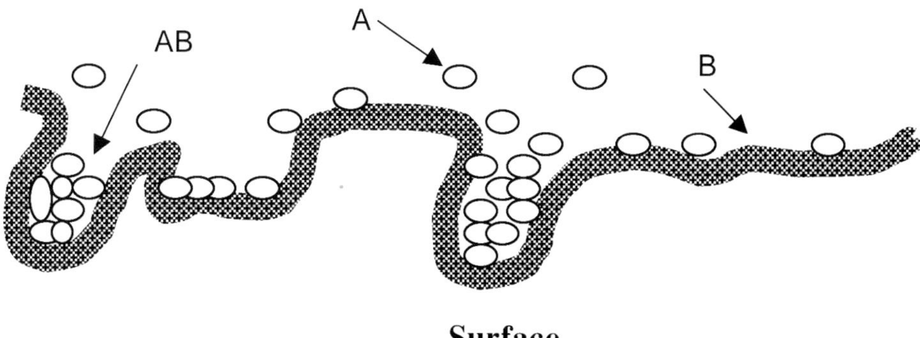

Figure 5.7 Partitioning between gas phase and adsorbed species (adapted from www.jhu.edu/~chem/fairbr/lang.html).

face coverage coefficient Φ, which is the ratio of actual moisture content to the monolayer moisture value, $\Phi = m/m_0$ and $0 < \Phi < 1$. Thermodynamically, the adsorption process is:

$$A + B \leftrightarrow AB \tag{5.42}$$

Assuming there is a fixed number of surface sites available, the equilibrium constant is:

$$K_m = \frac{[AB]}{[A][B]} \tag{5.43}$$

Assuming [AB] is proportional to the surface coverage of adsorbed molecules or Φ, [B] is proportional to the number of vacant sites or (1 − Φ), and [A] is proportional to the pressure of the gas p, then the equilibrium constant is b, and Equation 5.43 can be rewritten as:

$$K_m = b = \frac{\theta}{(1-\theta)p} \tag{5.44}$$

Rearranging Equation 5.44 gives the surface coverage $\Phi = m/m_0$ as:

$$\theta = \frac{m}{m_0} = \frac{bP}{1+bP} \tag{5.45}$$

Calculating a_w by dividing the actual pressure p by the vapor pressure of a pure system at 100% saturation in the vapor phase p_0, and using water as the adsorbent, Equation 5.45 becomes:

$$\frac{a_w}{m} = C + \frac{a_w}{m_0} \tag{5.46}$$

where a_w is the water activity, m is the moisture content, m_0 is the monolayer moisture content, and C is the constant equal to k/bm_0. Thus, a plot of a_w/m versus a_w is a straight line with a slope of $1/m_0$ and an a_w intercept of *C*, which is the temperature-dependent factor. Rarely, however, do we have enough data points for moisture sorption below m_0 to test this equation. One should note that at low p, Equation 5.45 becomes:

$$m = m_0 bp = S_0 p \tag{5.47}$$

This is Henry's law for adsorption of a gas of low pressure where S_0 is considered the solubility constant.

The Langmuir model is generally not applicable to most food systems because of the variability of the excess heat of adsorption for different sites of adsorption, the condensation of small pools of water in capillaries, and the existence of interaction between adsorbed water molecules on the surface. At higher pressure, and eliminating the restrictions of this model, the BET isotherm was developed from the Langmuir equation by Stephen Brunauer, Paul Emmet, and Edward Teller (1938) specifically for nonpolar gases on nonpolar surfaces (Karel 1975). One application where the Langmuir equation has been applied successfully is the adsorption and degassing of carbon dioxide in fresh roasted ground coffee (Shimoni and Labuza 2000, Anderson et al. 2003).

Brunauer-Emmet-Teller (BET) isotherm

The BET isotherm is one of the most successful ways to determine the monolayer moisture content for foods (Labuza 1968, 1975; Karel 1975). The two constants obtained from the BET model are the monolayer moisture content, m_0, and the energy constant, C_s, both of which come from the Langmuir equation at higher p. The model generally holds only for a limited range of a_w; however, the BET monolayer concept as a point for moisture stability of dry foods was found to be a reasonable guide to stability with respect to moisture content (Labuza 1970, Labuza et al. 1970). It should be noted that this a_w range of applicability is between 0 and 0.55; above the upper limit, the results deviate from the straight line portion when plotted as a linear equation. The general linear equation form for the BET isotherm equation is:

$$\frac{a_w}{(1\text{-}a_w)m} = \frac{1}{m_0} + \left[\frac{c-1}{m_0 c}\right] a_w \tag{5.48}$$

where *m* is the moisture in g/100 g solids at a_w and temperature T, and m_o is the monolayer value in same units, while *C* is the surface heat constant given by:

$$c = \text{constant} = e^{Q_s/RT} \tag{5.49}$$

where Q_s is the excess heat of sorption (cal/mol). The value of *C* depends on the sorption characteristics of the adsorbing site and decreases above the monolayer. For type II isotherms, the value varies from 2 to 50. When Q_s is very large, in the case of chemisorption (type I isotherm), the value varies from 50 to 200, whereas for type III isotherms representing crystals, Q_s approaches 0 and the constant *C* is less than 2. The monolayer concept is the main feature of the BET isotherm and can be calculated if Equation 5.48 is rearranged as:

$$\frac{a_w}{(1\text{-}a_w)m} = I + Sa_w \tag{5.50}$$

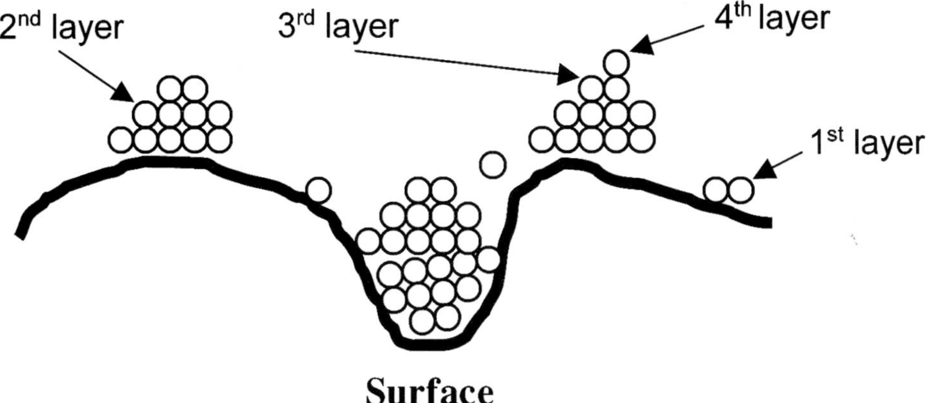

Figure 5.8 The schematic illustration of BET isotherm layers.

where I is the intercept and S is the slope of the straight line when $a_w/[(1 - a_w)m]$ versus a_w is plotted. Then the monolayer moisture value from this plot is:

$$m_0 = \frac{1}{I+S} \tag{5.51}$$

The main feature of the BET model is the monolayer concept, and that the sorption process is regulated by two mechanisms. One mechanism involves Langmuir kinetics, in which gas molecules are directly absorbed on selected sites at the solid surface until the latter is covered by a layer of gas one molecule thick, and the other mechanism is condensation in subsequent layers while all surface sites are assumed to have similar energy with no cross surface interaction (see Figure 5.8) (Van den Berg 1985). Note that the model does not predict the adsorption in the capillary region.

The failure of the BET model at higher a_w levels led to the development of the Guggenheim, Anderson, and de Boer (GAB) model. This three-parameter isotherm equation has been widely used to describe the moisture sorption behavior of foods.

Guggenheim-Anderson-de Boer isotherm

The GAB model was developed by Guggenheim Anderson and deBoer as an improved version of the BET model for multilayer adsorption (Van den Berg and Bruin 1981). Having a reasonably small number of parameters (three), the GAB equation has been found to adequately represent experimental data in the a_w range of 0 to 0.95 for most foods of practical interest. The GAB equation has been recommended by the European Project Group COST 90 on Physical Properties of Food (Wolf et al. 1985) and was accepted as the best equation to model the moisture sorption isotherm at the International Symposium on the Properties of Water (ISOPOW) in 1983. Both BET and GAB models are based on the same principles of monolayer coverage; however, the GAB model introduced an additional degree of freedom (an additional constant, k_b) by which the model has greater versatility. The monolayer value of the food product is obtained by:

$$m_0 = \frac{m_0 k_b c a_w}{\left[1 - k_b a_w\right]\left[1 - k_b a_w + c k_b a_w\right]} \tag{5.52}$$

where the value of k_b varies from 0.7 to 1, the value of C is between 1 and 20, m_0 is the monolayer value as grams of water per grams of solids, and is the water activity at moisture m. This equation (5.52) can be solved either by a nonlinear regression computer program in which at least five points are required or by converting to a polynomial to apply stepwise regression.

$$\frac{a}{m} = \frac{k_b}{m_0}\left[\frac{1}{c} - 1\right]a^2 + \frac{1}{m_0}\left[1 - \frac{2}{c}\right]a + \frac{1}{m_0 k_b c} \tag{5.53}$$

Thus, rearranging the GAB into a polynomial we have:

$$\frac{a}{m} = \alpha a^2 + \beta a + \varepsilon$$

$$\alpha = \frac{k_b}{m_0}\left[\frac{1}{c} - 1\right] \quad \beta = \frac{1}{m_0}\left[1 - \frac{2}{c}\right] \quad \varepsilon = \frac{1}{m_0 k_b c} \tag{5.54}$$

Finally using the binomial equation to find the GAB solution,

$$k_b = \frac{\sqrt{\beta^2 - 4\alpha\varepsilon} - \beta}{2\varepsilon} \qquad c = \frac{\beta}{\varepsilon k_b} + 2 \qquad m_0 = \frac{1}{\varepsilon k_b c} \tag{5.55}$$

This can easily be set up in an Excel spreadsheet and obtained at http://www.ardilla.umn.edu/Ted_Labuza/FScN%204312/4312-default.html.

The GAB model is basically similar to the BET equation in its assumption of localized physical adsorption in multilayers with no lateral interactions. Incorporation of the parameter k_b, however, assumes that the multilayer molecules have interaction with the sorbent that ranges in energy levels somewhere between those of the monolayer molecules and the bulk liquid. The major advantages of the GAB model are (1) it has a viable theoretical background; (2) it describes sorption behavior of most foods, from 0 to 0.95 a_w; (3) it has a mathematical form with only three parameters, making it very amenable to engineering calculations; (4) its parameters have physical meaning in terms of the sorption processes; and (5) it is able to describe some temperature effects on isotherms by means of Arrhenius-type equations (Van den Berg and Bruin 1981).

Lomauro et al. (1985a, 1985b) evaluated the fit of the GAB equation to 163 moisture sorption isotherms representing a variety of food products—45.5% of fruit isotherms, 54.5% of vegetable isotherms, 43% of meat isotherms, 44% of milk product isotherms, 81% of starchy foods isotherms, 100% of nut and oilseed isotherms, 61% of coffee and tea isotherms, and 91% of spice isotherms were reported to have good fit with the GAB model (Lomauro et al. 1985b).

Although the BET and GAB models and the monolayer concept on which they are based are apparently useful in explaining various stability mechanisms, they are not always compatible with other aspects of the moisture sorption phenomenon, which supports the use of empirical equations for modeling of moisture sorption. Note that the m versus a_w

data are used to determine the parameter constants in these models, so even though they fit, it does not prove they have theoretical meaning; they are, however, useful as a tool.

Empirical Models

Besides the availability of computer-aided curve-fitting models based on nonlinear regression, there are traditional empirical models such as the Henderson, Chirife, Smith, Oswin, and Kuhn models. These models are linearized with two or three fitting parameters used mostly as a complementary verification of sorption data combined with the BET or GAB models. We will briefly go through some of these multiparameter models, describing their applications and limitations in order to have a broader view of moisture sorption prediction models.

The Smith model was developed by Smith (1947) as a two-parameter empirical model to describe the final curved portion of the moisture sorption isotherms of biopolymers with high molecular weight. The experimental results showed that the equation is limited to an a_W level of 0.5 to 0.95 for wheat desorption (Becker and Sallans 1956) and applicable for a_W higher than 0.3 for peanuts (Young 1976). The Smith equation is:

$$M_w = A + B\ln(1 - a_w) \tag{5.56}$$

The Oswin model was developed by Oswin (1946) as another two-parameter empirical model in the form of a series expansion for sigmoid-shaped curves. The model gave fairly good fit for the moisture sorption isotherms of starchy foods, meats, and vegetables (Boquet et al. 1978). The Oswin equation is:

$$M_w = A\left[\frac{a_w}{1 - a_w}\right]^B \tag{5.57}$$

The Henderson model was developed by Henderson (1952) as a two-parameter empirical model. Despite contradictions in the model's true applicability to some foods, it fits in some cases, most likely due to being a log–log[1 − a] versus log[m] equation. The Henderson model is:

$$\ln\left[\ln\left[1 - a\right]\right] = \ln C + b\ln m \tag{5.58}$$

Thus, a plot of ln[ln[1 − a]] versus ln[m] should be a straight line, which in most cases it is not. The constant in the equation relates to C and m_0.

The Iglesias-Chirife model (1978) was developed as another two-parameter empirical model to describe the water sorption behavior of fruits and other high–sugar-containing products. The equation is:

$$\ln(M_w + (M_w^2 + M_{0.5w})^{1/2}) = Aa_w + B \tag{5.59}$$

Simpler models or models with fewer adjustable parameters are preferable to more elaborate models in general. Thus, if parameters for BET and GAB models could be determined by independent tests, there would be no need for empirical models, compared with the compactness and usefulness of the BET/GAB family. As long as the constants in a model have no physical significance, any three-parameter model can effectively represent a sigmoid-shaped moisture sorption isotherm, where the magnitude of parameters depends on the a_w range and number of points, as in the case of BET/GAB family. However, when the constants are used to calculate monolayer values, most empirical models are insufficient as this parameter does not exist. For these cases, it is safer to use a model that does not require, although it does not exclude, the existence of a monolayer, such as a four-parameter empirical model. The Peleg model derived by Peleg (1993) is the simplest four-parameter empirical model and is used for moisture sorption isotherms of both sigmoid and nonsigmoid shapes (Peleg 1993). The equation includes double power expansion as:

$$M_w = Aa_w^{\ C} + Ba_w^{\ D} \tag{5.60}$$

where $C < 1$ and $D > 1$.

Again, as noted above, the model may give a good fit but does not include the estimation of m_0 so it lacks a theoretical basis.

Effects on Isotherms

The moisture sorption properties of foods have been shown to be influenced by food composition, processing treatment, temperature, pressure, and relative humidity (Iglesias and Chirife 1976a, Lasekan and Lasekan 2000). The influence of these effects is of great importance in food processing. Considering that the a_w is an equilibrium concept, any single or combined processing effect might change the adsorbing sites, so these need to be accounted for. In this section, we review the effects of temperature, pressure, and composition on moisture sorption isotherms.

Temperature Effect on Isotherms

The effect of temperature on the sorption isotherm is of great importance given that foods are exposed to a range of temperatures during storage and processing, and a_w changes with temperature. In describing the moisture sorption isotherm, the temperature has to be specified and held constant because temperature affects the mobility of water molecules and the dynamic equilibrium between the vapor and adsorbed phases (Labuza et al. 1970, Kapsalis 1981). Therefore, moisture sorption isotherms are plotted with a specified and constant temperature. In general, the effect of temperature on increasing the a_w at constant moisture content is greatest at lower to intermediate water activities. Water activity increases as temperature increases for a constant moisture content. As an example, because of the nature of water binding, at constant a_w, foods that follow the type II isotherm hold less water at higher temperatures than at lower temperatures (see Figure 5.9). In addition, at high a_w and temperature, some new solutes may dissolve causing a crossover at high a_w, as noted in Figure 5.9.

Labuza (1968), Loncin (1980), and Iglesias and Chirife (1976b) have shown that the Clausius-Clapeyron equation can be applied to predict the isotherm value at any temperature if the corresponding excess heat of sorption is known at constant moisture content. The effect of temperature follows the Clausius-Clapeyron equation:

$$\ln \frac{a_{w2}}{a_{w1}} = \frac{Q_s}{R}\left[\frac{1}{T_1} - \frac{1}{T_2}\right] \tag{5.61}$$

where a_{w2} and a_{w1} are the a_w at temperature (K) T_2 and T_1, respectively; Q_s is the heat of sorption in J/mol (as a function of moisture content); and R equals 8.314 J/mol K. Q_s, the

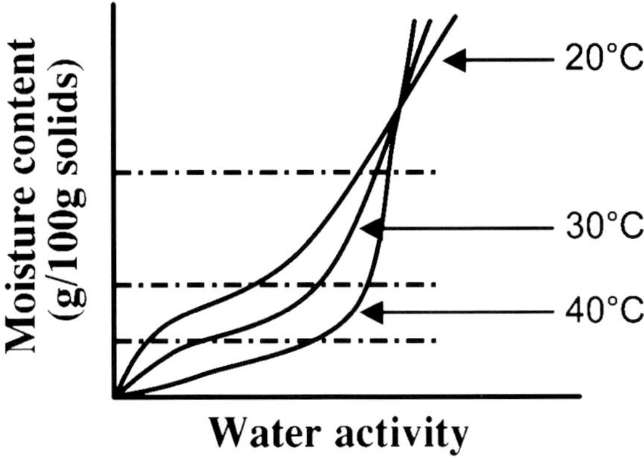

Figure 5.9 Water activity change for a food subjected to a temperature shift (adapted from Labuza, T.P. 1984. *Moisture Sorption: Practical Aspects of Isotherm Measurement and Use.* St. Paul, MN: AACC International Publishing).

excess binding energy for removal of water, is the only unknown and, unfortunately, there are no standard tables listing Q_s for different foods. Therefore, to predict the a_w of a food at any given temperature, moisture sorption isotherms must be determined for at least two temperatures. Then a plot of log a_w versus 1/T (K) will give a straight line at constant moisture content, as seen in Figure 5.10, and the a_w at any temperature for that moisture can be found. The slope of the line (Q_s/R) decreases to zero as moisture content increases. This is indicative of reduced water interactions (less binding energy) with the surface for adsorption, behaving more like pure water. The effect of temperature with respect to moisture content is shown to be greatest at low moisture contents. Above an a_w value of 0.8, no temperature effect is observed. Note that the above mentioned spreadsheet for the GAB and BET equations also has a worksheet for predicting the isotherm for any other temperature if data are available for two temperatures.

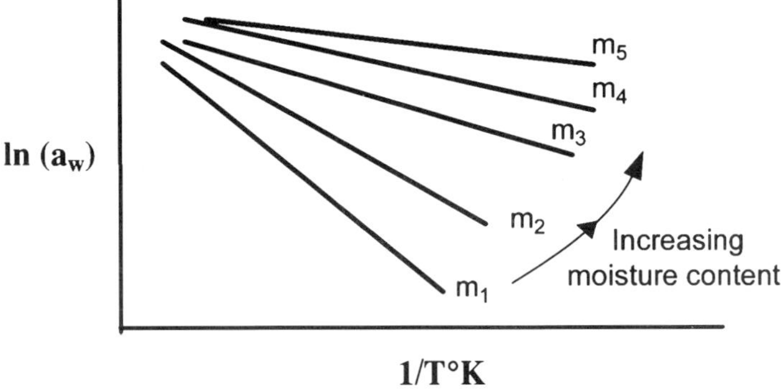

Figure 5.10 Plot of log a_w versus temperature for predicting a_w (from Labuza, T.P. 1984. *Moisture Sorption: Practical Aspects of Isotherm Measurement and Use,* St. Paul, MN: AACC International Publishing. Reprinted with permission).

Let us illustrate the effect of temperature on the moisture sorption isotherm with an example. We will determine the change in a_w of a packaged product affected by temperature change.

Example 5.9: Determination of Final Water Activity Affected by Temperature Change
A product packaged at an initial temperature of 25°C and initial a_w of 0.4 is temperature abused at 40°C in a truck. Moisture gain or loss is considered to be 0 as it is sealed inside an impermeable pouch. If the excess heat of sorption is 9 kJ/mol, the final a_w of the product is calculated by using the Clausius-Clapeyron equation (5.61), where inserting the data gives:

$$\ln\frac{0.4}{a_2}=\frac{-9000}{8.314}\left[\frac{1}{298.15}-\frac{1}{313.15}\right]=-0.174 \tag{5.62}$$

Thus, the final a_w is:

$$\frac{0.4}{a_2}=e^{-0.17}$$

and thus,

$$a_2=\frac{0.4}{0.84}=0.47$$

If the same example is considered for heat of sorption, with a Q_s equal to 18 kJ/mol, then the final a_w becomes:

$$a_2 = 0.562$$

This indicates that the a_w increases to 0.47 if the food is brought to 40°C. If the product has a critical a_w of 0.45, which could induce caking, the temperature abuse would be a substantial problem. As seen in the second part of the example with a higher value of heat of sorption, the change in a_w is more drastic.

Pressure effect on isotherms
Pressure also has an effect on the a_w of a food system, but the effect is small compared with the temperature effect. In most cases the pressure effect can be neglected unless elevated pressures are used, as in the case of an extrusion process. The thermodynamic effect of pressure on activity was discussed by Glasstone and Lewis (1960), who showed that a change in total pressure of the system will affect the vapor pressure. At equilibrium, any change in chemical potential of the liquid state will be equal to a change in chemical potential of the vapor. Thus, beginning with the equilibrium point of the chemical potential of water and the vapor state,

$$d\mu_L = V_L dP_T = d\mu_V = V_V dP_V \tag{5.63}$$

where V_L and V_V are the molar volume of the liquid and vapor, respectively, and dP_T and dP_V are the total pressure change and vapor pressure change, respectively. Rearranging Equation 5.63 gives the equation for pressure effect on a_w:

$$\ln \frac{a_2}{a_1} = \frac{\overline{V}_L}{RT}\left[P_2 - P_1\right] \tag{5.64}$$

At 20°C, this becomes:

$$\ln \frac{a_2}{a_1} = 9.6x10^{-7}\,\Delta P \tag{5.65}$$

where P_1 and P_2 are total pressures in mm Hg for initial and final pressure, respectively. Now we will illustrate the effect of pressure change on a_w with the following example.

Example 5.10: Determining Final Water Activity Affected by Pressure Change
A food product with an a_w value of 0.82 at sea level (760 mm Hg) is brought to a place where atmospheric pressure is 0.75 and the temperature is 20°C. The change in a_w is calculated by using Equation 5.64 as:

$P_2 = 0.75 \text{ atm} \times 760 \text{ mm Hg/atm} = 570 \text{ mm Hg}$

Then the pressure change ΔP is:

$\Delta P = [570\text{-}760] = -190 \text{ mm Hg}$

Plugging ΔP into Equation 5.65 yields:

$$\ln \frac{a_2}{a_1} = \left[9.6x10^{-7}\right](-190) \tag{5.66}$$

$$a_2 = a_1.e^{-1.82x10^{-4}}$$

Therefore, the final water activity, a_2, at 0.75 atm is:

$a_2 = (0.9998) \times 0.2 = 0.819$

As can be seen, for pressure change of 90 mm Hg, the decrease of a_w is negligible and probably cannot be measured.

Combined temperature-pressure effect
In some cases, a_w is influenced by the combination effect of temperature and pressure. Depending on the severity of each factor, neglecting the pressure effect could be an effective solution; however, in some cases, as in the extrusion process, the combined effect is calculated by summing up each factor using Equations 5.61 and 5.64. The combined effect of temperature and pressure on moisture sorption isotherms is estimated by using the following equation:

$$\partial \ln \frac{P}{P_0} = \left[\frac{\partial \ln \frac{P}{P_0}}{\partial T}\right]_P dT + \left[\frac{\partial \ln \frac{P}{P_0}}{\partial T}\right]_T dP \tag{5.67}$$

Integrating and rearranging the terms in the above equation yields:

$$\ln\frac{a_2}{a_1} = \frac{-Q_s}{R}\left[\frac{1}{T_2} - \frac{1}{T_1}\right] + \frac{\overline{V}_L \Delta P}{RT_2} \tag{5.68}$$

The combined effect of temperature and pressure will be illustrated in the following example by examining both the combined and single effects of each factor.

Example 5.11: Determining the Effect of Combined Temperature–Pressure Change
In an extrusion process, the food matrix enters the extruder with an initial a_w a_1 of 0.3 at 20°C. The exit pressure is 10,000 psi at 150°C, whereas the pressure in the processing room is 14.7 psi. The final a_w of the product at the exit can be determined by calculating the influence of each factor (pressure and temperature) and through substitution in Equation 5.68.

Temperature effect
Heat of sorption Qs for $a_1 = 0.3$ is equal to 2.1 kJ/mol. Therefore, from Equation 5.61 we obtain:

$$\ln\frac{a_2}{a_1} = \frac{-2100}{8.314}\left[\frac{1}{423.15} - \frac{1}{293.15}\right] = 0.2647$$

Water activity affected by temperature change is:

$$a_2 = a_1 e^{0.2647} = 0.391$$

Pressure effect
The final a_w affected by a pressure change is calculated from Equation 5.64:

$$\ln\frac{a_2}{a_1} = \frac{V_m}{RT_2}\left[P_2 - P_1\right] = 0.353$$

Water activity affected by pressure change is:

$$a_2 = a_1 e^{0.3527} = 0.556$$

Summing up the effects of both temperature and pressure change results in:

$\ln[a_2/a_1] = 0.2637+0.3527 = 0.616$

Thus, the final a_w, a_2, is:

$$a_2 = a_1 e^{0.616} = 0.556$$

At higher pressures, such as those occurring in an extruder, a_w can increase significantly with change in pressure even more dramatically than with a change in temperature.

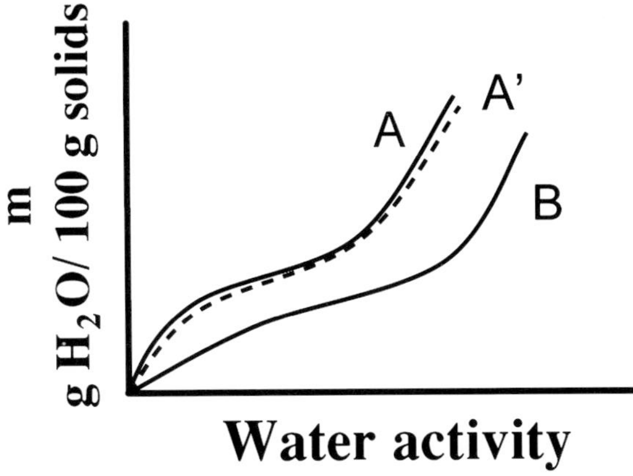

Figure 5.11 The effect of fat on moisture sorption (adapted from Labuza, T.P. 1984. *Moisture Sorption: Practical Aspects of Isotherm Measurement and Use*. St. Paul, MN: AACC International Publishing).

If the moisture isotherm as a function of temperature is known, equations for combined effects are used and the increase in a_W can be predicted assuming that the moisture content does not change. Similar to the last example, pressures in the order of 4000 psi (272 atm), which is generally used for extrusion of some semimoist foods, a dough at 0.8 a_W will increase by 1.22 times to 0.976 a_W. The higher a_W at the higher temperature of the extruder could significantly influence the reaction rates, microbial destruction rates, and product stability. Therefore the consequences of pressure and temperature change should be carefully investigated.

Food composition effect on isotherms

When working with complex food systems consisting of multidomain systems of ingredients, the effect of composition plays one of the most important roles affecting moisture sorption behavior. This section will review the influence of food composition on a_W.

Fats, among other food components, do not contribute significantly to moisture adsorption because they are generally hydrophobic. Lecithin, however, will adsorb water due to the choline and phosphate groups. The −OH groups on the glycerol will also adsorb moisture, but (1) the hydrophobic interactions between individual molecules block these groups and (2) the glycerol portion has a much lower mass than the fatty acid chain. It also is possible for the ester linkage in the acyl bond to adsorb, but, again, this is lower in weight compared with the rest of the molecule. Therefore, isotherms on a per gram nonfat solid basis do not differ significantly for a material into which fat is added. However, as a result of dilution a shift in moisture sorption curve (B) is observed, as shown in Figure 5.11. In Figure 5.11, A represents a system with no fat, e.g., a nonfat dry milk powder, and curve B represents the water adsorption with 20%. If the values of m for the fat added (B) isotherm are recalculated as g H_2O/100 g nonfat solids (A′), theoretically it should be the same as curve A.

Proteins are generally represented as type II moisture sorption isotherms due to their easily plasticized nature, resulting in increased availability of all polar groups. Heating at high temperatures generally increases adsorption.

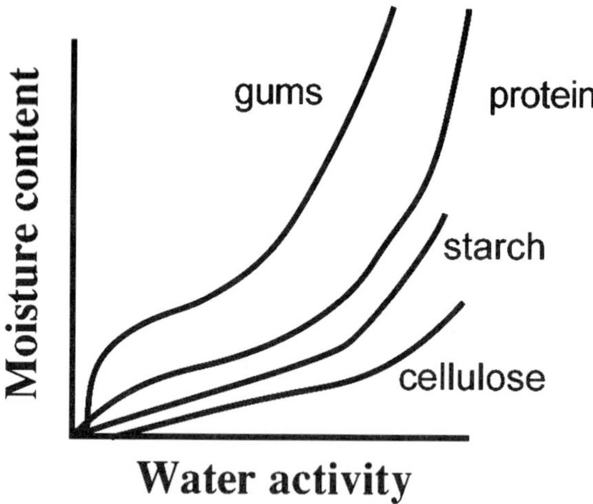

Figure 5.12 The effect of ingredient type on moisture sorption isotherms (adapted from Labuza, T.P. 1984. *Moisture Sorption: Practical Aspects of Isotherm Measurement and Use*. St. Paul, MN: AACC International Publishing).

Carbohydrates (e.g., starch and gum polymers) have more hydrogen bonds per monomer compared with proteins; however, plasticization is not easy due to their internal bonding structure. In the case of raw starch, high temperature and moisture are required for gelatinization to open up the structure to adsorption. Only surface adsorption is observed in the case of cellulose, which is crystalline in nature, whereas gums have more hydrogen bonds and more open chains, resulting in higher water-holding capacity. A general moisture sorption isotherm comparison of some carbohydrates to protein is given in Figure 5.12.

The effect of heating on the water-binding capacity of starch can be discussed based on either the plasticizing-gelatinization effect or the collapsing and/or recrystallization of the structure. Plasticizing of starch requires a high water content to gelatinize at temperatures above 60° to 65°C. In the case of high heat, such as in oil, the starch may recrystallize, or else there is a collapse of structure as in fried rice or on the surface of a potato chip. Sometimes, this is called resistant starch.

Mixtures of food components

Individual sorption isotherms can be used to estimate the equilibrium a_w in multicomponent food systems. In practice, it is assumed that the ingredients do not interact with each other in a closed system. The additive rule of mixtures is applied where the weight average isotherms are used to calculate the moisture sorption for a multicomponent food system based on the dry weight of each component (see Figure 5.13).

To create the weight average isotherm of a multicomponent mixture, the moisture content of each component at given a_w a_i is obtained; then the equation for multicomponent mixtures is used to calculate moisture content of the mixture. For a binary mixture of two components, m_A and m_B, from the isotherm at a_i, the moisture content on the line at each a_i is:

$$m_i = f_A m_A + f_B m_B = \text{g } H_20/\text{g solids} \tag{5.69}$$

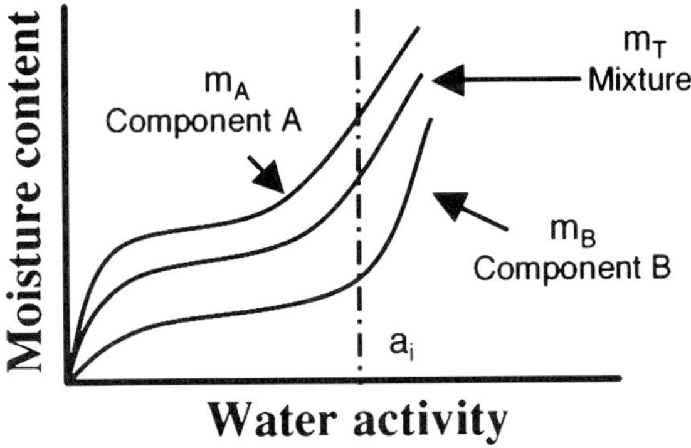

Figure 5.13 Weight average isotherms are used for multicomponent food systems (adapted from Labuza, T.P. 1984. *Moisture Sorption: Practical Aspects of Isotherm Measurement and Use*. St. Paul, MN: AACC International Publishing).

where m_A and m_B denote the moisture content, and f_A and f_B represent the weight percentage of components, respectively. The solid weight fraction of each component f is calculated as:

$$f_A = \frac{W_A}{W_T} \text{ and } f_B = \frac{W_B}{W_T} \text{ etc.} \quad (5.70)$$

where W_A and W_B represent the weight (g) of dry solids of A and B, respectively, and W_T represents the total weight of solids.

Equation 5.69 is repeated for five points on the isotherm and the GAB model is used to establish the whole isotherm. For total moisture of m_T (g water/g total A + B + X ... solids), the equilibrium a_w is read from the line produced. Note that the term $f_x m_x$ can be added to Equation 5.69 for other ingredients.

Hysteresis

As mentioned previously, the moisture sorption curve can be generated from an absorption process (starting with a dry system in which a_w is close to 0) or a desorption process (starting with a wet system in which a_w is >0.97). When food solids are exposed to conditions where vapor pressure of water is higher than vapor pressure of water in the solids, adsorption occurs. When conditions are reversed, lower vapor pressure of water in surrounding food solids becomes the driving force for desorption. Typical curves (see Figure 5.14) result in more water being held at the same a_w for the desorption curve than for the adsorption curve, above and below some closure points. At equal vapor pressure, the amount of adsorption and desorption for the same food may differ, which is referred to as sorption hysteresis. Hysteresis is in fact a thermodynamic impossibility because chemical potential or a_w is a state function, and thus the same composition and water content should always occur at a given a_w. One of the reasons for differences in moisture content between the two closure points is that, during drying (desorption), some solutes may supersaturate below

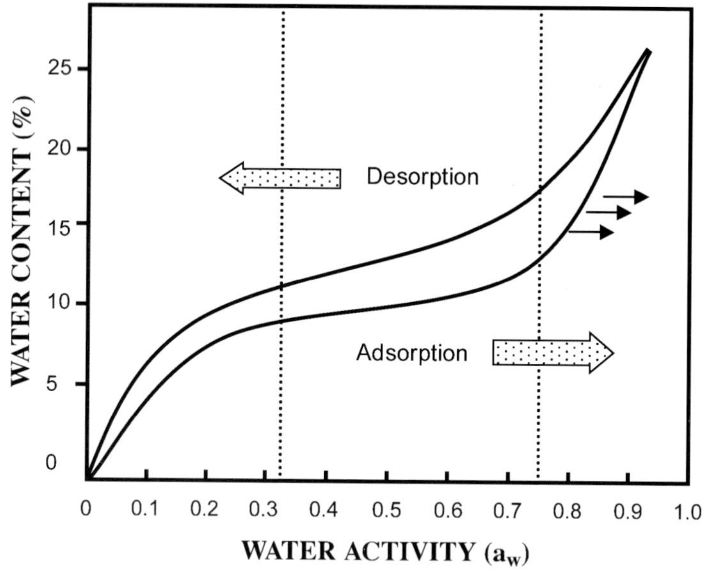

Figure 5.14 Generalized water sorption isotherm showing a hysteresis loop (adapted from www.lsbu.ac.uk/water/activity.html).

their crystallization a_W and thus hold more water as a_W is lowered. Foods with high sugar content usually exhibit this phenomenon. Second, the capillaries can empty differently upon desorption; the narrow ends of surface pores will trap and hold water internally below the a_W where it should have been released during adsorption; the narrow end prevents the body from filling. Last, the surface tension γ_s and the wetting angle Φ, from the Kelvin equation, differ between adsorption and desorption, resulting in a higher moisture content for desorption.

Hysteresis has some practical aspects. If a moist low a_W product is desired, a large desorption hysteresis would be beneficial (i.e., there would be more water at the same a_W). However, at the same a_W, the higher moisture content (desorption) also results in a greater rate of loss for some chemical reactions, reducing the shelf-life, which is undesirable (Kapsalis 1981). Extensive discussion on water sorption hysteresis is available in the literature. Acott (1976) showed this for microbial growth, whereas Chow et al. (1973) demonstrated this for lipid oxidation (Kapsalis 1981, 1987; Karel 1989).

Depending on the type of food and the temperature, a variety of hysteresis loop shapes can be observed (Wolf et al. 1972). The principal factors affecting hysteresis are composition of product, isotherm temperature, storage time before isotherm measurement, pretreatments, drying temperature, and number of successive adsorption and desorption cycles. Based on the composition of the food affecting hysteresis, three types of foods were defined: high-sugar foods, high-protein foods, and starchy foods. In high-sugar or high-pectin foods, hysteresis mainly occurs at below the monolayer region (Okos et al. 1992). In high-protein foods, the phenomenon is extended through an a_W of about 0.85 (Kapsalis 1981). Last, in the third group, starchy foods, a large hysteresis loop occurs with a maximum closure point a_W of about 0.70 (Okos et al. 1992). Overall, total hysteresis decreases

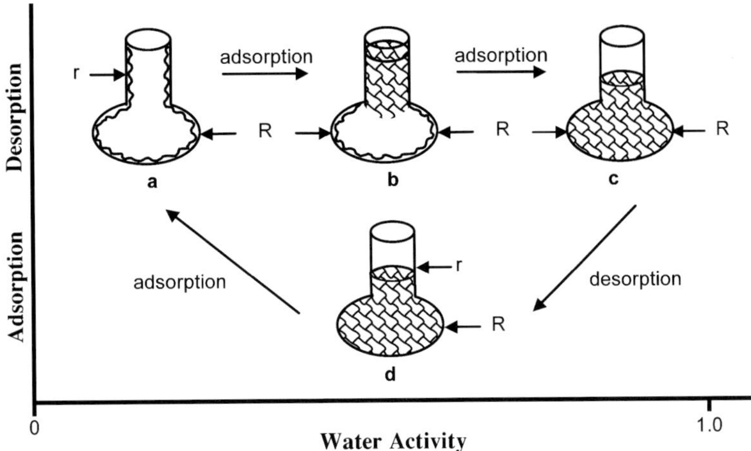

Figure 5.15 The schematic illustration of the hysteresis theory: (**a**) beginning of adsorption (very low a_W) where surface of the pore fills first, (**b**) intermediate a_W (radius *r* fills first), (**c**) high a_W (pressure fills the radius *R*), (**d**) beginning of desorption (a_W is very high) (adapted from Labuza, T.P. 1984. *Moisture Sorption: Practical Aspects of Isotherm Measurement and Use.* St. Paul, MN: AACC International Publishing).

as sorption temperature increases (Wolf et al. 1972). In general, the types of changes encountered upon adsorption and desorption will depend on the initial state of the sorbent, the transition taking place during adsorption, and the speed of desorption (Kapsalis 1981).

Theories of Sorption Hysteresis

Several theories have been proposed to explain the hysteresis phenomenon; however, no qualitative prediction is available in the literature. In general, theories for interpretation of hysteresis are based on either capillary condensation of porous solids, phase changes of non-porous solids, or structural changes of nonrigid solids (Kapsalis 1987). One of the major aspects is the condensation of water within capillaries as discussed earlier with smaller capillaries filling first on adsorption. Foods contain numerous pores or capillaries in which water exists (Bluestein and Labuza 1972, Karel 1973). The escaping tendency of water in capillaries or pores is less than that of pure water resulting in a lower a_W as well, as under the Kelvin equation. The effect of water condensed in the capillaries on hysteresis can be partially explained by the "ink-bottle theory" (McBain 1935, Rao and Das 1968). Considering the Kelvin equation, the theory is based on the shape of the capillaries, with narrow necks and large bodies, for which the radius of each dominates the rate of filling and emptying during the adsorption and desorption process. In adsorption, condensation first takes place on the surface of both the large- and small-diameter cavities. The subsequent steps are shown in Figure 5.15, where at some intermediate a_W the small capillaries fill (see Figure 5.15b). Finally, at high enough vapor pressure, the water is forced in to fill the whole system (see Figure 5.15c). On desorption, the whole system is filled (see Figure 5.15c). As one lowers a_W, the controlling factor is the small radius *r*. As the a_W is lowered (see Figure 5.15d), the system reaches an a_W in equilibrium with *r*, so the pores remained filled. Then, as the vapor pressure reaches the a_w in equilibrium with each small pore, the whole pore empties (see Figure 5.15a) except for the surface layers of pore.

As for other theories, partial sorption of chemical species, surface impurities, or phase changes can also lead to hysteresis. As a sugary material transitions from a glassy state to a rubbery state on adsorption, some sugars can crystallize and collapse may occur, both of which mean less water adsorption. Structural changes when polymeric materials swell during adsorption are also factors contributing to hysteresis. Hydration of proteins allows water to bind before desorption, while dehydrated proteins have some polar sites unavailable for water binding prior to adsorption (Cerefolini and Cerefolini 1980).

Supersaturation below crystallization a_w during desorption is a common phenomenon that occurs in some high-sugar and salt–containing solutions, causing them to hold more water due to decreased a_w.

Finally, to solve the dilemma of the same moisture at a given a_w, the above example indicates changes in state and structure so the two systems are really different materials.

Working Isotherm

Finally, there is the question as to which isotherm direction to follow, concerning a food product subjected to exposure to an external relative humidity and starting with an initial moisture content not equal to 0. In this case the initial state of the product allows for both adsorption or desorption based on difference between the external % RH and the initial a_w. The initial moisture content of the product can be either below the monolayer or somewhere between the monolayer and upper isotherm limit. It should be expected that if the initial a_w is below the external % RH (i.e., the a_w), the adsorption isotherm will be followed as it picks up moisture. However, if the initial a_w is greater than external % RH, desorption would occur but in a fashion depending on how the product was first made. If the product was first dried to below the lower closure point, and then water is added to obtain the initial value, any further gain or loss of moisture should be on the adsorption isotherm. However, if the product was initially made by drying it down to the wetted state, on rehumidification the product would begin to shift over to the adsorption isotherm. This latter phenomenon, which includes a crossover between desorption and adsorption curves, is in reality the true "working isotherm" for most fried, baked, or extruded food products, because they are usually dried to some level and then subjected to either moisture gain or loss (see Figure 5.16).

Final Remarks

Water activity prediction models and moisture sorption isotherms based on a broad perspective using fundamental concepts were reviewed in this chapter. In addition to empirical and theoretical models provided, there are various commercially available computer programs designed to predict a_w for different food systems.

Water Analyzer Series is a program that does all of these calculations and is available from T. P. Labuza (for MacIntosh computers with systems 9.2 or less). ERC CALC software offered by Campden and Chorleywood Food Research Association (United Kingdom) provides calculation for equilibrium relative humidity from product formulations, investigation of moisture migration between various food ingredients, and estimation of mold-free shelf-life of products. This program uses the Grover equation and further information about this

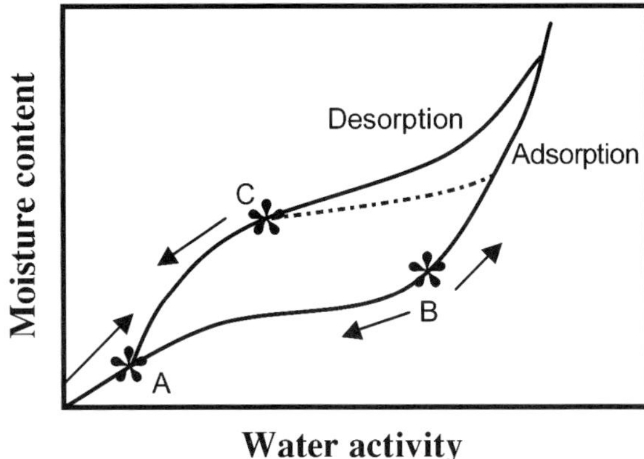

Figure 5.16 Adsorption-desorption hysteresis loop, showing paths taken by (**A**) a dry food during sorption, (**B**) a sample prepared by adsorption to an intermediate level, and (**C**) a sample prepared by desorption (from Labuza, T.P. 1984. *Moisture Sorption: Practical Aspects of Isotherm Measurement and Use*, St. Paul, MN: AACC International Publishing. Reprinted with permission).

software can be obtained from the company Website (sofa.dartnet.co.ukm). And a_W CALC software designed by American Institute of Baking (Manhattan, KS) provides calculation of a_w including estimation of mold-free shelf-life (Bell and Labuza 2000). Detailed information about this program is available online (www.aibonline.org). And as noted previously, there is an Excel spreadsheet available online for BET and GAB calculations.

References

Acott, K., and Labuza, T.P. 1976. Microbial growth response to water sorption preparation. *Food Technology (London)* 10(6):603–611.

Anderson, B.E., Shimoni, I., and Labuza, T.P. 2003. Degassing kinetics of carbon dioxide in packaging of fresh roasted ground coffee. *Journal of Food Engineering* 59(1):71–78.

Barbosa-Cánovas, G.V., and Vega-Mercado, H. 1996. Physical, chemical, and microbiological characteristics of dehydrated foods. In: *Dehydration of Foods*, ed. G.V. Barbosa-Cánovas and H. Vega-Mercado, pp. 29–95. New York: International Thomson Publishing.

Becker, H.A., and Sallans, H.R. 1956. A study of the relationship between time, temperature, moisture content, and loaf volume by the bromate formula in the heat treatment of wheat and flour. *Cereal Chemistry* 33:254–265.

Bell, L.N., and Labuza, T.P. 2000. *Moisture Sorption: Practical Aspects of Isotherms Measurements and Use*. St. Paul, MN: AACC International Publishing.

Boquet, R., Chirife, J., Iglesias, H. 1978. Equations of fitting water sorption isotherms of foods. II. Evaluation of various two parameters models. *Journal of Food Technology* 13:329–327.

Bromley, L.A. 1973. Thermodynamic properties of strong electrolytes in aqueous solutions. *AICHE Journal* 19(2):313–320.

Brunauer, S. 1945. *The Adsorption of Gases and Vapors*. Princeton, NJ: Princeton University Press.

Brunauer, S.P. Emmett, H., and Teller, E. 1938. Adsorption of gases in multimolecular layers. *Journal of the American Chemical Society* 60(2):309–319.

Cerefolini, G.F., and Cerefolini, C. 1980. Heterogenicity, allostericity, and hysteresis in adsorption of water by proteins. *Journal of Colloid and Interface Science* 78:65.

Chirife, J. 1987. Prediction of water activity in foods. In: *Food Preservation by Combined Methods Based on Lowering Water Activity*. Programa de Ciencia y Technologia para el Desarrollo, Mexico.

Chirife, J., Favetto, G., and Ferro-Fontan, C. 1982. The water activity of fructose solutions in the intermediate moisture range. *Lebensittel-Wissenschaft und-Technologie* 15(3):150–160.

Chirife, J., Favetto, G., and Scorza, O.C. 1982. The water activity of common liquid bacteriological media. *Journal of Applied Bacteriology* 53:219–222.

Chirife, J., and Ferro-Fontan, C. 1980a. Water activity of aqueous lactulose solutions. *Journal of Food Science* 45(6):1706–1707.

Chirife, J., and Ferro-Fontan, C. 1980b. A study of the water activity lowering behavior of polyethylene glycols in the intermediate moisture range. *Journal of Food Science* 45(6):1717–1719.

Chirife, J., and Fontan, C.F. 1982. Water activity of fresh foods. *Journal of Food Science* 47:661–663.

Chirife, J., Ferro-Fontan, C., and Benmergui, E.A. 1980. The prediction of water activity in aqueous solutions in connection with intermediate foods. IV. Water activity prediction in aqueous nonelectrolyte solutions. *Journal of Food Technology* 15(1):59–70.

Chirife, J., Fontan, C.F., and Scorza, O.C. 1980. A study of the water activity lowering behavior of some amino acids. *Journal of Food Technology* 15:383.

Chou, H.E., Acott, K.M., and Labuza, T.P. 1973. Sorption hysteresis and chemical reactivity: Lipid oxidation. *Journal of Food Science* 38:316–319.

Fontan, C.F., Chirife, C., and Benmergui, E.A. 1979. The prediction of water activity in aqueous solutions in connection with intermediate moisture foods. II. On the choice of the best a_w lowering single string electrolyte. *Journal of Food Technology* 14:639–646.

Glasstone, S. 1946. *Textbook of Physical Chemistry*. New York: Van Nostrand.

Glasstone, S., and Lewis, D. 1960. *Elements of Physical Chemistry*, pp. 251–253. Princeton, NJ: Van Nostrand Co., Inc.

Grover, D.W. 1947. The keeping properties of confectionary as influenced by its water vapor pressure. *J. Soc. Chem. Ind*. 66:201–205.

Iglesias, H.A., and Chirife, J. 1976. Isosteric heats of water sorption on dehydrated foods. Part I, Analysis of the differential heat curves. *Lebensittel-Wissenschaft und-Technologie* 9:116–122.

Iglesias, H.A., and Chirife, J. 1976a. BET monolayer values in dehydrated food components. *Lebensittel-Wissenschaft und-Technologie* 9:107–113.

Iglesias, H.A., and Chirife, J. 1976b. Isosteric heats of water sorption on dehydrated foods. Comparison with BET theory. *Lebensittel-Wissenschaft und-Technologie* 9:123–126.

Iglesias, H.A., and Chirife, J. 1982. *Handbook of Food Isotherms: Water Sorption Parameters for Food and Food Components*. New York: Academic Press.

Kapsalis, J.G. 1981. Moisture sorption hysteresis. In: *Water Activity: Influences on Food Quality*, eds. L.B. Rockland and G.E. Stewart, p. 143. New York: Academic Press.

Kapsalis, J.G. 1987. Influences of hysteresis and temperature on moisture sorption isotherms. In: *Water Activity: Theory and Applications to Food*, eds. L.B. Rockland and L.R. Beuchat, p. 173. New York: Marcel Dekker.

Karel, M. 1973. Recent research and development in the field of low-moisture and intermediate moisture foods. *Critical Reviews on Food Technology* 3:329.

Karel, M. 1975. Physico-chemical modifications of the state of water in foods: A speculative survey. In: *Water Relations of Foods*, ed. R.B. Duckworth, p. 639. London: Academic Press.

Karel, M. 1989. Role of water activity. In: *Food Properties and Computer-Aided Engineering of Food Processing Systems*, eds. R.P. Singh and A.G. Medina, pp. 135–155. Norwell, MA: Kluwer Academic Publishers.

Katz, E.E. and Labuza, T.P. 1981. The effect of water activity on the sensory crispness and mechanical deformation of snack food products. *Journal of Food Science* 46:403–409.

Labuza, T.P. 1968. Sorption phenomena in food. *Food Technology* 22(3):15.

Labuza, T.P. 1970. Properties of water as related to the keeping quality of foods. SOS symposium on physical and chemical properties of foods. *Proceedings of Third International Congress Food Science IFT*, pp. 618–635.

Labuza, T.P. 1971. Kinetics of lipid oxidation in foods. *Critical Reviews in Food Technology* 2:355–405.

Labuza, T.P. 1975. Sorption phenomena in foods: Theoretical and practical aspects. In: *Theory, Determination and Control of Physical Properties of Food Materials*, ed. C. Rha. Dordrecht, Holland: D. Reidel Publishing Company.

Labuza, T.P. 1978. The technology of intermediate moisture foods in the U.S. In: *Food Microbiology and Technology*. Parma, Italy: Medicina Viva Publishing.

Labuza, T.P. 1984. *Moisture Sorption: Practical Aspects of Isotherm Measurement and Use*. St. Paul, MN: AACC International Publishing.

Labuza, T.P., and Hyman, C.R. 1998. Moisture migration and control in multidomain foods. *Trends in Food Science and Technology* 9:47–55.

Labuza, T.P., Kaanane, A., and Chen, Y.J. 1985. Effect of Temperature on the moisture sorption isotherms and water activity shift of two dehydrated foods. *Journal of Food Science* 50:385.

Labuza, P.S., and Labuza, T.P. 2004. Cotton candy shelf-life. *Journal of Food Processing and Preservation* 28:274–287.

Labuza, T.P., Roe, K., Payne, C., Panda, F., Labuza, T.J., Labuza, P.S., and Krusch, L. 2004. Storage stability of dry food systems: influence of state changes during drying and storage. In: *Drying 2004,* eds. M. Silva and S. Rocha, pp. 48–68. Brazil: Ourograf Grafica Campinas.

Labuza, T.P., Tannenbaum, S.R., and Karel, M. 1970. Water content and stability of low moisture and intermediate moisture foods. *Journal of Food Technology* 24:543–550.

Lasekan, O.O., and Lasekan, W.O. 2000. Moisture sorption and the degree of starch polymer degradation on flours of popped and malted sorghum (sorghum bicolor). *Journal of Cereal Science* 31:55–61.

Leiras, M.C., Alzamora, S.M., and Chirife, J. 1990. Water activity of galactose solutions. *Journal of Food Science* 55(4):1174–1176.

Leung, H.K. 1986. Water activity and other colligative properties of foods. In: *Physical and Chemical Properties of Foods*, ed. M.R. Okos. St. Joseph, MI: American Society of Agricultural Engineers.

Lindsey, R.C. 1985. Food additives. In: *Food Chemistry*, 2nd Edition, ed. O.R. Fennema. New York: Marcel Dekker.

Lomauro, C.J., Bakshi, A.S., and Labuza, T.P. 1985a. Evaluation of food moisture sorption isotherm equations. Part I. Fruit, vegetable and meat products. *Food Science and Technology* 18:118.

Lomauro, C.J., Bakshi, A.S., and Labuza, T.P. 1985b. Evaluation of food moisture sorption isotherm equations. Part II. Milk, coffee, tea, nuts, oilseeds, spices and starchy foods. *Food Science and Technology* 18:118.

Loncin, M. 1980. Water adsorption isotherms of foods at high temperature. *Lebensittel-Wissenschaft und-Technologie* 13:182.

McBain, J.W. 1935. An explanation of hysteresis in the hydration-dehydration of gels. *Journal of the American Chemical Society* 57:699.

Miracco, J.L., Alzamora, S.M., Chirife, J., and Ferro-Fontan, C. 1981. On the water activity of lactose solutions. *Journal of Food Science* 46(5):1612–1613.

Money, R.W., and Born, R. 1951. Equilibrium humidity of sugar solutions. *J. Sci. Food Agric.* 2:180.

Norrish, R.S. 1966. An equation for the activity coefficients and equilibrium relative humidity of water in confectionary syrups. *Journal of Food Technology* 1:25–49.

Okos, M.R., Narsimhan, G., and Singh, R.K. 1992. Food dehydration. In: *Handbook of Food Engineering*, eds. R. Heldman and D.B. Lund, p. 437. New York: Marcel Dekker.

Oswin, C.R. 1946. The kinetics of package life. III The isotherm. *J. Chem. Ind. (London)* 65:419.

Payne, C., and Labuza, T.P. 2005a. The brittle-ductile transition of an amorphous food system. *Drying Technology* 23(4):1–16.

Payne, C., and Labuza, T.P. 2005b. Correlating perceived crispiness intensity to physical changes in an amorphous snack food. *Drying Technology* 23(4):17–36.

Peleg, M. 1992. On the use of the WLF model in polymers and foods. *CRC Critical Reviews on Food Science and Nutrition* 32:59–66.

Peleg, M. 1993. Assessment of a semi-empirical four parameter general model for sigmoid moisture sorption isotherms. *Journal of Food Processing and Engineering* 16(1):21.

Pitzer, K.S. 1973. Thermodynamics of electrolytes. I. Theoretical basis and general equations. *Journal of Physical Chemistry* 77(2):268–277.

Pitzer, K.S., and Mayorga, G. 1973. Thermodynamics of electrolytes. 2. Activity and osmotic coefficients for strong electrolytes with one or both ions univalent. *Journal of Physical Chemistry* 77:2300–2308.

Rahman, M.S. 2006. State diagrams of foods: Its potential use in food processing and physical stability. *Trends in Food Science and Technology* 17(3):129–141.

Rao, S.K., and Das, B. 1968. Varietal differences in gelatin, egg albumin, and casein in relation to adsorption-desorption hysteresis with water. *Journal of Physical Chemistry* 72:1233.

Robinson, R.A., and Stokes, R.H. 1956. *Electrolyte Solutions*. London: Butterworths.

Roos, Y., and Karel, M. 1991. Applying state diagrams to food processing and development. *Food Technology* 45(12):66–71.

Ross, K.D. 1975. Estimation of water activity in intermediate moisture foods. *Food Technology* 13(3):26–34.

Salwin, H. 1959. Defining minimum moisture contents for dehydrated foods. *Food Technology* 13:594–595.

Scott, W.J. 1957. Water relations of food spoilage microorganisms. *Advances in Food Research* 7:83–124.

Sherwin, C., and Labuza, T.P. 2006. Beyond water activity and glass transition: A broad perspective on the manner by which moisture can influence reaction rates in food. In: *Proceedings ISOPOW 9,* ed. P. Cano. Buenos Aires, Argentina.

Shimoni, I., and Labuza, T.P. 2000 Degassing kinetics and sorption equilibria of carbon dioxide in fresh roasted ground coffee. *Journal of Food Engineering* 23:419–436

Sloan, A.E., and Labuza, T.P. 1976. Prediction of water activity lowering ability of food humectants. *Journal of Food Science* 41:532.

Smith, S.E. 1947. The sorption of water vapor by high polymers. *Journal of American Chemical Society* 69:646.

Taoukis, P., Breene, W., and Labuza, T.P. 1988. Intermediate moisture foods. In: *Advances in Cereal Science and Technology*, Vol. IX, ed. Y. Pomeranz, pp. 91–128. St. Paul, MN: AACC Press.

Teng, T.T., and Seow, C.C. 1981. A comparative study of methods for prediction of water for multicomponent aqueous solutions. *Journal of Food Technology* 16:409–419.

Van den Berg, C., and Bruin, S. 1981. In: *Water Activity: Influence on Food Quality*, eds. L.B. Rockland and G.F. Stewart, p. 1. New York: Academic Press.

Van den Berg, C. 1985. Development of B.E.T.-like models for sorption of water on foods, theory and relevance. In: *Properties of Water in foods,* eds. D. Simatos and J.L. Multon. Dordrecht, the Netherlands: Martinus Nijhoff Publishers.

Van den Berg, C., and Bruin, S. 1981. Water activity and estimation in food systems. In: *Water Activity: Influences on Food Quality*, eds. L.B. Rockland and G.F. Stewart, pp. 1–61. New York: Academic Press.

Wolf, M., Walker, J.E., and Kapsalis, J.G. 1972. Water vapor sorption hysteresis in dehydrated food. *Journal of Agricultural and Food Chemistry* 20:1073–1077.

Wolf, W., Spiess, W.E.L., and Jung, G. 1985. Standardization of isotherm measurements. In: *Properties of Water in Foods: In Relation to Food Quality and Stability,* eds. D. Simatos and J.L. Multon, pp. 661–679. Dordrecht, the Netherlands: Martinus Nijhoff Publishers.

Young, J.H. 1976. Evaluation of models to describe sorption isotherms and desorption equilibrium moisture content isotherms of Virginia-type peanuts. *Transactions of ASAE* 19:146.

6 Measurement of Water Activity, Moisture Sorption Isotherms, and Moisture Content of Foods

Anthony J. Fontana, Jr.

This chapter describes the methods for measuring water activity (a_w), moisture sorption isotherms, and moisture content in foods. It is not intended as an extensive review but rather as an overview of the methods, along with their advantages and disadvantages, and the precautions required to make an informed selection of the appropriate technique. Reliable laboratory instrumentation is necessary for obtaining accurate and reproducible measurements and for guaranteeing the safety, quality, and shelf-life of foods. The methods vary in their accuracy, repeatability, speed of measurement, stability in calibration, linearity, and convenience of use. Methods for measuring a_w have been reviewed in detail by many authors (Troller and Christian 1978; Prior 1979; Troller 1983a; Rödel 1998, 2001; Reid et al. 2001a; Fontana and Campbell 2004).

Critical Parameters and Precautions

Obtaining the desired accuracy in an a_w measurement depends on several factors: measurement technique, calibration of the technique, temperature stability during the measurement process, vapor equilibration, and method of sample preparation. With so many factors affecting this measurement, accuracy may seem daunting. However, careful performance of the setup and measurement processes will greatly improve the chances for acceptable results.

Calibration/Validation

No matter what technique is chosen to measure a_w, instrument calibration must be verified before any samples are analyzed. Prior to reading the a_w standards, ensure that the instrument is clean and free of contamination. Most instrument manufacturers have detailed cleaning procedures for sensors and chambers that should be followed carefully. Care should be taken to not damage any surfaces in the measurement chamber, as damaged surfaces can absorb water and lead to inaccurate results. The instrument should be calibrated at the same temperature at which the sample a_w is measured.

A series of known a_w standards should be measured to ensure the accuracy of the instrument. Salt standards are excellent choices for calibration/validation. There are two types of salt standards, saturated or unsaturated, and both are easily made or acquired.

Saturated salt solutions in the form of salt slurries can be prepared by several methods (Stoloff 1978, AOAC 1995). Saturated salts are attractive because they can be made up without a precision balance. Start with reagent-grade salt and add distilled water in small increments, stirring well with a spatula after each addition, until the salt can absorb no more water as evidenced by free liquid (where it will take on the shape of the container but will not easily pour). Care must be taken to prevent moisture gain or loss by sealing the samples when not in use. When a salt standard is incorrectly prepared so that it consists mostly of liquid with a few crystals at the bottom, it can result in a layer of less-than-saturated solution at the surface, which will produce a higher reading than anticipated. Conversely, solid crystals protruding above the surface of the liquid can lower the readings.

Appendix A lists the a_w values of several saturated salts, and a more complete list is available in Greenspan (1977). Although saturated salt slushes cover most of the a_w spectrum, there are a limited number of choices. In addition, the a_w of saturated salts is highly dependent on temperature and care is needed to make proper solutions. The salt slurry should be made at, or preferably above, the temperature at which the testing is to be done because the solubility of many salts increases significantly with temperature and the excess salt may not be enough to yield a saturated solution (Bell and Labuza 2000). Most saturated salt solutions exhibit a decrease in a_w with increased temperature, because they have a negative heat of solution with increased solubility at higher temperatures. Water activities for saturated salts at temperatures other than those in Appendix A can be calculated using Equation 6.1:

$$a_w = \exp\left[\frac{k_1}{T - k_2}\right] \tag{6.1}$$

where T is temperature and k_1 and k_2 are empirical coefficients specific to each salt (Labuza et al. 1985).

Unsaturated salts are also an excellent choice for calibration or verification standards because they can be prepared to any value of a_w. Thus, if samples are measured at a certain a_w, salt standards can be prepared that verify measurement accuracy in that range. An additional benefit is that their a_w is relatively unaffected by temperature over a wide range (Chirife and Resnik 1984). Appendix B lists the a_w of various sodium chloride and lithium chloride solutions according to Robinson and Stokes (1965). The data in Robinson and Stokes can be used to compute a_w for other unsaturated salt solutions using Equation 6.2:

$$a_w = \exp(-v\phi c M_w) \tag{6.2}$$

where v is the number of ions in the salt (e.g., NaCl has two ions, Na^+ and Cl^-), ϕ is the osmotic coefficient, c is the molal concentration of salt, and M_w is the molecular mass of water. Errors are possible with improper preparation of salt solutions. To make correct calibration standards, both the solute (salt) and solution (water) must be weighed with good accuracy. The salt should be dried in an oven and cooled before use. Because unsaturated salt solutions can lose or gain water from the environment and change the solution's a_w, the salt solution must be sealed when not in use.

Temperature

The role of temperature in a_w measurement was discussed in Chapter 2. However, it is also significant in a practical sense when considering the state of the sample placed in the measurement chamber of an instrument. Samples warmer than the chamber temperature may condense water on the sensor and chamber surfaces (Reid et al. 2001a). In addition, samples that differ from the sensor temperature will have increased equilibration times unless sample temperature is measured directly.

Table 6.1 Vapor Pressure of Water as a Function of Temperature.

Temperature (°C)	P_o (kPa)
0	0.61
5	0.87
10	1.23
15	1.70
20	2.34
25	3.17
30	4.24
35	5.62
40	7.38
45	9.58
50	12.34

Source: *CRC Handbook of Chemistry and Physics*, 74th Edition. 1993. Cleveland, OH: The Chemical Rubber Company.

The saturation vapor pressure of water depends (to a good approximation) on temperature as seen in Table 6.1. Values can be computed from Equation 6.3:

$$p_o(T) = 0.611\exp\left(\frac{17.502T}{240.97 + T}\right) \tag{6.3}$$

where p_o is a function of Celsius temperature (T), so for any value of T supplied, a value of p_o is obtained. Now consider two samples, sample 1 with a vapor pressure of p_1 and sample 2 with a vapor pressure of p_2. If the samples are in vapor equilibrium with each other, $p_1 = p_2$, then from the definition it follows that $a_w = p/p_o$, $a_{w1} = p_1/p_o(T_1)$, and $a_{w2} = p_2/p_o(T_2)$. If $T_1 = T_2$ (isothermal), then $a_{w1} = a_{w2}$ because $p_o(T_1) = p_o(T_2)$. But if the samples differ in temperature, $p_o(T_1) \uparrow p_o(T_2)$, hence, $a_{w1} \uparrow a_{w2}$, even though $p_1 = p_2$. At high a_w, a temperature difference of 1°C between two samples produces an a_w difference of about 6% (Fontana and Campbell 2004). This is an extremely important consideration in the design of a_w measurement instrumentation. Without careful control or measurement of both the sample and sensor temperatures, no meaningful data can be collected (Reid 2001).

The effects just discussed relate to temperature differences between the sensor and the sample. There are, however, changes in a_w of foods with temperature just as there are in saturated salt solutions. The temperature coefficient for the a_w of foods range from 0.003 to 0.02°C^{-1} for high-carbohydrate or high-protein foods (Fennema 1985). Thus, a 10°C decrease in temperature can cause a lowering in a_w ranging from 0.03 to 0.2 a_w. Consequently, regulations often specify the temperature of a_w analysis (e.g., 25°C [AOAC 1995] and 20°C in Europe). Some a_w instruments have temperature control features to allow measurement of a_w at specific temperatures. Samples may need longer thermal equilibration time to equilibrate to the specific temperature or preequilibration of the samples is necessary to avoid the above errors.

Sample Preparation

The importance of consistency and repeatability in sample preparation cannot be overstated to obtain accurate and reproducible a_w readings. Because a_w is an intensive property, the sample only needs to be large enough to allow vapor equilibrium without changing its moisture content. Typical sample size is from 5 to 10 ml. Equipment should be used and maintained in accordance with the manufacturer's instruction manual and with good laboratory practice. If there are any concerns, the manufacturer of the instrument should be consulted.

For most foods, a_w is an important property of stability and safety. Thus, samples ranging from raw ingredients to finished goods need to be analyzed for a_w. All types of food samples (liquid, solid, powder, gel, syrup, emulsion, granule, etc.) may be measured for a_w, but they must be representative of the entire product. Many samples do not need any preparation; simply place the sample into the measurement chamber and measure the a_w. Typically, at least three samples of the product are taken to obtain a representative a_w (Voysey 1999). Care should also be taken to maintain cleanliness during sample preparation. For most commercial a_w systems, disposable sample cups are available. It is highly recommended they be used once and discarded. This practice eliminates the possibility of contamination in the sample cups, which could affect the a_w reading of the next sample. If, for some reason, cups are used for multiple samples, they should be carefully cleaned and thoroughly dried before each use. Cleanliness in all other parts of the sampling procedure is important as well, including sampling tools and storage containers (Fontana and Campbell 2004).

Water activity analysis should be performed on samples on the same day as they are collected. Any amount of exposure to ambient humidity can cause moisture loss or gain, so the time between collection and measurement should be kept to a minimum. Samples should also be collected in sealed containers to prevent moisture exchange. Often, it is assumed containers are sealed against water loss when they are not. For example, sample cups provided with some a_w measurement systems have tight-fitting plastic lids. Although the lids restrict moisture movement, they will not completely stop water loss or gain, and the sample a_w can change over time. For long-term storage, seal the plastic lid with Parafilm or tape. Polyethylene plastic bags (e.g., Whirl-Pak, Nasco, Inc.) are appropriate for short-term storage as long as the majority of the air is removed, but the product may gain or lose moisture over time. Glass jars or mylar bags should be used for long-term storage to prevent any moisture exchange with the environment. If long-term storage is required, sample containers should be stored in a constant temperature freezer with a minimum headspace to reduce water vapor and condensation effects. Samples removed from a freezer for a_w determination should be slowly warmed to ambient temperature to allow the sample to absorb any condensate (Voysey 1999).

Some precautions should be taken for multicomponent, coated, and glassy samples and/or water-in-oil emulsions to ensure a representative and homogeneous sample is tested. Multicomponent foods such as fruitcake should be sampled such that all constituents are proportionally represented. If the sample is to be sliced, crushed, or ground, then a consistent technique is necessary to ensure reproducible results. For example, if one sample is ground for 15 seconds and a second sample is ground for 1 minute, the second one will have smaller particles with more surface area and therefore a greater chance of gaining or losing moisture to the environment. In addition, there is the potential for loss of moisture from the heat generated during grinding. The grinding of meat products or cellular tissue can result in tissue maceration and may cause changes in a_w through the destruction of cellular tissue and release of intercellular water and solutes. However, if the final product contains a ground component, obviously, grinding is necessary to determine the a_w of the sample. The time between slicing, crushing, or grinding and a_w measurement should be consistent or additional precautions will be necessary to prevent moisture exchange with the room air (Fontana 2001).

Water-in-oil emulsions like butter or natural peanut butter, which may form an oil layer above the sample, contain lipids that act as barriers to vapor diffusion prolonging equilibration times dramatically. Adding glass beads to the emulsion will help decrease equilibration times by adding surface area to the sample. Other samples that require precaution include glassy products, such as hard candy, because they can absorb only small amounts of water from the air due to their physical state. Both sample types should be run in a chamber that has been dried to a_w below the sample before measurement, to ensure the sample releases water to the vapor phase to establish equilibrium.

Samples containing volatiles, such as propylene glycol, vinegar, or alcohol, may also need special measurement techniques, depending on the a_w sensor. High a_w samples obtained from fruits, vegetables, or meats and liquid samples from syrups or juices have water activities near 1.0. Therefore, they may need to be read in combination with a dry sample to ensure that the measurement sensor is not saturated (only for some measurement methods) and condensation does not adversely affect readings (Fontana and Campbell 2004).

In the case of multicomponent products, it may be necessary to separate the components or layers and measure each individually. Because a_w is a driving force for moisture migration, it is necessary to know the a_w values for the individual components or layers. For example, in a flake cereal with fruit pieces, it is important to know the individual component water activities. The a_w of cereal flakes and fruit pieces should be as close as possible to prevent moisture migration and, thus, to keep the flakes from becoming soggy and the fruit pieces hard and brittle (Fontana 2001).

In addition to equilibrium between the liquid phase water in the sample and the vapor phase, the internal moisture equilibrium of the sample is important. If a system is not at internal moisture equilibrium, one might measure a steady vapor pressure (over the period of measurement), which is not the true a_w of the system. An example of this might be a baked good or a multicomponent food. Initially, out of the oven, a baked good is not at internal equilibrium; the outer surface is at a lower a_w than the center of the baked good. One must wait a while for the water to migrate and the system to come to internal equilibrium. Other foods may be in a metastable state, like powdered milk, where transitions can occur from the amorphous to crystalline state causing a_w changes. To ensure accurate and reproducible a_w measurements, it is therefore important to remember the assumption underlying the definition of a_w of equilibrium (Fontana 2001).

Water Activity Measurement Methods

Hygrometers

Chilled mirror dew point

Chilled mirror dew point measurement is a primary method for determining vapor pressure that has been in use for decades (Harris 1995); it is based on fundamental thermodynamic principles and, as such, does not require calibration (Fontana 2001, Rödel 2001). Dew point instruments are accurate, fast, simple to use, and precise (Richard and Labuza 1990, Snavely et al. 1990, Roa and Tapia de Daza 1991, Voysey 1993). The measurement range of commercial dew point meters is 0.030 to 1.000 a_w with a resolution of ± 0.001 a_w and an accuracy of ± 0.003 a_w. Measurement time is typically less than 5 minutes. Vendors of commercial instruments based on this principle are listed in Table 6.2.

The basic principle involved in dew point determination of vapor pressure in air is that

Table 6.2 Vendor List for Commercial Instruments.

WATER ACTIVITY INSTRUMENT MANUFACTURERS

DECAGON DEVICES, INC.
950 N.E. Nelson Court
Pullman, WA 99163, USA
Phone: 800-755-2751
509-332-2756
Fax: 509-332-5158
Email: sales@decagon.com
Web: www.decagon.com

ROTRONIC INSTRUMENT CORP.
160 E. Main Street
Huntington, NY 11743, USA
Phone: 631-427-3898
Fax: 631-427-3902
Web: www.rotronic-usa.com

NOVASINA
A division of Axair Ltd.
Talstrasse 35-37
CH-8808 Pfäffikon, Switzerland
Phone: +41 55 416 66 60
Fax: +41 55 416 62 62
Web: www.novasina.com

TESTO INC.
35 Ironia Road
Flanders, NJ 07836, USA
Phone: 800-227-0729
973-252-1720
Fax: 973-252-1729
Web: www.testo.de

HAIR HYGROMETER MANUFACTURERS

LUFFT MESS-UND
Regeltechnik Gmbh
Gutenbergstrasse 20
Postfach 4252
D-70719 Fellbach, Germany
Phone: +49 (0)711/51822-0
Fax: +49 (0)711/51822-41
Email: info@Lufft.de
Web: www.lufft.de

FREEZE POINT DEPRESSION INSTRUMENT MANUFACTURERS

NAGY Messsysteme GmbH
Siedlerstr. 34
71126 Gaeufelden, Germany
Phone: +49-7032-7 66 70
Fax: +49-7032-7 21 89
Email: info@nagy-instruments.de
Web: www.NAGY-Instruments.de

ISOTHERM GENERATOR MANUFACTURERS

DVS
Surface Measurement Systems Ltd.
3 Warple Mews, Warple Way
London W3 0RF, UK
Phone: +44(0)208 749 4900
Fax: +44(0)208 749 6749
Web: www.smsuk.co.uk

HIDEN ISOCHEMA LTD.
231 Europa Boulevard
Warrington, WA5 7TN, UK
Phone: +44(0)1925 244678
Fax: +44(0)1925 244664
Email: info@hidenisochema.com
Web: www.isochema.com

TA INSTRUMENTS
109 Lukens Drive
New Castle, DE 19720, USA
Phone: 302-427-4000
Fax: 302-427-4001
Email: info@tainst.com
Web: www.tainstruments.com

RUBOTHERM
Präzisionsmeßtechnik
Universitätsstraße 142
44799 Bochum, Germany
Phone: +49 (0)234 70996-0
Fax: +49 (0)234 70996-22
Email: info@rubotherm.de
Web: www.rubotherm.de

Table 6.2 Vendor List for Commercial Instruments (*continued*).

QUANTACHROME INSTRUMENTS 1900 Corporate Drive Boynton Beach, FL 33426, USA Phone: 800-989-2476 561-731-4999 Fax: 561-732-9888 Web: www.quantachrome.com	**VTI Corporation** 7650 West 26th Avenue Hialeah, FL 33016, USA Phone: 305-828-4700 Fax: 305-828-0299 Web: www.vticorp.com
C.I. Electronics Ltd. Brunel Road Churchfields, Salisbury SP2 7PX, UK Phone: +44 (0)1722 424100 Fax: +44 (0)1722 323222 Web: www.cielec.com	

air can be cooled without a change in water content until it is saturated. Dew point temperature is the temperature at which the air reaches saturation, which is determined in practice by measuring the temperature at which condensation starts on a chilled mirror. With modern dew point instruments, a sample is equilibrated within the headspace of a sealed chamber containing a mirror, optical sensor, internal fan, and infrared thermometer. A thermoelectric (Peltier) cooler precisely controls the mirror temperature, and a thermocouple behind the mirror accurately measures the dew point temperature when condensation starts. The exact point at which condensation appears is detected using an optical reflectance sensor. This sensor emits infrared light onto the mirror, and reflected light is detected. When condensation occurs on the mirror, a change in reflectance is registered and the dew point temperature (T_d) is measured. At the same time, the sample temperature (T_s) is measured with an infrared thermometer (thermopile), and both temperatures are used to calculate a_w using Equation 6.4:

$$a_w = \frac{p_o(T_d)}{p_o(T_s)} \tag{6.4}$$

The purpose of the internal fan is for air circulation to reduce vapor equilibrium time and to control the boundary layer conductance of the mirror surface (Campbell and Lewis 1998).

Chilled mirror dew point instruments may have limitations with the use of certain volatile compounds, which can co-condense on the mirror surface with water vapor. The effect of volatiles on a dew point instrument is a function of volatile concentration and sample matrix. Instrument performance and reliability of measurements are minimally affected by the presence of low concentrations of less than 1% acetic acid or ethanol, and less than 3.5% propylene glycol. The presence of 2% acetic acid or 1.0% ethanol in foods or model systems produced statistically significant changes in a_w readings, although they are not considered "practically" important in most systems examined. Natural aromatic flavors and ground spices do not affect the dew point measurements. Thus, the chilled mirror dew point technique is an appropriate method for measuring foods with strong odors (aromatic organic volatile compounds) or foods containing low levels (i.e., up to 1%) of ethanol or acetic acid, and 3.5% propylene glycol (J. Chirife, unpublished data). Samples containing volatile compounds will not damage the sensor or instrument, but a sample cup of activated charcoal should be measured afterward to absorb and remove residual volatiles from the headspace.

Electric hygrometer

Many commercial instruments based on the electric hygrometer sensor are available for measurement of a_w. There are two types of electric hygrometers: capacitance- or resistance-type sensors. They both work on the same principle, in which a material (either a salt film or proprietary hygroscopic polymer film) changes its electrical response as a function of relative humidity. Depending on the water vapor pressure of the surrounding air, water will adsorb or desorb within the sensor and alter the electrical properties of the hygroscopic material. The sensor must be calibrated to convert the resistance or capacitance value to units of a_w. In these instruments, a sample is placed in a sealed chamber containing the sensor. The sample, air, and sensor must come to vapor and thermal (if not temperature compensated) equilibrium for accurate a_w measurements. Both types of sensors are reviewed in detail by Troller and Christian (1978), Prior (1979), Rahman (1995), Rahman and Sablani (2001), and Rödel (2001). Manufacturers of these instruments are listed in Table 6.2. Overall, these instruments measure the entire a_w range from 0 to 1.0 a_w with a resolution of ±0.001 a_w and an accuracy of between 0.01 and 0.02 a_w.

Electric hygrometers offer a reliable means for measuring a_w, provided some precautions are taken. Requirements for careful calibration, determining the equilibration time and the influence of temperature, must be considered to ensure accurate and reproducible readings. Stekelenburg and Labots (1991) found that the equilibration time required for various products to reach a constant a_w value (a change of <0.01 a_w) increased at higher a_w and mentioned that the following precautions must be taken for reliable measurement: (1) a_w value is taken when the reading (0.001 unit) has been constant for 10 minutes, (2) humidity sensors are calibrated regularly to compensate for drift, (3) a separate calibration curve is made for each sensor, and (4) sensors are calibrated at the same temperature at which samples are measured.

Difference in temperature between the sample and sensor can cause large errors. At high a_w, a 0.1°C temperature difference results in a 0.006 a_w error. If the a_w is high and sample temperature is above sensor temperature, water can condense on the sensor. Labuza et al. (1976) found that the equilibration time varied from 20 minutes to 24 hours depending on the humidity range and food material due to both thermal and vapor equilibration. Sample temperature must be known to obtain accurate a_w measurements (Rahman and Sablani 2001). Some instruments do not measure sample temperature; thus, they require careful temperature control or have long read times to allow thermal equilibration. Without careful control and measurement of temperatures in the regions occupied by the sample and the sensor (or allowing enough time for thermal equilibration), no meaningful data can be collected (Reid 2001). Product literature on modern electric hygrometers suggests short read times, although these claims have not been verified.

Recent improvements have been made to both the resistance and capacitance sensors such that several disadvantages discussed in older literature like hysteresis (differences in final a_w due to approaching equilibrium by either adsorption or desorption of water) and contamination by volatiles are less important. Mechanical and chemical filters are available to protect the sensor from contamination, although they may increase the equilibration time and do not protect against all volatile compounds (Pollio et al. 1986, Hallsworth and Nomura 1999). The chemical filters also have an a_w of their own and, depending on the type of sample, may influence the final a_w reading. The advantages and disadvantages of electric hygrometers depend on the needs of the end user. The portability of some electric hygrometers makes them well suited for at-line or in-process checks.

Hair or polymer hygrometer

It has been known for centuries that hair changes its dimensional length with atmospheric humidity. In a hair hygrometer, several strands of hair are attached between a fixed point and a lever arm to produce a deflection when relative humidity changes. More recently, braided strands of polyamide thread have replaced the hair (Jakobsen 1983, Gerschenson et al. 1984, Stroup et al. 1987, Rödel 2001). These instruments are called hair, polymer, filament, or fiber-dimensional hygrometers. Instruments using this method are listed in Table 6.2.

For measurement, the user simply places a sample in the hygrometer sample cup and seals the lid containing the built-in polymer hygrometer. Measuring a_w with this instrument requires about 3 hours of equilibration at constant temperature (Leistner and Rödel 1975), although Labuza et al. (1977) suggested longer equilibration times are necessary at high a_w values. The range of measurement is 0.3 to 1.0 a_w, with greatest accuracy between 0.3 and 0.8 a_w (Labuza 1975a). Typically, the accuracy of the method is limited to ±0.02 to 0.03 a_w, because this technique exhibits significant hysteresis, although Rödel et al. (1975) and Jakobsen (1983) reported accuracy and reproducibility of ±0.01 a_w at a constant temperature of 25°C, with sufficient sample temperature equilibration, and using an adsorptive procedure to prevent the hysteresis effect. Presencc of glycerol or other volatile substances may damage the sensitivity of the instrument during long equilibration, but they seem to be less sensitive to interfering vapors than the electric sensors (Gerschenson et al. 1984). Due to long read times and low accuracy, hair hygrometers have limited use in a_w measurement but may be useful for approximating a_w (Rahman 1995, Troller 1983a).

Thermocouple psychrometer

Thermocouple psychrometery is another method for directly determining the water vapor pressure above a food sample (Prior et al. 1977, Wiebe et al. 1981). A psychrometer measures the wet bulb temperature, which is related to the vapor pressure, using Equation 6.5:

$$p = p_o(T_w) - \gamma P_a(T_d - T_w) \tag{6.5}$$

where $p_o(T_w)$ is the saturated vapor pressure at the wet bulb temperature, γ is the psychrometer constant ($6.66 \times 10^{-4}\ C_{-1}$), P_a is the pressure of the air, and T_d and T_w are the dry bulb (ambient) and wet bulb temperatures, respectively. The wet bulb temperature is the temperature of a wet surface cooled only by the evaporation of water and should not be confused with the dew point temperature.

To measure the wet bulb temperature, a thermocouple psychrometer uses an extremely small thermocouple (a junction of two dissimilar wires producing a voltage proportional to the temperature differential). The thermocouple is positioned above a sample in a small (15 ml) sealed chamber. A small droplet of water is placed or condensed on the thermocouple junction surface. As the water evaporates, the latent heat of vaporization cools the thermocouple to the wet bulb temperature. Thus, a_w (p/p_o) is calculated using the above equation and the saturation vapor pressure (p_o) at the air temperature (see Table 6.1).

The most desirable feature of thermocouple psychrometry is its accuracy in the high range of a_w. At water activities near 1.0, the accuracy of the instrument is better than ±0.0005 a_w. The measurement range is limited to >0.93 a_w due to difficulty in condensing water on the thermocouple below this a_w level. Using a modification of the technique, it is possible to measure >0.78 a_w (Sharpe et al. 1991) and >0.55 a_w (Richards and Ogata 1958), but the accuracy decreases considerably below 0.90 a_w. Because the psychrometer measures a minute temperature difference ($T_d - T_w$), thermal gradients in the instrument due to ambient temperature fluctuations must be avoided. Measurement times are generally greater than 1 hour to allow for full thermal equilibrium. Steps have been taken recently to simplify this technique, but most users still find thermocouple psychrometry difficult, time consuming, and not as user-friendly as the electric and the dewpoint techniques (Fontana and Campbell 2004).

Isopiestic Method

The isopiestic method determines a_w of foods by relating gravimetric water content to a_w using a moisture sorption isotherm curve. A sample of known mass is stored in a closed chamber and allowed to reach equilibrium either with an atmosphere of known ERH or with a reference material of known sorption isotherm. Equilibrium between the environment and the sample is determined by weighing at intervals until constant weight is established. The method is described in detail by Troller and Christian (1978), Troller (1983a), Rahman (1995), and Sablani et al. (2001).

Water activity may be determined using the isopiestic method with either of two protocols. In the first protocol, a moisture sorption isotherm is created for the sample. Once the sorption isotherm is established for the sample, a_w can be determined from the moisture content using this isotherm curve. The isotherm curve is created using a series of controlled humidity chambers created using saturated salt standards. The water activities of saturated salt slurries as a function of temperature are listed in Appendix A. Samples are placed in the controlled chambers, allowed to equilibrate to the a_w of the salt slush, and then measured for gravimetric water content using an appropriate method (see later in chapter). Several days, or even weeks, may be required to establish equilibrium under static air conditions, but results can be hastened by evacuating the chamber and simultaneously equilibrating all relative humidity chambers. One has to be careful not to cause the salt solution to spatter during evacuation or to blow the sample out of the weighing dish when the vacuum is released. Once the isotherm curve is established for a product, the moisture content of any sample can be used to determine a_w by using the isotherm curve.

In the second protocol, the determination of a_w is accomplished by equilibrating the sample with a small amount of reference material such that the material equilibrates to the a_w of the sample. The reference material has a well-known moisture sorption isotherm. The gravimetric moisture content change of the reference material relates the a_w through the reference material's sorption isotherm curve. Examples of reference material include microcrystalline cellulose (Vos and Labuza 1974) and proteins (casein or soy isolates) (Fett 1973). Advantages of microcrystalline cellulose as a reference material are identified by Spiess and Wolf (1987b) as (1) its stable crystalline structure from −18° to 80°C, (2) its stable sorption properties after two or three repeat adsorption and desorption cycles, (3) its well-published sigmoid-shaped curve, and (4) its advantage of being readily available.

Food samples are equilibrated with the reference material usually in a vacuum desiccator for 24 to 48 hours depending on the a_w range. Comparison between this method and an electric hygrometer gave good correlations at a_w levels greater than 0.90 and superior precision at a_w levels of 0.90 or less.

The isopiestic method is attractive because it is a relatively straightforward method that is easy to set up and inexpensive. The main disadvantages are (1) lengthy equilibration times, (2) at high relative humidity, long equilibration time that can lead to microbial growth on samples and invalidation of results, (3) loss of conditioned atmosphere each time the chamber is opened to remove the sample for weighing causes delay in reaching equilibrium (Rahman 1995), and (4) differences in temperature between the sample and the standard cause large errors. Changes in the procedure have been suggested, such as allowing for continuous in situ weighing of the sample during equilibration (Spiess and Wolf 1987b), use of a fan to circulate air in the chamber, or increasing the surface area of the sample by slicing or grinding.

Vapor Pressure Manometer

Manometers determine a_w by directly measuring the vapor pressure above a food. The vapor pressure manometer method is reviewed in detail by Rahman (1995), Rizvi (1995), and Rahman et al. (2001). Makower and Myers (1943) were the first to describe the method and apparatus for foods. Method, apparatus, and design improvements are described by Taylor (1961), Sood and Heldman (1974), Lewicki et al. (1978), Nunes et al. (1985), Lewicki (1987), Lewicki (1989), and Benado and Rizvi (1987). Design and setup of a water vapor pressure capacitance manometer are discussed by Troller (1983b) and Zanoni et al. (1999).

The a_w of a sample after pressure and thermal equilibrium is calculated using Equation 6.6 (Lewicki et al. 1978):

$$a_w = \frac{(\Delta h_1 - \Delta h_2)}{P_w^v} \rho g \tag{6.6}$$

where P_w^v is the vapor pressure of pure water (in Pa), ρ is the density (kg/m^3) of manometric oil, g is the acceleration due to gravity (m/sec^2), and h_1 and h_2 are height readings of manometric oil (m). The accuracy of this method is ±0.005 a_w for $a_w < 0.85$ (Acott and Labuza 1975) and decreases to no better than ±0.02a_w at higher a_w values (Labuza et al. 1976). Temperature control is a critical factor in accurate a_w determinations. Lewicki et al. (1978) showed that the accuracy of the measurement can be improved by accounting for differences between the sample and headspace temperature. Nunes et al. (1985) reported an accuracy of ±0.009a_w in the 0.75 to 0.97 a_w range. For increased precision, Rizvi (1986) suggests maintaining the following conditions: (1) constant temperature for the whole system, (2) a ratio of sample volume to vapor space volume large enough to minimize changes in a_w due to loss of water by vaporization, and (3) a low density and low vapor pressure oil used as manometric fluid.

The vapor pressure manometer is not suitable for samples containing volatiles. Volatiles will contribute to the pressure above a food sample and contaminate the manometric fluid changing its properties with time. Respiration from high numbers of bacteria or mold prevents vapor pressure equilibrium. Manometric systems can be made useless by living products that respire, such as grains or nuts, by active fermentation, or by products that expand

Table 6.3 Water Activity in the Frozen State.

Temperature (°C)	p_{ice} (kPa)	p_{SCW} (kPa)	Water Activity
0	0.6104	0.6104	1.00
–5	0.4016	0.4216	0.953
–10	0.2599	0.2865	0.907
–15	0.1654	0.1914	0.864
–20	0.1034	0.1254	0.82
–25	0.0635	0.0806	0.79
–30	0.0381	0.0509	0.75
–40	0.0129	0.0189	0.68
–50	0.0039	0.0064	0.62

p_{ice} and $p_{supercooled\ water}$ are the vapor pressures above ice and supercooled water (scw), respectively, at the same temperature.
Source: CRC Handbook of Chemistry and Physics, 74th Edition. 1993. Cleveland, OH: The Chemical Rubber Company.

excessively when subjected to high vacuum (AOAC 1995). This method is considered a standard method but can only be used in the laboratory (Rahman 1995).

Freezing Point Depression

When an ice phase is present in foods, the ice exerts a vapor pressure that depends only on the temperature of the ice. In the frozen state, a_w is measured according to Equation 6.7:

$$a_w = \frac{p_{ice}}{p_{supercooled\ water}} \tag{6.7}$$

where p_{ice} and $p_{supercooled\ water}$ are the vapor pressures above ice and supercooled water, respectively, at the same temperature. Table 6.3 lists the vapor pressures above ice and supercooled water (CRC Handbook 1993) and a_w at various temperatures.

The point at which water freezes within a food product is based on the colligative properties (Raoult's law) of the sample. The freezing point corresponds to the temperature at which the vapor pressures of the solid and liquid phases of water are the same. The composition and concentration of solutes within a food determine the temperature at which water freezes, but the a_w of a frozen product is determined by the temperature. This method is described in Troller and Christian (1978), Rahman (1995), and Rödel (2001).

Although freezing point depression analysis is simple, precise, and unaffected by volatiles, it is limited to solutions and products with $a_w > 0.80$. The method has been used successfully under experimental conditions by some authors (Strong et al. 1970, Fontan and Chirife 1981, Rey and Labuza 1981, Lerici et al. 1983, Alzamora et al. 1994). In theory, the technique is relatively easy, but actually capturing the point at which the water freezes is rather difficult. Ice formation requires nucleation and water will supercool beyond its freezing point. The freezing point is found as the latent heat of freezing warms the sample back to the freezing point. Other difficulties include accurate temperature measurement for the calculation of a_w and the recording of the exact temperature at which the peak occurs. In addition, the sample size is small which makes choosing a representative sample somewhat difficult. It is also important to note that the a_w is at the freezing temperature and must be corrected to obtain the a_w at ambient temperatures (Fontana and Campbell 2004).

Commercial instruments based on freezing point depression that measure a_w are found in Table 6.2. The measuring range is between 0.8 and 1.0 a_w with an accuracy of up to

±0.001 a_w and measurement time is from 5 to 20 minutes. It is specifically designed for meat and meat product samples.

Moisture Sorption Isotherm Determination

The moisture sorption isotherm of a food is obtained from the equilibrium moisture contents determined at several a_w levels at constant temperature. The theory, use, and application of sorption isotherms are described in Chapter 5. A detailed review of sorption isotherm determination is found in Labuza (1975b), Wolf et al. (1985), Spiess and Wolf (1987a), and Bell and Labuza (2000). The moisture sorption isotherm is created in one of two ways: (1) food samples that are either dried (absorption), hydrated (desorption), or native (working) are placed in controlled humidity chambers at constant temperature and the weight is measured until equilibrium (constant weight) or (2) a series of samples with varied moisture contents are established by adding or removing moisture; then, a_w and water content are measured.

Controlled humidity chambers are established for the first method using saturated solutions of various inorganic salts (Appendix A), organic acids, bases, or alcohols to produce a constant vapor pressure in the atmosphere at constant temperature. Isotherms by definition are obtained at constant temperature (±1°C). Laboratory temperature is generally not constant enough because of changes during evenings and weekends. Temperature control and other precautions as described earlier are necessary when dealing with saturated salt solutions. Constant humidity can also be established through use of humidity generators. One needs six to nine different a_w levels (five minimum) to establish an isotherm. Several types of containers (desiccators, glass jars, or fish tanks) can be used for holding the samples at constant a_w. Triplicate samples weighed to ±0.0001 g should be used if enough space is available.

When the water vapor concentration in the headspace of the vacuum or air chamber is adequate, the gain or loss of moisture from the sample will depend primarily on the internal resistance to moisture change and heat transfer from the sample. If the amount of sample begins to approach the total air volume, then the vapor pressure of water in the space cannot be maintained adequately by evaporation from the surface of the saturated salt solution. Generally, a large ratio (>10:1) of salt slurry surface area to sample surface area and an air volume to sample volume ratio of 20:1 will prevent this problem (Bell and Labuza 2000). A general policy of a maximum of 3 weeks for any one study can be set unless there is a problem of recrystallization or nonequilibrium at high a_w.

The other method used to create a moisture sorption isotherm involves measuring a_w. This method is acceptable as long as the a_w device is properly calibrated and is sensitive enough for the entire a_w range (Bell and Labuza 2000). In this method, a set of samples of varying moisture content are prepared. Dry samples are moistened with water or placed in a sealed container over water while wet samples are dried down or equilibrated over desiccant. The a_w and moisture content are then measured by some appropriate method. Moisture content can be determined from the change in weight if the original moisture content is known. This method readily lends itself to doing both the absorption and desorption isotherms and allows a_w at different temperatures for constant moisture to determine isotherms as a function of temperature. Commercial systems (see Table 6.2) are available for determining the moisture sorption isotherms of small samples. These devices continuously monitor weight changes as humidified air passes over a sample.

Moisture Content Determination

Moisture content determination is an important analytical measurement in food research, manufacturing, and quality control. Knowledge of both moisture content and a_w gives the whole water story within a food. In addition, if these are known at different levels, i.e., the moisture sorption isotherm, many aspects of shelf-life, quality, and safety can be inferred. Methodologies of moisture determination have been reviewed by many authors (Pande 1974, 1975; Karmas 1980; Park 1996; Reid et al. 2001b).

Methods for measurement of moisture content in foods are classified into two groups: direct and indirect methods. In direct methods, moisture is removed by drying, distillation, extraction, or other physicochemical techniques and is measured gravimetrically or by any other direct means. Direct methods include oven (convection or vacuum) drying, thermogravimetric analysis, Karl Fischer titration, freeze-drying, azeotropic distillation, refractometry, and gas chromatography. Indirect methods do not remove water from a sample but rather measure some property or parameter of the water within the sample. These methods include spectroscopic (infrared, near infrared, NMR), dielectric capacitance, microwave absorption, sonic or ultrasonic absorption, and AC and DC conductivity.

These two categories of techniques have advantages and disadvantages. Direct methods are widely used for laboratory analysis and have high accuracy and even absolute values but are generally time-consuming and require manual operations. Indirect methods are dependent on the results of direct measurements, against which they are calibrated. They are faster and lend themselves to continuous and automated measurement in industrial processes.

Gravimetric or loss-on-drying methods are the most common and easiest procedures for analysis of moisture content. As an example, approximately 29 of the 35 moisture methods recognized by the Association of Official Analytical Chemists (AOAC) for nutritional labeling are gravimetric methods using some sort of drying oven (Sullivan and Carpenter 1993). Moisture content is determined by drying to constant weight in an oven. Moisture is calculated as the percent loss in weight after drying using Equation 6.8 for wet basis and Equation 6.9 for dry basis moisture content:

$$\%\,\text{Moisture}_{\text{Wet Basis}} = \frac{\text{Wet wt.} - \text{Dry wt.}}{\text{Wet wt.}} \times 100 \tag{6.8}$$

or

$$\%\,\text{Moisture}_{\text{Dry Basis}} = \frac{\text{Wet wt.} - \text{Dry wt.}}{\text{Dry wt.}} \times 100 \tag{6.9}$$

To obtain the true moisture content, it is important to be aware of factors involved in the gravimetric technique. Factors pertaining to sample weighing include balance accuracy, ambient relative humidity, weighing time with exposure to atmosphere, handling of drying dishes, sample homogeneity, and sample spillage. Factors pertaining to oven conditions include temperature, temperature gradients, pressure, air velocity, and oven humidity. Factors pertaining to sample drying include drying time, size and shape of sample and container, loss of other volatile material, chemical decomposition, scorching, and oven capacity. Factors pertaining to postdrying operations include loss of dry sample fines, efficiency of desiccator, and balance buoyancy effect. Some factors such as oven temperature and pressure are more important than others, and some factors such as ambient relative humidity are less important (Karmas 1980).

References

Acott, K.M., and Labuza, T.P. 1975. Inhibition of *Aspergillus* in an intermediate moisture food system. *Journal of Food Science* 40:137–139.

Alzamora, S.M., Chirife, J., and Gerschenson, L.N. 1994. Determination and correlation of the water activity of propylene glycol solutions. *Food Research International* 27:65–67.

Association of Official Analytical Chemists (AOAC). 1995. *Official Methods of Analysis of AOAC International.* Ed. T.R. Mulvaney. Arlington, VA: AOAC International. pp. 42–1 to 42–2.

Bell, L.N., and Labuza, T.P. 2000. *Moisture Sorption: Practical Aspects of Isotherm Measurement and Use,* 2nd Edition. St. Paul: American Association of Cereal Chemists.

Benado, A.L., and Rizvi, S.S.H. 1987. Water activity calculation by direct measurement of vapor pressure. *Journal of Food Science* 52:429–432.

Campbell, G.S., and Lewis, D.P., inventors. 1988. Decagon Devices, Inc., assignee. 1998 Oct 6. Water activity and dew point temperature measuring apparatus and method. U.S. patent 5,816,704.

Chirife, J., and Resnik, S.L. 1984. Unsaturated solutions of sodium chloride as reference sources of water activity at various temperatures. *Journal of Food Science* 49:1486–1488.

CRC Handbook of Chemistry and Physics, 74th Edition. 1993. Cleveland, OH: The Chemical Rubber Company.

Fennema, O.R. 1985. Water and ice. In: *Food Chemistry,* 2nd Edition, ed. O.R. Fennema, pp. 23–67. New York: Marcel Dekker.

Fett, H.M. 1973. Water activity determination in foods in the range 0.80 to 0.99. *Journal of Food Science* 38:1097–1098.

Fontan, C.F., and Chirife, J. 1981. The evaluation of water activity in aqueous solutions from freezing point depression. *Journal of Food Technology* 16:21.

Fontana, A.J. 2001. Dewpoint method for the determination of water activity. In: *Current Protocols in Food Analytical Chemistry,* ed. R.E. Wrolstad, pp. A2.2.1 to A2.2.10. New York: John Wiley & Sons.

Fontana, A.J., and Campbell, C.S. 2004. Water activity. In: *Handbook of Food Analysis*, 2nd Edition, ed. L.M.L. Nollet, pp. 39–54. New York: Marcel Dekker.

Gerschenson, L., Favetto, G., and Chirife, J. 1984. Influence of organic volatiles during water activity measurement with a fiber-dimensional hygrometer. *Lebensittel-Wissenschaft und-Technologie* 17:342–344.

Greenspan, L. 1977. Humidity fixed points of binary saturated aqueous solutions. *Journal of Research of the National Bureau of Standards: A Physics and Chemistry* 81A:89–96.

Hallsworth, J.E. and Nomura, Y. 1999. A simple method to determine the water activity of ethanol-containing samples. *Biotechnology and Bioengineering* 62:242–245.

Harris, G.A. 1995. Food water relations. *Food Technology Europe 2* Dec 1995/Jan 1996:96–98.

Jakobsen, M. 1983. Filament hygrometer for water activity measurement interlaboratory evaluation. *Journal of the Association of Official Analytical Chemists* 66:1106–1111.

Karmas, E. 1980. Techniques for measurement of moisture content of foods. *Food Technology* 34:52.

Labuza, T.P. 1975a. Interpretation of sorption data in relation to the state of constituent water. In: *Water Relations of Foods*, ed. R.B. Duckworth, pp. 155–172. New York: Academic Press.

Labuza, T.P. 1975b. Sorption phenomena in foods: Theoretical and practical aspects. In: *Theory, Determination and Control of Physical Properties of Food Materials*, ed. C. Rha, pp. 197–219. Dordrecht, the Netherlands: D. Reidel Publishing Co.

Labuza, T.P., Acott, K., Tatini, S.R., Lee, R.Y., Flink, J., and McCall, W. 1976. Water activity determination: A collaborative study of different methods. *Journal of Food Science* 41:910–917.

Labuza, T.P., Kreisman, L.N., Heinz, C.A., and Lewicki, P.P. 1977. Evaluation of the Abbeon cup analyzer compared to the VPM and Fett-Vos methods for water activity measurement. *Journal of Food Processing and Preservation* 1:31–41.

Labuza, T.P., Kaanane, A., and Chen, J.Y. 1985. Effect of temperature on the moisture sorption isotherms and water activity shift of two dehydrated foods. *Journal of Food Science* 50:385–391.

Leistner, L., and Rödel, W. 1975. The significance of water activity for micro-organisms in meats. In: *Water Relations of Foods*, ed. R.B. Duckworth, pp. 309–323. London: Academic Press.

Lerici, C.R., Piva, M., and Dalla, R.M. 1983. Water activity and freezing point depression of aqueous solutions and liquid foods. *Journal of Food Science* 48:1667–1669.

Lewicki, P.P. 1987. Design of water activity vapor pressure manometer. *Journal of Food Engineering* 6:405–422.

Lewicki, P.P. 1989. Measurement of water activity of saturated salt solutions with the vapor pressure manometer. *Journal of Food Engineering* 10:39–55.

Lewicki, P.P., Busk, G.C., Peterson, P.L., and Labuza, T.P. 1978. Determination of factors controlling accurate measurement of aw by the vapor pressure manometric technique. *Journal of Food Science* 43:244–246.

Makower, B., and Myers, S. 1943. A new method for the determination of moisture in dehydrated vegetables. Proceedings of Institute of Food Technologists, 4th Conference 156.

Nunes, R.V., Urbicain, M.J., and Rotstein, E. 1985. Improving accuracy and precision of water activity measurement with a water vapor pressure manometer. *Journal of Food Sci*ence 50:148–149.

Pande, A. 1974. *Handbook of Moisture Determination and Control.* Vol. 1. New York: Marcel Dekker.

Pande, A. 1975. *Handbook of Moisture Determination and Control.* Vol. 2–4. New York: Marcel Dekker.

Park, Y.W. 1996. Determination of moisture and ash contents of food. In: *Handbook of Food Analysis*, ed. L.M.L. Nollet, pp. 59–92. New York: Marcel Dekker.

Pollio, M.L., Kitic, D., Favetto, G., and Chirife, J. 1986. Effectiveness of available filters for an electric hygrometer for measurement of water activity in the food industry. *Journal of Food Science* 51:1358–1359.

Prior, B.A., Casaleggio, C., and vanVuuren, H.J.J. 1977. Psychrometric determination of water activity in the high a_w range. *Journal of Food Protein* 40:537–539.

Prior, B.A. 1979. Measurement of water activity in foods: A review. *Journal of Food Protein* 42:668–674.

Rahman, M.S. 1995. Water activity and sorption properties of food. In: *Food Properties Handbook*, ed. M.S. Rahman, pp. 1–86. Baca Raton, FL: CRC Press.

Rahman, M.S., and Sablani, S.S. 2001. Measurement of water activity by electronic sensors. In: *Current Protocols in Food Analytical Chemistry,* ed. R.E. Wrolstad, pp. A2.5.1 to A2.5.4. New York: John Wiley & Sons, Inc.

Rahman, M.S., Sablani, S.S., Guizani, N., Labuza, T.P., and Lewicki, P.P. 2001. Direct manometric determination of vapor pressure. In: *Current Protocols in Food Analytical Chemistry*, ed. R.E. Wrolstad, pp. A2.4.1 to A2.4.6. New York: John Wiley & Sons, Inc.

Reid, D.S. 2001. Factors to consider when estimating water vapor pressure. In: *Current Protocols in Food Analytical Chemistry*, ed. R.E. Wrolstad, pp. A2.1.1 to A2.1.3. New York: John Wiley & Sons, Inc.

Reid, D.S., Fontana, A.J., Rahman, M.S., Sablani, S.S., Labuza, T.P., Guizani, N., and Lewicki, P.P. 2001a. Vapor pressure measurements of water. In: *Current Protocols In Food Analytical Chemistry*, ed. R.E. Wrolstad, pp. A2.1.1 to A2.5.4. New York: John Wiley & Sons, Inc.

Reid, D.S., Ruiz, R.P., Chinachoti, P., and Krygsman, P.H. 2001b. Gravimetric measurements of water. In: *Current Protocols in Food Analytical Chemistry*, ed. R.E. Wrolstad, pp. A1.1.1 to A1.3.11. New York: John Wiley & Sons, Inc.

Rey, D.K., and Labuza, T.P. 1981. Characterization of the effect of solute in water-binding and gel strength properties of carrageenan. *Journal of Food Science* 46:786–789.

Richard, J., and Labuza, T.P. 1990. Rapid determination of the water activity of some reference solutions, culture media and cheese using a new dew point apparatus. *Sciences Des Aliments* 10:57–64.

Richards, L.A., and Ogata, G. 1958. Thermocouple for vapor pressure measurement in biological and soil systems at high humidity. *Science* 1089–1090.

Rizvi, S.S.H. 1986. Thermodynamic properties of foods in dehydration. In: *Engineering Properties of Foods*, eds. M.A. Rao and S.S.H. Rizvi. New York: Marcel Dekker, Inc.

Rizvi, S.S.H. 1995. Thermodynamic properties of foods in dehydration. In: *Engineering Properties of Foods,* 2nd Edition, eds. M.A. Rao and S.S.H. Rizvi, p. 123. New York: Marcel Dekker, Inc.

Roa, V., and Tapia de Daza, M.S. 1991. Evaluation of water activity measurements with a dew point electronic humidity meter. *Lebensittel-Wissenschaft und-Technologie* 24:208–213.

Robinson, R.A., and Stokes, R.H. 1965. *Electrolyte Solutions; The Measurement and Interpretation of Conductance, Chemical Potential, and Diffusion in Solutions of Simple Electrolytes*. London: Butterworth.

Rödel, W., Ponert, H., and Leistner, L. 1975. Verbesserter aw-Wert-Messer zur Bestimmung der Wasseraktivität (aw-Wert) von Fleisch und Fleischwaren. *Fleischwirtschaft* 55:557–558.

Rödel, W. 1998. Water activity and its measurement in food. In: *Instrumentation and Sensors for the Food Industry*, ed. E. Kress-Rogers, pp. 375–415. Cambridge: Woodhead Publishing Limited.

Rödel, W. 2001. Water activity and its measurement in food. In: *Instrumentation and Sensors for the Food Industry*, 2nd Edition, E. Kress-Rogers and C.B. Brimelow, pp. 453–483. Boca Raton, FL: CRC Press.

Sablani, S.S., Rahman, M.S., and Labuza, T.P. 2001. Measurement of water activity using isopiestic method. In: *Current Protocols In Food Analytical Chemistry*, ed. R.E. Wrolstad, pp. A2.3.1 to A2.3.10. New York: John Wiley & Sons, Inc.

Sharpe, A.N., Diotte, M.P., and Dudas, I. 1991. Water activity tester suited to compliance and high Aw work. *Journal of Food Protein* 54:277–282.

Snavely, M.J., Price, J.C., and Jun, H.W. 1990. A comparison of three equilibrium relative humidity measuring devices. *Drug Dev. Ind. Pharm.* 16:1399–1409.

Sood, V.C., and Heldman, D.R. 1974. Analysis of a vapor pressure manometer for measurement of water activity in nonfat dry milk. *Journal of Food Science* 39:1011–1013.

Spiess, W.E.L., and Wolf, W. 1987a. Critical evaluation of methods to determine moisture sorption isotherms. In: *Water Activity: Theory and Applications to Food*, eds. L.B. Rockland and L.R. Beuchat, pp. 215–233. New York: Marcel Dekker, Inc.

Spiess, W.E.L., and Wolf, W. 1987b. Water activity. In: *Water Activity: Theory and Applications to Food*, eds. L.B. Rockland and L.R. Beuchat, pp. 215–233. New York: Marcel Dekker.

Stekelenburg, F.K., and Labots, H. 1991. Measurement of water activity with an electric hygrometer. *International Journal Of Food Science and Technology* 26:111–116.

Stoloff, L. 1978. Calibration of water activity measuring instruments and devices: Collaborative study. *Journal of the Association of Official Analytical Chemists* 61:1166–1178.

Strong, D.H., Foster, E.M., and Duncan, C.L. 1970. Influence of water activity on the growth of *Clostridium perfrigens*. *Applied Microbiology* 19:980–987.

Stroup, W.H., Peeler, J.T., and Smith, K. 1987. Evaluation of precision estimates for fiber-dimensional and electrical hygrometers for water activity determinations. *Journal of the Association of Official Analytical Chemists* 70:955–957.

Sullivan, D.M., and Carpenter, D.E. 1993. *Methods of Analysis for Nutritional Labeling*. Arlington, VA: Association of Official Analytical Chemists (AOAC) International.

Taylor, A.A. 1961. Determination of moisture equilibria in dehydrated foods. *Food Tech.* 15:536–540.

Troller, J.A. 1983a. Methods to measure water activity. *J. Food Prot.* 46:129–134.

Troller, J.A. 1983b. Water activity measurements with a capacitance manometer. *J. Food Sci.* 48:739–741.

Troller, J.A. and Christian, J.H.B. 1978. Methods. *Water Activity and Food.* New York: Academic Press. pp. 13–47.

Vos, P.T. and Labuza, T.P. 1974. Technique for measurements of water activity in the high aw range. *J. Agric. Food Chem.* 22:326–327.

Voysey, P. 1993. An evaluation of the AquaLab CX-2 system for measuring water activity. *F M B R A Digest* 124:24–25.

Voysey, P.A. 1999. *Guidelines for the measurement of water activity and ERH in foods*. Gloucestershire: Campden & Chorleywood Food Research Association Group.

Wiebe, H.H., Kidambi, R.N., Richardson, G.H., and Ernstrom, C.A. 1981. A rapid psychrometric procedure for water activity measurement of foods in the intermediate moisture range. *J. Food Prot.* 44:892–895.

Wolf, W., Spiess. W.E.L., and Jung, G. 1985. Standardization of isotherm measurements. In: *Properties of Water in Foods*, eds. D. Simato and J.L. Multon, pp. 661–679. Dordrecht, the Netherlands: Martinus Nijhoff Publishers.

Zanoni, B., Peri, C., Giovanelli, G., and Pagliarini, E. 1999. Design and setting up of a water vapour pressure capacitance manometer for measurement of water activity. *J. Food Eng.* 38:407–423.

7 Moisture Effects on Food's Chemical Stability

Leonard N. Bell

The term *stability* refers to the ability of a substance to resist change over a specific period of time. For a food product, the length of time it remains acceptable (i.e., undergoes minimal changes) is known as its *storage stability* or *shelf-life.* A food product should be stable with respect to several modes of deterioration during distribution and storage so that its acceptability is maintained until consumption.

Shelf-life can be reduced or lost due to microbial, physical, and chemical changes in food. Microbial growth within a food product can cause spoilage, which is characterized by the development of sour flavors and off-aromas, as well as textural deterioration. The growth of pathogenic, or food-borne-illness-causing, microorganisms will also make the food unsafe. Thus, to optimize food shelf-life, the growth of both spoilage and pathogenic microorganisms should be prevented. In addition to microbial stability, the physical properties of a food should be preserved. Both the hardening of a soft food product (e.g., raisins) and the softening of a crunchy food product (e.g., ready-to-eat cereals) cause a reduction in shelf-life due to an unacceptable physical change. A third type of stability that affects the shelf-life of food is chemical stability. The chemical modification of food ingredients during storage can have deleterious effects on the flavor, color, and nutritional value of the food. For example, ascorbic acid degradation in orange juice causes a reduction in nutritional quality and can also cause undesirable darkening of the product. Understanding the chemical changes that occur in a food is a necessary component of maintaining and predicting its shelf-life.

Numerous types of chemical reactions can occur in foods, resulting in product deterioration over time. Many ingredients are susceptible to hydrolysis, or cleavage, with the addition of water. For example, the acidic environment of carbonated beverages promotes aspartame hydrolysis and consequently a loss of sweetness. Oxidation is another frequently encountered reaction type. Unsaturated pigments, fatty acids, flavors, and ascorbic acid can all be oxidized, yielding changes in the appearance, sensory characteristics, and nutritional value of the food. Carbonyl-amine reactions (i.e., Maillard-type reactions or nonenzymatic browning) cause changes in food color, flavor, and protein availability. Darkening of evaporated milk is an example of the Maillard reaction. Enzymes catalyze reactions that normally occur very slowly in food; desirable food ingredients may be lost or undesirable reaction products may be formed. Browning of fresh-cut apples and bananas is due to enzymatic activity. Depending on the food, one or more of these reaction types can occur during storage, causing a decrease in product quality and loss of shelf-life. Knowledge about the types of reactions occurring in food products provides a better understanding for controlling those reactions and extending food shelf-life.

Reaction Kinetics

Modeling Chemical Stability Data

Before attempting to control chemical reactivity in foods and predict product shelf-life, collection of chemical stability data under various environmental conditions (e.g., various temperatures, relative humidities) is required. Stability data will consist of some measurable attribute (e.g., vitamin concentration, color change), which is quantified as a function of time. Thus, the rate of change data, d[A]/dt, is obtained mathematically as follows:

$$\frac{d[A]}{dt} = k[A]^n \tag{7.1}$$

where [A] is the amount of the attribute being measured, n is the molecularity of the reaction, and *k* is the rate constant. The study of chemical reaction rates and the factors that influence the rates is known as reaction kinetics.

Strictly speaking, the molecularity comes from knowing the reaction mechanism. However, the exact mechanisms of most food-related reactions are unknown and a suitable value for n is selected in an empirical manner. Food chemical stability data are most commonly modeled by setting n equal to either 0 or 1, depending on which model gives the best linear fit (Labuza and Kamman 1983, Labuza 1984). When setting n to 0 gives the best linear fit, the reaction is said to behave as a pseudo zero order reaction. If the best fit is obtained from setting n to 1, then the reaction is modeled using pseudo first order kinetics. Additional information about molecularity and the order of reactions can be found in chapters on reaction kinetics in most physical chemistry textbooks (Alberty 1987, Levine 1988).

When n is replaced by 0 in Equation 7.1, integrating the resulting equation yields the following pseudo zero order kinetic equation:

$$[A] = [A]_o - k_{obs}t \tag{7.2}$$

where [A] is the amount of some attribute measured at time, t, $[A]_o$ is the amount of the attribute initially when t = 0, and k_{obs} is the pseudo zero order rate constant. The units of the zero order rate constant are concentration per time (e.g., mmol/day). The minus sign in Equation 7.2 represents a decrease in amount of attribute, A. If the amount of attribute A was increasing, the minus sign would be replaced by a plus sign. Based on Equation 7.2, a linear plot of [A] as a function of time should give a straight line with a slope equal to the rate constant. For example, Figure 7.1a shows a pseudo zero order kinetic plot of the formation of brown pigmentation due to a carbonyl-amine type reaction involving fructose and glycine at pH 3.5 and 60°C (Reyes et al. 1982). After a short lag period, the amount of browning increases linearly as a function of time, with a pseudo zero order rate constant of 4.8 O.D./day. A larger rate constant would indicate a faster extent of browning. Thus, rate constants are used as a measure of reactivity and in determining the amount of product change over a certain time period.

When a reaction follows pseudo first order kinetics (n=1), integration of Equation 7.1 results in:

$$\ln[A] - \ln[A]_o = -k_{obs}t \tag{7.3}$$

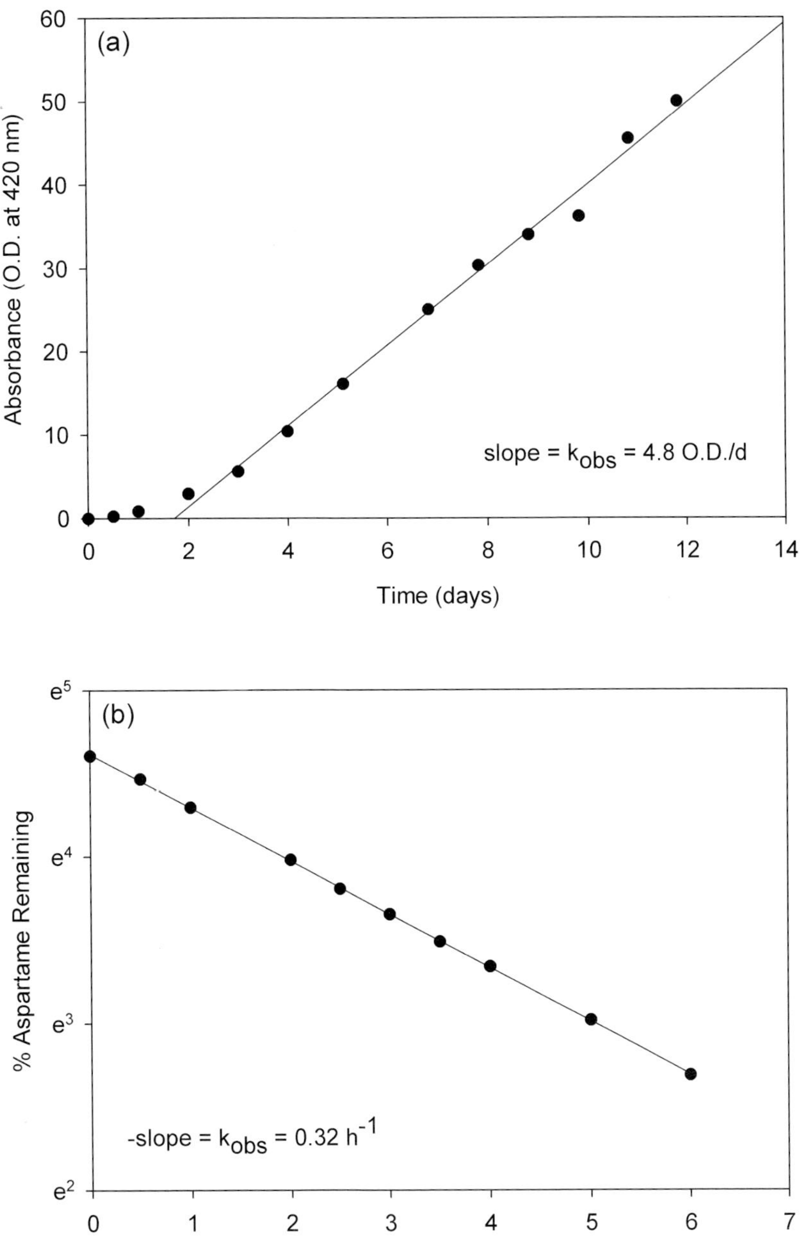

Figure 7.1 Modeling of stability data using (**a**) pseudo zero order kinetics for browning between glycine and fructose at pH 3.5 and 60°C (data from Reyes et al. 1982) and (**b**) pseudo first order kinetics for aspartame degradation in 0.1 M phosphate buffer at pH 7 and 37°C (data from Bell 1989).

where [A] is the amount of some attribute measured at time, t, $[A]_o$ is the amount of the attribute initially when t = 0, and k_{obs} is the pseudo first order rate constant. The units of the first order rate constant are reciprocal time (e.g., day^{-1}). Equation 7.3 can be rearranged as:

$$\ln(\%\,[A]\text{ remaining}) = -k_{obs}t \tag{7.4}$$

A plot of either the natural log of [A] or percent [A] remaining as a function of time would yield a straight line with a slope equal to the pseudo first order rate constant. Again, a larger rate constant translates into a faster reaction. Figure 7.1b shows an example of a pseudo first order kinetic plot of the degradation of aspartame in 0.1 M phosphate buffer at pH 7 and 37°C. The aspartame degradation rate constant in this solution was 0.32 hr^{-1} (Bell 1989).

It is frequently more convenient to express the kinetics in terms of reaction half-life. The half-life, $t_{1/2}$, is the time required for the amount of A to be reduced by 50%. For a first order reaction, the half-life is independent of initial reactant concentration and is calculated as:

$$t_{1/2} = \frac{0.693}{k_{obs}} \tag{7.5}$$

Using the aspartame example in Figure 7.1b, the rate constant of 0.32 hr^{-1} would equate to a half-life of approximately 2.2 hours. At pH 7 and 25°C, the aspartame half-life is approximately 6 hours (Bell 1989). The larger half-life at the lower temperature indicates greater aspartame stability.

Occasionally, a reaction may need to be modeled in a manner other than via pseudo zero or first order kinetics. One such case is the initial step of the Maillard reaction, where an amino acid reacts with a reducing sugar. A reaction in which two reactants participate, like the Maillard reaction, may be modeled via second order kinetics. If the reaction involves two molecules of the same substance or two different compounds at the same concentration (i.e., [A] = [B]), Equation 7.1 becomes:

$$\frac{d[A]}{dt} = -k_{obs}[A]^2 \tag{7.6}$$

This is integrated into

$$\frac{1}{[A]} - \frac{1}{[A]_o} = k_{obs}t \tag{7.7}$$

If concentrations of A and B are different, then

$$\frac{d[A]}{dt} = -k_{obs}[A]\,[B] \tag{7.8}$$

This integrates into

$$\frac{1}{([A]_o - [B]_o)} \ln \frac{[A]\,[B]_o}{[A]_o[B]} = k_{obs}t \tag{7.9}$$

Due to its complexity, Equation 7.9 is used infrequently to model chemical stability of foods.

Temperature Effects on Chemical Stability

Of the various factors that can influence chemical stability, the one universal variable that generally has the largest effect on chemical reaction rates is temperature. Simply stated, if a reaction needs to occur faster, temperature can be elevated. If the rate of reaction needs to decrease, temperature is reduced. Using refrigeration to lower the rate of food deterioration is the application of this basic concept. Thus, in order to control chemical reactivity and predict product shelf-life, it is important to understand how to model the effect of changing temperature on chemical reactions.

From a physical chemistry perspective, the effect of temperature is most appropriately modeled using the Arrhenius equation, as shown below:

$$\ln(k_{obs}) = \ln(A) - \frac{E_A}{RT} \tag{7.10}$$

where A is the preexponential factor, E_A is the activation energy, R is the ideal gas constant, and T is absolute temperature (i.e., in Kelvin). The activation energy is determined by plotting $\ln(k_{obs})$ as a function of 1/T; the slope of the plot multiplied by the ideal gas constant yields the activation energy. Activation energy is a measure of a reaction's temperature sensitivity. Reactions having large activation energies are more sensitive to temperature changes than reactions with low activation energies. Figure 7.2a shows the Arrhenius plot for ascorbic acid degradation in a model system humidified to water activity (a_w) 0.75 (Lee and Labuza 1975). The activation energy and preexponential factor of this reaction were determined to be 16.7 kcal/mol and 1.54×10^{10}, respectively. Using these values (E_A and A) and Equation 7.10, the rate for ascorbic acid degradation in this model system can be predicted at other temperatures. Under typical refrigerated conditions (i.e., 4°C), for example, the ascorbic acid half-life is predicted to be approximately 30 days.

Another method for evaluating the effect of temperature on a reaction is through the use of shelf-life plots, as shown in Figure 7.2b. By plotting the log of half-life as a function of temperature (in degrees Celsius), a term known as the Q_{10} value can be determined from the linear slope ($\log Q_{10} = -10 \times$ slope). The Q_{10} is also an indication of a reaction's temperature sensitivity, with higher Q_{10} values signifying greater temperature sensitivity. By definition, Q_{10} is the ratio of the half-life at one temperature with that at a temperature 10°C higher, as shown in Equation 7.11.

$$Q_{10} = \frac{t_{1/2\,(T)}}{t_{1/2\,(T+10)}} \tag{7.11}$$

From Figure 7.2b, a Q_{10} value of approximately 2.5 was determined, meaning that the ascorbic acid half-life (or shelf-life) would be reduced by 60% for each 10°C increase in temperature. Although the shelf-life plot is easier to use, it is applicable only to narrow temperature ranges. If a broad temperature range is required, the Arrhenius equation will predict the chemical stability more accurately.

Other Factors Affecting Chemical Reactions

In addition to temperature, many other factors also influence the rates of chemical reactions and thus food stability. Some of these variables are pH, light, oxygen, and other ingredients. Hydrolysis reactions typically occur faster at low pH values. For example, su-

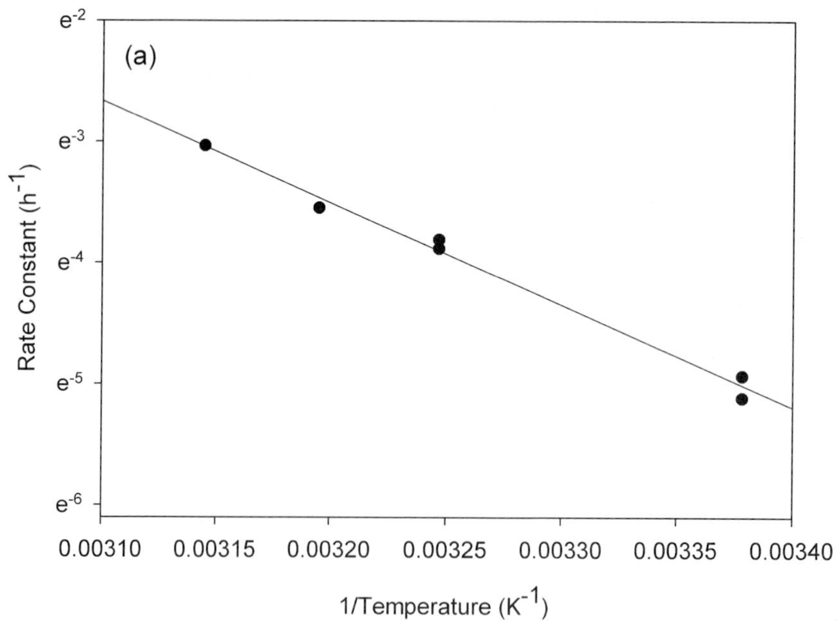

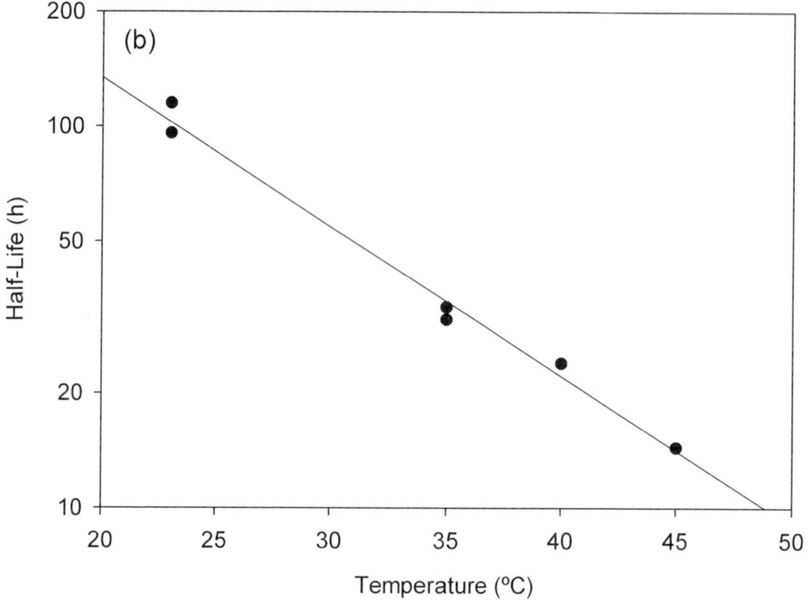

Figure 7.2 Modeling the temperature sensitivity of ascorbic acid degradation at a_w 0.75 using (**a**) Arrhenius kinetics and (**b**) shelf-life plot (data from Lee and Labuza 1975).

crose hydrolysis is catalyzed by acidic conditions (Kelly and Brown 1979). The degradation of thiamin, riboflavin, and ascorbic acid occurs faster at elevated pH values (Gregory 1996). Riboflavin is also sensitive to light exposure (Gregory 1996). Oxidation reactions (such as lipid oxidation) require aerobic conditions, and they are catalyzed by metal ions within the food system (Nawar 1996). Phosphate buffer salts also influence chemical stability by catalyzing the degradation of aspartame (Bell and Wetzel 1995), as well as the early steps of the Maillard reaction (Bell 1997a). Proper control of these variables can reduce chemical reactivity and extend product shelf-life.

One additional factor that frequently has a significant effect on food stability is water. As presented in Chapters 8 and 10, physical and microbial stability of food is strongly dependent on moisture. Chemical stability is no different. The remaining sections of this chapter address the various effects water can have on food chemical stability, including its effects on specific food-related reactions.

Possible Roles of Water in Chemical Reactions

Water is frequently the primary substance found in food products. Thus, the effects of water on food stability, and chemical stability in particular, need to be understood. Even when the amount of water in a food is low, it can impart significant effects on chemical reactions. The effect of water on chemical reactivity can be viewed from several perspectives. A food's moisture content and a_w are two parameters frequently evaluated with respect to food chemical stability. Some reactions are influenced by the dielectric properties of solutions, which are also affected by water. The molecular mobility of food components depends on the physical state of the food, which is also influenced by water. In addition, water can affect the reaction's sensitivity to temperature. Each of these effects of water is discussed below.

Moisture Content

The moisture content of a food has an effect on chemical reactivity. For example, ascorbic acid stability in a model food system at constant a_w was greater when the system was prepared via humidification (i.e., adsorption) than by dehydration (i.e., desorption); the system prepared via dehydration had a higher moisture content and less viscous reaction medium, which allowed for enhanced ascorbic acid degradation (Lee and Labuza 1975). In addition to influencing the viscosity of the reaction medium, water functions as a solvent for the dissolution of reactants. Sufficient water is necessary for solutes to dissolve, diffuse, and react. As the moisture content increases above that required for reactant dissolution, reactant dilution occurs, which consequently decreases the reaction rate, as shown by Equation 7.1. Numerous reactions, especially hydrolysis reactions, also use water directly as a reactant. If the concentration of this reactant is reduced, the reaction rate decreases. Thus, the effect of water on chemical reactivity depends on the reaction type and moisture content relative to reactant solubility. However, as discussed next, moisture content alone is not the sole indicator of chemical reactivity, nor is it the most appropriate stability indicator.

Water Activity

The general concept of a_w was presented in Chapter 2. Over three decades ago, it was recognized that chemical stability data correlated better with a_w than moisture content (Acker

1969, Rockland 1969). For example, it has been demonstrated that physical entrapment of water by gums does not change a_w values in comparison to aqueous solutions of similar composition (Rey and Labuza 1981, Wallingford and Labuza 1983). Thus, substances, such as hydrocolloids and insoluble fibers, can be added to reduce moisture content and physically contain water within a food without changing its a_W. Aspartame degradation in 0.1 M phosphate buffer at pH 7, a_w 0.99, and 30°C occurred at the same rate whether in solution (98.6% moisture) or a semisolid agar gel (92.3% moisture); the difference in moisture content or physical state did not affect aspartame reactivity (Bell 1989, 1992). In addition, aspartame degradation in a low-molecular-weight rubbery polyvinylpyrrolidone (PVP) system at a_w 0.54 was similar to that in a high-molecular-weight glassy PVP system at the same a_w (Bell and Hageman 1994). Thus, if the chemical behavior or properties of the water (e.g., a_w) in two foods are similar, chemical reactivity is often similar as well.

As mentioned in previous chapters, a_w is a thermodynamic term related to the chemical potential of water. Thermodynamics addresses equilibration phenomena rather than kinetic or time-dependent phenomena (Alberty 1987, Levine 1988). Thus, a_w dictates the direction of moisture migration until an equilibrium condition is achieved but gives no indication regarding the equilibration rate. Similarly, a_w does not appear directly in the kinetic equations presented earlier; these equations use reactant concentrations.

Because a portion of total moisture content is unavailable, a_w gives a better indication about water's ability to function as a reactant and solvent (Karel 1973, 1975; Labuza 1975). Thus, the concentration of water available to react, such as for hydrolysis reactions, is related more to a_w than to net moisture content. Likewise, the reactant concentration in solution is related to water's ability to behave as a solvent, which also depends on a_w (Gal 1975). As a_w (and moisture) increases, water's capacity for dissolving reactants also increases, which is partially due to a higher dielectric constant at higher moisture contents. As the a_W increases further, a moisture content is reached wherein the reactants completely dissolve. Once this a_w is obtained (all reactants are in solution), additional moisture sorption at higher water activities will cause the reactants to become increasingly dilute. This dilution occurs because, at high water activities, an additional a_w increase, however small, results in a large increase in the moisture content. Therefore, chemical reactivity typically increases with increasing a_w to a maximum followed by a decrease in reactivity (see Figure 7.3).

The relationship between a_w and equilibrium moisture content of a material is described by the moisture sorption isotherm, as introduced in Chapter 5. Moisture sorption data can be modeled by various equations, two of which are the Brunauer-Emmet-Teller (BET) and Guggenheim-Anderson-deBoer (GAB) equations (Bell and Labuza 2000). The GAB equation is shown below where m is the moisture content (on a dry weight basis), m_o is the monolayer moisture value, k is a multilayer factor, and *C* is a heat constant.

$$m = \frac{m_o \cdot k \cdot C \cdot a_w}{(1 - k \cdot a_w)(1 - k \cdot a_w + k \cdot C \cdot a_w)} \tag{7.12}$$

Nonlinear regression of moisture sorption data using the GAB equation yields values for m_o, k, and *C*. The ability to determine the monolayer moisture value (or simply, monolayer) is especially useful because the monolayer and its corresponding a_w have been shown to correlate with optimum food stability (Salwin 1959). Similarly, Duckworth and

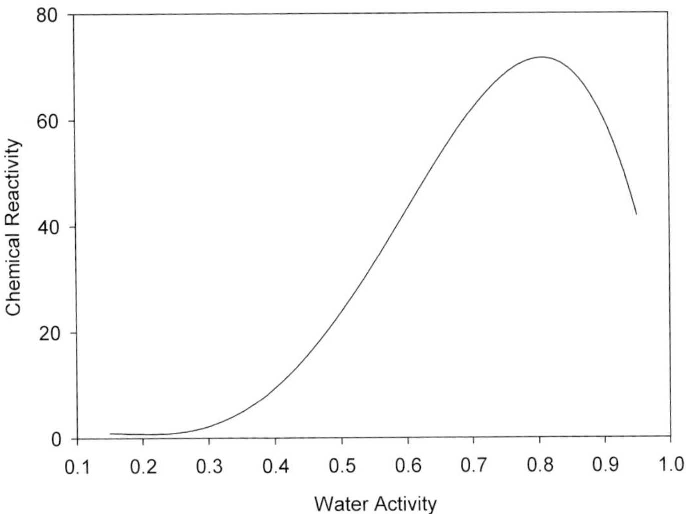

Figure 7.3 Generalized effect of water activity on chemical reactivity.

Smith (1963) demonstrated that solute movement was not detectable below the monolayer value but was detectable above it. Reactant mobility is a prerequisite for reactivity so lower mobility at the monolayer translates into greater product stability. As a_w is increased above the monolayer, chemical reactivity increases until reactant dilution occurs (see Figure 7.3).

Despite the fact that a_w is a thermodynamic term not directly linked with chemical kinetics, an empirical relation was found to exist between the rates of many chemical reactions and a_w from the monolayer up to the reactivity maximum. The Q_A concept, which is similar to the Q_{10} concept, provides a method for modeling the effect of a_w on chemical stability. The Q_A value represents the increase in reactivity (decrease in half-life) due to a 0.1-unit increase in a_w; Q_A is mathematically defined as follows.

$$Q_A = \frac{t_{1/2\,(a_w)}}{t_{1/2\,(a_w+0.1)}} \tag{7.13}$$

A Q_A value of 1.5 means that for each 0.1 a_w increase, the reaction half-life will decrease by 33% or the rate will increase by 50%. By plotting the natural log of reaction half-life versus a_w, Q_A can be calculated from the slope using the following equation:

$$Q_A = e^{-0.1 \cdot \text{slope}} \tag{7.14}$$

Figure 7.4 shows an example of Q_A plots for aspartame degradation. Using the Q_A value, the half-life (or shelf-life) can be estimated at any a_w between the monolayer and the reactivity maximum.

Dielectric Properties

The dielectric constant gives an indication of the polarity of a solvent and its ability to interact with or solvate ionic molecules (Connors et al. 1986). The solvent properties of

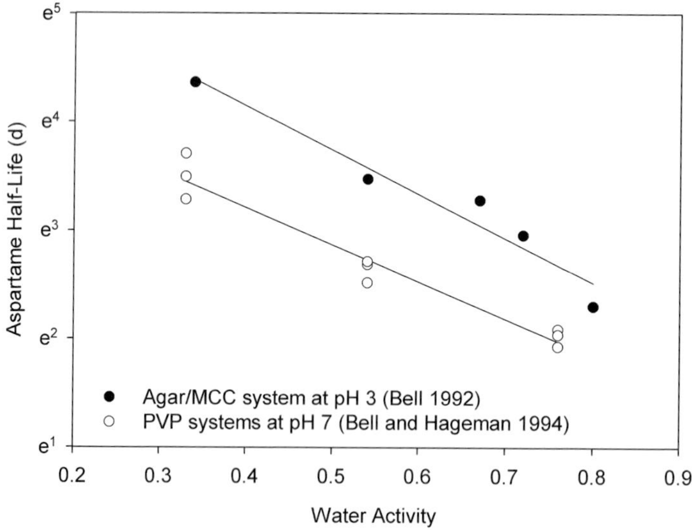

Figure 7.4 Q_A plots for aspartame degradation.

water are therefore attributed in part to its relatively high dielectric constant. Solvents having high dielectric constants are better able to solvate (i.e., hydrate) ions, thus promoting dissolution and preventing ion pairing. As the dielectric constant of a solution decreases, the attractive forces between the ion pairs increase (Chang 1981, Levine 1988). Because the dielectric properties of aqueous solutions influence the behavior of ionic substances, the dielectric constant would be expected to affect reactivity involving charged species. For an aqueous-based reaction of two oppositely charged ions, their hydration, associated with water's high dielectric constant, will reduce interaction between the ions and their reaction rate will be slower than in a solvent with a lower dielectric constant (Connors et al. 1986). A change in dielectric properties of a solution can also promote protein precipitation because differences in the hydration of the ionic amino acid residues result in a change in the net molecular configuration of the protein (Lehninger 1981). For neutral reactants that yield neutral products, the polarity of the solvent should have a minimal effect on the reaction (Connors et al. 1986).

The dielectric properties of water containing dissolved substances are between those of rigid ice and pure liquid water (Karel 1973), with the dielectric constant generally decreasing as the solute concentration increases. Bell and Labuza (1992) showed that 1 molal aqueous solutions of glycerol, glucose, and sucrose each had a different dielectric constant despite having the same a_w. However, at a concentration of 1 molal, these solutions each had different moisture contents; the dielectric constant correlated fairly well with moisture content. As a solution becomes increasingly dilute, the dielectric constant and a_w increase. The effect on a reaction could therefore be due to a change in the a_w, dielectric constant, or reactant concentration. This scenario demonstrates the difficulty in pinpointing exactly the effect water has on a chemical reaction. Water activity has received more attention with respect to food chemical stability than dielectric properties because a_w has been correlated with more food stability issues (e.g., microbial stability, physical stability, moisture transfer). In addition, a_w has become a standard determination made during product development, whereas measuring dielectric properties has not.

Reactant Mobility

In addition to the aforementioned effects of water, the presence of moisture can also influence the physical characteristics of solid foods. Water can act as a plasticizer converting amorphous glassy foods into amorphous rubbery foods. This conversion of a glass into a rubber, known as the glass transition, is accompanied by a large decrease in viscosity and increase in molecular mobility. If a glassy food immobilizes the reactants, their reactivity should be very low. Conversion to the rubbery state as a result of moisture uptake would increase reactant mobility and, consequently, it may be expected to promote an increase in reactivity. Although a_w effects have been investigated for four to five decades, the significance of the glass transition with respect to chemical reactivity has only been studied over the past 10 to 15 years. Additional information about glass transition theory as applied to foods appears in Roos (1995).

Water and Temperature Sensitivity

Another effect that water has on reaction kinetics is the change in temperature sensitivity of the reaction. Activation energies and Q_{10} values often decrease as a_w increases, indicating that lower moisture foods are more sensitive to temperature fluctuations. For example, the Q_{10} value for aspartame degradation at pH 5 and a_w 0.33 was 4.3, while at a_w 0.99 it was 2.6 (Bell 1989). In other words, an increase in temperature would promote a greater shelf-life loss at a_w 0.33 than 0.99. This larger temperature sensitivity is partially compensated by the general increase in stability at lower water activities. A potential rationale for the effect of water on activation energies has been provided previously based on enthalpy-entropy compensation theory (Labuza 1980a, Bell and Labuza 1994).

As demonstrated in the previous discussion, water can influence chemical reactions in a variety of ways. Different reaction mechanisms and types respond differently to the effects of water. A discussion of water's effect on specific food-related chemical reactions follows.

Aspartame Degradation

The high-intensity sweetener aspartame has been incorporated into a wide variety of high- and low-moisture food products. The stability of this dipeptide has been thoroughly investigated with respect to the effect of numerous environmental stresses, including the role of water. Before discussing the various effects of water on aspartame degradation, its degradation mechanism should be briefly introduced.

Mechanism of Aspartame Degradation

Aspartame consists of an aspartic acid residue linked to a phenylalanine residue, whose carboxylic acid group has been methylated. The degradation mechanism of this dipeptide depends on the pH of the reaction media. Aspartame has two primary degradation modes. One mechanism involves hydrolysis of the methyl ester group to produce α-aspartylphenylalanine (α-AP). This hydrolysis reaction occurs predominantly under acidic conditions. The other degradation mechanism is a cyclization reaction resulting from the nucleophilic attack of the phenylalanine carbonyl by the aspartic acid free amine, produc-

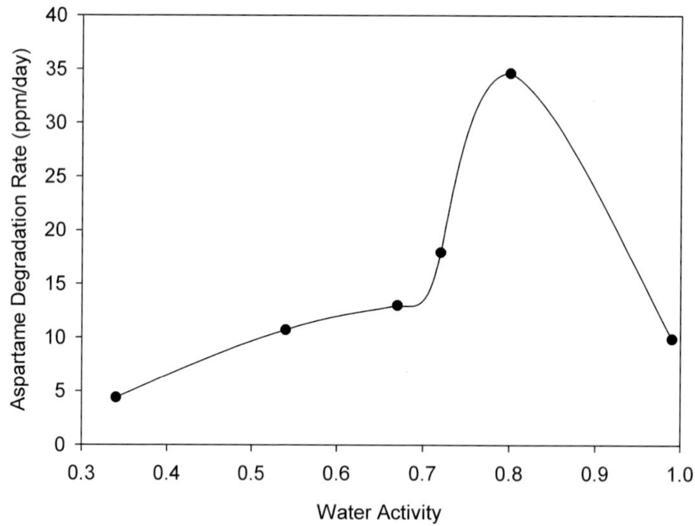

Figure 7.5 Aspartame degradation rates as a function of water activity in agar/microcrystalline cellulose model systems at pH 3 and 30°C, assuming an initial aspartame concentration of 500 ppm (data from Bell and Labuza 1991a, Bell 1992).

ing a diketopiperazine (DKP). The cyclization reaction predominates under neutral conditions. Both α-AP and DKP are produced at pH 3 to 7 but in differing amounts (Bell and Labuza 1991b). Although other secondary reactions may occur depending on environmental conditions, these are the two primary reaction pathways. The conversion of aspartame into α-AP and DKP is catalyzed by buffer salts, with phosphate buffer being especially effective (Bell and Wetzel 1995). Generally, between pH 4 and 7, degradation rates decrease with decreasing pH (Homler 1984, Bell and Labuza 1991b). Additional information about aspartame stability can be found in the literature (Bell 1997b).

Effect of Water on Aspartame Degradation

Aspartame degradation rates were evaluated as a function of a_w in an agar/microcrystalline cellulose/phosphate buffer model system (Bell and Labuza 1991a, Bell 1992). As shown in Figure 7.5, aspartame degradation was fastest around a_w 0.8. As a_w decreased below that level, phosphate buffer salts precipitated, causing a lowering of pH and a reduction in the degradation rate. In addition, the mobility of the reactants would decrease at the lower water activities, resulting in less interaction between the phosphate buffer salts and aspartame, which also lowers the reaction rate. At water activities above 0.8, buffer salts are less concentrated, which reduces their catalytic ability and decreases the rate of aspartame degradation.

Because reactant mobility could have a significant effect on aspartame degradation, a study was conducted to compare the effects of a_w with that of mobility issues associated with the glass transition. Bell and Hageman (1994) used PVP of different molecular weights to study the rearrangement of aspartame into DKP at pH 7. In PVP, a_w and moisture content can be held constant, whereas the glass transition and mobility change as a function of PVP molecular weight (Bell and Hageman 1995). At a_w 0.54, aspartame degradation rates were the same regardless of the PVP existing as a glassy or rubbery matrix. However, at a constant glass transition temperature, aspartame stability was greater in the lower a_w system. Thus, aspartame stability was found to correlate with a_w but not the glass transition temperature. In this study, the phosphate buffer concentration was better controlled by adding an amount of buffer proportional to the water content at each a_w. Thus, the buffer concentration and pH should be maintained better than in the Bell (1992) study, such that the increased reactivity cannot be explained by an increase in reactant concentration. Based on the glass transition data, reactant mobility did not play a major role with respect to increasing the reactivity at higher water activities. The greater reactivity (smaller half-life) observed at the higher water activities could be attributed to a change in the dielectric properties of the system, as explained next.

The isoelectric point of aspartame is at pH 5.2 (Homler 1984). At pH levels below the isoelectric point, aspartame carries a net positive charge, while at pH levels above the isoelectric point it carries a net negative charge. Thus, at pH 7, degradation occurs primarily from the enhanced proton transfer within the negatively charged aspartame by the negatively charged phosphate ion resulting in formation of cyclic DKP (Bell and Wetzel 1995). As a_w and moisture content increase, so, too, do the dielectric constant and polarity of the reaction medium. For a reaction involving two ions with the same charge, an increase in solvent polarity would increase the reaction rate (Connors et al. 1986). The expected increase in the polarity of the PVP system at pH 7 with increasing a_w could therefore explain the increase in aspartame degradation rates and the shorter half-life. The role of solvent polarity in the agar/microcrystalline cellulose systems is more difficult to assess due to the precipitation of buffer salts and the consequential change in pH as a_w decreases.

The above explanation is consistent with the findings of Sanyude et al. (1991), who specifically examined the effects of dielectric properties on aspartame degradation in a pH 4.5 solution. At this pH, aspartame has a net positive charge and its reaction with negatively charged buffer or hydroxyl ions would be enhanced by a decrease in the dielectric constant of the solution. This result was observed by Sanyude et al. (1991) where the aspartame degradation rate constant approximately doubled when the dielectric constant decreased from 65 to 45. In polar solutions having high dielectric constants, the solvation sphere around oppositely charged ions would limit their interactions and reactivity. Despite the apparent effect of the dielectric properties on aspartame degradation, this effect is relatively minor compared with that of pH and buffer concentration (Bell 1997b).

As mentioned earlier, a_w is a more commonly measured parameter than the dielectric constant. Increasing a_w can change the dielectric properties and promote solute dissolution, both of which affect aspartame stability. To empirically model the effect of a_w, Q_A plots for aspartame degradation in the agar/microcrystalline cellulose model system (Bell and Labuza 1991a, Bell 1992) and the PVP model system (Bell and Hageman 1994) were constructed (see Figure 7.4). Interestingly, the Q_A values for aspartame degradation in both model systems were 1.4 to 1.5, indicating that for each 0.1-unit increase in a_w, the aspartame degradation rate increases by 40% to 50%. This type of modeling is useful for predicting the effect an a_w change has on food stability, as demonstrated later in this chapter.

The other effect moisture has on aspartame degradation is with respect to the temperature sensitivity of the reaction. As the a_w and moisture content increased, the activation energies decreased (Bell and Labuza 1991a). For example, at pH 5 and a_w 0.33, the E_A and Q_{10} values for aspartame degradation were 27.1 kcal/mol and 4.3, while at a_w 0.99 they were 17.9 kcal/mol and 2.6. Thus, a change in temperature from 20° to 30°C would reduce the shelf-life of the a_w 0.33 system by 77%, whereas in the high-moisture system the shelf-life reduction would be 62%. However, the initial shelf-life in the low a_w system would be much longer than in the high a_w system, compensating for the greater temperature sensitivity. When estimating product shelf-life, both a_w effects and temperature effects need to be taken into account.

Vitamin Degradation

The shelf-life of food products can also decrease from the degradation of vitamins. The food does not typically appear or taste different, but from a nutritional perspective, it is not equivalent to the original product. Three vitamins of interest that can degrade during storage are ascorbic acid, thiamin, and riboflavin. Their degradation mechanisms are varied and complex; a detailed discussion of these has been presented by Gregory (1996). From the perspective of moisture effects, vitamin stability has most commonly been evaluated with respect to a_w.

Ascorbic Acid Stability

Ascorbic acid, or vitamin C, is a quite unstable vitamin, which degrades via oxidation. This oxidation reaction is catalyzed by trace metals (e.g., copper, iron) in the food. Ascorbic acid degradation increases as a_w increases (Lee and Labuza 1975, Riemer and Karel 1977, Laing et al. 1978, Dennison and Kirk 1982). Using the data of Lee and Labuza (1975) and Riemer and Karel (1977), the Q_A values for ascorbic acid degradation were determined to range from 1.6 to 1.7. In addition, at constant a_w, Lee and Labuza (1975) noted enhanced degradation in a system prepared by moisture desorption as compared to the same system prepared by moisture adsorption; greater reactivity occurred in the system having a larger moisture content (i.e., the desorption system). Lee and Labuza (1975) attributed this phenomenon to a decreased viscosity of the reaction medium, which would allow easier diffusion of oxygen and metal ions to the ascorbic acid. Dennison and Kirk (1982) also noted the significance of sufficient water for dissolving and mobilizing the metal ions that catalyze ascorbic acid destruction.

However, the above results could also be explained by a change in the dielectric properties. Again, as a_w and/or moisture content increases, the dielectric constant and polarity of the system would be expected to increase as well. The initial step of ascorbic acid degradation involves the neutral ascorbic acid molecule being converted into dehydroascorbic acid via a charged intermediate (Gregory 1996). An increase in solvent polarity would be expected to increase the reaction rate for this type of reaction (Connors et al. 1986). It is possible that both the higher dielectric constant and reduced viscosity associated with increasing a_w enhance ascorbic acid degradation in a synergistic manner.

The temperature sensitivity of ascorbic acid degradation can also depend on the a_w and moisture content. The activation energies for ascorbic acid degradation in dehydrated tomato juice decreased from 24.6 kcal/mol in the dry state to 16.2 kcal/mol at a_w 0.75 (Riemer and Karel 1977). This result is consistent with that found by Vojnovich and Pfeifer (1970) for ascorbic acid stability in mixed cereal. However, in a different cereal-type product, Kirk et al. (1977) found the opposite effect, where E_A increased with increasing a_w. Other studies in model food systems have not shown a definite pattern with respect to E_A and a_w (Lee and Labuza 1975, Laing et al. 1977). Due to the conflicting results, the potential effects of moisture on the temperature sensitivity of ascorbic acid degradation should be determined for each specific product formulation and packaging to ensure accurate shelf-life predictions.

Riboflavin Stability

Riboflavin, or vitamin B_2, degrades predominantly by a photochemical mechanism. An increase in a_w also enhances the rate of riboflavin degradation (Dennison et al. 1977). Riboflavin was incorporated into a cereal-type product at pH 6.8, which also contained ascorbic acid. Because the degradation of the neutral riboflavin produces neutral products, an increase in the dielectric constant should have minimal effects on the reaction rate. The increase in reactivity noted in this study could be explained by two possible factors, both of which relate to the pH of the cereal. Bell and Labuza (1994) noted that dehydration often resulted in the pH being lower than in the initially hydrated state. The lower a_w systems could have a lower pH than their higher a_w counterparts; riboflavin stability is greater at lower pH values (Gregory 1996). Similarly, the destruction of the ascorbic acid at higher water activities could cause a pH increase, which would consequently enhance riboflavin degradation in comparison to lower pH/a_w systems. Thus, a pH gradient may exist with the pH being lower than expected at low a_w values and higher than expected at high a_w values. Using the data of Dennison et al. (1977), the Q_A value for riboflavin degradation was determined to be 1.3, which means riboflavin is less sensitive to a_w changes than ascorbic acid. The lower sensitivity to changes in a_w is consistent with the earlier discussion where increasing solvent polarity was not expected to affect the rate of riboflavin degradation, but enhanced ascorbic acid degradation.

Thiamin Stability

The stability of thiamin, or vitamin B_1, has been evaluated in terms of both a_w and glass transition. Maximum reactivity has been shown to occur between a_w 0.3 and 0.5 (Dennison et al. 1977, Bell and White 2000). Kamman et al. (1981) showed an increase in thiamin degradation in pasta as a_w increased from 0.44 to 0.65. A decrease in reactivity was noted as a_w increased from 0.65 to 0.85 in a model food system (Arabshahi and Lund 1988). Depending on the food system and a_w range in the study, the a_w of maximum reactivity shifts.

Bell and White (2000) indicated that thiamin degradation in the glassy state generally increased as the system became less glassy (i.e., as T_g decreased), which was attributed to enhanced mobility of catalytic phosphate buffer ions. If the reaction was largely impacted by reactant mobility, conversion into the rubbery state would have continued to increase reactivity dramatically. However, on conversion of the glassy PVP into rubbery PVP at higher water activities, reactivity actually decreased. Because the phosphate buffer concentration was carefully controlled, dilution was not a possible explanation for the decrease in reactivity at higher water activities. Several possible explanations exist for the lower reactivity at higher water activities observed by Dennison et al. (1977), Arabshahi and Lund (1988), and Bell and White (2000). In studies by Dennison et al. (1977) and Arabshahi and Lund (1988), reactant dilution may indeed be responsible for the slight decrease in reactivity at higher a_w levels. In the Bell and White (2000) study, the explanation provided was that structural collapse of the rubbery systems at higher water activities actually reduced the reactant mobility by eliminating pore spaces where reactions were occurring more freely. A third explanation again involves the increase in polarity of the system at higher water activities and moisture contents. Thiamin degradation at pH 7 involves a positively charged thiamin molecule and negatively charged hydroxyl and phosphate buffer ions. As discussed previously, the more polar systems at higher moisture contents promote hydration of the oppositely charged ions, which reduces reaction rates. Similar results were reported by Arabshahi and Lund (1988), who noted enhanced thiamin loss in the presence of nonpolar humectants. Based on the data, it appears several water-related factors may be affecting thiamin degradation rates.

The temperature sensitivity of thiamin degradation also depends on the a_w and moisture content. The E_A values for thiamin degradation decreased with increasing a_w, decreasing from 30.8 to 26.6 kcal/mol as a_w increased from 0.44 to 0.65 (Kamman et al. 1981). Arabshahi and Lund (1988) found the opposite effect wherein increasing a_w resulted in slightly higher activation energies. Again, as in the case of ascorbic acid stability, there is not a single generalized effect due to water on the temperature sensitivity of thiamin degradation. Each formulation needs to be evaluated separately.

Maillard Reaction

Reaction Overview

The Maillard reaction is one of the most widely studied reactions in food science. This reaction is responsible for both desirable and undesirable aromas, flavors, and browning in food products. The initial step of the Maillard reaction involves an unprotonated amine that acts as a nucleophile by attacking a carbonyl group. Amino acids and reducing sugars are the most common sources of the amine and carbonyl groups, respectively. The nutritional value of a food can be reduced by the destruction of essential amino acids, especially lysine, via this reaction. The rate of the Maillard reaction is typically expressed in terms of reactant loss or brown pigment produced. The initial steps of the reaction are well characterized, but later steps leading to the development of brown pigments are not. The general scheme of the Maillard reaction, as outlined by Hodge (1953), indicates that some steps release water as a reaction product, while other steps require water as a reactant. For example, the initial step of the Maillard reaction between an amino acid and reducing sugar is often referred to as a condensation reaction, which produces a glycosylamine and water. The active participation of water as reactant and product in this reaction complicates interpretation of kinetic data collected as a function of a_w.

Effect of Water on the Maillard Reaction

The Maillard reaction is a bimolecular reaction requiring the interaction of two reactants. Thus, factors affecting reactant concentration and mobility would likely have a significant impact on the reaction rate. Similar to previously discussed reactions, the rate of brown pigment formation via the Maillard reaction increased as a_w increased up to a maximum, followed by a decrease at higher water activities (Loncin et al. 1968, Eichner and Karel 1972, Warmbier et al. 1976, Labuza 1980b, Karmas and Karel 1994). As moisture is adsorbed into the dry system, reactants begin to dissolve and mobilize. With greater moisture adsorption, more reactants dissolve and their mobilities increase, resulting in faster reaction rates. At some critical moisture level, additional moisture sorption results in dilution of the reactants. Lower concentrations of reactants at the higher water activities cause a reduction in the reaction rate, as described by the kinetic equations.

Because the Maillard reaction appeared to have some dependency on molecular mobil-

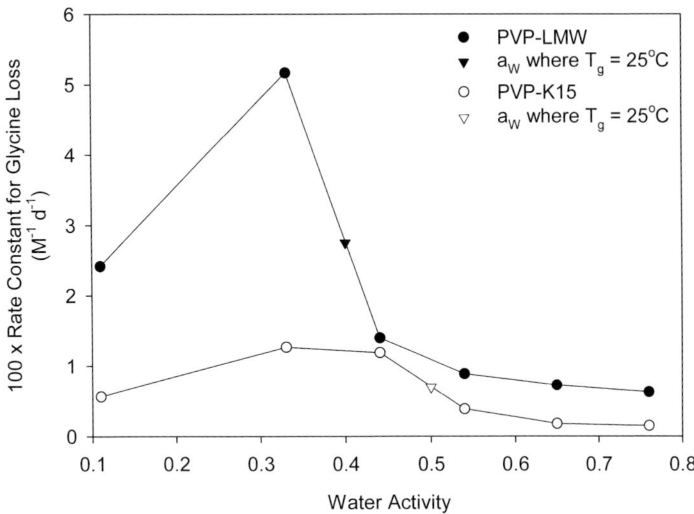

Figure 7.6 Glycine loss via the Maillard reaction as affected by water activity in polyvinylpyrrolidone (PVP) at pH 7 and 25°C (data from Bell et al. 1998).

ity, the glass transition theory was viewed as an alternative approach to describing the effect of water on this particular reaction. Various researchers collected data that suggested the mobility-related issues associated with the glass transition influenced the rate of the Maillard reaction (Karmas et al. 1992, Buera and Karel 1995, Lievonen et al. 1998). However, a change in moisture content or temperature was used in these studies to evaluate the effects of the glass transition, so it was unclear which variable was actually affecting the reaction. In addition, reactant concentration was not controlled, thus different reactant concentrations resulted at the different a_w/moisture combinations.

Bell et al. (1998) again used PVP of varying molecular weights to control the glass transition independent of moisture and temperature. Reactants (glycine and glucose) were incorporated in amounts proportional to the expected amount of adsorbed water so their concentration would not be a variable. Both reactant loss and brown pigment formation were measured. At a constant a_w, moisture content, and reactant concentration, the loss of glycine and development of brown pigment were faster in the PVP having the lowest glass transition temperature (i.e., the most mobile system). Thus, it appears the plasticization effect of water was a significant factor influencing the rate of this reaction.

Despite the evidence supporting the role of glass transition and molecular mobility on the Maillard reaction rate, contradictory results were also noted by Bell et al. (1998). As shown in Figure 7.6, for a given type of PVP, the rate constant for glycine loss increased with increasing a_w to a maximum and then decreased. When the browning rate was plotted as a function of a_w, the rate (in terms of optical density relative to the mass of dry solids) increased in the glassy state and plateaued in the rubbery state as shown in Figure 7.7. Sherwin (2002) suggested that the optical density readings should be corrected for excess pigment produced per gram of solid at the higher water activities, due to the higher net amount of reactants. This correction is accomplished by dividing the browning rate by

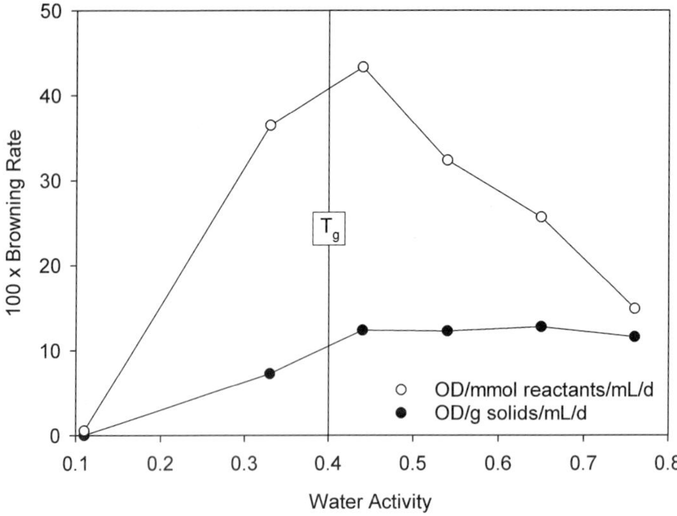

Figure 7.7 Formation of brown pigment via the Maillard reaction as affected by water activity in low-molecular-weight polyvinylpyrrolidone at pH 7 and 25°C (data from Bell et al. 1998).

the mmol of reactants per gram of solids. As shown in Figure 7.7, this correction results in a decrease in the browning rate at higher water activities, which was similar to that shown for reactant loss in Figure 7.6. The a_w associated with the glass transition temperature is noted in both Figures 7.6 and 7.7; PVP is glassy at water activities below this point and rubbery above this point.

Reactant dilution could not be used to explain these decreases in reactivity, as in previous studies, because the experimental design produced systems having the same reactant concentration regardless of the a_w at which they were stored. In addition, these rubbery systems should promote greater molecular mobility, and therefore reactivity, than their glassy counterparts. It was hypothesized that the collapse of the rubbery PVP eliminated pores in which the reaction was occurring, resulting in a decrease in the reactivity. This hypothesis was later verified by White and Bell (1999), who demonstrated that the Maillard reaction occurred slower in collapsed PVP than highly porous PVP at the same a_w and moisture content. Structural collapse is also a result of moisture plasticization and the conversion of a glassy matrix into a rubbery matrix.

Another factor may also partially explain the decrease in reactivity at higher moisture contents. As mentioned earlier, the initial step of the Maillard reaction involves the amino acid attacking the reducing sugar to produce the glycosylamine and water. This reaction step is reversible (Hodge 1953), which means an increase in the concentration of a reaction product would shift the equilibrium toward the reactants, thus decreasing the rate of the forward reaction. The additional water at the higher water activities could therefore cause a decrease in the rate of reactant loss via product inhibition. Thus, even if collapse was not occurring, a decrease in reactivity may still be observed. Data from our laboratory have indicated that the addition of glycerol to phosphate buffer solutions containing glucose and glycine resulted in an approximate doubling of the glucose loss rate constant in comparison to solutions without glycerol (Bell and Chuy 2005). As water hydrogen bonds to glycerol, it is less available to participate in the reverse reaction and actually shifts the equilibrium toward the forward reaction, enhancing its rate. This observation is consistent with excess water causing a decrease in the reaction rate.

The negatively charged amino acid reacting with the neutral sugar produces a negatively charged glycosylamine. Because the glycosylamine is a larger molecule than the amino acid, the negative charge is spread over a greater volume. An increase in solvent polarity, associated with an increase in the moisture content, would therefore be expected to lower the rate of this initial step (Connors et al. 1986). Thus, similar to other ionic reactions, the dielectric properties of the solvent system may partially explain the effect of water on this reaction.

The effect of water on the Maillard reaction can be summarized from two perspectives. In low-moisture systems (a_w below the reactivity maximum), water's roles in reactant solubility and mobility are important, with rates increasing with increasing a_w. Depending on the type of high-moisture food product, the reactant dilution, mobility restrictions from collapse, product inhibition from water, and greater solvent polarity can all reduce the rate of the Maillard reaction as a_w increases above the reactivity maximum.

Labuza and Saltmarch (1982) reviewed a variety of kinetic data regarding the effect of temperature on the Maillard reaction. They noted that the general trend was for activation energies and Q_{10} values for brown pigment formation to decrease as a_w increased. Thus, a small change in temperature has a larger effect on the browning reaction in low moisture foods than in solutions.

Lipid Oxidation

Reaction Overview

Unsaturated fatty acids can undergo oxidation producing a variety of off-flavors and off-aromas. Details of the mechanism and kinetic modeling of this complex reaction have been presented previously (Labuza 1971, Karel 1992). Briefly, the initial step of lipid oxidation involves formation of a lipid free radical, which is frequently catalyzed by trace metals. The propagation step involves oxygen reacting with the lipid free radial to produce a peroxyl free radical, which consequently reacts with another fatty acid to yield a hydroperoxide and a new lipid free radical. The hydroperoxide undergoes cleavage producing the volatile off-flavors and off-aromas. If two free radicals interact to form nonradical products, termination of the reaction occurs. The primary effects of water relate to the catalyzing ability of the metal ions and the diffusion of oxygen, as discussed below.

Effect of Water on Lipid Oxidation

Figure 7.8 shows the general pattern for lipid oxidation as a function of a_w. As a_w increases from the dry state, the rate of lipid oxidation decreases to a minimum near the monolayer. Additional moisture adsorption above the monolayer promotes an increase in reactivity. The reactivity trend shown in Figure 7.8 has been reported by numerous researchers (Maloney et al. 1966, Labuza et al. 1971, Quast and Karel 1972). As moisture is gained from the dry state, hydration spheres form around metal ions, decreasing their catalytic ability for initiating free radical formation. Water can also quench free radicals, which lowers the oxidation rate. During moisture sorption, water replaces air within the pores and capillaries of the food, reducing the rate of oxygen diffusion. The net effect is a slower oxidation rate at low water activities compared with that in the dry state. As moisture is

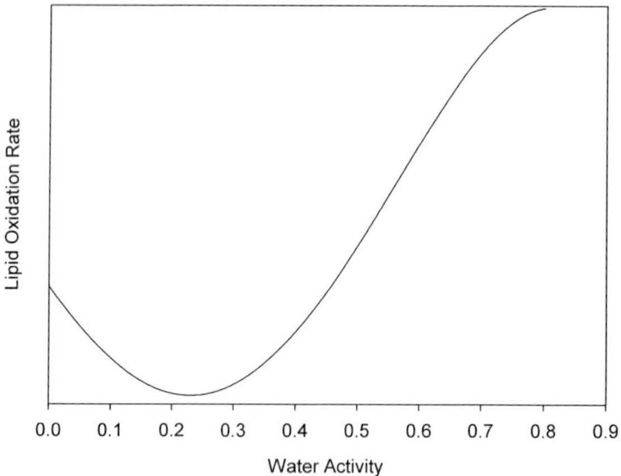

Figure 7.8 Generalized effect of water activity on the rate of lipid oxidation.

gained beyond this point, water solubilizes a larger concentration of catalytic metal ions. In addition, the mobility of oxygen and the metal ions increases, which enhances lipid oxidation. It should be pointed out that the effects of water on lipid oxidation are not solely due to a_W. Solute hydration and dissolution are related to dielectric properties, whereas reactant mobility is related to the glass transition.

Several studies have evaluated lipid oxidation with respect to the glass transition. Oxidation was prevented by encapsulating the lipids in a dense glassy matrix, due to limited oxygen diffusion (Gejl-Hansen and Flink 1977, Shimada et al. 1991, Labrousse et al. 1992). Conversion into the rubbery state without structural collapse may increase oxidation via increased diffusion of metal ions and oxygen. Structural collapse of a rubbery matrix that entrapped the lipids decreased oxidation rates due to a reduction in porosity and oxygen diffusion (Labrousse et al. 1992, Ma et al. 1992). However, if collapse or crystallization resulted in the lipids being exuded from the encapsulant, oxidation rates increased (Gejl-Hansen and Flink 1977, Shimada et al. 1991, Labrousse et al. 1992). Nelson and Labuza (1992) provide additional information about the effects of water on lipid oxidation in a detailed review.

Dehydration is a common technique for enhancing the stability of food products. As mentioned earlier, most reactions in foods cease at moisture contents below the monolayer. It is therefore tempting to simply dry foods to as low a moisture content as possible to guarantee this optimum stability. However, lipid oxidation data (as a function of a_W) demonstrate that simply drying to extremely low moisture contents, although stabilizing some reactions, usually enhances the rate of lipid oxidation. Not only will food quality suffer from extreme drying, the expense of food processing will be greater as well. Determining the monolayer from moisture sorption data can help prevent such problems.

Enzymatic Reactions

A wide variety of enzymatic reactions is possible within food systems, resulting in product deterioration. When discussing enzymatic reactions, two aspects that should be addressed are factors affecting enzyme stability and factors affecting enzyme activity. If an enzyme is not stable and experiences denaturation, its activity is typically lost. However, if an enzyme is not denatured, it will retain some, if not all, of its activity and other external factors will affect the reaction. Depending on the enzyme and type of reaction, water can affect the reaction in several ways. Water can be a reactant in enzymatic hydrolysis reactions. Water also acts as a solvent for the dissolution of reactants. Because reactants must diffuse into the enzyme's active site, molecular mobility issues are important. As mentioned earlier, protein hydration, as affected by the dielectric properties of the solution, will influence the enzyme's structure and functionality. Due to the tremendous number and types of enzymes in foods, the following discussion highlights only a few studies demonstrating the complexity of water's effect on enzyme reactions.

Enzyme Stability

Chen et al. (1999a) evaluated the effects of a_w and glass transition on invertase stability in PVP of differing molecular weights. Stability decreased with increasing a_w and moisture content. No clear relation to the glass transition was observed. The type of PVP (low-molecular-weight LMW versus high-molecular-weight K30) had a significant effect on invertase stability, with greater destabilization occurring in PVP-LMW. The Q_A values for invertase loss were 1.3 in PVP-LMW and 1.07 in PVP-K30. Monsan and Combes (1984) also reported that invertase stability was not simply related to a_w but that the polyol type also influenced its stability. They noted that stability was greater in the presence of polyols with greater chain length, which is similar to that observed in the PVP systems. Monsan and Combes (1984) added that the stabilization or destabilization effect could be due to a change in the enzyme's three-dimensional conformation as well as the water structure around the enzyme. Similarly, Larreta-Garde et al. (1987) concluded that the microenvironment around the enzyme controlled its stability, with a_w being one of the parameters. Although dielectric effects were not discussed in any of the above articles, the collective data enable one to hypothesize that differences in the dielectric properties may be affecting enzyme stability via changes in the aqueous environment surrounding the enzyme and thus its conformational structure.

Contrary to the above discussion, tyrosinase stability in PVP did correlate with the glass transition (Chen et al. 1999b). After short-term storage, the activity remaining decreased as the glass transition temperature decreased (i.e., PVP became more rubbery with greater molecular mobility). The rate constants for tyrosinase activity loss also correlated better with glass transition than with a_w. The rate constants increased to a maximum in the rubbery state, followed by a decrease as the PVP collapsed. Thus, immobilization of the enzyme from either a glassy matrix or a collapsed rubbery matrix appears to preserve its activity. The response of tyrosinase to water appears to be quite different from that of invertase, indicating that broad generalizations about enzyme stability should be avoided.

The stability of enzymes with respect to moisture cannot be explained using a single parameter. Generally, enzymatic stability is preserved at very low moisture contents near the monolayer. As moisture content increases, both the a_w and dielectric constant increase. In addition, molecular mobility increases as the glass transition temperature decreases. Other dissolved substances may affect the dielectric properties of the system, or interact directly with the enzyme, changing its stability as well. Depending on the specific enzyme, each of these factors can influence enzyme stability to some degree.

Enzyme Activity

Although enzyme stability is frequently optimized at low moisture contents, enzyme activity typically requires some amount of water, which is used to either solubilize the substrates, mobilize the reactants, or obtain the correct enzyme molecular conformation; water can also be used as a reactant. Acker (1963) showed that the extent of lecithin hydrolysis by phospholipase increased as a_w increased. Similarly, the activity of tyrosinase in organic solvent–water mixtures increased with increasing a_w (Tome et al. 1978). This latter paper concluded that a_w was probably a more important variable than changes in the viscosity and dielectric constant.

More recently, invertase-mediated sucrose hydrolysis in solids was found to occur very slowly until the a_w exceeded 0.7, above which reactivity increased dramatically (Silver and Karel 1981, Chen et al. 1999a, Kouassi and Roos 2000). Although Silver and Karel (1981) postulated that diffusion limitations could explain the results, Chen et al. (1999a) attributed the increase in reactivity above a_w 0.7 primarily to greater sucrose dissolution rather than to an increase in molecular mobility. Because hydrolysis was not perceptible in the glassy state and its rate increased dramatically in the rubbery state, Kouassi and Roos (2000) concluded that molecular mobility was an important factor at higher water activities. Regardless of data interpretation, four things are required for sucrose hydrolysis by invertase: the correct enzyme conformation, sucrose dissolution, sucrose diffusion to the active site, and sufficient water for the hydrolysis. Depending on the matrix, the relative importance of these factors may vary.

Due to the wide array of enzymes and systems containing enzymes, broad generalizations about the role of water on enzyme reactions are difficult to make. Enzyme stability increases as moisture content decreases. The exact reasons for this could relate to the dielectric properties, solute–enzyme interactions, and/or decreased molecular mobility. Enzyme activity generally increases as moisture content increases. The activity of an enzyme depends on sufficient water to dissolve the substrate and promote its diffusion into the active site, as well as having the enzyme in the proper molecular conformation. Dissolution and the enzyme structure are related to a_w and dielectric properties, whereas diffusion is related to molecular mobility and the glass transition.

Shelf-life Testing

General Requirements for Conducting a Shelf-life Test

To improve the stability of a food product and predict its shelf-life, stability testing must be conducted. The product tested should be as similar to the commercial product as possible, including its formulation, processing, and packaging. A reliable method for analyzing changes in the product should be used. The optimum data would include at least eight data points collected as a function of time until reaching a 50% change in the measured attribute responsible for loss of shelf-life (i.e., the reaction should advance to at least its half-life). The stability data should be collected at three temperatures and three water activities to enable more accurate shelf-life predictions at other environmental conditions.

Aspartame Stability as an Example

The shelf-life of an aspartame-sweetened product expires when 10% of the aspartame is lost. A food product development team needs to predict the product shelf-life at 20°C and a_w 0.15. Table 7.1 lists aspartame stability data at three temperatures and three water activities, as collected by Bell (1989). From this data, a shelf-life plot can be constructed and

Table 7.1 Aspartame Stability Data.

Conditions	Rate Constant (day^{-1})	$t_{10\% \text{ loss}}$* (days)
a_W 0.34, 30°C	0.00548	19.2
a_W 0.35, 37°C	0.01538	6.9
a_W 0.42, 45°C	0.04828	2.2
a_W 0.34, 30°C	0.00548	19.2
a_W 0.57, 30°C	0.01574	6.7
a_W 0.66, 30°C	0.02224	4.7

*$t_{10\% \text{ loss}}$ = time to 10% loss of aspartame.
Source: Bell, L.N. 1989. Aspartame degradation kinetics in low and intermediate moisture food systems. Master's thesis, University of Minnesota, St. Paul, MN. p. 150.

the Q_{10} determined. Similarly, the Q_A can be calculated from a Q_A plot. The values for the Q_{10} and Q_A were determined to be 4.25 and 1.55, respectively. To predict the shelf-life, θ_s, at 20°C and a_w 0.15, the following equation is used:

$$\theta_s = (t_{10\%,\, a_w\, 0.34,\, 30°C}) \times (Q_{10})^{(30-T)/10} \times (Q_A)^{(0.34-a_w)/0.1} \qquad (7.15)$$

Substituting in the shelf-life at a_W 0.34 and 30°C, the Q_{10} and Q_A values, and the target environmental conditions (i.e., T = 20°C and a_w = 0.15), the following solution results:

$$\theta_s = (19.2 \text{ days}) \times (4.25)^{10/10} \times (1.55)^{0.19/0.1} = 188 \text{ days} \qquad (7.16)$$

Thus, using the kinetic data in Table 7.1, a shelf-life of 188 days was predicted for an aspartame-sweetened product at 20°C and a_W 0.15.

Accelerated Shelf-life Testing

Stability testing is frequently a time-consuming process. To achieve shelf-life predictions in a shorter length of time, accelerated testing can be used. Accelerated shelf-life testing uses extreme water activities and temperatures to enhance the reaction rate, thereby reducing the time required for shelf-life predictions. Although time and money may be saved, several limitations are associated with accelerated shelf-life testing. At higher temperatures, the a_W, pH, and reactant solubility are all different compared with ambient conditions. Phase changes, such as the glass transition and crystallization, may result in a structurally different product at higher temperatures, which can be accompanied by differences in reactivity. Multiple reactions often occur in a food product, with one having a greater predominance with respect to shelf-life loss. As temperature changes, the relative importance of the reactions can change due to differences in their temperature sensitivities. The overall effect is a less accurate shelf-life prediction. Therefore, great care must be exercised when using accelerated shelf-life testing so that significant under- or over-estimation of product shelf-life can be avoided.

Summary

As discussed throughout this chapter, water has many potential roles with respect to its effect on food chemical stability. Empirical correlations between chemical stability and a_W

continue to be most frequently encountered. Plasticization of glassy systems into rubbery systems by water influences reactions where molecular mobility is required. The dielectric properties of aqueous systems appear to affect reactions involving charged species. Water acts as a solvent for reactants and as a reactant itself. However, as noted in the above discussion, no single physicochemical parameter describes the effect of water on chemical reactions. The effect of water on chemical stability depends on the specific reaction type and the matrix in which the reaction is occurring. The commonality that should be noted is that water in food systems cannot be ignored because food stability is significantly affected by even low amounts of water.

References

Acker, L. 1963. Enzyme activity at low water contents. In: *Recent Advances in Food Science—3*, eds. J.M. Leitch and D.N. Rhodes, pp. 239–247. London: Butterworths.

Acker, L.W. 1969. Water activity and enzyme activity. *Food Technology* 23:1257–1270.

Alberty, R.A. 1987. *Physical Chemistry,* 7th Edition. New York: Wiley. p. 934.

Arabshahi, A., and Lund, D.B. 1988. Thiamin stability in simulated intermediate moisture food. *Journal of Food Science* 53:199–203.

Bell, L.N. 1989. Aspartame degradation kinetics in low and intermediate moisture food systems. Master's thesis, University of Minnesota, St. Paul, MN. p. 150.

Bell, L.N. 1992. Investigations regarding the definition and meaning of pH in reduced-moisture model food systems. Doctoral dissertation, University of Minnesota, St. Paul, MN. p. 219. [Available from University Microfilms, Ann Arbor, MI.]

Bell, L.N. 1997a. Maillard reaction as influenced by buffer type and concentration. *Food Chemistry* 59:143–147.

Bell, L.N. 1997b. Stability of the dipeptide aspartame in solids and solutions. In: *Therapeutic Protein and Peptide Formulation and Delivery*, eds. Z. Shahrokh, V. Sluzky, J.L. Cleland, S.J. Shire, and T.W. Randolph, pp. 67–78. Washington, DC: American Chemical Society.

Bell, L.N., and Chuy, S. 2005. Catalytic effect of humectants on the Maillard reaction. Presented July 16-20, Institute of Food Technologists Annual Meeting, New Orleans, LA.

Bell, L.N., and Hageman, M.J. 1994. Differentiating between the effects of water activity and glass transition dependent mobility on a solid state chemical reaction: Aspartame degradation. *Journal of Agricultural and Food Chemistry* 42:2398–2401.

Bell, L.N., and Hageman M.J. 1995. A model system for differentiating between water activity and glass transition effects on solid state chemical reactions. *Journal of Food Quality* 18:141–147.

Bell, L.N., and Labuza, T.P. 1991a. Aspartame degradation kinetics as affected by pH in intermediate and low moisture food systems. *Journal of Food Science* 56:17–20.

Bell, L.N., and Labuza, T.P. 1991b. Potential pH implications in the freeze-dried state. *Cryo-Letters* 12:235–244.

Bell, L.N., and Labuza, T.P. 1992. pH of low moisture solids. *Trends in Food Science and Technology* 3:271–274.

Bell, L.N. and Labuza, T.P. 1994. Influence of the low-moisture state on pH and its implication for reaction kinetics. *Journal of Food Engineering* 22:291–312.

Bell, L.N., and Labuza, T.P. 2000. *Moisture Sorption: Practical Aspects of Isotherm Measurement and Use*, Second Edition. St. Paul, MN: American Association of Cereal Chemists. p. 122.

Bell, L.N., and Wetzel, C.R. 1995. Aspartame degradation in solution as impacted by buffer type and concentration. *Journal of Agricultural and Food Chemistry* 43:2608–2612.

Bell, L.N., and White, K.L. 2000. Thiamin stability in solids as affected by the glass transition. *Journal of Food Science* 65:498–501.

Bell, L.N., Touma, D.E., White, K.L., and Chen, Y.H. 1998. Glycine loss and Maillard browning as related to the glass transition in a model food system. *Journal of Food Science* 63:625–628.

Buera, M.P., and Karel, M. 1995. Effect of physical changes on the rates of nonenzymatic browning and related reactions. *Food Chemistry* 52:167–173.

Chang, R. 1981. *Physical Chemistry With Applications to Biological Systems,* 2nd Edition. New York: Macmillan. p. 659.

Chen, Y.H., Aull, J.L., and Bell, L.N. 1999a. Invertase storage stability and sucrose hydrolysis in solids as affected by water activity and glass transition. *Journal of Agricultural and Food Chemistry* 47:504–509.

Chen, Y.H., Aull, J.L., and Bell, L.N. 1999b. Solid-state tyrosinase stability as affected by water activity and glass transition. *Food Research International* 32:467–472.

Connors, K.A., Amidon, G.L., and Stella, V.J. 1986. *Chemical Stability of Pharmaceuticals*, 2nd Edition. New York: Wiley. p. 847.

Dennison, D.B., and Kirk, J.R. 1982. Effect of trace mineral fortification on the storage stability of ascorbic acid in a dehydrated model food system. *Journal of Food Science* 47:1198–1200, 1217.

Dennison, D., Kirk, J., Bach, J., Kokoczka, P., and Heldman, D. 1977. Storage stability of thiamin and riboflavin in a dehydrated food system. *Journal of Food Processing and Preservation* 1:43–54.

Duckworth, R.B., and Smith, G.M. 1963. Diffusion of solutes at low moisture levels. In: *Recent Advances in Food Science—3*, eds. J.M. Leitch and D.N. Rhodes, pp. 230–238. London: Butterworths.

Eichner, K., and Karel, M. 1972. The influence of water content and water activity on the sugar-amino browning reaction in model systems under various conditions. *Journal of Agricultural and Food Chemistry* 20:218–223.

Gal, S. 1975. Solvent versus non-solvent water in casein-sodium chloride systems. In: *Water Relations of Foods*, ed. R.B. Duckworth, pp. 183–191. London: Academic Press.

Gejl-Hansen, F., and Flink, J.M. 1977. Freeze-dried carbohydrate containing oil-in-water emulsions: Microstructure and fat distribution. *Journal of Food Science* 42:1049–1055.

Gregory, J.F. 1996. Vitamins. In: *Food Chemistry*, 3rd Edition, ed. O.R. Fennema, pp. 531–616. New York: Dekker.

Hodge, J.E. 1953. Chemistry of browning reactions in model systems. *Journal of Agricultural and Food Chemistry* 1:928–943.

Homler, B.E. 1984. Aspartame: Implications for the food scientist. In: *Aspartame—Physiology and Biochemistry*, eds. L.D. Stegink and L.J. Filer, pp. 247–262. New York: Dekker.

Kamman, J.F., Labuza, T.P., and Warthesen, J.J. 1981. Kinetics of thiamin and riboflavin loss in pasta as a function of constant and variable storage conditions. *Journal of Food Science* 46:1457–1461.

Karel, M. 1973. Recent research and development in the field of low-moisture and intermediate-moisture foods. *CRC Critical Reviews in Food Technology* 3:329–373.

Karel, M. 1975. Physico-chemical modification of the state of water in foods—A speculative survey. In: *Water Relations of Foods*, ed. R.B. Duckworth, pp. 639–657. London: Academic Press.

Karel, M. 1992. Kinetics of lipid oxidation. In: *Physical Chemistry of Foods*, eds. H.G. Schwartzberg and R.W. Hartel, pp. 651–668. New York: Dekker.

Karmas, R., Buera, M.P., and Karel, M. 1992. Effect of glass transition on rates of nonenzymatic browning in food systems. *Journal of Agricultural and Food Chemistry* 40:873–879.

Karmas, R., and Karel, M. 1994. The effect of glass transition on Maillard browning in food models. In: *The Maillard Reactions in Chemistry, Food, and Health,* eds. T.P. Labuza, G.A. Reineccius, V.M. Monnier, J. O'Brien, and J.W. Baynes, pp. 182–187. Cambridge, UK: Royal Society of Chemists.

Kelly, F.H.C., and Brown, D.W. 1979. Thermal decomposition and colour formation in aqueous sucrose solutions. *Sugar Technology Review* 6:1–48.

Kirk, J., Dennison, D., Kokoczka, P., and Heldman, D. 1977. Degradation of ascorbic acid in a dehydrated food system. *Journal of Food Science* 42:1274–1279.

Kouassi, K., and Roos, Y.H. 2000. Glass transition and water effects on sucrose inversion by invertase in a lactose-sucrose system. *Journal of Agricultural and Food Chemistry* 48:2461–2466.

Labrousse, S., Roos, Y., and Karel, M. 1992. Collapse and crystallization in amorphous matrices with encapsulated compounds. *Sciences Des Aliments* 12:757–769.

Labuza, T.P. 1971. Kinetics of lipid oxidation in foods. *CRC Critical Reviews in Food Technology* 2:355–405.

Labuza, T.P. 1975. Interpretation of sorption data in relation to the state of constituent water. In: *Water Relations of Foods*, ed. R.B. Duckworth, pp. 155–172. London: Academic Press.

Labuza, T.P. 1980a. Enthalpy/entropy compensation in food reactions. *Food Technology* 34(2):67–77.

Labuza, T.P. 1980b. The effect of water activity on reaction kinetics of food deterioration. *Food Technology* 34(4):36–41, 59.

Labuza, T.P. 1984. Application of chemical kinetics to deterioration of foods. *Journal of Chemical Education* 61:348–358.

Labuza, T.P., and Kamman, J.F. 1983. Reaction kinetics and accelerated tests simulation as a function of temperature. In: *Computer-Aided Techniques in Food Technology*, ed. I. Saguy, pp. 71–115. New York: Dekker.

Labuza, T.P., and Saltmarch, M. 1982. The nonenzymatic browning reaction as affected by water in foods. In: *Water Activity: Influences on Food Quality*, eds. L.B. Rockland and G.F. Stewart, pp. 605–650. New York: Academic Press.

Labuza, T.P., Heidelbaugh, N.D., Silver, M., and Karel, M. 1971. Oxidation at intermediate moisture contents. *Journal of the American Oil Chemists Society* 48:86–90.

Laing, B.M., Schlueter, D.L., and Labuza T.P. 1978. Degradation kinetics of ascorbic acid at high temperature and water activity. *Journal of Food Science* 43:1440–1443.

Larreta-Garde, V., Xu, Z.F., Biton, J., and Thomas, D. 1987. Stability of enzymes in low water activity media. In: *Biocatalysis in Organic Media*, eds. C. Laane, J. Tramper, and M.D. Lilly, pp. 247–252. Amsterdam: Elsevier.

Lee, S.H., and Labuza, T.P. 1975. Destruction of ascorbic acid as a function of water activity. *Journal of Food Science* 40:370–373.

Lehninger, A.L. 1981. *Biochemistry*. 2nd Edition. New York: Worth. p. 1104.

Levine, I.N. 1988. *Physical Chemistry*, 3rd Edition. New York: McGraw-Hill. p. 920.

Lievonen, S.M., Laaksonen, T.J., and Roos, Y.H. 1998. Glass transition and reaction rates: Nonenzymatic browning in glassy and liquid systems. *Journal of Agricultural and Food Chemistry* 46:2778–2784.

Loncin, M., Bimbenet, J.J., and Lenges, J. 1968. Influence of the activity of water on the spoilage of foodstuffs. *Journal of Food Technology* 3:131–142.

Ma, Y., Reineccius, G.A., Labuza, T.P., and Nelson, K.A. 1992. The stability of spray-dried microcapsules as a function of glass transition temperature. Presented June 21-24, Institute of Food Technologists Annual Meeting, New Orleans, LA.

Maloney, J.F., Labuza, T.P., Wallace, D.H., and Karel, M. 1966. Autoxidation of methyl linoleate in freeze-dried model systems. I. Effect of water on the autocatalyzed oxidation. *Journal of Food Science* 31:878–884.

Monsan, P., and Combes, D. 1984. Effect of water activity on enzyme action and stability. In: *Enzyme Engineering 7*, eds. A.I. Laskin, G.T. Tsao, and L.B. Wingard, pp. 48–63. New York: New York Academy of Sciences.

Nawar, W.W. 1996. Lipids. In: *Food Chemistry,* 3rd Edition, ed. O.R. Fennema, pp. 225–319. New York: Dekker.

Nelson, K.A., and Labuza, T.P. 1992. Relationship between water and lipid oxidation rates: Water activity and glass transition theory. In: *Lipid Oxidation in Food*, ed. A.J. St. Angelo, pp. 93–103. Washington, DC: American Chemical Society.

Quast, D.G., and Karel, M. 1972. Effects of environmental factors on the oxidation of potato chips. *Journal of Food Science* 37:584–588.

Rey, D.K., and Labuza, T.P. 1981. Characterization of the effect of solutes on the water-binding and gel strength properties of carrageenan. *Journal of Food Science* 46:786–789, 793.

Reyes, F.G.R., Poocharoen, B., and Wrolstad, R.E. 1982. Maillard browning reaction of sugar-glycine model systems: Changes in sugar concentration, color, and appearance. *Journal of Food Science* 47:1376–1377.

Riemer, J., and Karel, M. 1977. Shelf-life studies of vitamin C during food storage: Prediction of L-ascorbic acid retention in dehydrated tomato juice. *Journal of Food Processing and Preservation* 1:293–312.

Rockland, L.B. 1969. Water activity and storage stability. *Food Technology* 23:1241–1248, 1251.

Roos, Y.H. 1995. *Phase Transitions in Foods*. San Diego: Academic Press. p. 360.

Salwin, H. 1959. Defining minimum moisture contents for dehydrated foods. *Food Technology* 13:594–595.

Sanyude, S., Locock, R.A., and Pagliaro, L.A. 1991. Stability of aspartame in water: organic solvent mixtures with different dielectric constants. *Journal of Pharmaceutical Sciences* 80:674–676.

Sherwin, C.P. 2002. A molecular mobility approach to describing the role of moisture in the Maillard browning reaction rate. Doctoral dissertation, University of Minnesota, St. Paul, MN, p. 231. [Available from University Microfilms, Ann Arbor, MI.]

Shimada, Y., Roos, Y., and Karel, M. 1991. Oxidation of methyl linoleate encapsulated in amorphous lactose-based food model. *Journal of Agricultural and Food Chemistry* 39:637–641.

Silver, M., and Karel, M. 1981. The behavior of invertase in model systems at low moisture contents. *Journal of Food Biochemistry* 5:283–311.

Tome, D., Nicolas, J., and Drapron, R. 1978. Influence of water activity on the reaction catalyzed by polyphenoloxidase (E.C.1.14.18.1.) from mushrooms in organic liquid media. *Lebensittel-Wissenschaft und-Technologie* 11:38–41.

Vojnovich, C., and Pfeifer, V.F. 1970. Stability of ascorbic acid in blends of wheat flour, CSM and infant cereals. *Cereal Science Today* 15:317–322.

Wallingford, L., and Labuza, T.P. 1983. Evaluation of the water binding properties of food hydrocolloids by physical/chemical methods and in a low fat meat emulsion. *Journal of Food Science* 48:1–5.

Warmbier, H.C., Schnickels, R.A., and Labuza T.P. 1976. Effect of glycerol on nonenzymatic browning in a solid intermediate moisture model food system. *Journal of Food Science* 41:528–531.

White, K.L., and Bell, L.N. 1999. Glucose loss and Maillard browning in solids as affected by porosity and collapse. *Journal of Food Science* 64:1010–1014.

8 Water Activity and Physical Stability

Gaëlle Roudaut

Water and temperature are among the most important parameters controlling the physical properties of biological systems. Indeed, water (through water content or water activity [a_w]) and the interactions between water and the other food components control both thermodynamic and dynamic properties. The influence of water on molecular mobility, and thus its consequences on physical stability, is more marked in systems where moisture content is limited or where most of the water is immobilized in the frozen phase. Therefore, water–mobility relationships are particularly relevant for food systems in temperature and humidity domains in the vicinity of the glass transition temperature. The glass transition was thoroughly examined in an earlier chapter. The main focus of this chapter will be on the role of water in the physical stability of foods with an emphasis on texture (i.e., crispness, softness, stickiness) and structure characteristics (i.e., crystallization, caking, collapse, etc.) of the products. The practical consequences of the physical instability will be illustrated on real or model foods. When possible, prediction models adapted to the described phenomena will be presented.

The role of water in the molecular mobility of the matrix constituting the product is crucial, particularly with regards to the material's physical stability. Kinetic phenomena controlling the physical stability of foods, such as diffusion, depend on a wide range of parameters, such as relative molecular size of the diffusing molecule and medium, or the porosity of the matrix.

The influence of a_w on the physical properties of foods may be seen as indirect, solely resulting from the water content–a_w relation. There has often been discussion as to whether the effects are attributed to a_w or water content. Physical state changes drive many physical properties, a_w depends on temperature, and glass transition temperature (T_g) depends on water content. According to Chirife et al. (1999), it is very difficult to separate the effect of both water and physical state. Indeed, increasing water activity does decrease T_g, but there also might be some effects only attributable to an increase in water content. There are very few studies that simultaneously consider the effect of temperature and humidity on a given physical property that would permit discriminating a water (i.e., water-solute or water-polymer interactions) or temperature effect (i.e., physical state change). The physical stability of food products therefore cannot be separated from their sorption isotherm and phase diagram data.

Plasticization—Antiplasticization: Influence of Water on Mechanical Properties

As previously mentioned, T_g decreases with increasing a_w or water content. A decrease of T_g below ambient or working temperature results in a marked decrease in the rigidity of

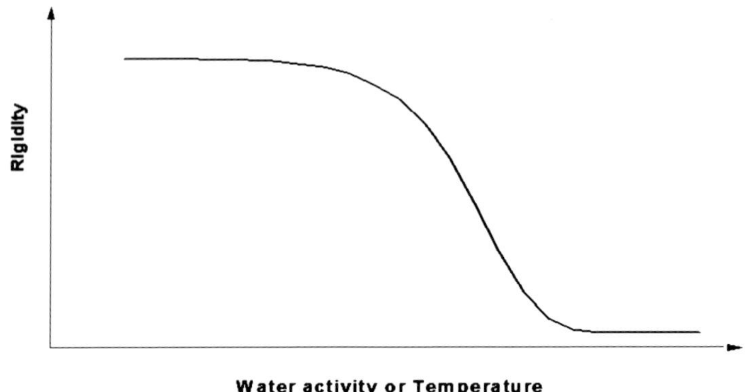

Figure 8.1 Schematic evolution of rigidity with water activity or temperature.

the hydrated product that is then plasticized (Ablett et al. 1986, Hutchinson et al. 1989). Among the mechanisms suggested to explain plasticization, the free volume concept is often used. At a given temperature, the smaller the molecules, the greater will be the associated free volume. Moreover, generally efficient plasticizers are also efficient solvents, reducing intramacromolecule and intermacromolecule interactions (Matveev et al. 2000) and replacing them by interactions involving water molecules. Sears and Darby (1982) stated that "water is the most ubiquitous plasticizer in our world." The efficiency of water as a plasticizer is based on water's affinity for other molecules and its ability to form a homogeneous mix without phase separation. The high dielectric constant of water also participates in the plasticizing action through a reduction of the electrostatic interactions between the groups of macromolecules. Generally, a plasticizer is described as a component favoring the mobility of the other component(s) in the material. The consequence of water plasticization on the material's mechanical properties has often been compared with the plasticization resulting from a temperature increase (and the associated thermal agitation). Changing the temperature at constant a_w or changing a_w at constant temperature may similarly affect the mechanical properties of a food material (see Figure 8.1).

However, in certain cases, water can exhibit an antiplasticizing effect. Instrumental and sensory measurements (Kapsalis et al. 1970, Harris and Peleg 1996, Roudaut et al. 1998, Suwonsichon and Peleg 1998, Valles Pamies et al. 2000, Waichungo et al. 2000) have shown for many products that in a limited humidity range (generally below 11% water), where water keeps its ability to decrease T_g, the rigidity increases with increasing a_w. As an example, Figure 8.2 shows the increase of sensory hardness of extruded bread between a_w 0.45 and a_w 0.6.

This mechanism, described as antiplasticizing for its opposite effect to plasticization, is of practical importance but its physical origin remains not fully understood. Two hypotheses are suggested:

- The initial water uptake may have increased the mobility, thus facilitating a possible structural rearrangement and/or a reinforcement of the interactions within the polymer constituting the product (Fontanet et al. 1997).

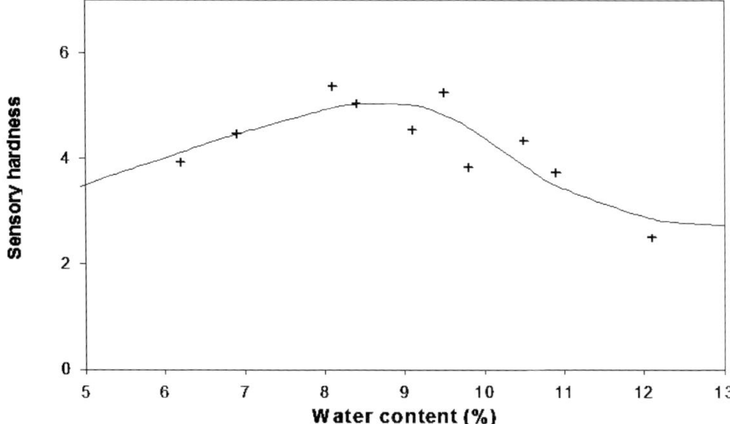

Figure 8.2 Influence of water activity on the sensory hardness of extruded bread.

- It could also result from a molecular densification of the product (decreasing free volume) (Kapsalis et al. 1970, Vrentas et al. 1988, Seow et al. 1999, Benczédi 2001).

Above a given moisture level (9% to 10% for extruded bread), the expected plasticization phenomenon would become dominant, and hardness would decrease (see Figure 8.2) (Roudaut et al. 1998).

The above-mentioned effects of water on mechanical properties play an important part in the influence of water on a food product's crispness.

Water and Crispness

The contribution of texture to the consumers' appreciation of a food product has been studied for nearly 40 years. In the early studies on the awareness of food texture, the importance of crispness was highlighted. For instance, when consumers were asked to generate attributes related to a list of specific foods, the term "crisp" was mentioned more often than any other attribute (Roudaut et al. 2002). This texture attribute may be, as a definition, associated with a combination of high-pitched sounds and the crumbling of the product as it is crushed through. Such behavior is generally encountered with puffed and brittle food products—more precisely, low-moisture systems (below 10% water or $a_W < 0.55$) rather than intermediate-moisture foods (IMF with a_W between 0.6 and 0.9 or water content between 10% and 40%), or high-moisture foods, with the exception of fruits or vegetables, which are sometimes described as crispy or crunchy.

This section will focus on cereal-based products. Many low-moisture (generally below 10% water), baked or extruded, cereal-based products such as breakfast cereals, wafers, biscuits, and snacks have a crispy texture. Generally, this texture is associated with a solid foam structure in which mechanical properties are determined by the foam's characteristics (cells' geometry and density) and the viscoelastic properties of the cell walls.

If the moisture content of these products increases, due to water sorption from the atmosphere or by mass transport from neighboring components or phases, a loss of crispness is observed (Nicholls et al. 1995). Because crispness is associated with freshness and

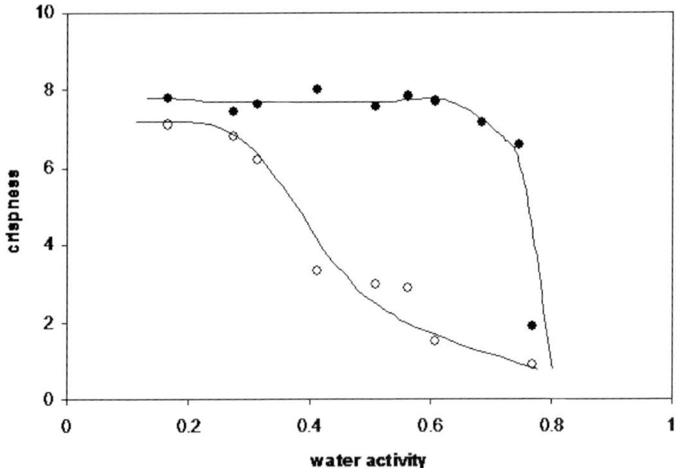

Figure 8.3 Influence of water activity on sensory crispness score (on a 1–to–10 scale) determined at first bite for extruded maize-based puffed snacks with (◦) 0% and (•) 20% sucrose (% dry basis).

quality, its loss is a major cause of consumer rejection. A great number of studies have been published on this topic (Peleg 1993, 1994, 1998; Harris and Peleg 1996; Tesch et al. 1996; Suwonsichon and Peleg 1998) with a view to characterizing and predicting the effect of water on crispness or to suggest the physical basis for such effects. The effect of hydration on crispness is sigmoidal in shape as illustrated in Figure 8.3 for maize extrudates characterized at first bite by a sensory panel. Although the different regions (before, during, and after the transition) can appear clearly separated, as with the 0% sucrose product in Figure 8.3, the changes generally observed seem to be the manifestation of a continuous process, i.e., with a smooth rather than an abrupt transition. The latter transition versus a_w can be described using the Fermi equation (8.1) (Peleg 1994) expressed as:

$$Y = \frac{Y_0}{1 + \exp\left(\frac{X - X_c}{b}\right)} \tag{8.1}$$

where Y is crispness, Y_0 crispness in the dry state, X_c the critical a_w or water content corresponding to $Y = Y_0/2$, and b is a constant expressing the steepness of the transition.

Similarly, mechanical properties featuring crispness (fracturability, amplitude of the acoustic emission at fracture) exhibit a similar evolution versus moisture (Peleg 1994, Roudaut et al. 1998). The pioneering work on the effect of water on crispness was presented by Brennan et al. (1974), followed by Katz and Labuza (1981) and Sauvageot and Blond (1991), in studies presenting sensory crispness and mechanical data for snacks and breakfast cereals equilibrated at different water activities.

Several explanations are suggested to explain such a change in texture: change in fracture mechanism (brittle to ductile behavior) (Kirby et al. 1993), glass transition (Ablett et al. 1986, Slade and Levine 1993), and localized motions preceding the glass transition

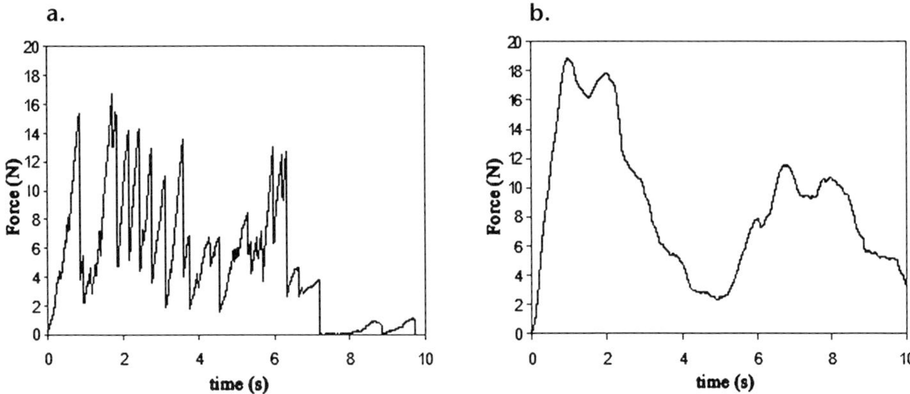

Figure 8.4 Force–deformation plots for maize extrudates equilibrated at a_w 0.28 (**a**) and a_w 0.75 (**b**).

(Roudaut et al. 1998, 2004). The fracture mechanism relates texture loss to changes in the force deformation curves obtained from compression or puncture tests of cellular foods. Each curve is characterized by two regions: a linear zone (from which stiffness modulus can be calculated) and an elasticity zone (which exhibits for brittle materials a series of peaks associated with the fracture of the individual cells constituting the product). As the material undergoes a transition from brittle to ductile, the curves change very markedly because deformation becomes dominant and the cell walls no longer break (see Figure 8.4). The overall force deformation plots are analyzed by describing their irregularities. Three main analyses have been developed: extracting parameters from each force peak, calculating the power spectrum, and determining the fractal dimension of the signal (Roudaut et al. 2002 and references therein).

As seen in Figure 8.4, when a_w increases from 0.28 to 0.75, the force–deformation curve smooths: the jaggedness (or peak numbers) is noticeably lower and has been shown to correlate to crispness (Peleg 1997, Van Hecke et al. 1998, Valles Pamies et al. 2000).

Baking or extrusion-cooking for a starch-based material leads to a major loss of the material's crystallinity. Thus, crispness has been associated with the amorphous state. Based on the previously described water sensitivity of T_g, the change from crispy (brittle, noisy) to deformable (ductile, silent) following rehydration has been attributed to a consequence of T_g decreasing below ambient temperature (Levine and Slade 1990). Very few published studies have reported the effect of temperature on this texture attribute using either the instrumental or the sensory approach. Only recently has one group developed texture tests at variable temperatures. However, such an approach appears particularly useful for verifying the role of the glass transition in crispness loss (Payne and Labuza 2005).

The effect of hydration (through a_w or water content) on crispness of cereal-based products varies with the formulation/recipe of the sample. As an example, extruded waxy maize exhibits a loss of crispness centered at a_w 0.40, whereas the critical a_w is 0.75 when the extruded waxy starch contains 20% sucrose (Valles Pamies et al. 2000) (see Figure 8.3). Such behavior cannot be explained by the glass transition because the sugar-rich sample has a lower Tg than the pure starch sample but a greater crispness at the highest a_w. It is noteworthy that at critical a_w, both samples were still in the glassy state.

Several authors report that crispness of cereal-based products could be affected by moisture pick-up (Hutchinson et al. 1989, Attenburrow et al. 1992, Kaletunc and Breslauer 1993, Nicholls et al. 1995, Li et al. 1998, Roudaut et al. 1998) at temperatures well below their T_g or the T_g measured for hydrated wheat starch (Zeleznak and Hoseney 1987).

Moreover, due to the complexity of the products, the existence of a unique glass transition and the role of the transition in the texture changes are questioned. Several authors (Kalichevsky et al. 1993, Slade and Levine 1993) have stressed that due to the products' complexity and heterogeneity, they may contain multiple phases with different T_g values. The textural changes could thus be caused by the glass transition of a minor phase (Nikolaidis and Labuza 1996), which may not be visible on differential scanning calorimetry thermograms. However, such a point may not be valid for single-component samples. For Peleg (1999), the contribution of glass transition to texture changes may not be relevant, because not all properties may change in unison as predicted by the glass transition theory. Indeed, different crispness-associated properties can be observed to vary within a wide range of water contents and thus may not result from a single event (Tesch et al. 1996).

While the contribution of the glass transition to texture changes is questioned, the importance of local motions is given attention. Indeed, localized movements of lateral groups or of small parts of the main polymer chain remain possible in the glassy state. These so-called "secondary" (β) relaxations, preceding the α-relaxation (onset of large amplitude cooperative movements), are also sensitive to water and thus could participate in the texture changes of glassy polymers (Wu 1992, Le Meste et al. 1996, Roudaut et al. 1999).

Water Activity and Softness

The influence of water on mechanical properties described for low-moisture cereal products with increasing hydration can also be considered for high- or intermediate-moisture (above 25% water) cereal products exposed to lower humidity atmospheres (below a_w 0.6).

Bread-like or sponge cake–like products exhibit a viscoelastic behavior similar to that of synthetic polymers (Slade and Levine 1991, Le Meste et al. 1992, Slade and Levine 1993). At their original moisture content (above 25%), in the rubbery plateau region, they are soft at room temperature and not very sensitive to changes in moisture content (see Figure 8.1). They become progressively leathery as moisture content decreases. Finally, below 20% moisture (this critical value depends of course on their composition), their Young's modulus increases sharply and they become brittle.

Stickiness, Caking, and Collapse

Caking and structure collapse processes generally concern products with high levels of soluble sugars, minerals, or protein hydrolysates (such as milk powders, instant coffee, dehydrated fruit juices, etc.). Generally used to describe powder behavior at a macroscopic scale, caking (formation of hard masses of greater size) might be observed as being a result of stickiness, when individual solids of a free-flowing powder stick to one another to ultimately form a mass of solids. This phenomenon can result in a range of structures—from small, easily breakable aggregates to rock-hard lumps of variable size or even a solidification of the whole powder mass (Peleg 1983). These structural changes will also be accompanied by changes in physical properties, such as an alteration of the visual aspect, flowability, handling properties, diminished wetability, and water dispersibility. Although caking may also result from fat melting, surface crystals solubilization, or electrostatic attraction, this discussion will only focus on moisture-induced caking in amorphous solids.

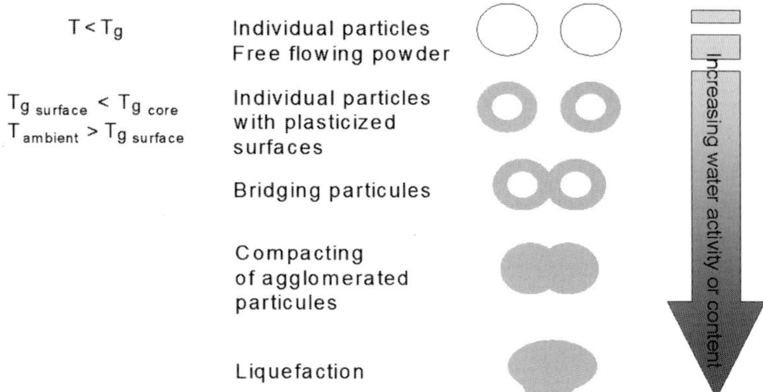

Figure 8.5 Schematic stages during caking versus a_w; water plasticization upon humidity exposure depresses T_g below ambient temperature and reduces the viscosity, allowing liquid bridging, compacting, and liquefaction of the particles.

Caking is known to occur under two major conditions, during drying or storage. However, the key parameter of the phenomenon is the same—a critical hydration level. When powder hydration increases (through water sorption, accidental wetting, moisture condensation), the particle's surface becomes plasticized (viscosity is lowcred) and tends to merge with the neighboring particle through an interparticle liquid bridging. The latter will take place if sufficient flow occurs during contact and the bridge is built strong enough to resist subsequent deformation (see Figure 8.5).

Caking occurs in several stages, and the extent of the consequences gradually increases, causing the free-flowing fine powder to turn first into an increasingly cohesive powder that is less flowing and finally into a solid mass. The early stages of caking may be desirable, when the powder made of fine particles has turned into larger particles through agglomeration. At this stage, agglomeration (known as wet granulation) is expected to improve handling and wetability for instantaneous dissolution. As previously mentioned, structural collapse is driven by the same mechanisms as caking and should be seen as an advanced stage of caking. Indeed, following the particle bridgings and their thickening, the interparticle space controlling the porosity of the material will be progressively lost, leading to a more-compact, collapsed material. .

The driving force for flow during contact is surface tension, and the resistance to flow is followed through the viscosity. Frenkel equation (8.2) (Downton et al. 1982, Wallack and King 1988) can be used to predict the order of magnitude of viscosity, which is critical for stickiness:

$$\eta = \frac{3\sigma t}{2R}\left(\frac{R}{x}\right)^2 \tag{8.2}$$

According to this equation, greater surface tension σ or longer contact time (t) will facilitate stickiness, whereas greater viscosity η (lower humidity, higher molecular weight,

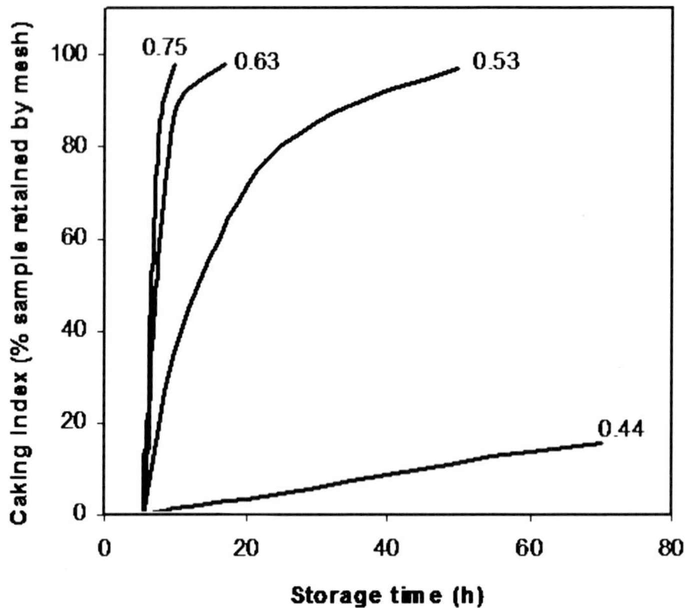

Figure 8.6 Caking index versus time for fish protein hydrolysates stored at various a_w at 30°C (from Aguilera et al. 1995).

lower temperature) or greater particle radius *R* will decrease the tendency of particles to stick (threshold ratio of × the bridge radius to *R* was estimated as 0.01). This mechanism predicts that efficient bridging should occur for combinations of temperature and moisture content corresponding to viscosity below a relatively constant value. For short-time contact, this isoviscosity should be within the range 10^6 to 10^8 Pa.s (Downton et al. 1982).

The time-scale of the observation is very important to the extent of the phenomenon studied; indeed, because particle size is time dependent, the conditions critical for stickiness to occur are different over the duration of a process (short times) or storage (long times).

The occurrence of powder stickiness is characterized by the determination of its sticky point. The *sticky point* is defined as the temperature at which the power needed to stir the powder in a tube increases sharply. Caking can be evaluated by a caking index corresponding to the amount of sample (%) retained by a given mesh. Figure 8.6 shows the effects of humidity on the caking kinetics of fish hydrolyzates: the higher the humidity (or temperature), the higher is the caking index (Aguilera et al. 1993). Similarly, the acceleration of powder caking versus relative humidity is shown in Figure 8.7 for powdered onions (Peleg 1993).

Both the collapse temperature and the sticky point have been shown to decrease when water content increases (To and Flink 1978). Above a certain humidity, when there is enough water to plasticize the system ($T > T_g$), the viscosity will be several decades lower than in the glass and the mobility will be high enough to allow flow against gravity.

The connection between caking and the glass transition is demonstrated by the parallel evolution of agglomeration temperature and T_g as a function of water content (Tsourouflis

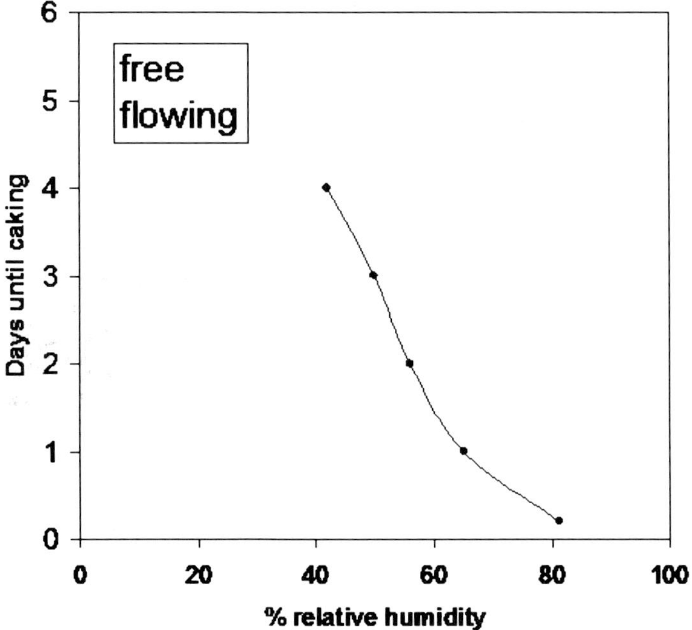

Figure 8.7 Caking kinetics (days until caking) of powdered onions versus relative humidity (from Peleg 1993).

et al. 1976, Downton et al. 1982). Roos and Karel (1991) reported, for sucrose–fructose model mixtures, a similar evolution in the sticky point (T_s) versus increasing moisture, with T_s located about 20°C above $T_{g\ onset}$. Both caking and stickiness rates increase with increasing $T - T_g$, and because many phenomena are controlled by the glass transition, they obey Williams Landel and Ferry kinetics as presented in Chapter 3 (Aguilera et al. 1995, Levi and Karel 1995).

Rather than a difficult, strict control of both relative humidity and temperature, caking may be prevented to a certain extent by the use of anticaking agents. Depending on the agent used, the operating mechanism is different (Peleg and Hollenbach 1984). The anticaking agent can:

- Prevent caking through competition for water (e.g., cellulose fibers in dry soup or cake mixes)
- Act as a physical barrier at the surface of the particle, decreasing liquid bridging (e.g., silicates in flavors, cocoa, dried soups, etc.), and then inhibiting crystal growth
- Increase T_g; in this case, anticaking agents are chosen among high-molecular-weight ingredients (e.g., starch or maltodextrins [in infant formulas]) (Hamano and Sugimoto 1978, Chuy and Labuza 1994, Aguilera 1995)

Crystallization

Given the threshold molecular mobility (both rotational and translational) that crystallization requires to take place, it is classically admitted that the mobility below T_g is not high enough to cause crystallization. A very detailed presentation of crystallization mechanisms and consequences in foods can be found in Hartel (2001).

When a_w increases, the solute molecules become increasingly mobile. Thus, the mobilized molecules collide in the proper orientation to form a nucleus. Such a nucleus is the starting point for crystallization that eventually spreads to the entire matrix. Moreover, this increased mobility facilitates the diffusion of crystallites and the overall crystal growth. Anhydrous crystal formation expels water from the matrix, and, depending on the close environment of the sample, the water molecules will diffuse toward the neighboring phase or to the atmosphere if the sample is exposed to open air. In some cases, crystallization can be a self-maintained process; indeed, as water is excluded from the crystals, it further plasticizes the system, "inducing a crystallization front" within the system. This water diffusion to the amorphous surrounding matrix may induce chemical degradations as described in Chapter 7 on chemical stability.

Among the products for which crystallization is particularly critical are spray-dried milk powders, ice cream, hard candies, soft cookies, and baked products. The a_w–crystallization relationship will be applied to several examples that follow.

Lactose and milk powders

Crystallization of lactose in products, such as ice cream or milk powders, has been studied since 1930 (Troy and Sharp 1930) as a result of its important practical consequences: decreased flowability of powders and granular texture development in ice cream. The rate of lactose crystallization in dairy powders increases when storage relative humidity is increased. Moreover, lactose will crystallize in the beta anhydride or in the alpha hydrate form depending on the a_w and temperature range (Saltmarch and Labuza 1980, Vuataz 1988).

The effect of water (as well as of temperature) on crystallization kinetics has been predicted by the Williams-Landel-Ferry (WLF) equation for lactose (Roos and Karel 1992), sucrose (Levi and Karel 1995), and lactose in milk powder (Jouppila and Roos 1994).

Soft cookies

Upon baking, soft cookies generally reach a moisture level where at room temperature they remain above their T_g and thus are perceived as soft. Under such conditions, sucrose is in the amorphous rubbery state (as well as in supersaturated conditions); thus it is in an unstable state. Its resulting crystallization (Levine and Slade 1993) is likely to cause the cookies to become firmer over storage time with an accompanying decreased acceptability of the products by consumers. The problem is partly solved with the use of high-fructose corn syrup, which maintains softness by inhibiting sucrose crystallization.

Hard candies

Hard candies are prepared at low a_w, between 0.2 and 0.4 (Bussiere and Serpelloni 1985), with temperature being the key technological parameter of the process. Water uptake on further handling or storage will cause two subsequent events. Depending on their composition (as mentioned in Chapter 3), T_g of the hard candies will be between 25° and 45°C; thus, in the first stage, the confectionery product undergoes a glass–to–supercooled liquid transition. Then, in a second stage, above T_g (high temperature exposure has similar effects), the product starts to flow, loses its shape (i.e., cold flow), and starts to crystallize. Such a phenomenon is described for confectionery as *graining*; indeed, it results in an

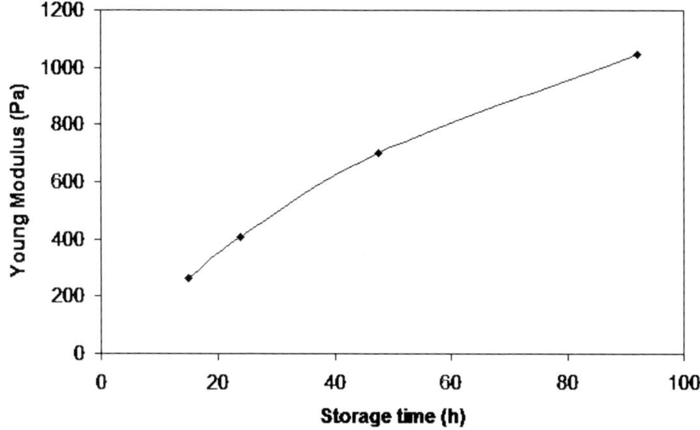

Figure 8.8 Elastic modulus versus aging time for bread crumbs (from Fessas and Schiraldi 1998).

opaque and glossless product, which may feel sandy (grained) in the mouth. The kinetics are controlled by a_w; below 0.12, there is no crystallization apparent after 3 years, whereas above 0.33, it appears within 3 days (Roos and Karel 1991).

Crystallization of starch during storage

As previously mentioned for soft cookies, soft freshly baked products at room temperature are mainly amorphous (as a result of the loss of crystallinity on baking) and above their T_g. In these conditions (above T_g but below melting temperature, T_m), long-range cooperative, structural reorganization such as crystallization might occur. This physical state change, called starch *retrogradation*, is accompanied by a well-known increase of product firmness over storage time (see Figure 8.8).

Crystallization rate versus temperature is known to exhibit a maximum between T_g and T_m—at low temperature, the high viscosity hinders the diffusion required for crystals growth, whereas the nucleation is limited at temperature approaching T_m. A similar bell-shaped behavior has been described for the evolution of crystallization rate as a function of water content. Indeed, isothermal retrogradation kinetics go through a maximum as a function of water content (see Figure 8.9); this maximum is shifted toward higher water content with decreasing temperature (Farhat and Blanshard 2001).

These results highlight that the role of water content on retrogradation kinetics is very different depending on the storage temperature. Indeed, whereas increasing moisture content above 35% (wb) accelerates retrogradation at 0°C, the effect is opposite at 40°C. This can be easily understood by a shift of both T_g and T_m to lower values due to the plasticizing effect of water.

Conclusion

This chapter has reviewed several aspects of the a_w or water content relationship with physical stability. Many properties of foods are affected by physical stability, especially where

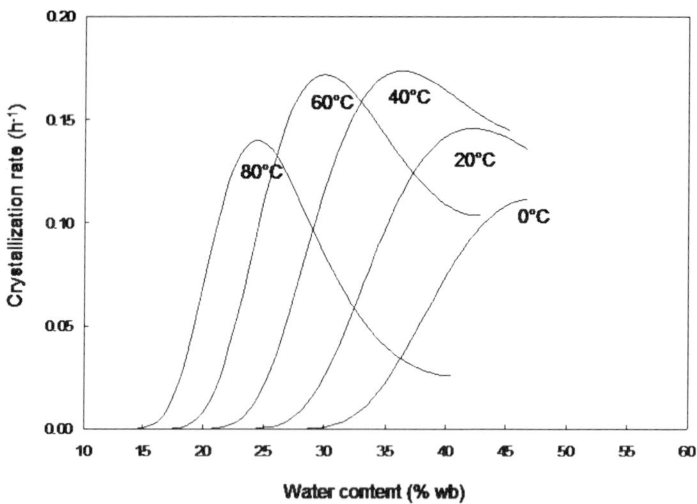

Figure 8.9 Simulation of effect of storage temperature on rate of recrystallization of starch–water systems as a function of water content (from Farhat and Blanshard 2001).

food structure is involved, whether at a molecular level (i.e., physical state) or at a macroscopic level (i.e., porosity).

If the role of water is mainly limited for sugar-rich products to a decrease of T_g below working temperature, an empirical rule based on the distance to T_g can be proposed (Colonna 2002).

a_w corresponding to T_g +10:	Adhesion to walls
a_w corresponding to T_g +20:	Caking
a_w corresponding to T_g +30:	Crystallization
a_w corresponding to T_g +40:	Collapse and mass caking (large scale)

If these relationships are in some cases closely related to the glass transition, the behavior of physical properties versus a_w or content is often more complex than a straightforward consequence of the glass transition at ambient temperature. Moreover, humidity-induced physical changes such as crystallization or densification will also affect chemical stability through a kinetic control of the diffusion. For example, crystallization of lactose has been shown to induce the release and oxidation of oil (methyl linoleate) encapsulated in lactose-gelatin food model (Shimada et al. 1991).

The physical stability of foods is of primary importance at different stages of a food product's shelf-life. Because the physical stability may affect several time-scales, an optimal control of the humidity will be required during processing, handling, transportation, storage, and consumption. When the effects of water cannot be overcome by the product's composition, technological solutions may lay in the design of moisture barriers or by a strict control of the temperature (e.g., refrigerated transportation). Therefore, future work should be carried out to obtain a better understanding of the mechanisms involved in physical instability and to further develop solutions to prevent the degradation of the physical properties of food systems.

References

Ablett, S., Attenburrow, G.E., and Lillford, P.J. 1986. The significance of water in the baking process. In: *Chemistry and Physics of Baking: Material, Process and Products*, eds. J.M.V. Blanshard, P.J. Frazier, and T. Galliard, pp. 30–41. London: Royal Society of Chemistry.

Aguilera, J.M., Del Valle, J.M., and Karel, M. 1995. Caking phenomena in amorphous food powders. *Trends in Food Science and Technology* 6:149–155.

Aguilera, J.M., Levi, G., and Karel, M. 1993. Effect of water content of the glass transition and caking of fish protein hydrolyzates. *Biotechnological Progress* 9:651–654.

Attenburrow, G.E., Davies, A.P., Goodband, R.M., and Ingman, S.J. 1992. The fracture behaviour of starch and gluten in the glassy state. *Journal of Cereal Science* 16:1–12.

Benczédi, D. 2001. Plasticization and antiplasticization of starch glasses by water. In: *Water Science for Food, Health, Agriculture and Environment*, eds. Z. Berk, R.B. Leslie, P.J. Lillford, and S. Mizrahi, pp. 273–281. Lancaster, PA: Technomic.

Bourne, M.C. 1986. Effect of water activity on texture profile parameters of apple flesh. *Journal of Texture Studies* 17:331–340.

Brennan, J.G., Jowitt, R., and Williams, A. 1974. Sensory and instrumental measurement of 'brittleness' and 'crispness' in biscuits. *Proceedings of the Fourth International Congress on Food Science and Technology* 11:130–143.

Bussiere, G., and Serpelloni, M. 1985. Confectionery and water activity: Determination of a_w by calculation. In: *Properties of Water in Foods,* eds. D. Simatos and J.L. Multon, pp. 635–647. Dordrecht, the Netherlands: Martinus Nijhoff.

Chirife, J., Buera, M.P., and Gonzalez, H.L. 1999. The mobility and mold growth in glassy/rubbery substances. In: *Water Management in the Design and Distribution of Quality Foods*, eds. Y.H. Roos, R.B. Leslie, and P.J. Lillford, pp. 285–298. ISOPOW 7. Lancaster, PA: Technomic.

Chuy, L.E., and Labuza, T. 1994. Caking and stickiness of dairy-based food powders as related to glass transition. *Journal of Food Science* 59(1):43–46.

Colonna, P. 2002. Rôle de l'eau dans les propriétés sensorielles des aliments. In: *L'eau Dans les Aliments*, eds. M. Le Meste, D. Lorient, and D. Simatos, pp. 21–47. Lavoisier, Paris: Tec&Doc.

Downton, G.E., Flores-Luna, J.L., and King, C.J. 1982. Mechanism of stickiness in hygroscopic, amorphous powders. *Industrial and Engineering Chemical Fundamentals* 21:447–451.

Duizer, L.M., Campanella, O.H., and Barnes, G.R.G. 1998. Sensory, instrumental and acoustic characteristics of extruded snack food products. *Journal of Texture Studies* 29, 397–411.

Farhat, I.A., and Blanshard, J.M.V. 2001. Modeling in the kinetics of starch retrogradation. In: *Bread Staling*, eds. P. Chinachoti and Y. Vodovotz, pp. 163–172. Boca Raton, FL: CRC Press.

Fessas, D., and Schiraldi, A. 1998. Texture and staling of wheat bread crumb: Effects of water extractable proteins and "pentosans." *Thermochim Acta* 323:17–26.

Fontanet, I., Davidou, S., Dacremont, C., and Le Meste, M. 1997. Effect of water on the mechanical behaviour of extruded flat bread. *Journal of Cereal Science* 25:303–311.

Hamano, M., and Sugimoto, H. 1978. Water sorption, reduction of caking and improvement of free flowingness of powdered soy sauce and miso. *Journal of Food Processing and Preservation* 2:185–196.

Harris, M., and Peleg, M. 1996. Patterns of textural changes in brittle cellular foods caused by moisture sorption. *Journal of Texture Studies* 73:225–231.

Hartel, R.W. 2001. *Crystallization in Foods*, p. 325. Gaithersburg, MD: Aspen.

Hutchinson, R.J., Mantle, S.A., and Smith, A.C. 1989. The effect of moisture content in the mechanical properties of extruded food foams. *Journal of Material Science* 24:3249–3253.

Jouppila, K., and Roos, Y.H. 1994. Water sorption and time-dependent phenomena of milk powders. *Journal of Dairy Science* 77:1798–1808.

Kaletunc, G., and Breslauer, K.J. 1993. Glass transition of extrudates: Relationship with processing-induced fragmentation and end-product attributes. *Cereal Chemistry* 70:548–552.

Kalichevsky, M.T., Blanshard, J.M.V., and Mash, R.D.L. 1993. Applications of mechanical spectroscopy to the study of glassy biopolymers and related systems. In: *The Glassy State in Foods*, eds. J.M.V. Blanshard and P.J. Lillford, pp. 133–156. Nottingham: Nottingham University Press.

Kapsalis, J.G., Walker, J.E., and Wolf, J.C. 1970. A physico-chemical study of the mechanical properties of low and intermediate moisture foods. *Journal of Texture Studies* 1:464–483.

Katz, E.E., and Labuza, T. 1981. Effect of water activity on the sensory crispness and mechanical deformation of snack food products. *Journal of Food Science* 46:403–409.

Kirby, A.R., Parker, R., and Smith, A.C. 1993. Effects of plasticizers on the fracture behaviour of wheat starch. Special publication of the Royal Society of Chemistry 133:250.

Le Meste, M., Huang, V.T., Panama, J., Anderson, G., and Lentz, R. 1992. Glass transition of bread. CFW Research of the American Association of *Cereal Chemists* 37:264–267.

Le Meste, M., Roudaut, G., and Rolée, A. 1996. Physical state and food quality—An example: The texture of cereal-based-foods. In: *Food Engineering 2000*, eds. P. Fito, E. Ortega-Rodriguez, and G.V. Barbosa-Cánovas, pp. 97–113. New York: Chapman and Hall.

Levi, G., and Karel, M. 1995. Volumetric shrinkage (collapse) in freeze-dried carbohydrates above their glass transition temperature. *Food Research International* 28:145–151.

Levine, H., and Slade, L. 1990. Influence of the glassy and rubbery states on the thermal, mechanical, and structural properties of doughs and baked products. In: *Dough Rheology and Baked Product Texture*, eds. H. Faridi and J.M. Faubion, pp. 157–330. New York: Van Nostrand Reinhold.

Levine, H., and Slade, L. 1993. The glassy state in applications for the food industry, with an emphasis on cookie and cracker production. In: *The Glassy State in Foods*, eds. J.M.V. Blanshard and P.J. Lillford, pp. 333–373. Nottingham: Nottingham University Press.

Li, Y., Kloeppel, K.M., and Hsieh, F. 1998. Texture of glassy corn cakes as function of moisture content. *Journal of Food Science* 63:869–872.

Matveev, Y.I., Grinberg, V.Y., and Tolstoguzov, V.B. 2000. The plasticizing effect of water on proteins, polysaccharides and their mixtures. Glassy state of biopolymers, foods and seeds. *Food Hydrocolloids* 14:425–437.

Nicholls, R.J., Appelqvist, I.A.M., Davies, A.P., Ingman, S.J., and Lillford, P.J. 1995. Glass transitions and fracture behaviour of gluten and starches within the glassy state. *Journal of Cereal Science* 25–36.

Nikolaidis, A., and Labuza, T.P. 1996. Glass transition state diagram of a baked cracker and its relationships to gluten. *Journal of Food Science* 61:803–806.

Payne, C.R., and Labuza, T.P. 2005. Correlating perceived crispness intensity to physical changes in an amorphous snack food. *Drying Technology* 23:887–905.

Peleg, M. 1983. Physical characteristics of food powders. In: *Physical Properties of Food*, eds. M. Peleg and E.B. Bagley, pp. 293–323. Westport, CT: AVI Publishing Co.

Peleg, M. 1993. Glass transitions and the physical stability of food powders. In: *The Glassy State in Foods*, eds. J.M.V. Blanshard and P.J. Lillford, pp. 435–451. Nottingham: Nottingham University Press.

Peleg, M. 1993. Mapping the stiffness-temperature-moisture relationship of solid biomaterials at and around their glass transition. *Rheologica Acta* 32:575–580.

Peleg, M. 1994. A mathematical model of crunchiness/crispness loss in breakfast cereals. *Journal of Texture Studies* 25:403–410.

Peleg, M. 1997. Measures of line jaggedness and their use in foods textural evaluation. *Critical Reviews of Food Science and Nutrition*. 37:491–518.

Peleg, M. 1998. Instrumental and sensory detection of simultaneous brittleness loss and moisture toughening in three puffed cereals. *Journal of Texture Studies* 29:255–274.

Peleg, M. 1999. Phase transitions and the mechanical properties of food biopolymers. In: *Biopolymer Science: Food and Nonfood Applications*, eds. P. Colonna and S. Guilbert, pp. 271–282. Paris: INRA Editions.

Peleg, M., and Hollenbach, A.M. 1984. Flow conditioners and anticaking agents. *Food Technology* March:93–102.

Roos, Y., and Karel, M. 1991. Plasticizing effect of water on thermal behavior and crystallization of amorphous food models. *Journal of Food Science* 56:38–43.

Roos, Y., and Karel, M. 1992. Crystallization of amorphous lactose. *Journal of Food Science* 57:775–777.

Roudaut, G., Dacremont, C., and Le Meste, M. 1998. Influence of water on the crispness of cereal based foods acoustic, mechanical, and sensory studies. *Journal of Texture Studies* 29:199–213.

Roudaut, G., Dacremont, C., Valles Pamies, B., Colas, B., and Le Meste, M. 2002. Crispness: A critical review on sensory and material science approaches. *Trends in Food Science and Technology* 13:217–227.

Roudaut, G., Maglione, M., and Le Meste, M. 1999. Sub-Tg relaxations in cereal-based systems. *Cereal Chemistry* 76:78–81.

Roudaut, G., Poirier, F., Simatos, D., and Le Meste, M. 2004. Can dynamical mechanical measurements predict brittle fracture behaviour? *Rheologica Acta* 44:104–111.

Sauvageot, F., and Blond, G. 1991. Effect of water activity on crispness of breakfast cereals. *Journal of Texture Studies* 22:423–442.

Sears, J.K., and Darby, J.R. 1982. Mechanism of plasticizer action. In: *The Technology of Plasticizers*, pp. 35–77. New York: Wiley Intersciences.

Seow, C.C., Cheah, P.B., and Chang, Y.P. 1999. Antiplasticization by water in reduced-moisture food systems. *Journal of Food Science* 64:576–581.

Shimada, Y., Roos, Y., and Karel, M. 1991. Oxidation of methyl linoleate encapsulated in amorphous lactose-based food model. *Journal of Agricultural and Food Chemistry* 39:637–641.

Slade, L., and Levine, H. 1991. A food polymer science approach to structure-property relationships in aqueous food systems: Non-equilibrium behavior of carbohydrate-water systems. In: *Water Relationships in Foods*, eds. H. Levine and L. Slade, pp. 29–102. New York: Plenum Publisher.

Slade, L., and Levine, H. 1993. The glassy state phenomenon in food molecules. In: *The Glassy State in Foods*, eds. J.M.V. Blanshard and P.J. Lillford, pp. 35–102. Nottingham: Nottingham University Press.

Suwonsichon, T., and Peleg, M. 1998. Instrumental and sensory detection of simultaneous brittleness loss and moisture toughening in three puffed cereals. *Journal of Texture Studies* 29:255–274.

Tesch, R., Normand, M.D., and Peleg, M. 1996. Comparison of the acoustic and mechanical signatures of two cellular crunchy cereal foods at various water activities levels. *Journal of the Science of Food and Agriculture* 70:347–354.

To, E.C., and Flink, J.M. 1978. "Collapse": A structural transition in freeze dried carbohydrates. *Journal of Food and Technology* 13:551–594.

Troy, H.C., and Sharp, P.F. 1930. Alpha and beta lactose in some milk products. *Journal of Dairy Science* 13:140–157.

Tsouflouris, S., Flink, J.M., and Karel, M. 1976. Loss of structure in freeze dried carbohydrates solutions: Effect of temperature, moisture content and composition. *Journal of the Science of Food and Agriculture* 27:509–519.

Valles Pamies, B., Roudaut, G., Dacremont, C., Le Meste, M., and Mitchell, J.R. 2000. Understanding the texture of low moisture cereal products: Mechanical and sensory measurements of crispness. *Journal of the Science of Food and Agriculture* 80:1679–1685.

Van Hecke, E., Allaf, K., and Bouvier, J.M. 1998. Texture and structure of crispy-puffed food products. II: Mechanical properties in puncture. *Journal of Texture Studies* 29:617–632.

Vrentas, J.S., Duda, J.L., and Ling, H.C. 1988. Antiplasticization and volumetric behavior in glassy polymers. *Macromolecules* 21:1470–1475.

Vuataz, G. 1988. Preservation of skim-milk powders: role of water activity and temperature in lactose crystallization and lysine loss. In: *Food Preservation by Moisture Control,* ed. C.C. Seow, pp. 73–101. Amsterdam: Elsevier Applied Science.

Waichungo, W.W., Heymann, H., and Heldman, D.R. 2000. Using descriptive analysis to characterize the effects of moisture sorption on the texture of low moisture foods. *Journal of Texture Studies* 15:35–46.

Wallack, D.A., and King, C.J. 1988. Sticking and agglomeration of hygroscopic, amorphous carbohydrate and food powders. *Biotechnology Progress* 4:31–35.

Wu, S. 1992. Secondary relaxation, brittle-ductile transition temperature and chain structure. *Journal of Applied Polymer Science* 46:619–624.

Zeleznak, K.J., and Hoseney, R.C. 1987. The glass transition in starch. *Cereal Chemistry* 64:121–124.

9 Diffusion and Sorption Kinetics of Water in Foods

Theodore P. Labuza and Bilge Altunakar

Shelf-life refers to the period of time that a food will retain an acceptable level of eating quality based on safety and organoleptic perspectives. Microbiological, enzymatic, and physicochemical reactions that simultaneously take place in any food are the major points of interest affecting this time period. To estimate a food's expected shelf-life using scientific models, identification of the main mechanisms that cause spoilage or loss of desirable characteristics, such as changes in texture, flavor, odor, or loss of key constituents, such as nutrients, is required.

Formulation, processing, packaging, and storage conditions, together with their relative impact as a function of food perishability, are the critical factors contributing to shelf-life estimation, while the effects of time, temperature, and humidity are crucial to food quality. At this point, knowledge of diffusion and sorption kinetics provides a scientific understanding of food stability as related to moisture gain or loss as well as a quantitative approach to maintaining food quality. If the changes taking place can be described mathematically, the way that these changes affect the quality of a product in any food process can easily be predicted. Therefore, this chapter is concerned with the kinetics of water diffusion and sorption in foods, including the possible use of a quantitative analysis of this process to predict shelf-life under different storage conditions. We often define food quality as a parameter, which can be any number of factors or combination of factors. This could be the result of chemical reactions, such as the reaction of fat with oxygen to produce off-flavor; it also could be due to physical changes—like loss of crispiness in a potato chip or formation of a visible defect such as black spots on shrimp. Combined, these can lower the acceptance or rejection scores, i.e., consumer perception (Labuza 2000).

Water vapor sorption by solids depends on many factors, among which the chemical composition, physicochemical state of ingredients, and physical structure are some of the most important ones contributing to this behavior. These parameters determine the quantity of moisture absorbed and the kinetics of the moisture sorption process (Spiess and Wolf 1986). The amount and the kinetics of moisture diffusion highly depend on the achievement of thermodynamical equilibrium, because moisture loss or gain from one region, or food component, to another region or component will occur continuously in order to reach thermodynamic equilibrium with the surrounding food components and environment. In the case of multidomain systems, for example, this includes food products that involve the partitioning of many components with different water activities and, thus, different chemical potentials such as granola bars with nuts, cereal pieces, and perhaps candy fruit bits. Moisture is exchanged because of the chemical potential difference between the components until the system finally reaches equilibrium water activity (a_w) throughout each domain. Diffusion and moisture migration kinetics play important roles in these types of dynamic systems. Similarly, in the case of shelf-life prediction, the shelf-life of a food product is determined by numerous complex interactions between parameters related to the product itself and/or associated with the external environment. These modifications could be chemical, physical, enzymatic, or microbiological and are mainly due to mass exchanges between foods and their environments. The finished product must be protected with effective packaging material such as metal cans, glass jars, or plastics. The performance of the material is estimated based on its efficiency to reduce mass transfer, which is measured in terms of the permeability to the specific component. The characteristics required depend on the deterioration reactions of the product being protected. For example, the fat in potato chips can oxidize to form off-odors, or off-flavors, and is controlled by the oxygen level in the package. The oxygen level is related to the difference between the rate at which oxygen is reacting with the lipid and the film permeability. At the same time, the rate is also controlled by the moisture content–a_w point on the moisture sorption isotherm, as pointed out in Chapter 5. Complicating this is the fact that if the a_w of the food is less than the external percent relative humidity (% RH), water will permeate into the package, raising the food's a_w and vice versa. The rate of gain or loss of moisture is a function of film permeability, and the area to volume (solid weight) of the food contained. Finally, each of these rates (permeance and reaction) is temperature dependent, giving a highly complex mathematical situation. In fact, mathematical solutions of this problem have only been done for potato chips (Quast and Karel 1972, Quast et al. 1972) and fresh roasted ground coffee (Cardelli and Labuza 2001).

Based on this, control of initial a_w and moisture migration is critical to the quality and possibly the "safety" of many foods. Ideally, food manufacturers develop products with defined processing conditions (e.g., 12 log cycles will kill *Botulinum* spores) and moisture contents in order to produce a safe finished product with optimum shelf-life. Quality and safety factors that manufacturers must consider are microbial growth, especially of pathogens, physical state, sensory properties, and the rate of chemical changes leading to loss of shelf-life. For dry and semimoist foods, shelf-life will depend on moisture content and a_w of each domain. Therefore, to achieve control of moisture migration, principles of mass transfer are used as the basis for preventive actions such as reconfiguration of the food to achieve similar water activities or adding edible barriers between the layers. An example is the inner chocolate layer used to reduce moisture transfer from frozen ice cream to the baked cereal cone used for frozen novelties and prefrozen and filled ice cream cones.

In this chapter, we examine the dynamics of moisture diffusion based on the basic principles of mass transfer and apply the fundamental relations for predicting moisture gain or loss through packaging. Finally, we review the methods and instrumentation used to measure moisture diffusion.

Diffusion of Water

Any particular component in a single-phase system can move, and it can be redistributed in absence of external mechanical or physical constraints, or pressure gradient. The concept of diffusion originates with this fact. In a given system, diffusional mass flow is proportional to the gradient of the concentration of the species with the defined matrix. Diffusion of water in simple gases and liquids can be analyzed and predicted by molecular dynamics. However, for the case of diffusion of water in solids and semisolids, which is of particular importance in foods and food processing systems, the mechanisms differ (Saravacos and Maroulis 2001).

Table 9.1 Typical Values for Effective Moisture Diffusivity (D_{eff}).

Product	a_w	Thickness (mm)	D_{eff} (m^2/sec) × 10^{-12}
Flour	0.11	3.6	3.86
Flour	0.75	8.3	32
Nonfat dry milk	0.75	3.1	21.3
Freeze-dried apple	0.75	2.7	4.06
Freeze-dried raw ground beef	0.75	10.9	30.6
Oatmeal cookie	0.75	10.1	3.97
Shredded wheat	0.75	2.9	5.52
Raisins	0.75	—	0.416

Source: Adapted from Tutuncu., M.A., and Labuza, T.P. 1996. Effect of geometry on the effective moisture transfer diffusion coefficient. *Journal of Food Engineering* 30:433-447.

It is generally assumed that transport of water in solids is controlled by molecular diffusion where the driving force can be a concentration gradient or moisture content gradient. Fick's law of diffusion refers to movement of a component through a binary mixture under a constant vapor pressure gradient at constant temperature. The driving force is therefore the chemical potential or vapor pressure difference between the two regions. For simplified analysis and calculations, one-dimensional diffusion is considered, applying Fick's first law of diffusion as:

$$J = -D\frac{dC}{dx}$$
$$J = \frac{m^2}{\sec}\frac{gH_2O}{m^3 m} = \frac{gH_2O}{\sec m^2} \tag{9.1}$$

where J is the flux defined as amount of moisture exchanged per unit of time per unit area (g H_2O/sec m^2), C is the concentration as mass per unit volume, x is the distance transversed by the concentration gradient, D is the diffusion coefficient (L^2/time), and dC/dx is the concentration gradient. The diffusion coefficient D of water in solids is thus usually defined as D_{eff} effective moisture diffusivity, which is an overall transport property incorporating all transport mechanisms and individually is hard to measure. Typical values for effective moisture diffusivity are given in Table 9.1.

Any solution for determination of moisture migration from one region or component to another, to reach equilibrium, requires information on the moisture diffusion properties of both components, which can be characterized by an effective moisture diffusion coefficient (D_{eff}). Water can be transported within solids by several transport mechanisms including capillary flow, Knudsen diffusion, transfers due to heat and pressure gradients, and external forces resulting in both vapor and liquid diffusion (Marousis et al. 1989).

Diffusion of Water in Foods

The transport of water in food materials is of fundamental importance as noted above. Although various mechanisms have been proposed to explain water transport in food materials, the effective diffusion model yields satisfactory results in engineering and technological applications, improving the quality and stability of the food (Simatos et al. 1981).

The equilibrium relationship between partial pressure of water in the gas phase and moisture content has been discussed extensively in Chapter 5 as being the basis of moisture sorption in foods. A number of empirical models have been proposed to characterize the adsorption and desorption process under equilibrium conditions. Depending on the sorption process, the activity of water at the surface of a food can be very different from the a_w inside, especially if dry air comes in contact with food containing large amounts of water or wet air in contact with a dry food. This difference might lead to evaporation of water with a corresponding decrease of a_w at the interface, transport of solutes toward the interface accompanying the migration of water, or diffusion of solutes from the surface toward the bulk of the solid (Loncin 1980). Evaporation of water decreases temperature at the surface of the solid, facilitating heat transfer between the air and interior of the matrix.

As far as transport properties are concerned, food structure plays a decisive role in water transport processes within food materials. According to various authors (Bruin and Luyben 1980), food materials can be classified into groups based on structural differences. The first group includes liquids and gels such as milk, fruit juices, and gelled products. The second group consists of dimensionally stable capillary-porous and hygroscopic-porous materials such as packed beds of materials; the last group includes dimensionally unstable capillary-porous and hygroscopic-porous deformable materials. Materials in the last group have matrices of a colloidal nature, exhibit shrinkage, and may develop a porous structure during drying as in vegetables and meat (Roques 1987). The structural collapse may be due to flow in the rubbery state as noted in Chapter 5. Model food materials based on granular and gelatinized starch are convenient experimental materials available for studying the mechanism of water movement in various food structures.

Factors Affecting Moisture Migration

Water activity equilibrium (thermodynamics) and rate of diffusion (dynamics of mass transfer) are the two main factors influencing moisture migration in multidomain foods. Multidomain foods with regions formulated to differentiate water activities cause the whole system to be in a nonequilibrium state. This will result in moisture migration from a higher a_w region (higher chemical potential) to the lower a_w region and can result in undesirable changes in the system described above.

The following example illustrates the condition of equilibrium. A single dry cracker with a_w 0.3 is put in a sealed chamber with a relative humidity of 75% at 25°C. The initial moisture of the cracker is 3% (w/w), and thus it gains water at this high humidity, i.e., a_w of air is 0.75, and for the cracker, 0.3, so there is a driving force toward the cracker. The final moisture of the cracker after equilibration is about 15% (w/w) when the cracker a_w equals the a_w of the humidified air at 75% RH. In a separate chamber, at the specified condition (i.e., 75% RH and 25°C), a piece of processed cheese food product (PCFP) with initial a_w 0.95 is introduced. Initially, the PCFP cheese has a moisture content of 60% (w/w) but loses its moisture because, opposite that of the cracker, its a_w of 0.95 is greater than that of the chamber at 0.75, i.e., the driving force causes evaporation into the air. After equilibration, the PCFP has a moisture content of about 25% (w/w) and a_w 0.75. The final a_w of both the cracker and the cheese food product is 0.75 because both were equilibrated at 75% RH (see Equation 9.1). This is the thermodynamic paradigm. The a_w in the two systems is the same, although moisture contents are quite different (15% versus 25%). If the cheese and cracker are now packaged together, no moisture exchange occurs because they are in thermodynamic equilibrium with the same chemical potential. As noted in Chapter 5, Equation 9.2 relates thermodynamic chemical potential of water vapor to a_w:

$$\mu = \mu_0 + RT \ln\left(\frac{p}{p_0}\right) = \mu_0 + RT \ln a_w \tag{9.2}$$

where μ is chemical potential of the water in each of the systems (the chamber, the cracker, and the cheese), μ_o is chemical potential of pure water vapor, *R* is ideal gas law constant, T is the temperature, p is the actual vapor pressure of water in equilibrium with each system, and p_o is the vapor pressure of water at saturation at temperature T. Using the above example, after equilibration the chemical potential of the water in each system is the same, i.e., $\mu_{cracker} = \mu_{chamber} = \mu_{cheese}$, and thus, $a_{w\ cracker} = a_{w\ chamber} = a_{w\ cheese}$; all have the same a_w.

To repeat, the chemical potentials of the water are the same when all the systems are brought to equilibrium at the same constant relative humidity. Even though the amount of moisture differed in the PCFP cheese and cracker, their water activities were the same; thus, no change in moisture content occurs when both are put together, unless chemical reactions or physical state changes take place. This shows that one approach to preventing moisture migration is the elimination of differences in a_w so as to maintain the original conditions. The question, of course, is whether one can achieve the desired textural characteristics at the final a_w. No one has yet found a way to make a crispy cracker or potato chip with high-moisture/a_w value. However, a soft product can be made at low a_w by using a liquid humectant like glycerol to lower the a_w. This was the basis of a patent developed by Kellogg to lower the a_w of raisins, such that when mixed with cereal (raisin bran) at 20% by weight, the raisins still maintained a soft texture and did not cause the breakage of teeth, a problem with untreated raisins. As seen, the difference in a_w explains the direction of change but has no influence on the rate of change to reach equilibrium. The rate is dependent on the matrix properties and geometries of each system, which can be modeled by Fick's first law of diffusion when applied to simple systems where vapor pressure remains constant on each side (not a real situation). For systems at constant temperature, the steady-state mass transfer for moisture can be expressed as:

$$\frac{dm}{dt} = \frac{k}{x} A(p_1 - p_2) \tag{9.3}$$

where dm/dt is the amount of moisture exchanged per unit of time (g H_2O/time), A is the area of transfer, *k* is the effective permeability of water, x is the path length for diffusion (meter), and p_1 and p_2 are the water vapor pressures in systems 1 and 2, respectively. It is obvious that in real systems, as moisture is exchanged, the difference between p_1 and p_2 becomes smaller and thus the rate of moisture transfer decreases. Thus, making the a_w of the systems as close as possible has some benefits.

Rate of Diffusion

Theoretically, diffusion of a small molecule, such as water vapor through a food domain, is controlled by the molecule's size, presence of other molecules in vapor phase that water molecules may collide with, and surrounding geometry, i.e., pore size distribution and how tortuous is the path. Because we are dealing with foods, generally at 1 atm total pressure (approximately 0.1 MPa), with air present in pores, the mean free path of a water molecule is very short because of the presence of nitrogen and oxygen molecules. Reported values of D_{eff} for materials are generally evaluated from either single particles or a bulk of material, where it is assumed both values are the same (Tutuncu and Labuza 1996). Example values were given in Table 9.1; further on we will see how D_{eff} values are measured. We can take a deeper look at Fick's law of diffusion in Equation 9.1, which describes unsteady-state diffusion as:

$$\frac{\partial m}{\partial t} = D_{eff} \frac{\partial^2 m}{\partial x^2} \tag{9.4}$$

where m is the moisture content (g H_2O/g dry solids), t is the time (seconds), D_{eff} is the effective moisture diffusivity (m^2/sec), and x is the distance (meters). According to Fick's law, D_{eff} is assumed as constant and diffusion as unidirectional where the external mass transfer resistance and volume change on moisture loss or gain are assumed negligible. Initial and boundary conditions are:

Initial condition	$m = m_0$	$0 \leq x \leq L_0$	$t = 0$
Boundary condition (1)	$m = m_e$	$t = 0,\ x = 0$	
Boundary condition (2)	$m = m_e$	$t = 0,\ x = 2L_0$	$t > 0$

The solution under these conditions for the average moisture content at any time t is a series solution obtained by the separation of variables method (Geankoplis 1972). In order to normalize the differences in initial sample weights, a dimensionless number Γ based on moisture content is defined and referred to as the unaccomplished moisture ratio. For moisture loss:

$$\Gamma = \frac{m - m_e}{m_i - m_e} \tag{9.5}$$

where m is the moisture at time t, m_e is the moisture reached at equilibrium with external atmosphere, m_i is the critical moisture and $m - m_e$ is defined as the unaccomplished moisture change, where at time $t = 0$, $\Gamma = (m_1 - m_e)/(m_i - m_e) = 1$.

As time approaches infinity (∞), then Γ approaches 0, so a plot of Γ versus time shows a decreasing slope.

For moisture gain:

$$\Gamma = \frac{m_e - m_i}{m_e - m} \tag{9.6}$$

Thus, at time $t = 0$ again $\Gamma = 1$ as m increases, Γ approaches ∞ (i.e., $m_e - m = 0$) with $m_e - m$ being the unaccomplished moisture change. So in this case the slope is upward starting out at $\ln\Gamma = \ln 1 = 0$.

Crank (1975) solves the problem of diffusional mass transfer in a plane sheet and an infinite cylinder for a given geometry of constant half-thickness and with a constant diffusion coefficient assuming molecular diffusion. Mathematical models for given geometries can be found in work by Geankopolis (1972) and Crank (1975). In a porous domain, molecules are allowed to transfer more quickly because of several mechanisms of moisture

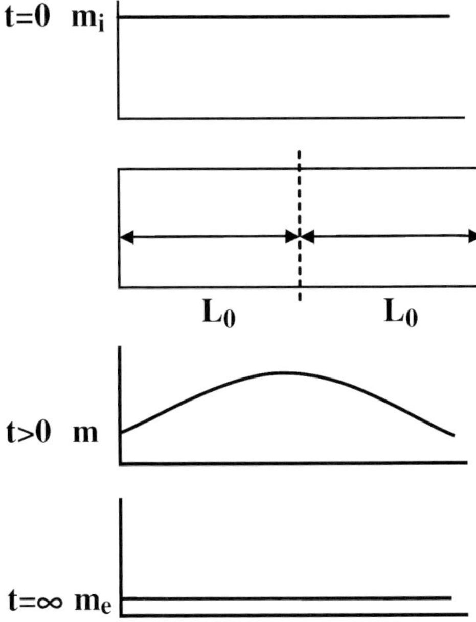

Figure 9.1 Unsteady-state diffusion for boundary conditions.

transfer (e.g., capillary action) together with liquid diffusion along the surface. In low-porosity materials, liquid diffusion is the main method of moisture transport. The smaller the pore size in the food domain matrix, the slower is the moisture migration. In addition, all the membranes, crystals, and lipids are barriers to moisture migration in a multidomain system. The more crystals or lipid interferences present in a system, the slower is the diffusion of water. It is important to note that the effective diffusion coefficient is dependent on both surrounding structure and solubility of water in the matrix, the latter of which is governed by the GAB equation. In Fick's first law (see Equation 9.1) the overall permeability of a matrix (k) is related to these factors through Equation 9.7:

$$k = D_{eff} S \tag{9.7}$$

where D_{eff} is the effective diffusivity in pore space [L^2/t (distance^2/time)] and S is the solubility of the diffusant in the material (g/cm^3). For food systems under consideration here, overall mass transfer is much more complex because it involves unsteady state diffusion (i.e., Δp or the a_w difference decreases over time), which has an exponential dependence on moisture, while the geometric dependence is L_0^2, not L_0. For example, if the thickness of a filling is doubled, equilibration will take four times longer. In the case of moisture loss from a slab transferred into the surroundings, i.e., an unpackaged food, overall transfer is usually modeled as Equation 9.8:

$$\ln \Gamma = \frac{m - m_e}{m_i - m_e} = \ln \frac{8}{\pi^2} - \frac{D_{eff}\pi^2}{4L_0^2} t \tag{9.8}$$

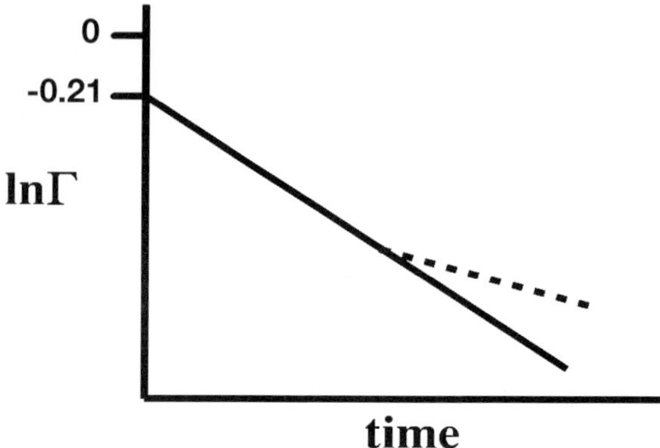

Figure 9.2 Plot of simple solution to Crank equation.

where m_e is final equilibrium moisture (dry basis), m_i is initial equilibrium moisture (dry basis), m is moisture at time t, D_{eff} is effective diffusivity in dimensions of L^2/t (e.g., m^2/sec where in this case m is meters), L_o is distance from the center to the surface, and Γ is the unaccomplished moisture ratio (moisture amount remaining to change over total moisture amount that potentially changes). Thus, a plot of lnΓ versus time (see Figure 9.2) gives a simple solution to the Crank equation.

Note that at $t = 0$, $\ln\Gamma = \ln[8/\pi^2] = -0.21$, not 1. Thus, the equation does not work well at the initial boundary condition. In fact, one usually obtains a broken line showing two effective diffusivities, one with a lower slope. Also note that many researchers ignore the $\ln[8/\pi^2]$ term and plot the lnΓ from the 0 value; i.e., they forget the true intercept. Tutuncu and Labuza (1996) showed that the above simple equation gives erroneous results when estimating D_{eff} and that at least 19 terms in the series expansion should be used to obtain the true D_{eff} (see Equation 9.9) within a single matrix:

$$\Gamma = \frac{8}{\pi^2} \sum_{n=0}^{n=19} \frac{1}{(2n+1)^2} \exp\left[\frac{-(2n+1)^2 \pi^2 D_{eff}}{4L_0^2}\right] t \tag{9.9}$$

Even this equation is difficult to use, as it assumes either a spherical or slab dimension and no effect from surrounding air space if the domain consists of particles like cereal flakes or raisins. In fact, Tutuncu and Labuza (1996) showed that measuring D_{eff} is very dependent on geometry, and thus any experimental design must take final product configuration into consideration. More recently, finite element and finite difference computer techniques have been used to model nonsimple geometric systems to obtain effective diffusivities (Hong et al. 1986).

The influence of pore structure on diffusion or permeability has been modeled by Geankopolis (1972). Unfortunately, there are few published values on the effective diffusivity of water vapor or liquid transport in pores of processed foods that would constitute a multidomain system. Most values are from drying experiments that are generally not applicable, as noted by Tutuncu and Labuza (1996). In general, effective diffusivity values for moisture range from 10^{-9} to 10^{-12} m^2/sec with higher values applicable to porous foods like cereal and lower values to dense foods like raisins or dates, as seen in Table 9.1. It should be noted the diffusion coefficient for water vapor in air is 2.4×10^{-5} m^2/sec and the self-diffusion coefficient in liquid water is about 2.4×10^{-9} m^2/sec. Thus, the structure of food can slow down diffusion significantly, which is an advantage in the creation of multidomain foods.

Another way to slow diffusion between two domains is to create a diffusion barrier between them. Obviously, this has to be an edible barrier, and therefore available barrier material is a limiting agent. There are several practical applications in the food industry to slow down moisture diffusion between different domains. Biquet and Labuza (1988) showed that chocolate is a good barrier against moisture because of the dispersion of fat (cocoa butter) throughout the matrix. This has limited applications in the case of candies, confectioneries, frozen bakery goods, and frozen desserts. For example, a chocolate layer placed on the inside of an ice cream cone helps reduce the moisture pickup from the ice cream filling in the cone and thereby maintains crispness of the cone during storage. Similarly, a mixed starch–lipid layer can be used between the sauce and the crust to prevent sogginess of crust in frozen pizza. Additionally, par-frying the crust clogs the pores with oil and expands the structure to increase path length "L," thereby reducing moisture pick-up. Biquet and Labuza (1988) summarized the basic properties needed for a good moisture barrier and described three major properties of importance for moisture barrier selection: pore size, structure, and adhesiveness. A good moisture barrier should have pores small enough to prevent water from passing through or to slow down the diffusion process. It should not contain structures similar to the permeant molecules, and it must be adhesive to the food surface. Fennema et al. (1993) and Miller and Krochta (1997) have given good reviews on the limitations of edible based films.

Diffusion of Water in Polymers

Understanding the mechanism of water sorption and diffusion in amorphous polymers is of particular importance in our interpretation of related transport phenomena in foods (Peppas and Peppas 1994). Water diffusion in polymers often deviates from the predictions of Fick's law, leading to non-Fickian diffusional behavior. The mechanism of transport is generally indicated with sorption kinetics by using a polymeric (usually a thin film) sample:

$$\frac{m}{m_e} = kt^n \tag{9.10}$$

and

$$\ln\frac{m}{m_e} = \ln k + n\ln t \tag{9.11}$$

where m and m_e are the moisture contents after sorption time t and at equilibrium, respectively, k is a constant, and n is the diffusional index. The diffusional index characterizes the type of diffusion, where for the value n, it equals 1.0, and the fraction sorbed is proportional to the square root of time. Here, the constant k acts as an effective diffusivity term but with units of inverse time to the 1/n power. Then best fit is determined by plotting $\ln[m/m_e]$ versus lnt for different values of n. For a sample of thickness L and effective diffusivity D_{eff}, the constant k for $n = 1$ can be derived as:

$$k = 4\frac{D_{eff}}{\pi x^2} \tag{9.12}$$

The physical structure of the polymeric material affects the effective moisture diffusivity D_{eff}. For glassy materials, values of D can be in the order of 10^{-14} m^2/sec, but the value can increase by 10^3 above the glass transition temperature for rubbery materials (Saravacos and Maroulis 2001).

As water penetrates the polymeric system, mechanical stresses are relaxed, resulting in a significant volume increase in the polymer structure, allowing for faster diffusion.

Film Moisture Diffusivities

As a water vapor barrier, a film can extend shelf-life and raise the quality of foods by limiting moisture migration, which could accelerate deteriorative reactions (Labuza 1980, Rockland and Nishi 1980). Permeability is a simple measure of water vapor transmission, which is defined as the ability of a substrate to transfer water vapor from one side to the other. When a film is maintained at a constant temperature, and there is a constant relative humidity on each side of the film, then the amount of water vapor passing through a given area of film for a given time is expressed as water vapor transmission rate (WVTR):

$$WVTR = \frac{dw}{Adt} = \frac{k}{x}\Delta p = \frac{\text{g water}}{\text{day.area}} \tag{9.13}$$

where x is film thickness in appropriate units, A is area (m^2), Δp is water vapor pressure difference across film (mm Hg H_2O), and k/x represents film permeance where k is film permeability in units, which vary depending on the way things are reported. For example, in the United States, polymer films are usually measured in thousandths of an inch and sold at 100 square feet, so the permeability k would be:

$$k = \frac{\text{grams water. } 1000^{\text{ths}} \text{of an inch}}{\text{day.100'' sf. mm Hg}}$$

which is a confusing mix of English and metric units. The permeance k/x is found by dividing k by the film thickness and the WVTR is found by multiplying k/x by the area and by Δp, the driving force. To summarize, WVTR is water vapor transmission amount per time area. Relative permeabilities of other gases in comparison to water vapor and water vapor permeance values are given in Table 9.2. As expected, water has the greatest permeability because it is small. CO_2, although larger than N_2 and O_2, has a greater permeability than those molecules, mainly because it has a zero dipole moment and thus easily dissolves in the plastic. Because the permeance is a product of the diffusion coefficient multiplied by solubility, this makes CO_2 faster but not as fast as water vapor.

Generally, the experimental method for measuring the WVTR of a film with uniform thickness (x) maintained at an external constant temperature and relative humidity is to use special cups filled with desiccant (Thwing-Albert cups). The dry and wet cup methods described in American Society for Testing and Materials (ASTM) E96-95 are commonly used as two standard dish methods. The dry cup method requires a desiccant, usually anhydrous calcium chloride, to maintain the inside of the dish at 0% RH. The wet cup method specifies distilled water in the dish to maintain 100% RH. When the weight of the cup is plotted as a function of time, the y-axis represents the gain of water; taking the slope and dividing by the exposed area gives the WVTR, as shown in Figure 9.3.

Table 9.2 Relative Permeabilities of Other Gases Compared to Water Vapor.

		$k = \frac{cc\ mil}{day\ m^2\ mmHg}$		
Film	H_2O	CO_2	O_2	N_2
Saran (PVC)	3.4×10^4	680	170	34
Mylar (PET)	5.1×10^6	3400	680	136
Nylon-6	3.4×10^7	2000	1000	340
Cellulose acetate	3.4×10^8	1.36×10^5	2×10^4	6800
Polyethylene (LD)	3.4×10^5	6.8×10^4	2×10^4	6800
Polystyrene	5.1×10^7	2.7×10^5	3.4×10^4	—

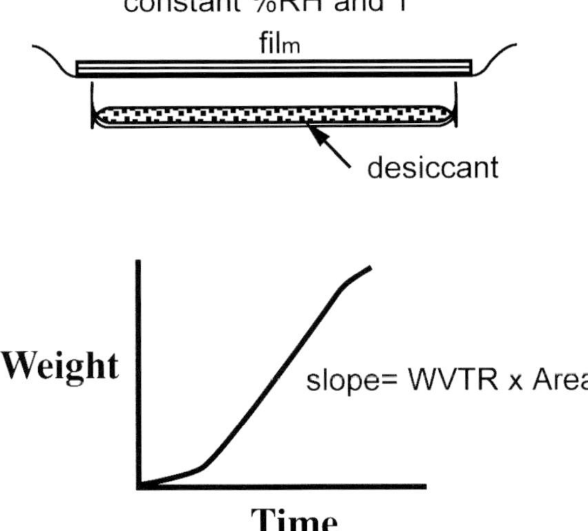

Figure 9.3 Water vapor transmission rate (WVTR) determination.

Generally, a 7- to 10-day period is used such that the desiccant does not reach its limit. From this WVTR, we can then calculate *k* and/or *k/x* as follows:

$$WVTR = \frac{slope}{A} = \frac{amount}{time.area} \quad (9.14)$$

$$k = permeability = \frac{WVTR}{\Delta p} x = \frac{slope}{A\ \Delta p} x \quad (9.15)$$

$$\frac{k}{x} = permeance = \frac{WVTR}{\Delta p} = \frac{slope}{A.\Delta p} \quad (9.16)$$

Various units for these are shown in Table 9.3.

Table 9.3 Units Used in Reporting Permeance Values

Amount of Vapor Transfer (Δn)	Thickness of Film (l)	Area (A)	Time (Δt)	Pressure Difference (Δp)
cc (STP)	mm	cm^2	sec	cm Hg
moles	cm	cm^2	sec	cm Hg
cc (STP)	cm	cm^2	sec	cm Hg
grams	cm	cm^2	sec	mm Hg
cc (STP)	mil	100 in^2	24 hr	atm
grams	cm	cm^2	24 hr	atm
cc (STP)	mil	m^2	24 hr	atm
cc (STP)	cm	cm^2	min	atm
grams	mil	100 m^2	hr	atm
cc (STP)	mil	100 in^2	100 hrs	atm

Table 9.4 Typical Film Moisture Permeability Values of Films.

Film	ASTM E96 WVTR Value 100°F/90%RH (g/day • m^2)	g/day • m^2 • mm Hg *k/x*
Cellophane		
Lacquered	8–15	0.18–0.34
Polymer coated	8–14	0.18–0.32
Cellulose acetate	200	4.52
Nylon	200	4.52
Pliofilm	8	0.18
Polyester	15	0.34
Polyethylene		
Low density	18	0.40
Medium density	8–15	0.18–0.34
High density	5–10	0.11–0.23
Polypropylene	8–10	0.18–0.23
Polystyrene	100	2.3
Saran	1.5–5	0.03–0.11
Vinyl	50	1.13
Foil laminate	0.1	0.0023

Table 9.4 gives the values for film permeabilities to moisture for a variety of common packaging films.

The key with films is that they have a resistance to moisture transfer greater than that of the food matrix; the same types of functions apply assuming a constant outside % RH and constant temperature, where:

$$\ln \Gamma = \ln \frac{m_c - m}{m_c - m_i} = \frac{k}{x} \frac{p_0}{b} \frac{A}{w_s} t = \phi t \tag{9.17}$$

Thus, a plot of lnΓ versus time is a straight line of slope ϕ with the constants *k* [film permeability in (g water • film thickness)/area of film • time • mm Hg H_2O vapor pressure], *x* (film thickness in similar units as in *k*), p_o (saturation water vapor pressure at temperature T of the experiment), *b* (isotherm slope in g water/100 g solids in the region of concern), A [film package area (same units as used in *k*)], W_s (dry weight solids/100 contained in package), m (moisture content as a f(t) in g water/100 g solids), m_i (initial moisture content), and m_c (maximum critical moisture content allowed on the basis of the straight line assumption for the isotherm in the region of concern).

From the above, when $m = m_i$ at time = 0/ $\ln\Gamma = \ln 1 = 0$; thus, as time progresses, we get a straight line of slope ϕ. This equation has been shown to work in most conditions given that the isotherm is known and a real *k* value is determined using dry desiccant as the food in a similar package and at the same test external % RH (Bell and Labuza 2000). This will be shown in greater detail below.

Multilayer Packaging

Use of multilayered packaging films or several films in series or parallel provides enhanced resistivity to oxygen and moisture. Series of flexible packaging film inserted into waxed paper box, such as used for cereals, is one example commonly used in food industry. For this case, the resistivity of films used in series is calculated as sum of the resistances:

$$\frac{1}{\Sigma\ \text{resistance}} = \frac{1}{\dfrac{x_T}{k_0 A}} = \frac{1}{\dfrac{x_1}{k_1 A} + \dfrac{x_2}{k_2 A} + \dfrac{x_3}{k_3 A}} \tag{9.18}$$

When multiple films are used in parallel, as in blister packages for drugs and candies, resistivity of the film is calculated as the sum of permeabilities for each film's area:

$$\Sigma\ \text{resistance} = \left[\frac{k}{x}\right]_0 A_T = \left[\frac{k}{x}\right]_1 A_1 + \left[\frac{k}{x}\right]_2 A_2 + \left[\frac{k}{x}\right]_3 A_3 \tag{9.19}$$

where *k/x* is permeability of each film layer with a surface area of A.

Nuclear Magnetic Resonance and Magnetic Resonance Imaging in Diffusion of Water in Foods

Most studies of moisture transport in foods and related polymeric materials involve experiments that yield integral functions. Data obtained from these functions are often insufficient for investigating the physics of moisture transport, especially for complex systems such as foods. Magnetic resonance imaging (MRI) and nuclear magnetic resonance imaging (NMR) are two of the concise techniques available for such investigation, which makes it possible to resolve coefficients used in determining effective transport coefficients, material structure, and material properties (MacCarthy et al. 1994). Recent work using NMR and MRI technology (Chen et al. 1997) is leading to a better understanding of moisture transport in foods.

NMR and MRI can be used to measure transport of mass by measuring molecular diffusion coefficients and/or by measuring internal gradients in component saturations. With MRI, actual rates of diffusion can be visualized during storage of foods and diffusivities estimated. Ruan and Litchfield (1992) used MRI to follow moisture mobility and distribution of corn kernels. Schmidt and Lai (1991) reviewed applications of NMR and MRI to study of water relations in foods. Umbach et al. (1992) calculated water self-diffusion coefficients in starch-gluten-water systems using NMR, while Heil et al. (1993) studied water migration in baking biscuits using MRI. Jeffrey et al. (1994) showed the capability of NMR microscopy to give radial profiles of water velocity in a model system. From these studies, both MRI and NMR methods seem to be useful in gaining a further understanding of rates and paths of moisture migration. Theoretically, self-diffusion coefficients for water from NMR data are higher than effective diffusion coefficients calculated from drying data. Differences between macroscopic and microscopic data are presumed to be a result of changes in sample dimension during drying experiments as well as due to impermeable barriers within the matrix. Extensive discussion on this subject can be found in Chapter 4, "Water Mobility in Foods."

Prediction of Food Shelf-Life Based on Moisture Barriers

The aim of food packaging is to prevent degradation in food, a condition that results in lower quality and makes the food unsuitable for consumption. Therefore, selecting the proper packaging is of great importance in considering transmission properties. A number of packaging models exist to predict moisture transfer and shelf-life of packaged foods (Heiss 1958, Karel 1976, Mizrahi et al. 1970a). These models assume that water vapor transmission through the packaging film is the rate of the limiting step. Many foods gain or lose moisture during long-term storage because commonly used packaging materials are permeable to moisture, where the water vapor pressure difference between the inside and outside of package under constant temperature is the driving force behind vapor diffusion. Steady-state mass transfer is assumed as the major mechanism of moisture transfer through the package, wherein WVTR is derived by substituting Henry's law into Fick's first law of diffusion, as illustrated in the sample calculation (see the "Diffusion of Water" section earlier in this chapter).

Packaging Predictions

Establishing sufficient product shelf-life is the primary goal of packaging. The package is designed to ensure shelf-life of product and is limited to within the time required to reach a critical moisture content. Microbiological activity and chemical stability based on composition of foods may be considered in determining a critical moisture content. If a food is susceptible to oxidation of unsaturated lipids, such as processed cereals, dry meats, fish, and vegetables, the rate of shelf-life loss increases as a_w decreases below the monolayer value (see Chapter 5). This monolayer value can be viewed as critical moisture content, which is associated with the critical a_w value. Figure 9.4 represents the relationship between moisture content and critical a_w based on quality, texture, and microbial growth limits.

The basic principle is to retain a food material's original condition after packaging by preventing moisture gain or loss based on values of critical moisture contents. A critical moisture content of a food material is generally used to determine maximum allowable moisture gain or loss from initial moisture content of the food. This critical moisture content is determined from the moisture sorption isotherm and the stability map. Simple equations have been derived to estimate moisture gain or loss in a food held in a semipermeable package using isotherm and the film permeability to moisture (Labuza and Cotreras-Medellin 1981, Taoukis et al. 1988).

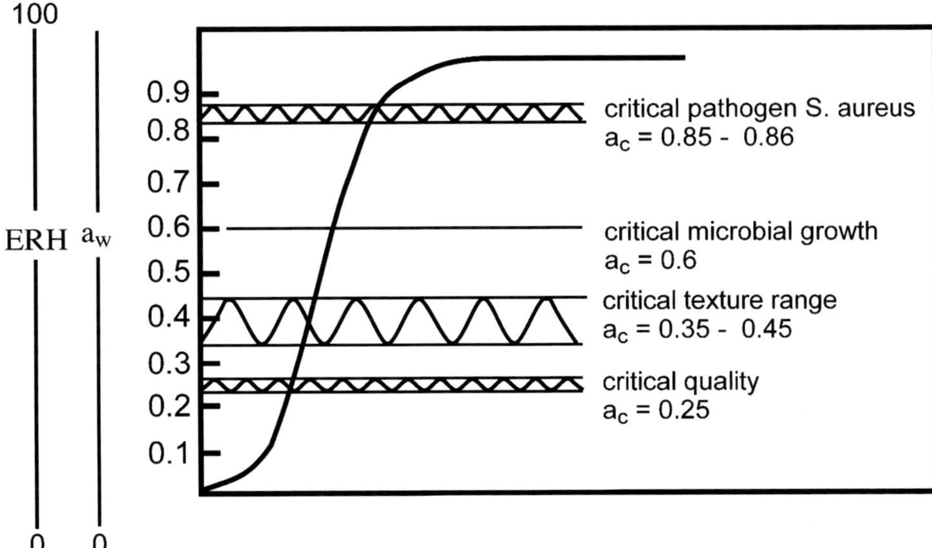

Figure 9.4 Water activity (a_w) versus moisture content plot showing critical water activities (a_c). ERH, equilibrium relative humidity.

Moisture Gain and Loss in Packaging

Moisture gain or loss from the food inside the package can be predicted by starting with a set of assumptions. The simplest is that the rate of gain or loss is constant over the whole shelf-life:

$$\frac{dm}{Adt} = t\frac{k}{xw_s}\left[p_{out} - p_{in}\right] \tag{9.20}$$

where w_s is the weight of solids contained and p_{out} (vapor pressure outside) and p_{in} (vapor pressure inside the package) are constant. This would be true only for a short period of time for moisture gain or loss for a high-moisture food as was shown in Figure 9.3, where a_w is essentially constant over a large moisture range. Then, the weight change would follow a straight line as in Figure 9.5, as noted earlier. For dry and semimoist foods, a change in moisture would affect the vapor pressure inside the package as determined by the mass of solids contained and the local slope of the moisture sorption isotherm. For a gain, the driving force would decrease over time, thereby resulting in "unsteady state mass transfer" with a decrease in rate of gain or loss over time. This is indicated by the lower line in Figure 9.5.

Several assumptions need to be made to help estimate the moisture gain or loss under unsteady conditions. First, it is assumed the food inside a package follows a linear moisture sorption isotherm with slope b (see Figure 9.6) for the range of moisture change that takes place. In this case, the portion of moisture sorption isotherm used takes the form:

$$m = b(a_w) + I \tag{9.21}$$

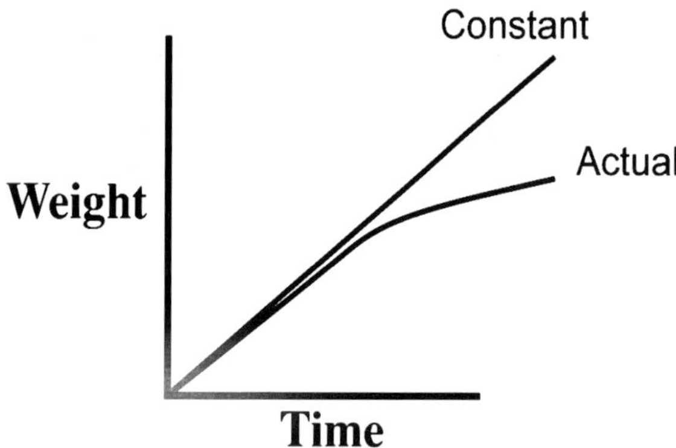

Figure 9.5 Moisture gain by a system under steady-state (constant) versus unsteady-state conditions.

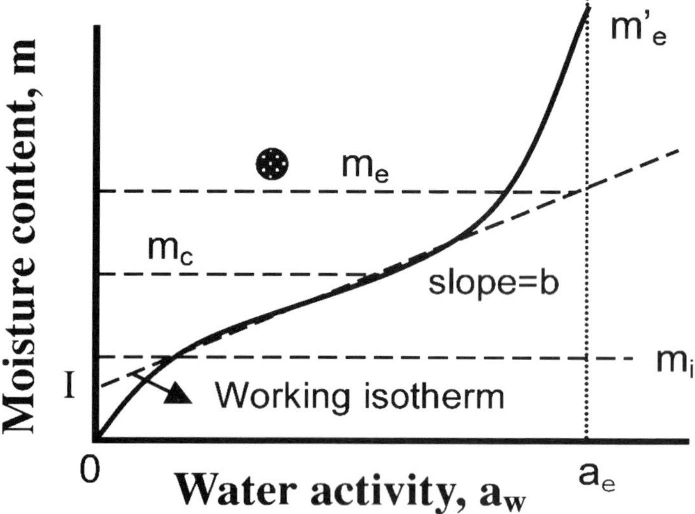

Figure 9.6 Working moisture isotherm used for packaging predictions, superimposed on true isotherm with the initial (m_i), critical (m_c), and equilibrium (m_e) moistures indicated (adapted from Bell, L.N., and Labuza, T.P. 2000. *Moisture Sorption: Practical Aspects of Isotherms Measurements and Use.* St. Paul, MN: AACC International Publishing).

where m is moisture content on the assumed straight line determined as a function of a_w, *b* is slope of line g H_2O/g solid, and I is the intercept at a = 0. Second, it is assumed that the packaging film is the major barrier to moisture transfer, which requires the assumption that instantaneous equilibrium of moisture is reached inside the package over time.

As shown in Figure 9.6, the straight line position chosen goes between the initial moisture content m_i and the critical moisture content m_c chosen as the upper limit for shelf-life.

The slope *b* in Equation 9.20 is the slope between these points. A dilemma exists in most situations because there are different intersections of the moisture content that the product would reach if placed with no packaging at the external humidity and an a_w of a_e. As seen, there is an m_e for the assumed straight line assumption and another m_e' on the real moisture sorption isotherm. Because of the mathematics, the former is chosen, and the solution then is to solve the following:

$$\frac{dm}{dt} = \frac{k}{x}\frac{A}{w_s}\left[p_{out} - p_{in}\right] \tag{9.22}$$

$$m_e = \mathrm{ba}_e + I = \frac{bp_{out}}{p_0} + I \tag{9.23}$$

$$m = \mathrm{ba}_{in} + I = \frac{bp_{in}}{p_0} + I \tag{9.24}$$

Substituting in the next equation for p_{out} and p_{in}, we get:

$$\frac{dm}{dt} = \frac{k_x}{x}\frac{A}{w_s}\left[\frac{p_0}{b}\left[m_e - m\right]\right] \tag{9.25}$$

Rearranging gives

$$\int_{m_i}^{m}\frac{dm}{m_e - m} = \int_0^t \frac{k_x}{x}\frac{A}{w_s}\frac{p_0}{b}dt \tag{9.26}$$

$$\ln\Gamma = \ln\left[\frac{m_e - m_i}{m_e - m}\right] = \frac{k}{x}\frac{A}{w_s}\frac{p_0}{b}t \tag{9.27}$$

Thus, at time 0, $\ln\Gamma = 0$ because $m = m_i$ and Γ increases linearly with time at slope = ϕ:

$$\Phi_{ext} = \frac{k}{x}\frac{A}{w_s}\frac{p_0}{b} \tag{9.28}$$

Thus, very easily, the time to reach some critical moisture content, m_e, can be determined as long as we know the permeability (*k/x*), the package area A, the weight of solids contained w_s, the isotherm slope *b*, and the vapor pressure of pure water p_0 at the test temperatures. Critical a_w values are 0.35 for caking, 0.4 for loss of crispiness, and 0.6 for microbial spoilage. Again, note that the solution is based on moisture content traversing the "linear working isotherm" represented by the dashed line in Figure 9.6; the time t required to reach a certain critical moisture content is derived by rearranging Equation 9.27 as:

$$t = \frac{\ln\Gamma}{\dfrac{kAp_0}{xw_s b}} = \frac{\ln\Gamma}{\Phi_{ext}} \tag{9.29}$$

Obviously, the isotherm must be determined at the same temperature at which moisture gain is predicted. The same assumptions are held for predicting moisture loss within a package.

The equation for moisture loss is similar to Equation 9.26 as follows:

$$\ln \Gamma = \ln \left[\frac{m_i - m_e}{m - m_e} \right] = \frac{kAp_0}{xw_s b} t \tag{9.30}$$

where Γ is unaccomplished moisture content, m_i is initial moisture, m_e is the equilibrium moisture on the linear equation for the isotherm, *k/x* is film permeance in g H_2O/day • m^2 • mm Hg, A is area of package, p_0 is vapor pressure of pure water at temperature of test, *b* is moisture sorption isotherm slope, and m is moisture content at any time *t*. A plot of $\ln\Gamma$ versus time will give a straight line, and a plot of m versus *t* shows a curved line. This curvature is because the driving force behind moisture gain decreases as moisture is picked up, because internal vapor pressure increases. The following example illustrates prediction of shelf-life for potato chips as a result of moisture gain up to a critical a_w, which yields loss of crispiness in potato chips.

Example 9.1

Potato chips, which become organoleptically unacceptable (i.e., become rubbery) at an a_w value of 0.45, were stored in mylar and polyethylene bags under abuse conditions at 35°C and 100% RH. Two replicates (A and B) of mylar and polyethylene bags sealed with chips at an initial moisture content of 1 g H_2O/100 g solids showed the following gains in weight over time (see Table 9.5).

To solve the problem, we need to know the film permeabilities that have been reported in industrial literature, as follows:

Polyethylene film:

WVTR = 9.3 g/m^2 day at 100°F/90% RH

Mylar film:

WVTR = 17 g/m^2 day at 39.5°C/90% RH

Note that permeabilities are in WVTR units and must be converted to *k/x* to solve the equation. For polyethylene, the WVTR test was reportedly done at 100°F/90% RH, which gives external water vapor pressure (p_{out}) of 49.2 (p_o value at 100°F) × 0.9 (% RH/100), or 44.28 mm Hg. The WVTR test (ASTM E96) is carried out by sealing film over dry desiccant in a special test cup. Thus, water vapor pressure inside (p_{in}) is 0. The driving force behind moisture transfer across the film to the desiccant is constant. Rate of moisture gain is:

$$\frac{\Delta W}{\Delta \theta} = \frac{gH_2O}{day} = \frac{k}{x} A[p_{out} - p_{in}] \tag{9.31}$$

for p_{out} constant and p_{in} equal to 0. Then,

$$\frac{\Delta W}{\Delta t} = \frac{k}{x} Ap_{out} \tag{9.32}$$

where p_{out} is outside vapor pressure. WVTR is expressed as g H_2O/day m^2, so

Table 9.5 Moisture Gains in Potato Chips Packaged in Mylar and Polyethylene Bags Stored at 35°F and 100% RH (A and B Represent Two Replicates).

	Package Weight in Grams*			
	Polyethylene		Mylar	
Time (days)	**A**	**B**	**A**	**B**
0	11.84	13.42	10.00	9.58
1.93	12.02	13.61	10.27	9.83
3.76	12.09	13.69	10.38	9.91
5.78	12.22	13.84	10.55	10.07
7.66	12.23	13.87	10.56	10.08
9.99	12.32	13.95	10.66	10.17
Bag weight, g	3.22	3.22	0.92	0.92
Total area, both sides, m^2	0.02908	0.02904	0.03017	0.0264

*Sealed with chips at mi = 1 g H_2O/100 g solids.

$$\frac{\Delta W}{\Delta t} = (WVTR)A \tag{9.33}$$

which is the slope of weight gain versus time for straight-line portion of test data. WVTR is slope dW/dt divided by package area. Thus,

$$\frac{k}{x} = \frac{WVTR}{p_{out}} = \frac{\Delta W / \Delta t}{Ap_{out}} \tag{9.34}$$

Therefore, for given conditions of polyethylene,

$$\frac{k}{x} = \frac{WVTR}{44.28} = \frac{9.3}{44.28} = 0.21 \text{ g H}_2\text{O/day m}^2 \text{ mm Hg} \tag{9.35}$$

The same calculation for the mylar bag under given conditions at 39.5°C/90% RH,

$$\frac{k}{x} = \frac{17}{44.28} = 0.384 \text{ g H}_2\text{O/day m}^2 \text{ mm Hg} \tag{9.36}$$

Because actual values of *k/x* can vary due to stretching of film, end seals, etc., it is best to take actual measurements using bags suspended over a vapor source. Bags are filled with desiccant, sealed, and then weighed periodically for up to 7 to 10 days. Results of this study at 35°C/100% RH (p_{out} = 42.2 mm Hg) show that manufacturer's reported permeance value is about three times higher than actual measured values.

The next step is to model the moisture sorption isotherm. A straight line is drawn on the isotherm that best fits data for the region the product moisture will traverse. We know from data that the initial moisture content (m_i) is 1 g H_2O/100 g solids and the critical moisture (m_c) for loss of crispiness is 4.6 g H_2O/100 g solids at a_w of 0.45. Thus, the equation of a straight line drawn through the two points gives the slope b = 0.0882 g H_2O/100 g solids.

If product is left in external conditions, then moisture (m_e) is calculated from Equation 9.16 under abuse conditions at 35°C/100% RH.

To obtain moisture content m_i, the initial weight at time zero is subtracted from total weight at any time and divided by weight of dry solids, and then added to initial moisture content. Thus,

$$m = \frac{(W)_t - (W)_i}{W_{drysolids}} + m_i \tag{9.37}$$

and

$$W_{drysolids} = (W_{initial} - W_{packaged}) \cdot \left[\frac{\%solids}{100} \right] \tag{9.38}$$

For example, for the polyethylene package at 1.93 days:

$$\%solids = 100 \left[1 - \frac{0.01}{1 + 0.01} \right] = 99.01 \tag{9.39}$$

$$W_{drysolids} = (11.84 - 3.22).9901 = 8.535\text{g} \tag{9.40}$$

$$m = \frac{12.02 - 11.84}{8.535} + 0.01 = 0.03109 \text{ g } H_2O\text{/g solids} \tag{9.41}$$

Therefore, at 1.93 days:

$$\Gamma = \frac{9.95 - 1}{9.5 - 3.11} = 1.40 \tag{9.42}$$

For every data point calculation of unaccomplished moisture gain data, Γ gives moisture data for potato chips. When the obtained unaccomplished moisture gain data at 100% RH and 35°C versus time for both packaging materials is plotted, the intercept of lines (where Γ crosses critical moisture content) corresponds to time (days) of loss of crispiness.

This example shows that in fact no long-term moisture gain studies are required. Measurement of the moisture sorption isotherm, the permeability of the proposed package type (*k/x*) and knowing the area–to–weight of dry solids ratio (A/W_S), along with the expected external conditions are the only parameters needed. The time required to reach a critical moisture m_c can then be quickly calculated or graphed via the above steps.

In the above example, polyethylene would give a shelf-life of only 4 to 5 days, while with mylar, only 3 days are allowed. Considering the fact that the worst conditions are typically tested in shelf-life studies, the actual shelf-life for the two types of packaging would be longer at 50% RH compared with the calculated values above.

Final Remarks

Moisture migration within the domains of food or through packaging material is an important concept leading to both physical and chemical changes in food systems, thereby reducing shelf-life, as reviewed in this chapter. Understanding and quantifying the factors involved in moisture migration such as thermodynamic and dynamic relations are essential to reduce moisture transfer. Moisture migration within a food system can be prevented by formulating the domains with the same a_w (as close as possible) through the use of selected solutes (humectants) and ingredients. Additionally, treating the product mechanically to create the smallest pore size is considered to be a method of prevention. The addition of ingredients such as sorbitol or other sugar alcohols increases local viscosity, thereby inhibiting diffusion and mobility, or the addition of an edible barrier between domains or on the surface of a product may help control moisture migration. For future studies dealing with moisture migration in multidomain food systems, several issues still need to be addressed. State diagrams as described in Chapter 5 for individual components of multidomain systems would be useful in industrial applications. At the molecular level, NMR and MRI studies of water binding and flow should continue to be researched for a clearer understanding of moisture migration during storage. Importantly, flavor masking of humectants is important in keeping the food system's flavor acceptability high while maintaining shelf-life. Finally, continued research into the development of edible films will result in safe, effective, and organoleptically acceptable moisture barriers.

References

Alfrey, T., Jr., Gurnee, E.F., and Lloyd, W.G. 1966. Diffusion in glassy polymers. Journal of Polymer Science 12:249–261.

American Society for Testing and Materials. 1996. ASTM E96. Standard test methods for water vapor transmission of materials. Available at www.astm.org.

Bell, L.N., and Labuza, T.P. 2000. Moisture Sorption: Practical Aspects of Isotherms Measurements and Use. St. Paul, MN: AACC International Publishing.

Biquet, B., and Labuza, T.P. 1988. Evaluation of the moisture permeability characteristics of chocolate films as an edible barrier. Journal of Food Science 53(4):989–997.

Bruin, S., and Luyben, K. C. 1980. Drying of food materials: A review of recent developments. In: Advances in Drying, edited by A.S. Mujumdar. New York: Hemisphere.

Cardelli, C., and Labuza, T.P. 2001. Application of Weibull hazard analysis to the shelf life of roasted and ground coffee. Lebensittel-Wissenschaft und-Technologie 34:273–278.

Chen, H. 1995. Functional properties and applications of edible films made from milk proteins. Journal of Dairy Science 78:2563–2583.

Chen, P.L., Long, Z., Ruan, R.R., and Labuza, T.P. 1997. Nuclear magnetic resonance studies of water mobility in bread during storage. Lebensittel-Wissenschaft und-Technologie 30:178–183.

Crank, J. 1975. Mathematics of diffusion, 2nd Edition. Oxford, UK: Clavedon Press.

Fennema, O.R., ed. 1996. Chemistry. New York: Marcel Dekker.

Fennema, O., Donhowe, I.G., and Kester, J.J. 1993. Edible films: Barriers to moisture migration in frozen food. Food Australia 45(11):521–525.

Geankoplis, C.J. 1972. Mass Transport Phenomena. New York: Holt, Reinhart, and Winston.

Gontard, N., Marchesseau, S., Cug, J.L., and Guilbert, S. 1995. Water vapor permeability of edible bilayer films of wheat gluten and lipids. International Journal of Food Science and Technology 30:49–56.

Greener Donhowe, I., and Fennema, O. 1994. Edible films and coatings: Characteristics, formation, definitions and testing methods. In: Edible Coatings and Films to Improve Food Quality, eds. J.M. Krochta, E.A. Baldwin, and M.O. Nisperos-Carriedo, pp. 1–24. Lanchester, PA: Technomic.

Heil, J.R., Ozilgen, M., and McCarthy, M.J. 1993. Magnetic resonance imaging analysis of water migration and void formation in baking biscuits. In: AICHE Symposium Series, ed. L.G. Elmer, 89(297):39–45.

Heiss, R. 1958. Shelf life determinations. Modern Packaging 31(8):119.

Hong, Y.C., Bakshi, A.S., and Labuza, T.P. 1986. Finite element modeling of moisture transfer during storage of mixed multi component dried foods. Journal of Food Science 51(3):554–558.

Jeffrey, K.R., Callaghan, C., and Xia, Y. 1994. The measurement of velocity profiles: An application of NMR microscopy. Food Research International 27(2):199–201.

Karel, M. 1976. Technology and application of new intermediate moisture foods. In: Intermediate Moisture Foods, eds. R. Davies, G.G. Birch, and K.J. Parker, pp. 4–31. London: Applied Science Publication.

Kester, J.J., and Fennema, O. 1986. Edible films and coatings: A review. Food Technology 40(12):47–59.

Kinsella, J.E., and Morr, C.V. 1984. Milk proteins: Physicochemical and functional properties. CRC Critical Reviews in Food Science and Nutrition 21(3):197–262.

Krochta, J.M. 1992. Control of mass transfer in foods with edible coatings and films. In: Advances in Food Engineering, eds. R.P. Singh and M.A. Wirakartakusumah, pp. 517–538. Boca Raton, FL: CRC Press.

Krochta, J.M., and de Mulder-Johnston, C. 1997. Edible and biodegradable polymer films: Challenges and opportunities. Food Technology 51(2):61–77.

Labuza, T.P., and Contreras-Medellin, R. 1981. Prediction of moisture protection requirements for foods. Cereal Foods World 26:335–343.

Labuza, T.P., Tannenbaum, S.R., and Karel, M. 1969. Water content and stability of low-moisture and intermediate-moisture foods. Food Technology 24:35.

Labuza, T.P. 1971. Kinetics of lipid oxidation in foods. Critical Reviews in Food Science (2)3:355–405.

Labuza, T.P. 1980. The effect of water activity on reaction kinetics of food deterioration. Food Technology 34(4):36–41.

Labuza, T.P. 1985. An integrated approach to food chemistry: Illustrative cases. In: Food Chemistry, edited by O.R. Fennema, pp. 913–938. New York: Marcel Dekker.

Labuza, T.P., and Schimdl, M.K. 1985. Accelerated shelf life testing of foods. Food Technology 39(9):57–62.

Labuza, T.P. 2000. Functional foods and dietary supplements: Safety, good manufacturing practice (GMPs) and shelf life testing. In: Essentials of Functional Foods, eds. M.K. Schmidl and T.P. Labuza. London: Aspen Press.

Loncin, M. 1980. Diffusion phenomena in solid. In: Food Process Engineering, eds. Y. Linko, J. Malkki, and J. Larinkari, pp. 354–363. London: Applied Science Publishers.

MacCarthy, M.J., Lasseux, D., and Maneval, J.E. 1994. NMR imaging in the study of diffusion of water in foods. Journal of Food Engineering 22:211–224.

Marousis, Z.B., Karathanos, V.T., and Saravacos, G.D., 1989. Effect of sugars on the water diffusivity in hydrated granular starches. Journal of Food Science 54:1496–1500.

Miller, K.S., and Krochta, J.M. 1997. Oxygen and aroma barrier properties of edible films. Trends in Food Science and Technology 8(7):228–237.

Mizrahi, S., Labuza, T.P., and Karel, M. 1970a. Computer aided predictions of extent of browning in dehydrated cabbage. Journal of Food Science 35:799–803.

Mulvihill, D.M., and Fox, P.F. 1989. Physico-chemical and functional properties of milk proteins. In: Developments in Dairy Chemistry, vol. 4, ed. P.F. Fox, pp. 131–172. London: Elsevier Applied Science.

Morillon, V., Debeaufort, F., Blond, G., and Voilley, A. 2000. Temperature influence on moisture transfer through synthetic films. Journal of Membrane Science 168(1-2):223–231.

Parkins, K.L., and Brown, W.D. 1982. Preservation of seafood with modified atmospheres. In: Chemistry and Biochemistry of Marine Food Products, eds. R.E. Martin, G.J. Flick, C.E. Hebard, and D.R. Ward, pp. 453–465. Westport, CT: AVI Publishing.

Peppas, N.A. 1995. Controlling protein diffusion in hydrogels. In: Trends and Future Perspectives in Peptide and Protein Drug Delivery, eds. V.H.L. Lee, M. Hashida, and Y. Mizushima, pp. 23–27. Chur: Harwood Academic Publishers.

Peppas, N.A., and Peppas, L.B. 1994. Water diffusion and sorption in amorphous macromolecular systems and foods. Journal of Food Engineering 22:189–210.

Quast, D.G., Karel, M., and Rand, W. 1972. Development of a mathematical model for oxidation of potato chips as a function of oxygen pressure, extent of oxidation and equilibrium relative humidity. Journal of Food Science 37:673–678.

Quast, D.G., and Karel, M. 1972. Computer simulation of storage life of foods undergoing spoilage by two interacting mechanisms. Journal of Food Science 37:679–683.

Rockland, L.B., and Nishi, S.K. 1980. Influence of water activity on food product quality and stability. Food Technology 34(4):42–51.

Roques, M.A. 1987. Diffusion in Foods: the work of COST 90bis subgroup. In: Physical Properties of Foods—2, eds. R. Jowitt, F. Escher, M. Kent, B. McKenna, and M. Roques, pp. 13–25. London: Elsevier.

Ruan, R., and Litchfield, J.B. 1992. Determination of water distribution and mobility inside maize kernels during steeping using magnetic resonance imaging. Cereal Chemistry 69(1):13–17.

Simatos, D., Le Meste, M., Petroff, D., and Halphen, B. 1981. Use of electron spin resonance for the study of solute mobility in relation to moisture content in model food systems. In: Water Activity: Influences on Food Quality, eds. L.B. Rockland and G.F. Stewart, pp. 391–346. New York: Academic Press.

Saravacos, G.D., and Maroulis, Z.B. 2001. Transport of water in food materials. In: Transport Properties of Foods. New York: Marcel Dekker.

Schmidt, S.J., and Lai, H.M. 1991. Use of NMR and MRI to study water relation in foods. In: Water Relations in Food, eds. H. Levine and L. Slade, pp. 405–522. New York: Plenum Press.

Spiess, W.E.L., and Wolf, W. 1986. The results of the COST 90 project on water activity. In: Physical Properties of Foods, eds. R. Jowitt, F. Escher, M. Kent, B. McKenna, and M. Roques, pp. 65–87. London: Applied Science Publishers.

Taoukis, P.S., El Meskine, A., and Labuza, T.P. 1988. Moisture transfer and shelf life of packaged foods. In: Food and Packaging Interactions, ed. J.H. Hotchkins, pp. 243–447. Washington, DC: American Chemical Society.

Thomas, N.L., and Windle, A.H. 1982. A theory of case II diffusion. Polymer 23:529.

Thompson, D. 1983. Response surface experimentation. Journal of Food Processing and Preservation 6:155–188.

Tsoubeli, M.N., Davis, E.A., and Gordon, J. 1995. Dielectric properties and water mobility for heated mixtures of starch, milk, protein, and water. Cereal Chemistry 72:64–69.

Tutuncu, M.A., and Labuza, T.P. 1996. Effect of geometry on the effective moisture transfer diffusion coefficient. Journal of Food Engineering 30:433–447.

Umbach, S.L., Davis, E.A., Gordon, J., and Callaghan, P.T. 1992. Water self-diffusion and dielectric properties determined for starch-gluten-water mixtures heated by microwave and by conventional methods. Cereal Chemistry 69(6):637–642.

Vrentas, J.S., and Duda, J.L., 1977. Diffusion in polymer-solvent systems, I, re-examination of the free volume theory. Journal of Polymer Science, Polymer Physics Edition 15(I):403–407.

Yang, L., and Paulson, A.T. 2000. Mechanical and water vapor barrier properties of edible gellan films. Food Research International 33:563–570.

10 Effects of Water Activity (a_w) on Microbial Stability: As a Hurdle in Food Preservation

María S. Tapia, Stella M. Alzamora, and Jorge Chirife

Water activity (a_w) is a major factor in preventing or limiting microbial growth. In several cases, a_w is the primary parameter responsible for food stability, modulating microbial response, and determining the type of microorganisms encountered in food. Of all the factors affecting microbial growth, death, and survival in food products (temperature, oxygen, nutrient availability, acidity and pH, presence of natural, or added inhibitors, etc.), the influence of a_w on vegetative microorganisms and spores is one of the most complex and fascinating, and for this reason it has been extensively studied by food microbiologists. Adverse environmental conditions, such as a_w changes that cause osmotic stress, can elicit the sporulation response in spore-forming microorganisms, but bacterial endospores and some fungal spores have special requirements, such as optimum a_w values for initiating germination, and outgrowth (minimal a_w for germination is usually higher than the minimum a_w for sporulation). In addition, the production of secondary metabolites (toxins) is affected by $a_{w.}$ Therefore, sporulation, germination, and toxin production are affected by a_w along with other environmental factors (Beuchat 1987, 2002).

There is no doubt, then, that control of a_w for preservation purposes was instinctively recognized by humans. Throughout history, a_w alone or in combination with other environmental parameters has been the base of many food preservation methods. Food dehydration, for instance, is an ancient technology based on the relatively modern principle of a_w reduction. Evolution of food drying has passed from traditional osmotic dehydration, based on the use of sugars and salt, to reduce a_w and the application of hot air in a continuous and batch process (cabinet, tunnel, spray drying, fluidized bed) and freeze-drying, to "new" concepts like vacuum application to the fluidized bed process or the possible substitution of air by superheated steam (Welti-Chanes et al. 2004). Currently, dry and dehydrated products exhibit a high level of popularity among modern consumers because these products are economical and convenient, having increased shelf-life, decreased cost, reduced packaging, and improved handling, transport, and shipping properties. Even in the face of trends like minimal processing that try to keep fresh-like characteristics in final products, a_w reduction can be accomplished by the addition of humectants at a minimum level maintaining the product in a high-moisture state and through refined techniques of dehydration like pulsed vacuum osmotic dehydration.

Whichever process is used, preservation of all food attributes is of major concern, because total quality refers not only to the delay or prevention of growth of food spoilage and poisonous organisms but also to the inhibition or delay of physicochemical and biochemical reactions deleterious to texture, color, flavor, and nutritive value of foods. However, microbial preservation in food processing is viewed as a primary consideration due to public health implications.

The concept of a_w flourished in the light of initially empirical and inconsistent observations between total moisture content and product stability, to current knowledge and understanding of the influence of this concept on safety and shelf-life in controlling microbial growth, death, survival, sporulation, and toxin production. This has resulted in the inclusion of the a_w principle in various government regulations (Food and Drug Administration [FDA] and U.S. Department of Agriculture [USDA] regulations, Good Manufacturing Practice [GMP] and Hazard Analysis and Critical Control Points [HACCP] requirements, NSF International [formerly the National Sanitation Foundation] Draft Standard 75), being an important critical control point for risk analysis, as defined by the HACCP concept (Fontana 2000). It is important to note that a_w inhibits growth but does not kill bacteria (or it does this very slowly). Therefore, a_w reduction is quite an effective method to eliminate the risk of toxigenic bacteria (i.e., bacteria that grow in food and produce toxin) such as *Staphylococcus aureus, Clostridium botulinum,* and *Bacillus cereus.* However, it does not eliminate the risk posed by infectious bacteria, i.e., *Salmonella,* some *Escherichia coli* strains, etc.

This chapter covers aspects that attempt to explain why the concept of a_w has been successfully applied to achieve microbial preservation of foods, its mode of action at the cellular level, and its application in food preservation along with other preservative factors, as a hurdle for microbial growth and metabolic activity.

Water Activity and Metabolic Activities of Microorganisms—Minimal Water Activity for Growth, Toxin Production, and Sporulation

Water activity has become one of the most important intrinsic properties in predicting the survival of microorganisms in food due to its direct influence on product quality and stability. William James Scott's classic demonstration (1953)—that it is not the water content but the a_w of a food system that governs microbial growth and toxin production—showed that microorganisms have a limiting a_w level below which they will not grow or produce toxins. Scott's work was expanded by his colleague J. H. B. Christian (1963) and had a profound impact on food technology. Thus, the minimal a_w value for growth emerged as one of the most investigated parameters that determine the water relations of microorganisms in food. This minimal a_w value defines in theory, the level below which a microorganism or group of microorganisms can no longer reproduce, even if others more resistant and adaptable to a_w reduction can grow and spoil or compromise product safety. Because in a food it is extremely difficult to isolate responses to a_w alone, minimal a_w values should be obtained in laboratory model systems in which all other factors that influence microbial response (pH, redox potential, temperature, nutrient availability, etc.) are at their optimum (Troller 1987). The concept should be used carefully when applied to food systems where other inhibitory factors are present and a_w is not generally at or below minimal levels for growth. It should serve as a reference and as a security level in food product development. It is also important to note that minimal a_w for growth varies depending on the type of a_w-depressing solute used in the growth media. This phenomenon is known as the "solute effect" and is discussed elsewhere in this chapter.

Accepted generalizations claim that the majority of spoilage bacteria will grow down to about 0.95 a_w, being the reason why bacteria are the dominant flora of most high-moisture foods. Other bacteria, many of public health concern, may reach values of 0.90 or even 0.85 a_w. With the exception of moderate and high halophilic ones (e.g., those that spoil

Table 10.1 Minimum a_w for Growth of Microorganisms.

Range of a_w	Microorganisms Inhibited by Lowest a_w in This Range
1.00–0.95	*Pseudomonas, Escherichia, Proteus, Shigella, Klebsiella, Bacillus, Clostridium perfringens, C. botulinum* E, G, some yeasts
0.95–0.91	*Salmonella, Vibrio parahaemolyticus, Clostridium botulinum* A, B, *Listeria monocytogenes, Bacillus cereus*
0.91–0.87	*Staphylococcus aureus* (aerobic), many yeasts (*Candida, Torulopsis, Hansenula), Micrococcus*)
0.87–0.80	Most molds (mycotoxigenic penicillia), *Staphyloccocus aureus*, most *Saccharomyces* (bailii) spp., *Debaryomyces*
0.80–0.75	Most halophilic bacteria, mycotoxigenic aspergilli
0.75–0.65	Xerophilic molds (*Aspergillus chevalieri, A. candidus, Wallemia sebi), Saccharomyces bisporus*
0.65–0.61	Osmophilic yeasts (*Sacharomyces rouxii*), a few molds (*Aspergillus echinulatus, Monascus bisporus*)
< 0.61	No microbial proliferation

Source: Adapted from Beuchat, L.R. 2002. Water activity and microbial stability. *Fundamentals of Water Activity*. IFT Continuing Education Committee, June 14-15, Anaheim, CA.

Table 10.2 Approximate a_w Values of Some Foods in the Range of 0.86 to 0.99.

Food Product	Approximate a_w Value
Fresh foods: Milk, vegetables, fruits, and meats	0.97–0.99
Canned products	0.97–0.98
Yogurt	0.98
Tomato paste (double)	0.98
Lebervhurst	0.97
Mozzarella cheese	0.97
Processed cheese	0.97
White bread (sliced)	0.96
Pâté de foie gras	0.95– 0.96
Mortadella (Italian)	0.95
Mayonnaise	0.95
Margarine	0.94
Salted olives	0.93–0.95
Fresh pasta (MA packaged)	0.92–0.94
Tomato ketchup	0.93
Soybean sauce	0.92
French dressing	0.92
Parmesan cheese	0.91
Salami	0.90
Intermediate-moisture cat food	0.88
Chocolate syrup	0.86

brines and salt-rich foods), bacteria do not compete well in "high osmotic" (low a_w) environments. In the case of high-sugar foods, osmophilic yeasts are favored instead. In general, at lower a_w values down to 0.61, yeast and molds take over, with filamentous fungi being by far the predominant microflora (Beuchat 1981, Hocking 1988, Fontana 2002). The interactions among a_w, pH, temperature, and other environmental factors are also determinant in selecting the flora that would prevail on particular foods.

Table 10.1 and Appendix D present the a_w limits for growth of microorganisms, and Table 10.2 gives examples of some foods with a_w values in the range of 0.86 to 0.99.

Table 10.3 Minimum a_w for Growth and Toxin Production by Bacteria and Molds.

Organism	Minimum a_w: Growth	Toxin Production
Bacteria		
Staphylococcus aureus	0.86	0.87 Enterotoxin A
		0.90 Enterotoxin B
Salmonella	0.93–0.95	
Vibrio parahaemolyticus	0.94	
Clostridium botulinum	0.94 A, B	0.94 A, B
	0.97 E	0.97 E
	0.965 G	0.965 G
Clostridium perfringens	0.93–0.95	
Bacillus cereus	0.92–0.93	0.95 (diarrhea)
Molds		
Aspergillus flavus	0.78–0.84	0.84 Aflatoxin
		0.83–0.87
A. parasiticus,	0.82	0.87 Ochratoxin
A. ochraceous	0.77–0.81	0.83–0.87
Penicillium cyclopium	0.82–0.85	0.87–0.90 Penicillic acid
P. viridicatum	0.80–0.81	0.83–0.86
A. ochraceous	0.77	0.80–0.88
P. cyclopium	0.82– 0.85	0.97 Patulin
P. marensii	0.79	0.99
P. patulum	0.81–0.85	0.95
P. expansum Stachybotrys atra	0.82–0.84	0.99 Stachybotryn
	0.94	0.94

Sources: Adapted from Beuchat, L.R. 1987. Influence of water activity on sporulation, germination, outgrowth and toxin production. In: *Water Activity: Theory and Applications to Food,* eds. L.B. Rockland and L.R. Beuchat, pp. 137–151. New York: Marcel Dekker; and Beuchat, L.R. 2002. Water activity and microbial stability. *Fundamentals of Water Activity*. IFT Continuing Education Committee, June 14-15, Anaheim, CA.

It is quite interesting to compare the minimal values for growth of different microorganisms and minimal values for toxin production (see Table 10.3). For example, *S. aureus* (an exceptionally tolerant organism to NaCl), with reported minimal a_w for growth as low as 0.85 to 0.86 (aerobically) and 0.90 to 0.91 (anaerobically), has demonstrated that Enterotoxin A–producing strains can produce the toxin at more adverse conditions (0.87 a_w) than strains producing Enterotoxin B (0.97 a_w) (Troller 1971, 1972; Lotter and Leistner 1978; Lee et al. 1981; Notermans and Heuvelman 1983). As for toxigenic spore-forming bacteria like *C. botulinum*, the minimal a_w for toxin production is similar to that required for germination and outgrowth of spores. Sporulation often occurs when the lower a_w limit for growth is approached. Type of solute, temperature, pH, nutrient availability, etc. strongly affect these responses. Spores of *C. botulinum* types A and B are capable of germination, growth, and toxin production at lower a_w (0.93 to 0.94) than are spores of type E (0.97) (Baird-Parker and Freame 1967, Beuchat 1987) or type G (0.965) (Briozzo et al. 1983). The lower limit for growth of mycotoxigenic molds is reported at ~0.78 a_w. Because growth and mycotoxin production do not always take place simultaneously, minimal a_w values for mycotoxin production are generally higher than minimal values for growth. This represents a natural safety margin because a_w decrease inhibits micotoxin production when mold growth may still be occurring (Northolt et al. 1995).

Bacterial and fungal spores are generally characterized by their extreme dormancy, associated with a dehydrated state of the protoplast, and by their requirement for heat shock or other severe treatment to initiate germination and outgrowth (Beuchat 1987). Water activity affects these microbial events, and generally optimal conditions of temperature, pH, oxygen tension, and nutrient availability are necessary to permit sporulation, germination, and toxin production at reduced a_w (Beuchat 1987, 2002). The mechanisms of germination and outgrowth are also influenced by the type of solutes used, as demonstrated by Jakobsen et al. (1972) with *B. cereus* spores when performing studies to determine minimal a_w at which heat-shocked spores germinated and grew. The test solutes (NaCl, KCl, glucose, fructose, sorbitol, glycerol, erythritol, and dimethylsulfoxide) greatly affected the germination response, with germination stopping at a_w values of 0.95 or less and outgrowth at a_w less than 0.93. Sinigaglia et al. (2002) developed a response surface model for the effects of temperature (20° to 40°C), pH (4.5 to 6.5), and a_w (0.94 to 0.99) in an attempt to obtain predictions of germination of *B. cereus* American Type Culture Collection (ATCC) 11778 spores. Germination depended, to varying extents, on the interactions among the independent variables and complexity of the medium, being affected by interactions of a_w with temperature and pH.

In the case of molds, minimal a_w values for sporulation have been determined (Beuchat 1987). Molds require higher a_w values for spore formation than for germination. The type of solute used to adjust a_w influences growth and sporulation of fungi. At any given temperature, a reduction in a_w will cause a decrease in the rate of germination. The presence of appropriate nutrients will tend to broaden the range of a_w and temperature at which spore germination and growth may occur, with the a_w range permitting germination, being greater at an optimum temperature.

The Challenge of Minimal a_w Limits for Growth

Minimal a_w values reported in Tables 10.1 and Table 10.3 for growth of different microorganisms have been mostly determined in liquid laboratory media, using sodium chloride to depress the a_w. However, one may question whether these values also apply to real reduced-moisture foods, and this question should be addressed. It has been well established that the inhibitory a_w for growth of *C. botulinum* types A and B in liquid broth media, adjusted with NaCl, is between 0.94 and 0.95 (Ohye and Christian 1966, Baird-Parker and Freame 1967). Glass and Doyle (1991) confirmed this minimum value of a_w for solid foods in a study on the relationship between a_w of fresh pasta (meat- or cheese-filled tortellini and flat noodle linguine or fettuccini) and toxin production by *C. botulinum*. Four types of fresh pasta with different a_w were inoculated with *C. botulinum*, packaged under modified atmosphere, and stored at 30°C for 8 to 10 weeks. The pH of all samples was favorable to *C. botulinum* growth. No toxin was detected in tortellini at 0.94 a_w; toxin was produced at 2 weeks in linguine at 0.96 a_w, whereas linguine or fettuccini at 0.93 or 0.95 a_w did not become toxic. Glass and Doyle (1991) concluded that a_w of the fresh pasta was the principal factor in preventing botulinal toxin in temperature-abused products. Dodds (1988) reported a study on the effects of a_w on toxin production by *C. botulinum* in cooked vacuum-packaged potatoes; a_w was controlled by the addition of NaCl and potatoes were incubated for 60 days at 25°C. Toxin was produced at 0.96 a_w, but no toxin was detected at 0.955 a_w, in good agreement with predictions made from the behavior of the bacterium in liquid broth. Valik and Gorner (1993) studied the growth of *S. aureus* in pasta dough in relation to its a_w and found that the bacterium multiplied until the a_w was below 0.86, at which point it ceased; this is in good agreement with the minimal a_w for growth of *S. aureus* determined in liquid broth media (Vaamonde et al. 1982, Chirife 1994). Giannuzzi and Parada (1991) studied the behavior of *S. aureus* in dehydrated milk, beef, and pork equilibrated at a_w values of 0.84 and 0.90 and incubated at 30°C. No growth was observed in any of the foods at a_w 0.84. However, growth occurred in all systems at a_w 0.90; this agrees with the known behavior of *S. aureus* in liquid broth media. Silverman et al. (1983) reported that the limiting a_w for growth of *S. aureus* in bacon sealed in cans and stored at 37°C was 0.87, in good agreement with known behavior in liquid broth media. King et al. (1984) studied the effects of a_w on mold growth in stored almonds. No mold growth was observed after 18 months on almonds stored at a_w 0.70, which is in good agreement with predictions made from the behavior of molds in laboratory media of controlled a_w (Beuchat 1983). Thus, in the absence of other hurdles, the minimum a_w values for growth found in Table 10.1 may be used for safety specifications.

The Safety of Water Activity Adjustment/Measurement for the Control of Bacterial Growth in Foods

For foods in which a_w is the main factor controlling the development of microbial hazards or spoilage, one must be certain that samples do not exceed a specified a_w. In the past 40 years or so, isopiestic equilibration, freezing point, hair or polymer, electrolytic, capacitance, or dew point hygrometers have been used to measure a_w in foods (Favetto et al. 1983, Aguilera et al. 1990, Rahman 1995). The accuracy of a_w determinations improved through those years up to the present time, where, for example, chilled mirror dew point instruments are accurate to about ±0.003 a_w (Fontana 2002). It should be noted that accurate measurements not only depend on the a_w measurement method used but also on standards used for verification (usually saturated salt solutions) and proper temperature control (Fontana 2002).

For decades, researchers determined the minimal water activities for microbial growth; however, these a_w levels were somewhat imprecise in various cases (i.e., disagreement was observed in the results reported). Lack of accurate measurement of a_w could have been one important reason for some disagreements. The importance of accurate measurement of a_w for microbial growth need not be emphasized and perhaps may be illustrated with reference to Figures 10.1 and 10.2. Figure 10.1 (Briozo et al. 1983) shows that *C. botulinum* cannot grow and produce toxin in a model cheese system at a_w 0.949, but at a_w 0.960 it does do so. The same happens with *S. aureus* (Silverman et al. 1983) inoculated in bacon, as shown in Figure 10.2—absence of growth in bacon is observed at a_w 0.90 but not at a_w 0.91 (at 20°C); at 37°C, growth is observed at a_w 0.86, but at a_w 0.87, there is growth inhibition. Thus, relatively small incremental margins in a_w (i.e., ±0.01) may lead to a growth–no growth response. It follows that it is advisable to consider a margin of safety in the selection of the working a_w, to take into account the error involved in the measurement of a_w in foods.

Effects of a_w on Thermal Resistance of Microorganisms

Microbial growth in foods is a complex event strongly affected by environmental factors that determine if growth occurs, continues to occur, or detains. Factors that affect growth also influence death and survival. The basis for survival and death as influenced by a_w is also very

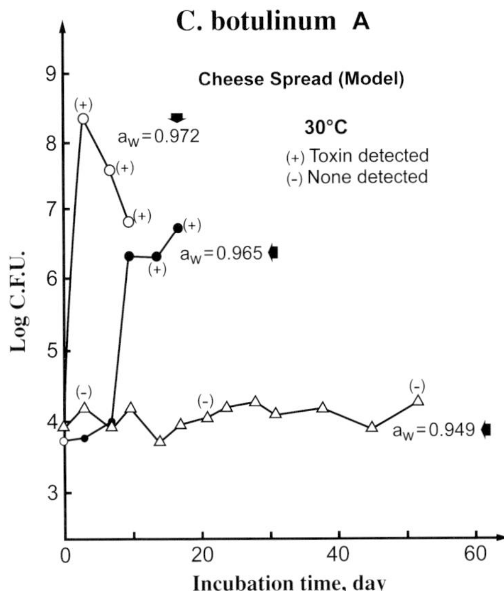

Figure 10.1 Effect of a_w on growth and toxin production by *C. botulinum* inoculated in processed cheese model system (adapted from Briozzo, J., Amato de Lagarde, E., Chirife, J., and Parada, J.L. 1983. Effect of water activity and pH on growth and toxin production by *C. botulinum* type G. *Applied Environmental Microbiology* 51:844–848).

complex. Multiple factors of intrinsic and extrinsic nature affect this relationship, differing with types of food, preservation factors, and the microflora involved (Lenovich 1987).

The increased resistance to temperature of vegetative cells and spores in low a_w environments has been extensively studied in terms of microbial inactivation (Lenovich 1987). When partially dehydrated cells are exposed to high temperatures, greater thermal resistance is displayed by the microorganism than when cells are grown at higher a_w values. Death curves are not always linear, and interpolation of *D*-values (and *z*-values) for thermal processes may not always be safe. Proteins and other essential cell constituents become more resistant to thermal damage in the partially dehydrated state. Particular attention has been given to spore-forming organisms due to their natural heat resistance. Heat resistance of these microorganisms has been reported to increase as their a_w decreases, with optimum resistance found at values between 0.2 and 0.5 a_w depending on the organism (Brown and Melling 1971). In a study performed by Corry (1974), the effects of sugars and polyols on the thermoresistance of *Salmonella enterica* serovar typhimurium was investigated, and maximum resistance was reported at 0.2 a_w.

Many of these studies have been performed in liquid media using different solutes to reduce a_w. The effect of solutes on the D-value differs for different microorganisms. The D-values in low a_w solutions and reduced-moisture foods must take into account not only the a_w but also the actual solute used for controlling a_w. In general, vegetative cells and spores are more resistant as the a_w of the heating menstrum is reduced, but as noted above, the type of solute used to adjust a_w may result in significant differences in the heat resistance of a given microorganism. Also, composition of the recovery media for detection of

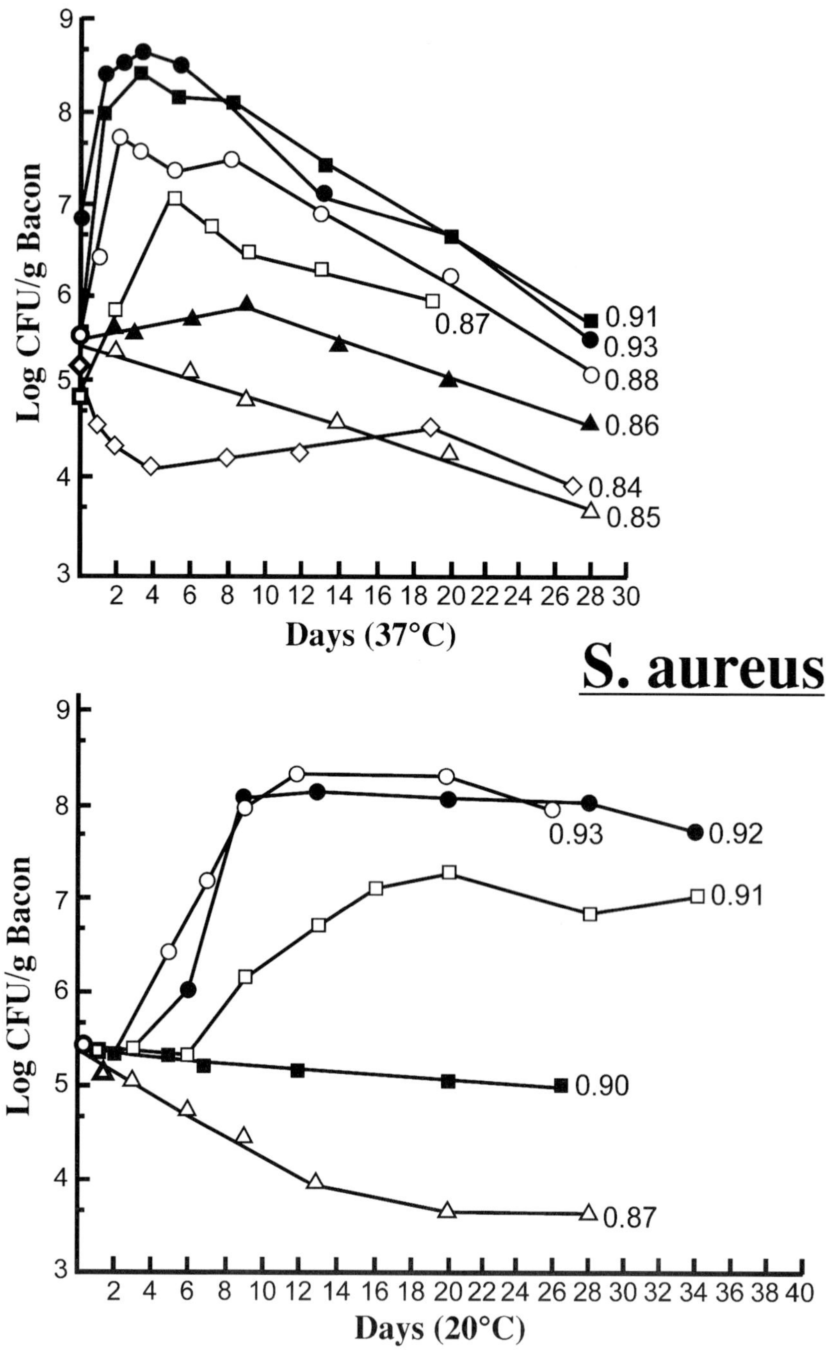

Figure 10.2 Effect of water activity on *S. aureus* growth in bacon incubated at 37°C and 20°C (reproduced with permission from Silverman, G.J., Munsey, D.T., Lee, C., and Ebert, E. 1983. Interrelationship between water activity, temperature and 5.5% oxygen on growth and enterotoxin secretion by *Staphylococcus aureus* in precooked bacon. *Journal of Food Science* 48:1783–1786).

Table 10.4 Effects of Solutes on D Values for *Salmonella* spp. at Various Water Activities.

Solute	% w/w (a_w)	D_{65} Values (min) *Salmonella typhimurium*	*Salmonella senftemberg*
Sucrose	30 (0.975)	0.7	1.4
	70 (0.824)	53	43
Glucose	30 (0.955)	0.9	2.0
	70 (0.748)	42	17
Fructose	30 (0.955)	0.5	1.1
	70 (0.748)	12	1.5
Glycerol	30 (0.915)	0.2	0.95
	70 (0.602)	0.9	0.7

Source: Adapted from Corry, J.E.L. 1974. The effect of sugars and polyols on the heat resistance of *Salmonellae*. *Journal of Applied Bacteriology* 37:31–43.

Table 10.5 Effect of Sucrose Concentration on Thermal Resistance of *L. monocytogenes* Scott A in Culture Media.

	D-Value (min) at Three Different Temperatures		
a_w	60°C	62.8°C	65.6°C
0.98	2.0	0.74	0.36
0.96	2.9	0.97	0.52
0.94	5.6	3.0	1.1
0.92	7.6	5.3	3.1
0.90	8.4	5.9	3.8

Source: Adapted from Sumner, S.S., Sandros, T.M., Harmon, M.C., Scott, V.N., and Bernard, D.T. 1991. The heat resistance of *Salmonella typhimurium* and *L. monocytogenes* in sucrose solutions of various water activities. *Journal of Food Science* 56:1741–1743.

the viable population after heat treatments may affect study results. Ionic solutes may decrease heat resistance at low levels but afford considerable protection at high concentration. Nonionic solutes have a variable effect; larger-molecular-weight solutes such as sucrose exert a protective effect against heat inactivation, while glycerol causes only a small increase in heat resistance. For instance, *S. aureus* heated in skim milk has a $D_{60°C}$ value of 5.3 minutes, while in skim milk (plus 57% sugar) the $D_{60°C}$ value is about 22 minutes (Mossel et al. 1991). Corry (1975) reported the dependence of decimal reduction times of *S. typhimurium* and *S. senftemberg* on a_w, with thermal inactivation being highest at high a_w (>0.95), decreasing as a_w decreased until 0.6 to 0.8 a_w, and increasing again until a_w approached very low values. Kirby and Davies (1990) reported increased thermal resistance of dehydrated cultures (on hydrophobic membranes) for *S. typhimurium* LT2 at high temperatures (135°C for 30 minutes). Results also showed that little or no death occurred during heat challenges of 1 hour at temperatures of up to 100°C. The survival of *S. typhimurium* (ATCC 13311) heated and recovered in media with 0%, 1%, 2%, 3%, 4%, or 5% (w/w) added sodium chloride was investigated by Mañas et al. (2001). A protective effect in the heating medium and an inhibitory effect in the recovery medium were observed. When the sodium chloride concentration was the same in both media, the protective effect exerted in the heating media dominated over its inhibitory effect in the recovery media.

Table 10.4 shows the effects of solutes on D-values of *Salmonella* spp. at various water activities. Other pathogens like *Listeria monocytogenes* have been also investigated in this regard, as shown in Table 10.5.

It is generally accepted that in dried foods, microorganisms do not grow because of the lack of enough available water to sustain growth. However, microorganisms may still be viable and capable of reproduction on rehydration, posing a potential risk for public health and focusing attention on decontamination processes for dried products like powders that are to be rehydrated to high water contents (i.e., milk powder). Laroche and Gervais (2003) studied the effect of high hot- (150°, 200°, and 250°C) and cold-temperature shocks (as a possible decontamination procedure for dried foods) on the viability of dried (glass bead immobilized) vegetative cells of *Saccharomyces cerevisiae* and *Lactobacillus plantarum* in low a_w environment. They observed an unexpected range of a_w values between 0.30 and 0.50 in which microorganisms were more resistant to the various treatments, with maximum viability at 0.32 a_w for *L. plantarum* and 0.50 a_w for *S. cerevisiae*. Because it is known that cells are able to acquire resistance to a severe stress condition when previously exposed to a mild form of the same or a different stress (Siderius and Mager 1997) and viable microorganisms in food powders suffer thermal, hydric, and ionic stresses, the increased thermal resistance in low a_w environments may be explained by the adaptation mechanisms that are developed. It has been suggested that water in close contact with the proteins could be a factor in determining cell inactivation. As the cell is heated, water molecules begin to vibrate, causing the disulfide and hydrogen bonds in the surrounding proteins to weaken and break, altering the final three-dimensional configuration, and possibly preventing the protein from functioning (Earnshaw et al. 1995). When only a small amount of water is present, these vibrations are reduced, decreasing the protein denaturation. This mechanism can explain, in some cases, the high viability obtained for cells at low a_w values.

For bacterial and fungal spores, the resistance to the lethal effects of heat may increase a 1000 times or more at low a_w values, usually showing a maximum in the range of 0.2 to 0.5 a_w (Mossel et al. 1991). Also, solute type and the nature of the fungal structure affect heat resistance of yeasts and molds. Table 10.6 (Beuchat 2002) shows the influence of a_w and type of solute on the D-values of some conidia, ascospores, and vegetative cells of molds and yeasts.

Osmotic and Specific Solute Effects

It is known that water is the single most important factor governing microbial spoilage in foods, and the concept of a_w has been very valuable in physiological studies of microorganisms, principally because measured values generally correlate well with the potential for growth and metabolic activity (Gould 1985). The a_w concept has assisted food scientists and microbiologists in their efforts to predict the onset of food spoilage, as well as to identify and control food borne diseases (Leistner and Rodel 1975, Silverman et al. 1983, Dodds 1989, Glass and Doyle 1991). It has been shown repeatedly in the literature that each microorganism has a critical a_w below which growth cannot occur (Brown 1974, Gould and Measures 1977, Beuchat 1983). For example, pathogenic bacteria cannot grow below 0.85 a_w; yeast and molds are more tolerant to a reduced a_w, but usually no growth exists below 0.62 a_w (Scott 1953, Hocking and Pitt 1979). A fundamental requirement for growth of microorganisms on substrates of high osmolality is the intracellular accumulation of solutes, by either transport or synthesis, in concentrations that counterbalance the osmolality of the external medium (Chirife et al. 1981, Prior et al. 1987, Hocking 1988).

Growth restriction due to environmental stresses is a common situation found by mi-

Table 10.6 Influence of a_w and Type of Solute on D-Values for Some Molds and Yeasts Conidia, Ascospores, and Vegetative Cells.

Molds/Yeasts	a_W	Temperature (°C)	Solute	D- Value (min)
Aspergillus flavus (conidia)	0.99	55	None	3
	0.90		NaCl	70
	0.90		Sucrose	66
	0.85		Glucose	66
Aspergillus parasiticus (conidia)	0.99	55	None	8
	0.90		NaCl	230
	0.90		Sucrose	199
	0.85		Glucose	214
Aspergillus niger (conidia)	1.00	55	None	6
	0.60		None	100
	0.30		None	216
	0.00		None	100
Penicillium puberulum (conidia)	0.99	48	None	31
	0.89		Sucrose	30
	0.93		NaCl	30
Byssochlamys nivea (ascospores)	0.98	75	Sucrose	60
	0.92		Sucrose	260
	0.84		Sucrose	470
	0.99	80	Control	39
	0.93		NaCl	48
	0.89		Sucrose	49
Zygosaccharomyces bailii (ascospores)	0.999	80	Control	8.5
				10
	0.975		NaCl	9.4
				20
	0.975		Sucrose	11
				16
	0.95		Sucrose	21
				20
Kluyveromyces marxianus (ascospores)	0.999	60	Control	24
	0.975		NaCl	30
	0.95		Sucrose	40
			NaCl	36
			Sucrose	54
Geotrichum candidum (vegetative cells)	0.99	52	None	30
	0.97		NaCl	21
	0.93		NaCl	10
	0.97		Sucrose	57
	0.89		Sucrose	59
Saccharomyces cerevisiae (vegetative cells)	0.99	51	None	21
	0.97		NaCl	24
	0.93		NaCl	13
	0.97		Sucrose	49
	0.89		Sucrose	53
Debaryomyces hansenii (vegetative cells)	0.99	48	None	12
	0.97		NaCl	17
	0.93		NaCl	18
	0.97		Sucrose	40
	0.89		Sucrose	43

Source: Adapted from Beuchat, L.R. 2002. Water activity and microbial stability. *Fundamentals of Water Activity*. IFT Continuing Education Committee, June 14-15, Anaheim, CA.

croorganisms in nature. Therefore, they have evolved different mechanisms to resist the adverse effects of these stresses. As internal media stability (composition and volume of fluids) is vital for survival and growth, these mechanisms, called "homeostatic mechanisms," act to ensure that key physiological activities and parameters in the microorganisms remain relatively unchanged, even when the environment around the cell is different and greatly perturbed (Gould 1996, Leistner and Gould 2002).

The response of microorganisms to lowered a_w is essentially a response to osmotic stress and is therefore often referred to as "osmoregulation" or "osmoadaptation" (Gould 2000). This reaction is most developed in microorganisms, particularly in the most osmotolerant of the yeasts and molds, but it is also widespread in animals and plants. Although the specific details of how each organism responds to a hyperosmotic shock are different and the organisms differ widely in the range of osmolarity over which they grow, several common features, both physiologically and genetic, have arisen (O'Byrne and Booth 2002). Bacterial response to hyperosmolarity includes two aspects. The first one (the most readily observable) concerns the ability of bacteria to accumulate osmoprotective compounds for turgor and growth restoration. The second one concerns the osmotic induction of general stress systems, with the consequent development of multitolerances toward other environmental stresses when subjected to hypertonic environments (Pichereau et al. 2000, O'Byrne and Booth 2002). Therefore, there is a mixed strategy used across the bacterial genera in response to an osmotic shock: existing transport systems and enzymes may be activated/inhibited by cues and/or the genes encoding their structural components may be induced/repressed (O'Byrne and Booth 2002).

Microbial cells have an internal osmotic pressure that is higher than that of the surrounding medium, resulting in a turgor pressure exerted outward on the cell wall, providing the mechanical force necessary for expansion of the cell and growth (Gutiérrez et al. 1995). When a microorganism is put into an environment of reduced a_w, water migrates from the cytoplasm of the cell (in a passive way or possibly mediated via water channels) and membrane turgor is lost. The homeostasis (or internal equilibrium) is disturbed and the organism will not multiply but will remain in the lag-phase until the equilibrium is reestablished. A universal and major response of cells to a reduced a_w is the accumulation of low-molecular-weight solutes in their cytoplasm at concentrations sufficient to just exceed the osmolality of the external medium. In this way, the cells regain or avoid loss of water via osmosis and maintain the turgor in the cell membrane that is essential for its proper functioning. Chirife et al. (1981) used experimental data from literature to theoretically calculate the intracellular a_w from the solute composition of various bacterial cells grown in media of a_w between 0.85 and 0.993. They found that the intracellular a_w was generally equal or slightly lower than that of the growth medium; their results are shown in Figure 10.3. The general reaction therefore appeared to be a homeostatic mechanism with respect to cell water content (Gould 1989). Compatible solutes (so-called because, even at very high relative concentrations, they do not appreciably interfere with the metabolic and reproductive functions of the cell) are generally nonionic solutes, since many enzymes will start to lose activity in presence of high salt concentration (Gutiérrez et al. 1995). While amino acids (proline, α-keto glutarate, γ-aminobutyric acid, glutamic acid) appear to be the most common compatible solutes in bacteria, polyols of various types (mannitol, cyclohexanetetrol, arabitol, sorbitol, glycerol, erythritol, etc.) are the predominant protoplasmic solutes in many fungi (Troller 1987). These compatible solutes have the following common properties (Gutiérrez et al. 1995): (1) they are soluble to high concentration and

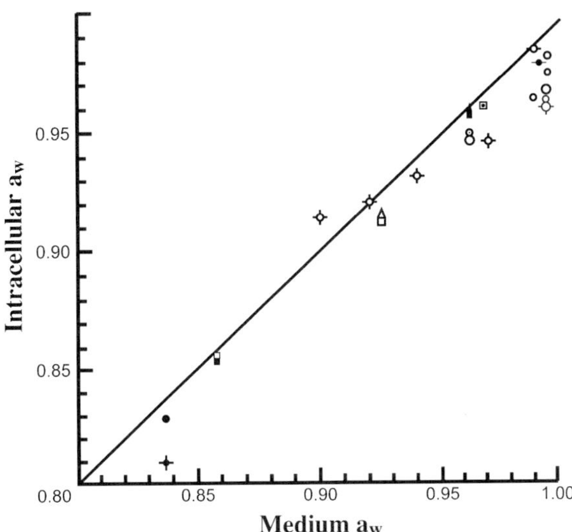

Figure 10.3 Comparison of intracellular and growth medium a_w for various halophilic and nonhalophilic bacteria (reproduced with permission from Chirife, J., Ferro-Fontán, C., and Scorza, O.C. 1981. The intracellular water activity of bacteria in relation to the water activity of the growth medium. *Journal of Applied Bacteriology* 50(3):475–479).

can be accumulated to very high levels in the cytoplasm of the cells; (2) they do not modify enzyme activity and can even protect enzymes for denaturation by salts; (3) they have relatively small molecular weight and are usually neutral or zwitterionic molecules; and (4) the cell membrane exhibits controlled permeability to them.

Depending on the nature of the growing media, compatible solutes either can be transported from the environment or can be synthesized de novo in the cytoplasm. Some solutes are only available from the environment (e.g., choline, betaine); others can be either synthesized or transported (e.g., proline), while others are only available by synthesis (e.g., trehalose). Accordingly, the availability of these compounds in the environment can influence the growth rate of the organisms under conditions of hyperosmotic stress. In particular, many foods contain a wide range of substances that will act as compatible solutes or are their precursors (e.g., quaternary amine, glycine betaine, and proline in plant materials; various types of meat, taurine in fish, and crabs, etc.) and thereby facilitate growth at lowered a_W, increasing the limit of tolerance to hyperosmolarity of the bacterium (O'Byrne and Booth 2002).

The level of compatible solute accumulation is set by the environmental osmolarity (O'Byrne and Booth 2002). The pool of accumulated solutes is also influenced by the degree of osmotic stress. Salts (usually potassium glutamate) are accumulated at low osmolarity of the environment while, as this salt is increased, the initial response is also the accumulation of glutamate. But this accumulation is transient as the cell initiates the accumulation of other compatible solutes because high concentrations of potassium glutamate are inimical to enzyme activity.

One other major microbial response to change in a_W is the adaptation of the membrane composition (Russell et al. 1995). For a wide range of bacteria, the most common alteration is the increase in the membrane proportion of anionic phospholipids and/or glycolipid as a means to preserve the proper bilayer phase and maintain its vital functions.

Because of their central role, research has concentrated on the compatible solutes, and much of the genetic basis of osmoregulation has been elucidated (Gould 2000). Adaptive strategies involve the osmotic regulation of the expression of a number of genes to optimize growth under the stress condition, allowing cells to modulate the rate of acquisition of compatible solutes. Many of these genes are under the control of alternative stress and stationary phase sigma factors, σ^S, in the Gram-negative and σ^B in the Gram-positive species (Pichereau et al. 2000). For example, the growth of *E. coli* in absence of other compatible solutes from the growth medium occurs by the accumulation via its synthesis of trehalose. Trehalose synthetic enzymes are under the control of the *rpoS* sigma factor, which accumulates when cells are grown at high osmolarity (O'Byrne and Booth 2002). In *E. coli,* the σ^S regulon includes over 50 different genes and the products of these genes confer resistance to a wide range of stress conditions. As presently known, there is a general response mechanism (the "global response") underlying many of the apparent distinct responses of microorganisms to different stresses imposed on them in foods (e.g., low a_w, low pH, low or high temperature, oxidative stress, starvation, etc.). This global response is mediated by the stationary-phase regulator *RpoS*, which regulates the expression of many important stationary-phase stress resistance genes linked to survival under starvation conditions and to survival in the stationary phase. As Gould (2000) stated, this fact would explain the cross-resistances to different stresses that have usually been found to occur in response to a single stress.

The usefulness of a_w has been somewhat diminished by the fact that measured a_w levels required for microbial safety and stability depended on the "specific solute effect." That is, the microbial response may differ at a particular a_w when the latter is obtained with different solutes. It has been established that the a_w of the medium is not the only determining factor regulating the biological response but that the nature of the a_w-controlling solute also plays a role (Christian 1981, Gould 1988, Ballesteros et al. 1993). Glycerol, for example, readily permeates the membrane of many bacteria and therefore does not initiate the same osmoregulatory response as nonpermeant solutes, such as sodium chloride and sucrose, and therefore has a different inhibitory a_w. While growth in the presence of sodium chloride and sucrose induced the accumulation of proline and glycine betaine by *S. aureus* cells subjected to osmotic stress, growth in the presence of the permeant humectant glycerol did not (Vilhelmsson and Miller 2002).

It has been shown that many pathogenic bacteria such as *C. botulinum* E, *C. botulinum* A and B, *Vibrio parahaemolyticus*, *B. cereus*, *L. monocytogenes, E. coli*, and *C. perfringens* are somewhat more tolerant to glycerol (lower minimal a_w) than to sodium chloride, and this is the norm in most situations. Troller and Stinson (1981) studied the effects of a_w reduction on growth by three lactic streptococci and found that, in general, sucrose was somewhat more restrictive of growth than was glycerol. An exception is the important food-borne pathogen *S. aureus*, for which the reverse is true (see Table 10.7); this suggests that the influence of the solute itself may be complex. The "solute effect" is particularly noticeable when solutes such as ethanol and various glycols (propylene glycol, polyethylene glycols, etc.) are used to adjust a_w (Chirife 1994), as shown in Table 10.8 for *S. aureus*. For solutes such as sodium chloride and sucrose, the minimal a_w supporting growth of *S. aureus* is in the vicinity of 0.86; however, the minimal a_w allowing growth was well above 0.86 when ethanol or glycols were used to control a_w. It has been demonstrated that ethanol

Table 10.7 Effect of Glycerol on Minimal a_w Supporting Growth of Pathogenic Bacteria in Laboratory Media.

Bacteria	**a_w Adjusted With NaCl**	**a_w Adjusted With Glycerol**
Clostridium botulinum E	0.966	0.943
Escherichia coli	0.949	0.940
Clostridium perfringens	0.945	0.930
Clostridium botulinum A and B	0.940	0.930
Vibrio parahaemolyticus	0.932	0.911
Bacillus cereus	0.930	0.920
Listeria monocytogenes	0.920	0.900
Staphylococcus aureus	0.860	0.890

Sources: Adapted from Chirife, J. 1994. Specific solute effects with special reference to *Staphylococcus aureus*. *Journal of Food Engineering* 22: 409–419; and Tapia, M., Villegas, Y., and Martínez, A. 1991. Minimal water activity for growth of *Listeria monocytogenes* as affected by solute and temperature. *International Journal of Food Microbiology* 14:333–337.

Table 10.8 Minimal a_w for Growth of *S. Aureus* in Media With a_w Adjusted With Different Solutes.

Solute	**Minimal a_w**
Ethanol	0.975
1,3-Butylene glycol	0.97
Propylene glycol	0.96
Polyethyleneglycol 200	0.93
Polyethyleneglycol 400	0.93
Glycerol	0.89
NaCl	0.860
Sucrose	0.867

Source: Adapted from Chirife, J. 1994. Specific solute effects with special reference to *Staphylococcus aureus*. *Journal of Food Engineering* 22:409–419.

and glycols showed specific antibacterial activity against *S. aureus,* which may be compatible with cell wall attack (Ballesteros et al. 1993).

The fact that ethanol and glycols have a marked "solute effect" has limited practical consequences in food preservation because these solutes (ethanol, glycols) have at present little chance to be used for a_w control in human foods, because of either regulatory or consumer demands for the "green label" foods (i.e., foods without chemical additives).

Vittadini and Chinachoti (2003) studied the effects of physicochemical and molecular mobility parameters on *S. aureus* growth in high-moisture, liquid, and homogeneous media. They found that growth parameters correlated better with a_w (regulated by the addition of NaCl) than with the physicochemical and molecular mobility of the media. Ballesteros et al. (1993) reported that in liquid growth media, there is no clear relationship between *S. aureus* response to modification of certain physical properties of the growth medium (viscosity, dieletric constant, oxygen solubility, and oxygen diffusivity). Inhibitory effects of sodium chloride and sucrose were primarily ascribed to their a_w-lowering abilities. Nevertheless, the influence of various physical properties of a_w-controlling solutes (membrane permeability, ionic and nonionic properties, etc.) on the microbial response deserves further study.

Although minimal a_w for *S. aureus* growth certainly depended, in various cases, on the solute used to adjust the a_w, the bacterium was not able to grow below the current accepted

Table 10.9 Minimal Water Activity for Growth of Pathogenic Bacteria in Laboratory Media With a_w Adjusted With Salts or Sugars.

Bacteria	NaCl	KCl	Sucrose	Glucose
Listeria monocytogenes	0.92	—	0.92	—
Vibrio parahaemolyticus	0.936	0.936	0.940	—
Clostridium botulinum G	0.965	—	0.965	—
Clostridium botulinum E	0.972	0.972	0.972	0.975
Clostridium perfringens	0.945	—	—	0.945
Staphylococcus aureus	0.864	—	0.867	—

Source: Adapted from Chirife, J. 1993. Physicochemical aspects of food preservation by combined factors. *Food Control* 4:210–215.

minimal a_w 0.86, and this result is of major importance regarding food safety (Chirife 1994). When the solutes most often present in reduced a_w-preserved foods (e.g., sodium chloride, potassium chloride, sucrose, glucose) are used to control a_w, specific solute effects are much less evident, as shown in Table 10.9 for various pathogenic bacteria.

A practical application of the "specific solute effect" may be found in the utilization of sodium lactate. During the past several years, natural sodium lactate (considered to be a "GRAS" additive) has been recognized as an effective food ingredient to extend product shelf-life and control pathogens, mainly in cooked meat products (Weaver and Selef 1993, Miller and Acuff 1994, Blom et al. 1997). Sodium lactate is a normal constituent of muscle tissue, but when added at higher levels (2% to 3%), foods demonstrated antimicrobial activity against a broad range of microorganisms. Although sodium lactate is quite effective in lowering a_w, it has been demonstrated that its antimicrobial activity is mainly due to a "specific solute (lactate ion) effect" rather than to lowering of a_w (Chen and Shelef 1992; Houtsma et al. 1996a, 1996b).

Microbial Growth in Reduced Moisture Foods: Water Activity and "Mobility"

As noted by Karel (1999), despite some limitations, a_w is an excellent stability factor for microbial growth, because the creation of an osmotic pressure difference between the hydrated microbial cell and the surrounding food leads to cessation of microbial growth, through several mechanisms that are fairly well understood. He also added that the concept of a_w has been a dominating factor in the technology of food preservation for the past 50 years; despite this, some researchers interested in the stability of foods point out that the a_w concept is not adequate as an index of microbial growth stability and have focused instead on the mobility aspects of food systems. Chirife and Buera (1996), Slade and Levine (1987, 1991) and Franks (1991) challenged the concept of a_w and its utility to predict microbial stability. These authors stated that in many products, equilibrium thermodynamic descriptors such as a_w are inappropriate because the measured physical properties are time dependent. In many foods, the solids (either biopolymers or low-molecular-weight carbohydrates) are in an amorphous, metastable state (Slade and Levine 1987, Roos and Karel 1991, Levine and Slade 1992). This amorphous matrix may exist either as a very viscous glass or as a more liquid-like rubbery structure. The characteristic temperature range over which the glass rubber transition occurs (glass transition temperature [T_g]) has been proposed as a physicochemical parameter that can define stability and safety of foods (Slade and Levine 1987, 1991; Levine and Slade 1992). These authors proposed replacing the concept of a_W with a water dynamics approach, to better predict the microbial stability of intermediate-moisture foods. High viscosity states such as glasses would greatly interfere with the growth of microorganisms, because mobility may be needed in reduced-moisture solid foods for transport of nutrients and metabolites within the food matrix. However, from an extensive review of available knowledge, Chirife and Buera (1996) demonstrated that these expectations were not supported by experimental evidence (Le Meste et al. 2002). Among the arguments they presented was the observation that many foods are in the rubbery state at moisture contents where these foods are known to be resistant to microbial growth. To the contrary, molds were reported to grow in wheat flour in moisture conditions where these products were likely to be glassy. Buera et al. (1998) showed that samples of stored glassy white bread were spoiled by xerophilic molds, suggesting that T_g (measured by differential scanning calorimetry [DSC]) cannot be considered as an absolute threshold for mold growth inhibition. Similarly, Chirife et al. (1999) showed that mold growth in maltodextrins may be possible below the T_g, and spoilage may not be prevented by keeping the product below the T_g.

A research of potential interest was performed by Lavoie et al. (1997), who tried to correlate the molecular mobility of water (NMR water mobility) with the lag phase and growth rate of *S. aureus*. Media consisted of solute–brain-heart infusion (BHI) mixtures of various moisture contents and three different solutes were used—NaCl, glycerol, and raffinose. Inspection of experimental results revealed some inconsistencies regarding maximum cell densities and variation of growth rates with moisture content that, in the present author's opinion, limited the validity of their conclusions. These authors related the lag phase of *S. aureus* to NMR signal intensity and a_W and moisture content and reported that relationships were linear on a semilog plot. Unfortunately, only three points were used to demonstrate such correlation, and several of their experimental data were not included in the correlation. Kou et al. (1999) used water and biopolymer mobility (as monitored by NMR relaxation and other techniques) and T_g of sucrose–starch model food systems to correlate those parameters with the conidia germination time of *Aspergillus niger*. They concluded that the translational mobility of water and T_g (overall system mobility) could provide alternative measures to supplement a_W for predicting mold germination. Although the use of NMR can help monitor the water availability factor as related to the microbial activity in foods (Lavoie and Chinachoti 1995), results obtained at present mostly applied to the specific situation of *A. niger* germination in starch–sucrose systems. Certainly, much more experimental data are needed.

Stewart et al. (2002) hypothesized that the decrease in mobility between glass-forming systems may partially explain the varied responses of microorganisms in systems with matching a_W values but in which different glass-forming humectants were used to achieve the targeted a_W. They studied the behavior of *S. aureus* at reduced a_W using sucrose plus fructose or glycerol to lower a_W in the growth media and reported that limiting a_W was 0.88 (at neutral pH) when sucrose-fructose was used and 0.86 when glycerol was used (which did not agree with literature data where limiting value using glycerol was 0.89, as reported by Marshall et al. 1971). According to Stewart et al. (2002), as T_g of the system increases, the viscosity increases, which in turn decreases the mobility of molecules in the system (including the mobility of water molecules). However, reality does not always appear to follow this scheme. Using available mathematical models for a_W prediction and literature data on viscosity of fructose, glucose, glycerol, and sucrose, Chirife et al. (1984) calcu-

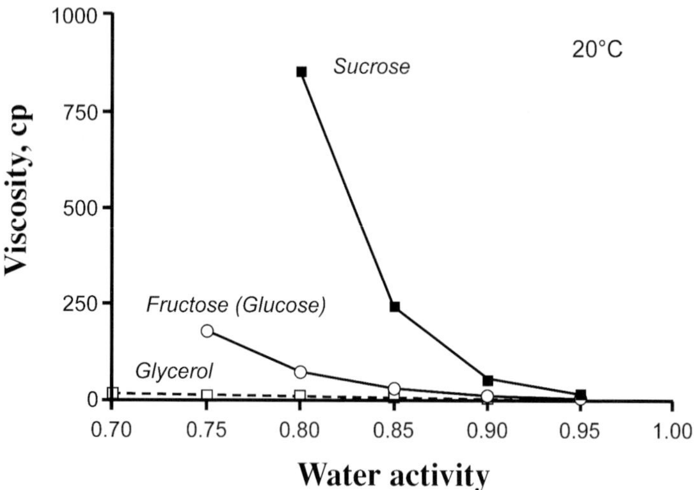

Figure 10.4 Calculated viscosities of sucrose, glucose, fructose, and glycerol solutions at various water activities at 20°C (curves for sucrose and glucose include some points corresponding to supersaturation) (adapted from Chirife J., Favetto, G., and Ferro-Fontán, C. 1984. Microbial growth at reduced water activities: Some physico-chemical properties of compatible solutes. *Journal of Applied Bacteriology* 56:259–264).

lated "iso-a_w viscosity" curves as shown in Figure 10.4. It can be seen that at any a_w, the order of viscosity is sucrose > glucose = fructose > glycerol, with sucrose being the most viscous solute. Pitt and Hocking (1977) studied the influence of glycerol or a glucose/fructose mixture on the growth rate of *Xeromyces bisporus* and *Chrysosporium fastidium* and found that rates were higher in glucose-fructose than in glycerol in a wide range of lowered water activities. González (1983) reported that radial growth rate at 31°C of *A. flavus* Link (a strain) was higher in sucrose media than in glycerol at water activities between 0.80 and 0.95. Chirife and Buera (1994) showed that germination time of *A. niger* at 28°C was shorter in sucrose media than in glycerol at 0.85 or 0.90 a_w. These experimental findings showed that the growth rates or germination time did not correlate with type of solute in the way predicted by mobility considerations (as noted above).

Karel (1999) noted that most foods are heterogeneous in their structure and that different "microregions" in a food material may have widely different physical and chemical properties. One of the advantages of a_w is that in most food products a_w equilibration is readily attainable (or *cuasi*) by water diffusion through the food, because water is a small and polar molecule that diffuses readily in most food structures. Chirife and Buera (1994) acknowledged that some reduced-moisture foods may be in a nonequilibrium situation; however, as far as the prediction of microbial growth is concerned, changes in nonequilibrium semimoist foods are generally very slow (within the time-frame of food's shelf-life) and/or small, so they do not seriously affect the application of a_w as a predictor of microbial growth. Nevertheless, the authors noted that the use of a_w should always consider the possible influences of some important nonequilibrium situations, such as foods containing amorphous crystallizable sugars.

Stewart et al. (2002) correctly noted that when Scott first published his work on the water relations of bacteria, he was well aware that a_w would not necessarily be adequate to describe all of the properties of solutions that influence microbial growth, metabolism, and survival. In recent years, evidence has suggested that molecular mobility (translational or rotational motion) may be related to some diffusion-limited properties of foods (Fennema 1996). However, based on present available information, glass transition/mobility concepts do not seem useful for predicting (with confidence) the microbial stability of foods and do not offer a better alternative than a_w to predict microbial stability of foods (in terms of growth–no-growth situations). Further research to assess the influence of solutes and water mobility on growth parameters (i.e., lag phase, growth rate) and metabolic activity of microorganisms in reduced moisture foods is needed, to discover whether a_w could be replaced by something more meaningful. As noted by Chirife and Buera (1996), the a_w and "mobility" approaches are not contradictory but complementary; one is concerned with the properties of the solvent (water) and the other with the structure of the food matrix. Thus, the role played by the combined effects of a_w, "mobility," and physicochemical properties of the media in modulating microbial response deserves to be investigated.

Water Activity as a Hurdle in Combined Preservation Techniques

As stated by Leistner in 1987 and earlier in 1978, it soon became apparent that in most foods where a_w is important for quality and stability, other factors (called "hurdles" by Leistner) contribute to the desired product shelf-life. Therefore, interest taken initially in a_w by food manufacturers was extended to other factors (e.g., Eh, pH, temperature, incorporation of additives, etc.).

Preservation systems, to be effective, must overcome microbial homeostatic resistance. Homeostatic mechanisms of vegetative microorganisms are energy dependent because the cell must expend energy to resist the stress factors, e.g., to repair damaged components, to synthesize new cell components, etc. In the case of spores, homeostatic mechanisms do not consume energy, but they are built into the cell before being exposed to an environmental stress (Leistner and Gould 2002). Combined preservation techniques interfere with the active homeostatic mechanisms that operate in the vegetative microbial cell and the passive refractory homeostatic mechanisms that operate in microbial spores at a number of sites or in a cooperative manner (Gould 1988, 1996). According to Leistner (1999, 2000), in foods preserved by hurdle technology, the possibility exists that different hurdles in a food will not just have an additive effect on stability but could act synergistically. A synergistic effect could be obtained if the hurdle in a food hits different targets (e.g., cell membrane, DNA, enzyme systems, pH, a_w, Eh) within the microbial cell and thus disturbs the homeostasis of the microorganisms present in several aspects. Therefore, using different hurdles in the preservation of a particular food should be an advantage, because microbial stability could be achieved with a combination of gentle hurdles. In practical terms, this could mean that it is more effective to use different preservative factors at small levels in a food than only one preservative factor at a large level, because different preservative factors might hit different targets within the bacterial cell and thus act synergistically.

A slight reduction in food a_w means that microorganisms surviving the processing of such foods, or contamination of the food after processing, will be osmoregulating at some level or other. However, if osmoregulation diverts resources away from normal cell biosynthetic processes, then it is clear that the stressed cells may be more vulnerable to other stresses in combined preservation systems, particularly when the other stresses also divert away from the synthesis of new cell material. Because homeostatic responses often require the expenditure of energy by the stressed cells, restriction of the availability of energy therefore is a sensible target to pursue. According to Gould (2000), this probably forms the basis of many of the successful, empirically derived, mild combination preservation procedures, exemplified by the "hurdle technology" and "multitarget preservation" approaches of Leistner.

Water activity continues to be one of the main hurdles to be manipulated, and the use of combinations of extrinsic and intrinsic factors together with lowered a_w levels is common in the food industry. Improved knowledge about the role of water in foods led to the rediscovery and optimization of old preservation techniques and to a renewed interest in foods that are shelf-stable with control of a_w. This applies to fully dehydrated, intermediate, and high-moisture traditional foods with inherent empiric hurdles and also to novel products, especially high-a_w foods, for which the hurdles are intelligently selected and intentionally applied. Traditional fully dehydrated and intermediate-moisture foods can be regarded as some of the oldest preserved foods by humans. However, in the quest for quality, the importance of considering the combined action of decreased a_w with other preservation factors as a way to develop new improved foodstuffs coincided almost simultaneously with modern food processing, to the point that currently, consumers are searching for fresh-like characteristics in many processed products. The food industry has responded to these demands with the so-called minimally processed foods, which have become a widespread industry that is receiving a lot of attention lately.

Therefore, the control of a_w for food design is being used in many ways according to needs (Alzamora et al. 2003):

- At various stages of the food distribution chain, during storage, processing, and/or packaging as a "back-up" hurdle in existing minimally processed products with short shelf-life to diminish microbial pathogenic risk and/or increase shelf-life (i.e., a slight reduction in a_w in addition to refrigeration and heating in "ready-to-eat foods")
- Traditionally, for obtaining long–shelf-life products (fully dehydrated and intermediate moisture ones)—actual trends in these applications are to obtain very high sensory quality products by utilizing advances in the knowledge of water sorption phenomena, a_w prediction, deleterious physicochemical reactions, and polymer science, as well as more controlled and/or sophisticated drying techniques
- As one of the preservative factors (together with other emerging and/or traditional preservative factors) to obtain high-moisture novel foods by hurdle techniques

In industrialized countries with ready availability of energy and infrastructure and widespread use of refrigeration, the control of a_w has been mainly applied to develop a great variety of mild thermally processed, chilled, and frozen-distributed foods. Topical applications include fermented meats (sausages, raw hams) and shelf-stable mild heat–treated meats (ready-to-eat fresh-like meats); *sous vide* and cook-chill dishes and healthful foods (low-fat and/or low-salt and functional foods); and foods processed via emerging techniques (e.g., hydrostatic high pressure) (Leistner and Gould 2002).

In contrast to the above scenario, in many developing countries, refrigeration is expensive and not always available. Thus, emphasis on the lowering of a_w approach has affected the development of ambient-stable foods, which require minimal energy, machinery, and infrastructure for processing, storage, and distribution (Leistner and Gould 2002). Common applications entail foods with reduced a_w (achieved by partial drying or addition of salt or sugar), usually combined with acidification (i.e., a reduction of pH) and addition of preservatives (e.g., fermented foods and fully dehydrated foods).

Most of the traditional foods that remain stable, safe, and tasty during long-term storage without refrigeration in developing countries such as Africa, Asia, and Latin America are intermediate-moisture foods, in which lowering of a_w is one of the main preservative factors or hurdles (Leistner and Gould 2002). Many of the manufacturing processes for intermediate-moisture foods were empirically developed, but now the hurdles and their specific roles are better understood and can be rationally selected to design or to optimize the preservation system.

There are actually two categories of foods with reduced a_w whose stability is based on a combination of factors: intermediate- and high-moisture foods. Intermediate-moisture foods generally range from 0.60 to 0.90 a_w and 10% to 50% water by weight (Jayaraman 1995). Additional hurdles provide the margin of safety against spoilage by microorganisms resistant to a_w (mainly molds and yeasts, which can grow at a_w as low as 0.61) and against some bacterial species that are likely to grow when the a_w value of the intermediate-moisture food is near the upper limit of water activities (i.e., 0.90 a_w). With these targets, the lowering of a_w is often combined with chemical preservatives (i.e., nitrites, sorbates, sulfites, benzoates, antimicrobials of natural origin, smoke components) and a reduction of pH (which usually inhibits or decreases bacterial growth, accentuates the action of preservatives, and increases the minimum a_w values for bacterial growth), and sometimes with competitive microorganisms. Other intermediate-moisture food products receive a thermal treatment during the manufacturing process that inactivates heat-sensitive microorganisms, while the subsequent hot filling in sealed containers further improves the microbial stability (Leistner and Gould 2002).

Most intermediate-moisture foods are designed to be stored for several months at ambient temperature even in tropical climates and to be eaten "as is" without rehydration. They are moist enough to be ready-to-eat without giving rise to a sensation of dryness but dry enough to be ambient-stable (Karel 1973, 1976; Jayaraman 1995). Many intermediate-moisture food products, because of the addition of very high amounts of solutes (such as sugar or salt) to reduce a_w to the desired level, are too sweet or salty, becoming undesirable from the nutritional and sensory point of view. Therefore, this category of products has been subjected in the past decade to continuous revision and discussion.

On the other hand, high-moisture foods have a_w values well above 0.90. Thus, in this category the reduction of a_w is a hurdle of less relative significance because most of the microorganisms are able to proliferate (Leistner and Gould 2002). Stability at ambient temperature is reached by applying intentional and carefully designed hurdle technology. High-moisture fresh-like fruits and cooked meat products, preserved by the a_w–mild heat treatment–pH–preservatives interaction and storable without refrigeration, represent a rational application of the combined approach (Alzamora et al. 1995, 2000; Leistner and Gould 2002).

In 1994, within the Science and Technology for Development (CYTED) Program, a project entitled Development of Intermediate Moisture Foods From Iberoamerica conducted a survey in 11 countries and collected information on 260 traditional intermediate- and high-moisture foods. Table 10.10 shows the main factors used in Spain and Latin America for preservation of the traditional foods that were evaluated (Tapia et al. 1994; Welti et al. 1994). Many of these products, also common in different parts of the world, are safe and

Table 10.10 The Main Factors Used in Iberoamerican Countries for Preservation of Traditional Foods by Combined Methods Technology.

Preservation Factors[a] Product Category	a_w	pH	F	t	Smoke	Preserv.	CF
Fruits and vegetables	X	X	X	—	—	X	X
Meat	X	X	X	—	X	X	X
Fish	X	X	X	X	X	—	—
Dairy	X	X	X	X	—	X	X
Bakery	X	—	X	—	—	X	—
Miscellaneous	X	X	X	—	—	X	—

[a]F = mild heat treatment; t = mild refrigeration; Preserv. = preservatives; CF = competitive flora.

storable without refrigeration and require inexpensive packaging. Selected representative products from each category, the process parameters involved, and their contribution as microbial stability factors (hurdles) are consigned in Table 10.11. Most shelf-stable foods do not rely solely on a_w for microbial control but on other preservation factors. The binary combination of a_w and pH acts as a relevant hurdle in many of these products, preventing proliferation of pathogenic microorganisms, while the rest (antimicrobials, thermal treatment, etc.) play a secondary role, mainly against spoilage flora (Tapia et al. 1994).

Different approaches have been explored for obtaining shelf-stability and fresh-likeness in fruit products. Commercial, minimally processed fruits are fresh (with high moisture) and are prepared for convenient consumption and distribution to the consumer in a fresh-like state. Minimum processing includes minimum preparation procedures such as washing, peeling, and/or cutting, packing, etc., after which the fruit product is usually placed in refrigerated storage, where its stability varies depending on the type of product, processing, and storage conditions. However, product stability without refrigeration is an important issue not only in developing countries but in industrialized countries as well. The principle used by Leistner for shelf-stable high-moisture meats ($a_w > 0.90$), where only mild heat treatment is used and the product still exhibits a long shelf-life without refrigeration, can be applied to other foodstuffs. Fruits would be a good choice. Leistner states that for industrialized countries, production of shelf-stable products is more attractive than intermediate-moisture foods because the required a_w for shelf-stable products is not as low and less humectants and/or less drying of the product is necessary (Leistner 2000).

If fresh-like fruit is the goal, dehydration should not be used in processing. Reduction of a_w by addition of humectants should be kept to a minimum level to maintain the product in a high-moisture state. To compensate for the high moisture left in the product (in terms of stability), controlled blanching can be applied without affecting the sensory and nutritional properties; pH reductions can be made that will not impair flavor; and preservatives can be added to alleviate the risk of potential spoilage microflora. In conjunction with these factors, slight thermal treatment, pH reduction, slight a_w reduction, and the addition of antimicrobials (sorbic or benzoic acid, limited amounts of sulfite), all placed in context with the hurdle technology principles applied to fruits, make up an interesting alternative to intermediate-moisture preservation of fruits, as well as to commercial minimally processed refrigerated fruits. Considerable research effort has been made within the CYTED Program and the Multinational Project on Biotechnology and Food of the Organization of American States (OAS) in the area of combined methods, geared to the development of shelf-stable high-moisture fruit products. Over the past two decades, use of

Table 10.11 Preservation Factors in Selected Food Products With Reduced a_w.

Preservation Factors							Level of Hurdle Relevance	
Product	a_w	pH	Antimicrobial (A)	Thermal Treatment (T)	Refrigeration Requirement (R)	Comp. Flora[a] (CF)	Most Relevant	Secondary
Meat products								
Sausage	0.92	5.6	Sodium nitrite	No	No	Yes	a_w	A, CF
Dry sausage	0.74	4.5	Sodium nitrite	No	No	Yes	a_w, pH	A, CF
Spanish ham	0.85	6.2	Sodium nitrite	No	No	No	a_w	A
Beef *foie gras*	0.87	6.3	Sodium nitrite	Yes	Yes	No	a_w, R	T
Vegetable products								
Ketchup	0.94	3.8	Potassium sorbate	Yes	No	No	pH, a_w	A, T
Garlic cream	0.84	4.0	Essential oils (naturally occurring)	No	No	No	a_w, pH	A
Garlic sauce	0.96	3.7	Essential oils (naturally occurring)	Yes	No	No	pH, a_w	T
Chili cream dip	0.84	4.2	Essential oils (naturally occurring)	Yes	No	No	a_w, pH	A, T
Fruit products								
Candied papaya	0.70	4.6	No	Yes	No	No	a_w	pH, T
Candied pineapple	0.80–50.87	4.5–5.6	No	Yes	No	No	a_w	pH, T
Dehydrated plum	0.77	3.9	No	No	No	No	a_w	pH
Dehydrated banana	0.62	5.1	No	No	No	No	a_w	
Peach jam	0.83	3.1	No	Yes	No	No	a_w, pH	T
Mango jam	0.81	5.0	Sodium benzoate	Yes	No	No	a_w, pH	T, A
Guava paste	0.78–50.88	3.5	Sulfite	Yes	No	No	a_w, pH	A, T
Sweet potato paste	0.84	3.4	Sodium benzoate	Yes	No	No	a_w, pH	A
Fishery products								
Brined anchovies	0.75	6.2	No	No	No	No	a_w	
Dry-salted anchovies	0.71–50.74	5.6–5.8	No	No	Yes	No	a_w	R
Cod-type dry fish	0.74–50.75	7.5–8.6	No	No	No	No	a_w	
Anchovies in oil	0.76–50.80	6.1–6.2	No	No	Yes	No	a_w	R
Smoked trout	0.96	5.4	Smoke	No	Yes	No	a_w	R
Smoked salmon	0.96–50.98	5.7–6.2	Smoke		No	Yes	No	R

(*continues*)

Table 10.11 Preservation Factors in Selected Food Products With Reduced a_w (*continued*).

Preservation Factors							Level of Hurdle Relevance	
Product	a_w	**pH**	**Antimicrobial (A)**	**Thermal Treatment (T)**	**Refrigeration Requirement (R)**	**Comp. Flora[a] (CF)**	**Most Relevant**	**Secondary**
Dairy products								
Sweet condensed milk	0.84	6.6	No	Yes	No	No	a_w, T	
Melted cheese	0.97–50.98	5.7–6.0	No	Yes	Yes	No	T, R	
Milk jam	0.81–50.85	5.6–6.0	No	Yes	No	No	a_w	Maillard products
Goat cheese	0.91	5.6	No	Yes	Yes	No	a_w	T, pH, R
Reggianito cheese	0.86	5.5	No	Yes	No	No	a_w	T, pH
Miscellaneous products								
Mayonnaise	0.93–50.94	3.8-3.9	Potassium sorbate	Yes	No	No	pH, a_w	A
Honey	0.62–50.69	3.1-3.3	No	No	No	No	a_w, pH	
Soy sauce (type 1)	0.79	4.7	No	Yes	No	No	a_w, pH	T
Soy sauce (type 2)	0.79	4.8	Sodium benzoate	Yes	No	No	a_w, pH	A, T

[a]Comp. flora = competitive flora.

Source: Adapted from Tapia, M.S., Aguilera, J.M., Chirife, J., Parada, E., and Welti-Chanes, J. 1994. Identification of microbial stability factors in traditional foods from Iberoamerica. *Revista Española de Ciencia y Tecnología de Alimentos* 34:145–163.

this approach has led to important developments of innovative technologies for obtaining shelf-stable "high-moisture fruit products" that are storable for 3 to 8 months without refrigeration. These new technologies are based on a combination of inhibiting factors to combat the deleterious effects of microorganisms in fruits, including additional factors to diminish major quality loss. Slight reduction of a_w (0.94 to 0.98 a_w), control of pH (pH 3.0 to 4.1), mild heat treatment, addition of preservatives (concentrations (1,500 ppm), and antibrowning additives were the factors selected to formulate the preservation procedure (Alzamora et al. 1989, 1993, 1995; Guerrero et al. 1994; Cerrutti et al. 1997).

There exist novel (in their application) and refined impregnation techniques for developing minimal processes. Pulsed vacuum osmotic dehydration, a new method of osmotic dehydration that takes advantage of the porous microstructure of vegetable tissues, is a technique that uses vacuum impregnation to reduce process time and improve additive incorporation. During vacuum impregnation of porous materials, important modifications in structure and composition occur as a consequence of external pressure changes. Vacuum impregnation shows faster water loss kinetics in short-time treatments compared with time-consuming atmospheric "pseudo-diffusional" processes, due to the occurrence of a specific mass transfer phenomenon, the hydrodynamic mechanism, and the result produced in the solid–liquid interface area. Many fruits and vegetables have a great number of pores and offer the possibility of being impregnated by a determined solution of solute and additives. Thus, product composition as well as its physical and chemical properties may be changed to improve its stability. An important advantage of using low pressures (approximately 50 mbar) in minimal preservation of fruit is that equilibration times are shorter than at atmospheric pressure (e.g., 15 minutes under vacuum versus a few hours in forced convection at atmospheric conditions, or a few days in media without agitation, reducing a_w to 0.97) (Alzamora et al. 2003). This process could be appropriate in the development of new minimally processed fruit products or in the development of improved pretreatments for traditional preservation methods such as canning, salting, freezing, or drying and also in high-quality jam processing (Alzamora et al. 2000).

At present, the physical nonthermal processes (high hydrostatic pressure, manothermo-sonication, oscillating magnetic fields, pulsed electric fields, light pulses, etc.) are receiving considerable attention. In combination with other conventional hurdles, they are of potential use for the microbial stabilization of fresh-like food products with little degradation to nutritional and sensory properties. With these novel processes, the goal intended is often not a sterile product but only a reduction of the microbial load, whereas growth of the residual microorganisms is inhibited by additional conventional hurdles. Interesting results have been reported by the research group at Universidad de las Américas (Mexico) in obtaining minimally processed avocado sauce, avocado purée, and banana purée. These fruit products were preserved by the interaction of blanching–high pressure–pH–a_w–preservatives, and the combination of heat treatment and high pressure significantly decreased browning reactions (Alzamora et al. 2000). Another group of hurdles, which at present is of special interest in industrialized as well as developing countries, includes "natural preservatives" (spices and their extracts, hop extracts, lysozyme, chitosan, pectine hydrolysate, etc.) (Leistner 2000). As an example, high-moisture strawberries can be preserved for at least 3 months by combining mild heat treatment, 3000 ppm vanillin (instead of synthetic antimicrobials), 500 ppm ascorbic acid, and adjustment of a_w to 0.95 and pH to 3.0 (Cerrutti et al. 1997).

Finally, we need to mention the excellent recompilation of traditional and artisanal combined methods used around the world (many of them involving the control of a_w) by the two world's leading authorities on hurdle technology, Professor Lothar Leistner and Dr. Grahame Gould (Leistner and Gould 2002). This overview covers hurdle techniques applied in developed countries and in Latin America, India, China, and Africa. Basic principles underlying preservation procedures are critically discussed for many popular products. Among them, it is interesting to cite the following:

- Paneer, a cottage cheese–type Indian product (hurdles: 0.97 a_w; pH 5, F_0 value 0.8), stable for several weeks without refrigeration
- Dudh churpi, an Indian dairy product (preparation: heating, acid coagulation, addition of sugar and potassium sorbate, smoking, drying)
- Meat (preparation: marination in salt, glycerol, nitrite, acidulants and ascorbate, cooking and packaging; 0.70 or 0.85 a_w, pH 4.6) storable at room temperature for 1 month or at 5–C for more than 4 months
- Rabbit meat, quite popular in China, marinated and cooked; fried; brined and cooked; or smoked (hurdles: 0.92 to 0.98 a_w, refrigeration).

Predictive Microbiology in Assessing the Effects of a_w in Combination With Other Hurdles

The value of predictive microbiology models is becoming increasingly recognized in the design of combined technologies. Until recently, food microbiologists have relied on nonkinetic empirical data obtained by challenge testing with specific microorganisms to predict the safety of foods. The traditional approach, rather than the kinetic one, was used to fix intrinsic and extrinsic factors governing microbial growth and then establish maximum and minimum limits in which organisms will grow or survive. While the hurdle concept is widely accepted as a food preservation strategy, its potential has yet to be fully realized, as there is a lack of quantitative data available to allow prediction of the adequate and necessary level of each hurdle. If the growth kinetics (lag phase duration, growth rate) or the interface between microbial growth and no growth for an identified target microorganism under several combinations of intrinsic and extrinsic factors can be predicted, the selection of such factors can be made on a scientific basis, and the selected hurdles can be kept at their minimum levels. Synergistic and additive interactions can be identified and sensory selection of hurdles and their levels may be done between several "safe" equivalent combinations of interactive factors determined by the models (McMeekin et al. 2000, McMeekin and Ross 2002, Alzamora et al. 2003).

Classification of predictive models is based on the population behavior that they describe and comprises growth models, limits of growth (growth–nongrowth interface), and inactivation models. Within each category, models are classified as being at the primary, secondary, or tertiary level (Buchanan 1993, Whiting 1995). Primary-level models describe the changes in the microbial response with time. Secondary-level models describe the responses by parameters of primary models to change in environmental factors. Tertiary-level models refer to application of software and expert systems developed from primary- and secondary-level models and constitute a user-friendly form in software for personal computers (McMeekin et al. 1993, McClure et al. 1994, McMeekin and Ross 2002). Among the large number of factors undoubtedly affecting microorganisms, the reduction in a_w is one of the few that exerts the most control over an organism's growth or decline and deserves particular attention in predictive modeling studies. Some examples of mathematical models applied to describe and understand the likely behavior of biological agents in environments with lowered a_w are addressed next.

Buchanan and Bagi (1997) explored the effects of four nonionic humectants (NaCl, mannitol, sorbitol, and sucrose) in combination with four pH levels (4.5, 5.5, 6.5, and 7.5) and three incubation temperatures (12°, 19°, and 28°C) on the growth kinetics of a three strain mixture of *E. coli* O157:H7 in BHI broth. Growth curves were fitted using the Gompertz equation. Increasing humectant concentrations interacted with decreasing temperatures and pH values to decrease lag phase duration and generation times. Solute identity had relatively little impact on the growth kinetics of *E. coli* O157:H7 at low solute concentration; however, differences among the solutes were noted as a_w became more restrictive. Accuracy and bias indices of the model indicated reasonable predictions for combinations of a_w, pH, and temperature that support the growth of the microorganism.

The growth response of *V. parahaemolyticus* activity was modeled with a form of the square root type model, incorporating a novel term for the effects of superoptimal water activities, which can be used to predict generation times for the temperature range 8° to 45°C and a_w range 0.936 to 0.995, permitting growth of halophilic organisms like *V. parahaemolyticus* (Miles et al. 1997). The predicted generation times were successfully compared with observed responses for both laboratory media and food (prawn, crab, cod, meat, milk, oyster, beef, etc.) in these temperature (8° to 45°C) and a_w (0.936 to 0.995) ranges.

Rosso and Robinson (2001) studied the effects of a_w on the radial growth of *Aspergillus flavus, A. nomius, A. oryzae, A. parasiticus, A. candidus, A. sydowii, Eurotium amstelodami, E. chevalieri*, and *Xeromyces bisporus.* Mold growth was described by a model derived from the Cardinal Model family. The model had an excellent descriptive ability for different mold species, different pH values, and different control solutes. The results could be used to control the quality of the product from formulation to storage, especially for long shelf-life products (cakes, biscuits, soft beverages, cheese, baked goods, etc.).

The growth responses of *L. monocytogenes* as affected by CO_2 concentration (0% to 100% v/v, balance nitrogen), NaCl concentration (0.5% to 8.0% w/v), pH (4.5 to 7.0), and temperature (4° to 20°C) were studied in laboratory medium, and growth curves were fitted using the model of Baranyi and Roberts (1994). Predictions of the model were compared with published data on food (raw beef, corned beef, cold-smoked salmon, cooked beef, chicken nuggets, etc.). The model was suitable for predicting growth of *L. monocytogenes* in various foods packaged under a modified atmosphere (Fernández et al. 1997).

Graham et al. (1996) studied the combined effect of temperature (4° to 30°C), pH (5.0 to 7.3) and NaCl (0.1% to 5.0%) on growth from spores of nonproteolytic *C. botulinum.* They used the data to construct a growth model and compared predictions from the model with growth reported in the literature. Predictions of growth from two models by Baranyi and Gompertz showed that these were suitable for use with fish, meat, and poultry products. These models should allow food processors to reduce the amount of challenge testing necessary to ensure food safety with regard to nonproteolytic *C. botulinum* and to be useful in the development of minimally processed foods, which rely on storage at low temperatures for preservation (i.e., refrigerated processed foods of extended durability [REPFEDs] and *sous vide* products).

A logistic regression model was first proposed by Ratkowsky and Ross (1995) to model the boundary between growth and no growth for bacterial strains in the presence of one or more growth controlling factors such as temperature, pH, and additives (e.g., salt and sodium nitrite). The model was illustrated with *Shigella flexneri* (Ratkowsky and Ross 1995) and subsequently used by many authors for assessing the influence of a_w in combination with other hurdles (temperature, antimicrobials, pH, etc.) on many pathogenic and spoilage microorganisms (McMeekin et al. 2000, Alzamora et al. 2002). A number of important scientific and practical implications of knowing the sharp cutoff between growth and no growth conditions were pointed out by McMeekin et al. (2000) as follows: (1) events occurring in one region may be reversed as the interface is crossed, and the study of physiological mechanisms close to either side may contribute to the knowledge of the mode of action of preservative factors; (2) for a_w and other stress factors, the interface can be closely defined (for a_w and pH, the definition is 0.1 to 0.2 pH unit and 0.01 to 0.03 a_w unit, respectively) even for different values of growth probability, showing that there is a sharp cutoff between growth and no growth conditions; and (3) the definition of growth–no growth interface provides an accurate set of conditions on which mild combined preservation techniques may be based, allowing the control of a process or formulation to ensure nongrowth of a dangerous food-borne pathogen or occurrence of a spoilage organism.

References

Aguilera, J.M., Chirife, J., Tapia, M.S., and Welti-Chanes, J. 1990. *Inventario de Alimentos de Humedad Intermedia Tradicionales de Iberoamérica.* Ciudad de México: Instituto Politécnico Nacional.

Alzamora, S.M., Cerrutti, P., Guerrero, S., and López-Malo, A. 1995. Minimally processed fruits by combined methods. In: *Food Preservation by Moisture Control—Fundamentals and Applications*, eds. J. Welti-Chanes and G. Barbosa-Cánovas, pp. 463–492. Lancaster: Technomic Publishing Company.

Alzamora, S.M., Fito, P., López-Malo, A., Tapia, M.S., and Parada-Arias, E. 2000. Minimally processed fruit using vacuum impregnation, natural antimicrobial addition and/or high hydrostatic pressure techniques. In: *Minimally Processed Fruits and Vegetables. Fundamental Aspects and Applications*, eds. S.M. Alzamora, M.S. Tapia, and A. López Malo, pp. 293–315. Gaithersburg: Aspen Publishers.

Alzamora, S.M., Gerschenson, L.N., Cerrutti, P., and Rojas, A.M. 1989. Shelf stable pineapple for long-term non-refrigerated storage. *Lebensittel-Wissenschaft und-Technologie* 22(5):233–236.

Alzamora, S.M., Tapia, M.S., Argaíz, A., and Welti-Chanes, J. 1993. Application of combined methods technology in minimally processed fruits. *Food Research International* 26:125–130.

Alzamora, S.M., Tapia, M.S., López-Malo, A., and Welti-Chanes, J. 2003. The control of water activity. In: *Food Preservation Techniques*, eds. L. Bogh-Sorensen, P. Zeuthen, and H. Lelieveld, pp. 126–153. Cambridge, England: CRC Press.

Baird-Parker, A.C., and Freame, B. 1967. Combined effects of water activity, pH and temperature on the growth of *Clostridium botulinum* from spore and vegetative cell inocula. *Journal of Applied Bacteriology* 30(3):420–429.

Ballesteros, S.A., Chirife, J., and Bozzini, J.P. 1993. Specific solute effects on *Staphylococcus aureus* cells subjected to reduced water activity. *International Journal of Food Microbiology* 20(2):51–66.

Baranyi, J., and Roberts, T.A. 1994. Predictive microbiology: Quantitative microbial ecology. *Food Microbiology* 10:43–59.

Beuchat, L.R. 1981. Microbial stability as affected by water activity. *Cereal Foods World* 26(7):345–349.

Beuchat, L.R. 1983. Influence of water activity on growth, metabolic activities and survival of yeasts and molds. *Journal of Food Protection* 46:135–141.

Beuchat, L.R. 1987. Influence of water activity on sporulation, germination, outgrowth and toxin production. In: *Water Activity: Theory and Applications to Food,* eds. L.B. Rockland and L.R. Beuchat, pp. 137–151. New York: Marcel Dekker.

Beuchat, L.R. 2002. Water activity and microbial stability. Fundamentals of Water Activity, IFT Continuing Education Committee, June 14-15, Anaheim, CA.

Blom, H., Nerbrink, E., Dainty, R., Hagtvedt, T., Borch, E., Nissen, H., and Nesbakken, T. 1997. Addition of 2.5% lactate and 0.25% acetate controls growth of *Listeria monocytogenes* in vacuum-packed, sensory acceptable servelat sausage and cooked ham stored at 4°C. *International Journal of Food Microbiology* 38:71–76

Briozzo, J., Amato de Lagarde, E., Chirife, J., and Parada, J.L. 1983. Effect of water activity and pH on growth and toxin production by *C. botulinum* type G. *Applied Environmental Microbiology* 51(4):844–848.

Brown, A.D. 1974. Microbial water relations: Features of the intracellular composition of sugar-tolerant yeasts. *Journal of Bacteriology* 118(3):769–777.

Brown, A.D., and Melling, J. 1971. Inhibition and destruction of microorganisms by heat. In: *Inhibition and Destruction of Microbial Cells*, ed. W.B. Hugo, pp. 1–37. London: Academic Press.

Buchanan, R.L. 1993. Developing and distributing user(friendly application software. *Journal of Industrial Microbiology* 12(3-5):251–255.

Buchanan, R.L., and Bagi, L.K. 1997. Effect of water activity and humectant identity on the growth kinetics of *Escherichia coli* O157:H7. *Food Microbiology* 14(5):413–423.

Buera, M.P., Jouppila, K., Roos, Y.H., and Chirife, J. 1998. Differential scanning calorimetry glass transition temperatures of white bread and mold growth in the putative glassy state. *Cereal Chemistry* 75(1):64–69.

Cerrutti, P., Alzamora, S.M., and Vidales, S.L. 1997. Vanillin as antimicrobial for producing shelf-stable strawberry purée. *Journal of Food Science* 62:608–610.

Chen, N., and Shelef, L.A. 1992. Relationship between water activity, salts of lactic acid, and growth of *Listeria monocytogenes* in a meat model system. *Journal of Food Protection* 55(8):574–578.

Chirife, J. 1993. Physicochemical aspects of food preservation by combined factors. *Food Control* 4:210–215.

Chirife, J. 1994. Specific solute effects with special reference to *Staphylococcus aureus*. *Journal of Food Engineering* 22:409–419.

Chirife, J., and Buera, M.P. 1994. Water activity, glass transition and microbial stability in concentrated and semimoist food systems. *Journal of Food Science* 59:921–927.

Chirife, J., and Buera, M.P. 1996. Water activity, water glass dynamics and the control of microbiological growth in foods. *Critical Reviews in Food Science and Nutrition* 36:465–513.

Chirife, J., Buera, M.P., and González, H.H. 1999. The mobility and mold growth in glassy/rubbery substances. In: *Water Management in the Design and Distribution of Quality Foods. ISOPOW 7*, eds. Y.H. Roos, R.B. Leslie, and P.J. Lillford, pp. 285–298. Basel, Switzerland: Technomic Publishing.

Chirife J., Favetto, G., and Ferro-Fontán, C. 1984. Microbial growth at reduced water activities: Some physicochemical properties of compatible solutes. *Journal of Applied Bacteriology* 56:259–264.

Chirife, J., Ferro-Fontan, C., and Scorza, O.C. 1981 The Intracellular water activity of bacteria in relation to the water activity of the growth medium. *Journal of Applied Bacteriology* 50(3):475–479.

Christian, J.H.B. 1963. Water activity and the growth of microorganisms. *Recent Advances in Food Research* 3:248–55.

Christian, J.H.B. 1981. Specific solute effects on microbial water relations. In: *Water Activity: Influences on Food Quality*, eds. L.B. Rockland and G.F. Stewart, p. 825. New York: Academic Press.

Corry, J.E.L. 1974. The effect of sugars and polyols on the heat resistance of *Salmonellae*. *Journal of Applied Bacteriology* 37:31–43.

Corry, J.E.L. 1975. The effect of water activity on the heat resistance of bacteria. In: *Water Relations of Foods*, ed. Ro, B. Duckworth. London: Academic Press.

Dodds, K.L. 1989. Combined effect of water activity and pH on inhibition of toxin production by *Clostridium botulinum* in cooked, vacuum-packaged potatoes. *Applied Environmental Microbiology* 55:656–60.

Earnshaw, R.G., Appleyard, J., and Hurst, R.M. 1995. Understanding physical inactivation processes: Combined preservation opportunities using heat, ultrasound and pressure. *International Journal of Food Microbiology* 28(2):197–219.

Favetto, G.J., Resnik, S.L., Chirife, J., and Ferro-Fontán, C. 1983. Statistical evaluation of water activity measurements obtained with the Vaisala Humicap humidity meter. *Journal of Food Science* 48:534–538.

Fennema, O. 1996. Water and ice. In: *Food Chemistry*, ed. O. Fennema, pp. 17–94. New York: Marcel Dekker.

Fernández, P.S., George, S.M., Sills, C.C., and Peck, M.W. 1997. Predictive model of the effect of CO_2, pH, temperature and NaCl on the growth of *Listeria monocytogenes*. *International Journal of Food Microbiology* 37:37–45.

Fontana, A.J. 2000. Water activity's role in food safety and quality. Second NSF International Conference on Food Safety, October 11-13, 2000, Savannah, GA.

Fontana, A.J. 2002. Measurement of water activity. Fundamentals of Water Activity, IFT Continuing Education Committee, June 14-15, Anaheim, CA.

Franks, F. 1991. Water activity: A credible measure of food safety and quality? *Trends in Food Science and Technology* 68–72.

Giannuzzi, L., and Parada, J.L. 1991. Sobre el crecimiento de *Staphylococcus aureus* en medios de actividad acuosa inferior a 0.86. *Revista Argentina de Microbiología* 23:79–85.

Glass, K., and Doyle, M.P. 1991. Relationship between water activity of fresh pasta and toxin production by proteolitic *Clostridium botulinum*. *Journal of Food Protection* 54:162–165.

Gould, G.W. 1985. Osmoregulation: Is the cell just a simple osmometer? The microbiological experience. In: *A Discussion Conference: Water Activity: A Credible Measure of Technological Performance and Physiological Viability*? Faraday Division, Royal Society of Chemistry, Girton College, Cambridge, England, July 1-3.

Gould, G.W. 1988. Interference with homeostasis—Foods. In: *Homeostatic Mechanisms in Microorganisms,* eds. J.G. Banks, R.T.G. Board, G.W. Gould, and R.W. Mittenbury. Bath, England: Bath University Press.

Gould, G.W. 1989. Drying, raised osmotic pressure and low water activity. In: *Mechanisms of Action of Food Preservation Procedures,* ed. G.W. Gould, pp. 97–118. New York: Elsevier Applied Science.

Gould, G.W. 1996. Methods for preservation and extension of shelf-life. *International Journal of Food Microbiology* 33:51–64.

Gould, G.W. 2000. Induced tolerance of microorganisms to stress factors. In: *Minimally Processed Fruits and Vegetables. Fundamental Aspects and Applications,* eds. S.M. Alzamora, M.S. Tapia, and A. López-Malo, pp. 29–37. Gaithersburg: Aspen Publishers.

Gould, G.W., and Measures, J.C. 1977. Water relations in single cells. *Philosophical Transactions of the Royal Society of London* 278:151.

Graham, A.F., Mason, D.R., and Peck, M.W. 1996. Predictive model of the effect of temperature, pH and sodium chloride on growth from spores of non(proteolytic *Clostridium botulinum. International Journal of Food Microbiology* 31:69–85.

Guerrero, S., Alzamora, S.M., and Gerschenson, L.N. 1994. Development of a shelf-stable banana purée by combined factors: Microbial stability. *Journal Food Protection* 57:902–907.

Gutiérrez, C., Abee, T., and Booth, I.R. 1995. Physiology of the osmotic stress response in microorganisms. *International Journal of Food Microbiology* 28:233–244.

Hocking, A.D. 1988. Strategies for microbial growth at reduced water activities. *Microbiology Science* 5:280–284.

Hocking, A.D., and Pitt, J.I. 1979. Water relations of some *Penicillum* species at 25°C. *Transactions of the British Mycological Society* 73:141–145.

Houtsma, P.C., Dewit, J.C., and Rombouts, F.M. 1996b. Minimum inhibitory concentration (MIC) of sodium lactate and sodium chloride for spoilage organisms and pathogens at different pH values and temperatures. *Journal of Food Protection* 59:1300–1304.

Houtsma, P.C., Kant-Muermans, M.L., Rombouts, F.M., and Zwietering, M.H. 1996a. Model for the combined effects of temperature, pH, and sodium lactate on growth rates *of Listeria monocytogenes* in broth and bologna-type sausages. *Applied Environmental Microbiology* 62:1616–1622.

Jakobsen, M., Filtenborg, O., and Bramsnaes, F. 1972, Germination and outgrowth of the bacterial spore in the presence of different solutes. *Lebensittel-Wissenschaft und-Technologie* 5:159–172.

Jayaraman, K.S. 1995. Critical review on intermediate moisture fruits and vegetables. In: *Food Preservation by Moisture Control—Fundamentals and Application*, eds. J. Welti-Chanes and G. Barbosa-Cénovas, pp. 411–442. Basel, Switzerland: Technomic Publishing.

Karel, M. 1973. Recent research and development in the field of low moisture and intermediate moisture foods. *Critical Reviews in Food Technology* 3:329–373.

Karel, M. 1976. Technology and application of new intermediate moisture foods. In: *Intermediate Moisture Foods,* eds. R. Davies, G.G. Birch, and K.J. Parker, pp. 4–31. London: Applied Science Publishers.

Karel, M. 1999. Food research tasks at the beginning of the new millennium—A personal vision. In: *Water Management in the Design and Distribution of Quality Foods. ISOPOW 7*, eds. Y.H. Roos, R.B. Leslie, and P.J. Lillford, pp. 535–549. Basel, Switzerland: Technomic Publishing.

King, D., Jr., Pitt, J.I., Beuchat, L.R., and Corry, J.E.L. 1984. Comparison of enumeration methods for molds growing in flour stored in equilibrium with relative humidities between 66 and 68%. In: *Methods for the Mycological Examination of Food*, pp. 143–145. New York: Plenum Press.

Kirby, R.M., and Davies, R. 1990. Survival of dehydrated cells of *Salmonella typhimurium* LT2 at high temperatures. *Journal of Applied Bacteriology* 68(3):241–246.

Kou, Y., Molitor, P.F., and Schmidt, S.J. 1999. Mobility and stability characterization of model food systems using NMR, DSC and conidia germination techniques. *Journal of Food Science* 64:950–959.

Laroche, C., and Gervais, P. 2003. Unexpected thermal destruction of dried, glass bead-immobilized microorganisms as a function of water activity. *Applied and Environmental Microbiology* 69(5):3015–3019.

Lavoie, J., and Chinachoti, P. 1995. The role of water mobility in promoting *Staphylococcus aureus* and *Aspergillus niger* activities. In: *Magnetic Resonance in Food Science*, eds. P.S. Belton, I. Delgadillo, A.M. Gil, and G.A. Webb, pp. 33–42. Cambridge, UK: The Royal Society of Chemistry.

Lavoie, J.P., Labbe, R.G., and Chinachoti, P. 1997. Growth of *Staphylococcus aureus* as related to ^{17}ONMR water mobility and water activity. *Journal of Food Science* 62:861–866.

Lee, R.Y., Silverman, G.J., and Munsey, D.T. 1981. Growth and enterotoxin A production by *Staphylococcus aureus* in precooked bacon in the intermediate moisture range. *Journal of Food Science* 46:1687.

Leistner, L. 1978. Hurdle effect and energy saving. In: *Food Quality and Nutrition* ed. W.K. Downey, pp. 553–557. London: Applied Science Publishers.

Leistner, L. 1987. Shelf-stable products and intermediate foods based on meat. In: *Water Activity: Theory and Applications to Food*, eds. L.B. Rockland and L.R. Beuchat, pp. 295–327. New York: Marcel Dekker.

Leistner, L. 1999. Combined methods for food preservation. In: *Food Preservation Handbook,* ed. M.S. Rahman, pp. 457–485. New York: Marcel Dekker.

Leistner, L. 2000. Hurdle technology in the design of minimally processed foods. In: *Minimally Processed Fruits and Vegetables. Fundamental Aspects and Applications*, eds. S.M. Alzamora, M.S. Tapia, and A. López-Malo, pp. 13–27. Gaithersburg: Aspen Publishers.

Leistner, L., and Gould, G.W. 2002. *Hurdle Technologies. Combination Treatments for Food Stability, Safety and Quality.* New York: Kluwer Academic/Plenum Publishers.

Leistner, L., and Rodel, W. 1975. The significance of water activity for microorganisms in meats. In: *Water relations of Foods,* ed. R.B. Duckworth, pp. 309–323. New York: Academic Press.

Le Meste, M., Champion, D., Roudaut, G., Blon, G., and Simatos, D. 2002. Glass transition and food technology: A critical appraisal. *Journal of Food Science* 67:2444–2458.

Lenovich, L.M. 1987. Survival and death of microorganisms as influenced by water activity. In: *Water Activity: Theory and Applications to Food*, eds. L.B. Rockland and L.R. Beuchat, pp. 119–136. New York: Marcel Dekker.

Levine, H., and Slade, L. 1992. Glass transitions in foods. In: *Physical Chemistry of Foods*, eds. H.G. Schwartzberg and R.W. Hartel, pp. 83–220. New York: Marcel Dekker.

Lotter, L.P., and Leistner, L. 1978. Minimal water activity for enterotoxin A production and growth of *Staphylococcus aureus. Applied Environmental Microbiology* 36:377.

Mañas, P., Pagan, R., Leguérinel, I., Condón, S., Mafart, P. and Sala, F. 2001. Effect of sodium chloride concentration on the heat resistance and recovery of *Salmonella typhimurium. International Journal of Food Microbiology* 63(3):209–216.

Marshall, B.J., Ohye, D.F., and Christian, J.H.B. 1971. Tolerance of bacteria to high concentrations of NaCl and glycerol in the growth medium. *Applied Microbiology* 21:363–364.

McClure, P.J., Blackburn, C. de W., Cole, M.B., Curtis, P.S., Jones, J.E., Legan, J.D., Ogden, L.D., Peck, M.W., Roberts, T.A., Sutherland, J.P., and Walker, S.J. 1994. Modelling the growth, survival and death of microorganisms in foods: The UK Micromodel approach. *International Journal of Food Microbiology* 23:265–275.

McMeekin, T.A., Olley, J.N., Ross, T., and Ratkowsky, D.A. 1993. *Predictive Microbiology: Theory and Application.* Taunton: Research Studies Press.

McMeekin, T.A., Presser, K., Ratkowsky, D., Ross, T., Salter, M., and Tienungoon, S. 2000. Quantifying the hurdle concept by modelling the bacterial growth/no growth interface. *International Journal of Food Microbiology* 55:93–98.

McMeekin, T.A., and Ross, T. 2002. Predictive microbiology: Providing a knowledge(based framework for change management. *International Journal of Food Microbiology* 78:133–153.

Miles, D.W., Ross, T., Olley, J., and McMeekin, T.A. 1997. Development and evaluation of a predictive model for the effect of temperature and water activity on the growth rate of *Vibrio parahaemolyticus. International Journal of Food Microbiology* 38:133–142.

Miller, R.K., and Acuff, G.R. 1994. Sodium lactate affects pathogens in cooked beef. *Journal of Food Science* 59:15–19.

Mossel, D.A.A., Corry, J.E.L., Struijk, C.B., and Baird, R.M. 1991. *Essentials of the Microbiology of Foods. A Text for Advanced Studies*, pp. 91–93. Chichester: John Wiley and Sons.

Northolt, M.D., Frisvad, J., and Samson, R.A. 1995. Occurrence of food-borne fungi and factors for growth. In: *Introduction to Foodborne Fungi,* eds. R.A. Samson, E.S. Hoekstra, J.C. Frisvad, and O. Filtenborg, pp. 243–250. Wageningen: Ponsen & Looyen.

Notermans, S., and Heuvelman, C.J. 1983. Combined effect of water activity, pH and sub-optimal temperature on growth and enterotoxin production of *Staphylococcus aureus. Journal of Food Science* 48:1832.

O'Byrne, C., and Booth, I.R. 2002. Osmoregulation and its importance to food(borne microorganisms. *International Journal of Food Microbiology* 74:203–216.

Ohye, D.F., and Christian, J.H.B. 1966. Combined effects of temperature, pH and water activity on growth and toxin production by *Clostridium botulinum* types A, B and E, in Proceedings of the Fifth International Symposium on Food Microbiology, Moscow, p. 217.

Pichereau, V., Hartke, A., and Auffray, Y. 2000. Starvation and osmotic stress induced multiresistances influence of extracellular compounds. *International Journal of Food Microbiology* 55:19–25.

Pitt, J., and Hocking, A. 1977. Influence of solute and hydrogen ion concentration on the water relations of some xerophilic fungi. *Journal of General Microbiology* 101:35–40.

Prior, B.A., Kenyon, C.P., van der Veen, M., and Mildenhall, J.P. 1987. Water relations of *Pseudomonas fluorescens*. *Journal of Applied Bacteriology* 62:119–128.

Rahman, S. 1995. *Food Properties Handbook*, pp. 1–14. Boca Raton, FL: CRC Press.

Ratkowsky, D.A., and Ross, T. 1995. Modelling the bacterial growth/no growth interface. 1995. *Letters in Applied Microbiology* 20:29–33.

Roos, Y.H., and Karel, M. 1991. Plasticizing effect of water on thermal behavior and crystallization of amorphous food models. *Journal of Food Science* 56:38–43.

Rosso, L., and Robinson, T.P. 2001. A cardinal model to describe the effect of water activity on the growth of moulds. *International Journal of Food Microbiology* 63:265–273.

Russell, N.J., Evans, R.I., ter Steeg, P.F., Hellemons, J. Verheul, A., and Abee, T. 1995. Membranes as a target for stress adaptation. *International Journal of Food Microbiology* 28(2):255–261.

Scott, W.J. 1953. Water relations of *Staphylococcus aureus* at 30°C. *Australian Journal of Biological Science* 6:549–556.

Siderius, M.H., and Mager, W.H. 1997. General stress response: In search of a common denominator. In: *Yeast Stress Responses*, eds. S. Hohmann and W.H. Mager, pp. 213–230. Berlin: Springer.

Silverman, G.J., Munsey, D.T., Lee, C., and Ebert, E. 1983. Interrelationship between water activity, temperature and 5.5% oxygen on growth and enterotoxin secretion by *Staphylococcus aureus* in precooked bacon. *Journal of Food Science* 48:1783–1786.

Sinigaglia, I., Corbo, M.R., Altieri, C., and Massa, S. 2002. Response surface model for effects of temperature, water activity and pH on germination of *Bacillus cereus* spores. *Journal of Food Safety* 22:121–134.

Slade, L., and Levine, H. 1991. Beyond water activity: Recent advances based on an alternative approach to the assessment of food quality and safety. *Critical Reviews in Food Science and Nutrition* 30:115–360.

Slade, L., and Levine, H. 1987. Structural stability of intermediate moisture foods—A new understanding. In: *Food Structure—Its Creation and Evaluation*, eds. J.R. Mitchell and J.M.V. Blanshard, pp. 115–147. London: Butterworths.

Stewart, C.M., Cole, M.B., Legan, J.D., Slade, L., Vandeven, M.H., and Schaffner, D.W. 2002. *Staphylococcus aureus* growth boundaries: Moving towards mechanistic predictive models based on solute-specific effects. *Applied and Environmental Microbiology* 68(4):1864–1871.

Sumner, S.S., Sandros, T.M., Harmon, M.C., Scott, V.N., and Bernard, D.T. 1991. The heat resistance of *Salmonella typhimurium* and *L. monocytogenes* in sucrose solutions of various water activities. *Journal of Food Science* 56:1741–1743.

Tapia, M., Villegas, Y., and Martínez, A. 1991. Minimal water activity for growth of *Listeria monocytogenes* as affected by solute and temperature. *International Journal of Food Microbio*logy 14:333–337.

Tapia, M.S., Aguilera, J.M., Chirife, J., Parada, E., and Welti-Chanes, J. 1994. Identification of microbial stability factors in traditional foods from Iberoamerica. *Revista Española de Ciencia y Tecnología de Alimentos* 34:145–163.

Troller, J.A. 1971. Effect of water activity on enterotoxin B production and growth of *Staphylococcus aureus*. *Applied Microbiology* 21:435.

Troller, J.A. 1972. Effect of water activity on enterotoxin A production and growth of *Staphylococcus aureus*. *Applied Microbiology* 24:440.

Troller, J.A. 1987. Adaptation and growth of microorganism in environments with reduced water activity. In: *Water Activity: Theory and Applications to Food,* eds. L.B. Rockland and L.R. Beuchat, pp. 111–117. New York: Marcel Dekker.

Troller, J.A., and Stinson, J.V. 1978. Influence of water activity on the production of extracellular enzymes by *Staphylococcus aureus*. *Applied Environmental Microbiology* 35(3):521–526.

Vaamonde, G., Chirife, J., and Scorza, O.C. 1982. An examination of the minimal water activity for *Staphylococcus aureus* ATCC 6538 P growth in laboratory media adjusted with less conventional solutes. *Journal of Food Science* 47:1259–1262.

Valik, L., and Gorner, F. 1993. Growth of *Staphylococcus aureus* in pasta in relation to its water activity. *International Journal of Food Microbiology* 20:45–48.

Vilhelmsson, O., and Miller, K.J. 2002. Humectant permeability influences growth and compatible solute uptake by *Staphylococcus aureus* subjected to osmotic stress. *Journal of Food Protection* 65:1008–1015.

Vittadini, E., and Chinachoti, P. 2003. Effect of physico-chemical and molecular mobility parameters on *Staphylococcus aureus growth. International Journal of Food Science and Technology* 38:841–849.

Weaver, R.A., and Shelef, L.A. 1993. Antilisterial activity of sodium, potassium or calcium lactate in pork liver sausage. *Journal of Food Safety* 13:133–146.

Welti-Chanes, J., Bermúdez, D., Mújica-Paz, H., Valdez-Fragoso, A., and Alzamora, S.M. 2004. Principles of freeze concentration and freeze-drying. In: *Handbook of Frozen Foods*, eds. Y.H. Hui, P. Cornillon, I. Guerrero Legarreta, M. Lim, K.D. Murrell, and W.-K. Nip. New York: Marcel Dekker.

Welti-Chanes, J., Tapia, M.S., Aguilera, J.M., Chirife, J., Parada, E., López-Malo, A., López, L.C., and Corte, P. 1994. Classification of intermediate moisture foods consumed in Ibero-America. *Revista Española de Ciencia y Tecnología de Alimentos* 34:53–63.

Whiting, R.C. 1995. Microbial modeling in foods. *Critical Reviews in Food Science and Nutrition* 35(6):467–494.

11 Principles of Intermediate-Moisture Foods and Related Technology

Petros S. Taoukis and Michelle Richardson

Water Activity and Stability of Foods

The water activity (a_w) of a food describes the degree of "boundness" of water contained in a food and is a measure of its availability to act as a solvent and to participate in chemical or biochemical reactions (Labuza 1977). The rate of these reactions can thus be expressed as a function of a_w as illustrated in Figure 11.1. At constant temperature, a unique relationship exists between the moisture content and a_w of a specific food, depending on its method of preparation (i.e., adsorption versus desorption). This relationship is described by the moisture sorption isotherm of the food (Bell and Labuza 1998). At very low a_w, water is tightly associated with surface polar sites by chemisorption and is generally unavailable for reactions and solutions. The upper limit of this region has been traditionally characterized as the "monolayer value," which for most foods occurs at 0.2 to 0.3 a_w. Above the monolayer, water is involved, in varying degrees, in multilayers and capillaries and is possibly entrapped in structural components. Its ability to act as a solvent, as a reaction medium, and as a reactant increases with increasing a_w (Labuza 1975). As a result, the rates of many deteriorative reactions increase exponentially with increasing a_w. At high a_w, a small change in a_w corresponds to a large increase in moisture content, leading to dilution of the reacting species; the rate of some reactions may level off or decrease (Labuza 1980). However, at low a_w, increased water content lowers the reaction-phase viscosity, which facilitates the diffusion of reacting species and thus tends to increase the reaction rate or offset the decrease that would otherwise result from the dilution effect. Several theories attempt to explain the effect of a_w on food deterioration reactions, as well as ways to systematically approach and model this effect (Taoukis et al. 1997). The moisture content and a_w can influence the kinetic parameters of the Arrhenius equation (k_A, E_A), the concentrations of the reactants, and in some cases, even the apparent reaction order, n. Most relevant studies have modeled either k_A as a function of a_w, related to the change of mobility of reactants due to the dependent changes of a_w on viscosity, or E_A as a function of a_w. The inverse relationship of E_A with a_w (increase in a_w decreases E_A, and vice versa) could be theoretically explained by the proposed phenomenon of enthalpy–entropy compensation discussed by Labuza (1980b). Additionally, moisture content and a_w directly affect the glass transition temperature of the system, T_g. With increasing a_w, T_g decreases. The transverse of T_g and change into a rubbery state has pronounced effects, especially on texture- and viscosity-dependent phenomena, but also on reaction rates and their temperature dependence. It has been proposed for dehydrated systems that a critical moisture content/a_w alternative to the monolayer value of the BET theory is the value at which the dehydrated system has a T_g of 25°C (Roos 1993). In complex systems, matrix porosity, molecular size, and phenomena such as collapse and crystallization occurring in the rub-

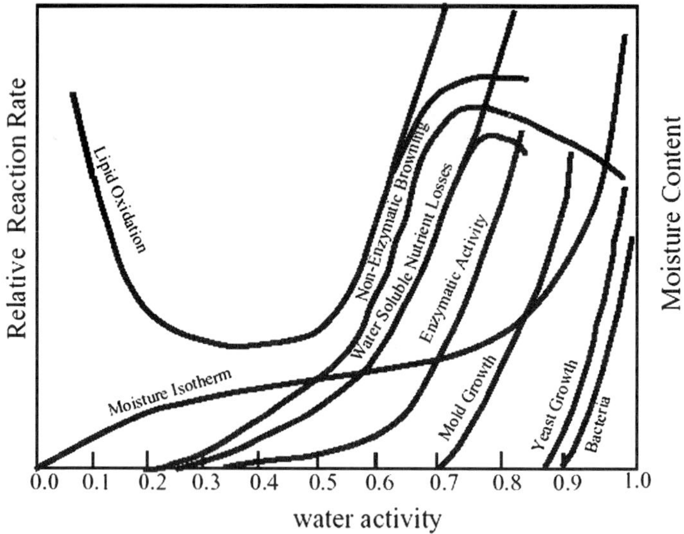

Figure 11.1 Water activity map (adapted from Labuza, T.P. 1970. Properties of water as related to the keeping quality of foods. In: *Proceedings of the Third Annual Meeting of the International Congress of Food Science and Technology*, pp. 618–635. Chicago: Institute of Food Technology).

bery state result in more complicated behavior. Both a_w and glass transition theory help explain the relationship between moisture content and deteriorative reaction rates.

Water activity has a marked effect on the growth of microorganisms, as shown in the stability map (see Figure 11.1). It was the search for an effective measure to describe the behavior of microorganisms in environments with reduced moisture that contributed to the establishment of the term "water activity" (Scott 1957) in food technology. The main response of microorganisms to a shift in environmental a_w, corresponding to a change in the concentration of nonpenetrant solutes, is the compensatory regulation of their internal solution contents. This adjustment tends to be at an internal solute concentration level that just exceeds the external osmolality, protecting the organisms from excessive water loss.

This osmotic stress suppresses microbial growth. If the environmental osmolality exceeds the osmoregulatory capacity of a microbial cell, then water is irreversibly lost, and growth ceases (Gould 1985). The ability of a microorganism to grow in a given environment is not determined exclusively by the a_w but also depends on the complex interactions of several other hurdles, including temperature, pH, oxidation–reduction potential, preservatives, and competitive microflora (Leistner and Rodel 1976, Troller 1980, Leistner and Gould 2002). According to the hurdle concept, all of these factors are regarded as barriers to cell growth. Growth can be inhibited by an appropriate combination of hurdles under conditions where each hurdle, individually, would be inadequate. For set values of other factors, a minimum a_w for growth can be defined. Minimum values of a_w for growth of several common microorganisms of significance to health and for toxin production are given in Table 11.1.

The U.S. Food and Drug Administration (FDA), in the Good Manufacturing Practice

Table 11.1 Water Activity (a_w) Requirements for Growth and Toxin Production of Microorganisms as Related to Food Safety.

Microorganism	Minimum a_w for Growth	Minimum a_w for Toxin Production
Aeromonas hydrophila	0.97	
Bacillus cereus	0.93–0.95	
Campylobacter jejuni	0.98	
Clostridium botulinum		
Type A	0.93–0.95	0.94–0.95
Type B	0.93–0.94	0.94
Type E	0.95–0.97	0.97
Clostridium perfringens	0.93–0.95	
Eschericia coli	0.95	
Listeria monocytogenes	0.92–0.94	
Salmonella spp.	0.92–0.95	
Staphylococcus aureus	0.86	0.87–0.90 (Enterotoxin A)
Vibrio parahaemoliticus	0.94	
Yersinia enterolytica	0.95	
Aspergillus clavatus	0.85	0.99 (Patulin)
Aspergillus flavus	0.78–0.80	0.83–0.87
Aspergillus ochraceus	0.77–0.83	0.83–0.87 (Ochratoxin)
	0.76–0.81	0.80–0.88 (Penicillin acid)
Aspergillus Parasiticus	0.82	0.87 (Aflatoxin)
Byssochlamys nivea	0.84	
Penicillium cyclopium	0.82–0.87	0.97 (Penicillin acid)
	0.81–0.85	0.87–0.90 (Ochratoxin)
Penicillium expansum	0.83–0.85	0.99 (Patulin)
Penicillium islandicum	0.83	
Penicillium martensii	0.79–0.83	0.99 (Penicillin acid)
Penicillium patulum	0.81–0.85	0.85 0.95 (Patulin)
Penicillium 0viridicatum	0.83	0.83–0.86 (Ochratoxin)
Stachybotrys atra	0.94	0.94
Trichothecium roseum	0.90	

Sources: Adapted from Beuchat, L.R. 1981. Microbial stability as affected by water activity. *Cereal Foods World* 26:345-349; Christian, J.H.B. 1981. Specific solute effects on microbial water relations. In: *Water Activity: Influences on Food Quality*, eds. L.B. Rockland and G.F. Stewart, pp. 825-854. New York: Academic Press.

regulations, under 21 CFR 110 (51 FR 22458; 1986), defines a *safe moisture level* as being a level of moisture low enough to prevent the growth of undesirable microorganisms in a finished product under the intended conditions of manufacturing, storage, and distribution. The maximum safe moisture level of a food is based on its a_w. The a_w of a food is considered safe if adequate data are available to demonstrate that the food's a_w is at or below a value not supportive of the growth of undesirable microorganisms (21 CFR 110.3).

Intermediate-Moisture Foods: Definition

The importance of the role of water in food preservation and in controlling food quality was recognized early in history. The reduction of available water by sun-drying or the addition of sugars and salt is the basis of some of the earliest preservation techniques (Labuza and Sloan 1981). The aforementioned new understanding and knowledge of water–food interactions has contributed in the past four decades to a systematic optimization of old preservation technologies and improved food products that are shelf stable, all based on the principle of reduced a_w. New production techniques designed to use state-of-the-art technology have led to several generations of new food products that require less energy to produce and distribute. Such foods are generally classified as intermediate-moisture foods (IMFs). IMF products are foods with moisture content higher than that of dry foods and are edible without rehydration. Despite their higher moisture content, they are designed to be shelf stable without needing refrigeration during distribution and storage. Thermal processing to the extent needed for canning is not required, although some IMF products may be pasteurized (Kaplow 1970).

IMFs have no precise definition based on water content or a_w. Generally, their moisture content is in the range of 10% to 40% and a_w is 0.60 to 0.90 (Karel 1973, Erickson 1982, Gould 1996).

The Committee for Intermediate-Moisture Foods at the National Center for Coordination of Research on Food and Nutrition (Centre National de Coordination des Etudes et Recherches sur la Nutrition et l'Alimentation) in France introduced a comprehensive definition (Multon 1981):

> Intermediate-moisture foods are food products of soft texture, subjected to one or more technological treatments, consumable without further preparation and with a shelf stability of several months assured without thermal sterilization, nor freezing or refrigeration, but by an adequate adjustment of their formulation: Composition, pH, additives and mainly a_w which must be approximately between 0.6 and 0.84 (measured at 25°C).

The characteristic properties of IMFs offer a number of advantages over "conventional" dry or high-moisture foods. IMF processing, as well as distribution, generally is substantially less energy intensive than drying, refrigeration, freezing, or canning. Additionally, IMF technology can potentially lead to a higher retention of nutrients and quality than that achieved with more vigorous processes, such as certain dehydration and thermal processing methods. IMFs can be stored without special precautions for several months. Although appropriate packaging is a factor in prolonged shelf-life, packaging requirements for IMF products are not as strict as it is for many other food categories. Completely impervious packages are not necessary, and loss of package integrity would not pose a health hazard, especially in environments with average humidity. Plasticity and chewability, without an oral sensation of dryness, make IMFs suitable for direct consumption with no preparation, resulting in convenience and further energy savings. Because of the plastic texture of these foods, they can be easily shaped into pieces of uniform geometry for easy packaging and storage, and they can be formed conveniently into individual servings. Due to their relatively low moisture, IMF products are concentrated in weight and bulk and have high nutrient and caloric density. These advantageous characteristics of IMFs are particularly compatible with the needs of modern consumers for convenient foods of high nutrient density. IMFs are especially attractive in situations where the food supply load, ability to resupply, and preparation time are limiting factors, as in military settings (Brockman 1970), space travel (Smith et al, 1971), geographical exploration, and mountaineering. IMF technology could be an alternative to energy-intensive methods of drying for preservation and storage. Thus, in tropical climates and Third World Countries, where refrigeration is limited and food spoilage is a vital problem, IMF technology could be a high value alternative (Obanu et al. 1975, Obanu 1981, Narashima Rao 1997). Relevant international projects aimed at the systematic exploitation of IMF principles have been sponsored to support technology upgrades in the production of traditional food items in developing countries, such as Indonesia and Malaysia (Seow 1988).

The legislatory status regarding the application of a_w and other hurdles and their required intensity to achieve food product stability is reviewed in Leistner and Gould (2002). The FDA Good Manufacturing Practice regulations include a provision that applies to IMFs. According to 21 CFR 110.80(b)(14), foods (such as IMFs) that rely on the control of a_w for preventing the growth of undesirable microorganisms should be processed to maintain a safe moisture level. Compliance with this requirement may be accomplished effectively by means of (1) monitoring the a_w of the food, (2) controlling the ratio between water and soluble solids in the finished food, and (3) protecting the finished food from moisture pick-up, so that the a_w does not increase to an unsafe level. This can be achieved by appropriate IMF technology.

Intermediate-Moisture Food Technologies

Water Activity Reduction

The principal requirement for the production of an IMF is the reduction of a_w in the product to an a_w value in the IMF zone. Additionally, the method used should result in products that are organoleptically acceptable without any further preparation steps. This is usually achieved with the addition of humectants, materials that lower a_w but also may impart a plastic texture and allow foods to retain their moist properties. In general, a_w reduction by simple drying of the food results in a texture that is too dry for direct consumption. Thus, most vegetables and meats would have to be dried until their moisture content was below 15% to achieve an a_w less than 0.85. At this moisture level, rehydration would be required before consumption. Therefore, the use of humectants is a fundamental and characteristic step in the production of IMFs, regardless of the specific manufacturing process applied.

A number of hygroscopic chemical compounds have been used or can be considered for use by the food industry as humectants. An effective humectant would ideally provide the following properties:

- Show no toxicity
- Exhibit adequate solubility in water
- Effectively lower the mole fraction of water, i.e., have a low molecular weight, possibly combined with ionic dissociation
- Show compatibility with the organoleptic characteristics, i.e., improve flavor or impart no flavor to IMF products
- Have low or no caloric value
- Reinforce dietetic image of the food by adding positive functional properties, e.g., act as a dietary fiber, a prebiotic or antioxidant component
- Have low cost

An ideal humectant would also exhibit synergism with other humectants. Most used or candidate humectants can be classified in one of four general categories: (1) sugars, (2) low-molecular-weight polyols, (3) protein derivatives, and (4) mineral and organic salts (see Table 11.2).

Table 11.2 Common Humectants, Conventional and Unconventional, and Their Principal Roles in Foods.

	Lowering of a_w	Plasticizing	Antimicrobial Action	Rehydration Ability	Delay of Crystallization	Sweetening	Other Roles[a]	FDA Status
Monosaccharides, disaccharides, and polysaccharides								
Pentoses	XX[b]	XX						GRAS
Hexoses (glucose, fructose)	XX	XX			XX	XX		GRAS
Mannose, galactose, etc.	XX	XX						GRAS
Disaccharides (sucrose, lactose, maltose)	XX	XX				XX		GRAS
Various oligosaccharides	XX		XX			X	XX	GRAS
Natural or industrial products: honey, invert sugar, high-fructose corn syrup, glucose syrup, maple syrup	XX	XX			XX			GRAS
Maltodextrins (dextrose equivalents 3–20)	X				XX			GRAS
Hydrolysates; gums and hydrolysates; and hydrolysates	XX	XX			XX	X		GRAS
Alcohols and polyols								
Ethanol	XX		XX					21 CFR 184. 1293
Sorbitol	XX	XX	XX		XX	XX		21 CFR 184. 1835
Mannitol, xylitol, erythritol	XX	X				XX		21 CFR 172.395
Glycerol	XX	XX	X		XX	X	X	21 CFR 182.1320
1,2-Propanediol, 1,2-butanediol (propylene glycol)	XX	XX		XX		X		21 CFR 182.1666
1,3-Butanediol, 1,3-pentanediol (1,3-butylene glycol)	XX	XX		XX				21 CFR 172.220
1,3,5-Polyols (4–12 carbon atoms)	XX	XX						21 CFR 172.864
Polyethylene glycols (mol wt 400, 600, 1500, 2400, etc.)	XX	XX						21 CFR 172.820
Mineral salts (NaCl, KCl, $CaCl_2$)	XX	X		X			XX	GRAS
Phosphates, polyphosphates	X						XX	CFR 182 & 184
Certain carbonates, sulfates	XX						XX	CFR 182 & 184
Salts of milk serum	XX	X						CFR 182 & 184

Table 11.2 Common Humectants, Conventional and Unconventional, and Their Principal Roles in Foods (*continued*).

	Lowering of a_w	Plasticizing	Antimicrobial Action	Rehydration Ability	Delay of Crystallization	Sweetening	Other Roles[a]	FDA Status
Organic acids								
Food acids and Na, K, Ca salts	XX	X	XX			X	XX	CFR 182 & 184
Ascorbic acid	X						XX	CFR 182 & 184
Proteins and derivatives								
Amino acids and salts	XX	X					XX	21 CFR 172.320
Oligopeptides	X						XX	GRAS
Protein hydrolysates	XX	XX					XX	GRAS

Sources: Adapted from Guilbert, S. 1984. Additifs et agents d'presseurs de l'aw. In: *Additifs et Auxiliaires de Fabrication dans les Industries Agro-alimentaires*, ed. J.L. Multon, pp. 199–227. Aprix, Paris: Tec et Doc Lavoisier.

[a] Other roles include pH regulation, protein solubilization and antioxidation, and increase in nutritional and dietetic value (e.g. prebiotics or soluble fibers).

[b] XX denotes a strong effect, and X denotes a moderate effect.

It was mentioned above that the effectiveness of a humectant in lowering a_w depends on its ability to lower the mole fraction of water as well as to interact with and alter the structure of the water in the food system. The a_w of an ideal solution is a direct function of the mole fraction of the solved component. According to Raoult's law (Raoult 1888),

$$a_w = X_w = \frac{n_w}{n_w + n_s} \tag{11.1}$$

where X_w is the mole fraction of water, n_w is the total moles of water, and n_s is the total moles of solute or solutes. Thus, the a_w of the aqueous ideal solution depends only on the total number of solute molecules (kinetic units) and not on the nature of the solutes. The smaller the molecular weight of the humectant (solute), the greater is its a_w-lowering effect (humectancy) per unit of weight dissolved. Dilute aqueous solutions of nonelectrolytes exhibit ideal solution behavior. For example, the a_w for solutions of glucose and glycerol in concentrations up to 4 mol/L and for sucrose in concentrations up to 2 mol/L is predicted accurately (with a deviation of less than 1%) by Raoult's law (Reid 1976). Equation 11.1 also applies to dilute salt solutions, in concentrations up to 1 mol/L, with each ionic species regarded as a distinct kinetic unit.

Substantial deviations from ideality are observed in more-concentrated systems. These deviations can be attributed to several factors—the most prevalent are interactions between solute molecules; the unavailability of some of the water in the food, to act as a solvent; and the binding of solutes to insoluble food components (e.g., proteins), which prevents solute molecules from entering actual solution. Nonideality of mixtures is generally described in thermodynamics by excess functions and is incorporated into the equations for a_w through the use of activity coefficients (Van Den Berg and Bruin 1981). Thus, the a_w of a real aqueous solution is:

$$a_w = \gamma_w X_w = \frac{n_w}{n_w + n_s \gamma_s} \tag{11.2}$$

where γ_w is the activity coefficient of water and γ_s is that of the solute. For ideal solutions, γ_s is unity. Generally, for humectants, γ_s is greater than 1 and γ_w is less than 1. Several theoretical approaches have led to equations for the computation of activity coefficients of liquid components (Prausnitz 1969).

Use of the commonly used equations in chemical engineering, the Margules, Van Laar, nonrandom two-liquid, and Wilson equations, has been limited in food science (Van Den Berg and Bruin 1981), although in principle they can be applied to humectant–water systems. The thermodynamics of concentrated electrolyte solutions is more complicated (Robinson and Stokes 1970). Activity coefficients and a_w of electrolyte solutions can be estimated through the use of generalized isotherms (Kusik and Meissner 1973) or equations (Pitzer and Mayorga 1973, Benmergui et al. 1979).

The aforementioned theoretical approaches, although useful for predicting the a_w of solutions, can become cumbersome for multisolute mixtures and can give only an approximate estimate for a real food system. To minimize development time of IMF products, a prediction equation is needed for the a_w-lowering effect of humectants in a complex food system containing solids that do not go completely into solution (Sloan and Labuza 1976).

Table 11.3 Binary Interaction Constants for Norrish Equation.

Humectant	K	Humectant	K
Alanine	−2.59 ± 0.37	Lactulose	−8.0 ± 0.3
α-Aminobutyric acid	−2.57 ± 0.37	Lysine	−9.3 ± 0.3
1,3-Butylene glycol	−3.47	Malic acid	−1.82 ± 0.13
2,3-Butylene glycol	−4.78	Maltose	−4.54 ± 0.02
Citric acid	−6.17 ± 0.49	Mannitol	−0.91 ± 0.27
Corn syrups		NaCl (0%–2%)	7.91
DE 33	−5.97	NaCl (2%–26%)	−6.26
DE 42	−5.31	Ornithine	−6.4 ± 0.4
DE 55	−5.18	Polyethylene glycol	
DE 64	−4.57	mol wt 200	−6.1 ± 0.3
DE 83	−3.78	mol wt 400	−26.6 ± 0.8
DE 91	−2.99	mol wt 600	−56 ± 2
Erythritol	−1.34	Proline	−3.9 ± 0.1
Galactose	−2.24	Propylene glycol	−1.0
Glucose, fructose	−2.25 ± 0.02	Sorbitol	−1.65 ± 0.14
Glycerol	−1.16 ± 0.01	Sucrose	−6.47 ± 0.06
Glycine	0.87 ± 0.11	Tartaric acid	−4.68 ± 0.5
KCl (0%–2%)	10.81	Urea	2.02
Lactic acid	−1.59 ± 0.2	Xylitol	−1.66
Lactose	−10.2	Xylose	−1.54 ± 0.02

Sources: Chirife, J., Ferro Fontan, C., and Scorza, O.C. 1980a. A study of the water activity lowering behavior of some amino acids. *Journal of Food Technology* 15:383–387; Chirife, J., Ferro Fontan, C., and Bemergui, E.A. 1980b. The prediction of water activity in aqueous solutions in connection with intermediate moisture foods IV. Aw prediction in aqueous non electrolyte solutions. *Journal of Food Technology* 15:59–70; Leiras, M.C., Alzamora, S.M., and Chirife, J. 1990. Water activity of galactose solutions. *Journal of Food Science* 55:1174–1176; Alzamora, S.M., Chirife, J., and Gerschenson, L.N. 1994. Determination and correlation of the water activity of propylene glycol solutions. *Food Research International* 27(1), 65–67; Bell, L.N., and Labuza, T.P. 2000. *Moisture Sorption: Practical Aspects of Isotherm Measurement and Use*, 2nd Edition. St. Paul, MN: American Association of Cereal Chemists.

This need has led to the development of a number of semitheoretical and empirical equations for predicting a_W.

Norrish (1966), by using the Margules–Van Laar equations, expressed the a_W of a binary system as:

$$\ln a_w = \ln X_w + K_1 X_{s1}^{\ 2} \tag{11.3}$$

where K_1 is the interaction constant between the solute (S_1) and water; this constant is determined by the slope of $\ln(a_W/X_W)$ versus X_{S1}^2. He generalized the equation for multicomponent solutions, neglecting ternary interactions, as follows:

$$\ln a_w = \ln X_w - \left[(-K_1)^{1/2} X_{s1} + (-K_2)^{1/2} X_{s2} + \ldots\right]^2 \tag{11.4}$$

where K_1, K_2,... are the binary interaction constants of solutes S_1, S_2, ..., respectively, in water, and Xs_1, Xs_2, ... are the mole fractions of the solutes, respectively. Interaction constants for numerous solutes are given in the literature. Values of K for common humectants are given in Table 11.3, compiled from various sources (Chirife et al. 1980, Chirife and Fontan 1980, Leiras et al. 1990, Alzamora et al. 1994, Bell and Labuza 2000).

Table 11.4 Grover Equation Constants.

Compound	E_j
Acids	2.5
Egg white	1.4
Ethanol	8.0
Glucose syrup 28 DE	0.7
Glucose syrup 42 DE	0.8
Glucose syrup 60 DE	1.0
Glycerol	2.0
Gums	0.8
Invert sugar	1.3
Lactose	1.0
Propylene glycol	4.0
Protein	1.3
Salts	9.0
Sorbitol	2.0
Starch	0.8
Sucrose	1.0

Source: Labuza, T.P. 1984. *Moisture Sorption: Practical Aspects of Isotherm Measurement and Use*. St. Paul, MN: American Association of Cereal Chemists.

Equation 11.4 can be modified to account for the positive values of K for inorganic salts as well as some organic acids:

$$\ln a_w = \ln X_{H_2O} + \frac{\sum_{i=1}^{j} K_i [X_i]^2}{\sum_{i=1}^{j} [X_i]^2} + \left[1 - X_{H_2O}\right]^2 \tag{11.5}$$

Chuang and Toledo (1976) proposed a modification of the Norrish equation that can be applied to multicomponent systems containing solids with molecular weights that are not well defined.

Grover (1947) derived the following equation, on an empirical basis, for predicting the a_w of sugar–water solutions:

$$100a_w = 104 - 10E_s + 0.45E_s^2 \tag{11.6}$$

where $E_s = \Sigma(E_j/M_j)$, E_j is a constant for compound j, and M_j is the number of grams of the component per gram of water in the system. This formula predicts the final a_w from the composition of the system and the ratios of the weight of each component to the total weight of the water in the food. Empirical constants for various components are given in the literature (Labuza 1984), with some listed in Table 11.4. The Grover equation has been successfully applied in the confectionery industry (Bussiere and Serpelloni 1985), but it is not accurate for high concentrations of humectants.

The Ross (1975) equation was derived theoretically based on the Gibbs-Duhem equation. It assumes that the effects on the activity coefficient of each solute, resulting from interactions between solutes, cancel each other. Thus, the same activity coefficients are used in a multicomponent system as in a binary water–solute system. This assumption leads to an easily applicable equation:

$$a_{wf} = a_{wo} a_{w1} \ldots a_{wn} \tag{11.7}$$

where a_{wf} is the final a_w of a food system after the addition of humectants; a_{wo} is the initial a_w of the system; and a_{wi} ($i = 1, \ldots, n$) is the a_w of a binary solution of humectant *i* in water, where the total water content of the system is considered. The value of a_{wo} can be measured or estimated from published sorption isotherm data for the food, and a_{wi} can be calculated from Equations 11.3 through 11.6 or from published sorption isotherm databases (Iglesias and Chirife 1982). It was proposed that the Ross equation would give more accurate predictions for systems containing nonsolute solids, if, for the estimation of a_{wi}, the water strongly bound to the solids (monolayer moisture) is not considered available for dissolving humectant *i* (Chirife et al. 1985).

A modification of the Ross equation was proposed by Ferro Fontan et al. (1981):

$$a_{wf} = a_{wo} \prod_i \left[a_{wi}(m) \right]^{mi/m} \tag{11.8}$$

In Equation 11.8, the a_w of each humectant is estimated at the total molality (m) of the mixture, instead of the molality of the humectant itself (m_i), and then it is raised to the power of m_i/m. Based on the results of an experimental study, Teng and Seow (1981) showed that the Ross equation usually overestimates the a_w of a mixture, whereas Equation 11.8 performs better than the Ross equation at high concentrations. Lilley and Sutton (1991) proposed yet another refinement of the Ross equation. Their model took into account the effect of size, solvation, and solute–solute interaction. For a binary system, the proposed equation has the following form:

$$a_{wf} = a_{w1} \bullet a_{w2} \bullet \exp[-(2\lambda_w J_{12} m_1 m_2)] \tag{11.9}$$

where J_{12} is the coefficient that accounts for nonideality caused by the heterotactic interactions between solutes 1 and 2 in the mixture. It can be approximately estimated from the coefficients of homotactic interactions between molecules of the same solute that are experimentally determined:

$$J_{12} = (J_{11} J_{22})^{1/2} \tag{11.10}$$

Equation 11.10 takes the form of the Ross equation when the heterotactic coefficient is 0. It was concluded by Lilley and Sutton that although the proposed equation improves a_w prediction, the accuracy of the Ross equation is sufficient for most applications, especially when the added requirements of the refined equations for background data is considered.

The equations discussed above provide useful tools for prior estimation of the effect of adding a humectant to a multicomponent system. Nevertheless, their accuracy depends on the particular system. The more concentrated the system (i.e., the lower the moisture of the system), the greater is the discrepancy to be expected between actual a_w and predicted values.

Salts, glycerol, and sugars are the most commonly used humectants. Having low molecular weights and high solubilities, they are very effective in lowering a_w. Polyols and sodium chloride exhibit higher humectancy than sugars (Sloan and Labuza 1975). However, the use of these humectants at levels required to achieve an a_w value within the IMF zone results in undesirable flavor. In order to achieve an acceptable flavor profile, combinations of these humectants and the addition of less effective or less common humectants are the usual approaches in product design and formulation.

Polymeric humectants, such as high-molecular-weight polyols and water-soluble gums, result in a_w that deviates strongly from that predicted by Raoult's law (i.e., they have a large γ_S). This is because they are more effective per unit of hydrophilic groups than are compounds with lower molecular weights, as was shown for polyethylene glycol with different degrees of polymerization (Bone 1969). The problem with high-molecular-weight humectants is the high viscosity of their solutions. For example, coconut skim milk, a natural humectant in this category (Hagenmaier et al. 1975) at a_w 0.7, has a viscosity of 20,000 cp.

Earlier research was focused on compounds showing strong deviations from Raoult's law, that is, compounds having high activity coefficients, and humectant combinations with pronounced synergistic activity. Sodium lactate has extraordinary humectancy (Karel 1975, Loncin 1975) and also exhibits significant synergism with sodium chloride but shows no synergism with sugars or polyols and has a negative antagonistic effect with citrates (Karel 1975).

Several researchers have explored the applicability and the relative effectiveness of alternative humectants that do not exhibit the flavor problems caused by high concentrations of salt, sugars, and glycerol. Amino acids and protein hydrolysates have been shown to be very effective humectants (Chirife et al. 1980, Guilbert et al. 1981, Anderson and Witter 1982). Guilbert (1981) added several humectants to water–glycerol solutions with a_w in the range of 0.8 to 0.9 and measured how efficiently they lowered a_W. Sodium chloride and organic salts were more efficient than the others (amino acids and highly hydrolyzed protein hydrolysates > glycerol and sorbitol > sucrose and glucose syrups > glucidic macromolecules and ovalbumin). A similar humectancy order was found when some of the humectants were added to a model IMF. Among the tested amino acids, the most effective in depressing a_w were L-arginine, L-lysine, L-ornithine, L-proline, and glycine. Equally effective was a mixture of 18 amino acids. The tested amino acids had no undesirable effect on flavor, except for arginine, which contributed a bitter flavor. A degree of bitterness also resulted from the addition of some protein hydrolysates. Other factors to be taken into account, when amino acids are considered for use as humectants, are their solubility and cost. A supply of fairly low-cost amino acids can be predicted for the near future, because of rapid advances in genetic engineering and fermentation technology. The added nutritional value of amino acids would be an additional benefit. However, amino acids are not allowed to be added to foods except for nutritional purposes.

Gee et al. (1977) studied alternative ways of producing soft, pliable textured foods, especially fruits and vegetables, at a shelf-stable a_w of 0.6 to 0.7. Their approach was to build up or alter the polymeric matrix of plant material by adding materials that could interact with the natural matrix and lead to increased water retention at a given a_w. Vegetable pieces (green snap beans), which would normally have a poor texture when dried to a_w levels of IMFs, were cooked in a number of solutions of naturally occurring substances, equilibrated overnight, drained, and dried to an a_w of 0.6 to 0.7. The water retention of these IMF pieces was compared with that of control samples of water-cooked pieces. Pieces treated with algin, gum arabic, tapioca starch, glycerol and glycerol oligomers, lactic acid, aloe vera gel, or yucca extract had increased water retention and plasticity. Other starches and gums had a negative effect. A combination of such naturally occurring humectants, at concentrations below their off-flavor threshold, can be considered for use in the production of IMFs. In the same study, mild lactic acid fermentation was shown to have a very beneficial effect on the texture of further-processed intermediate-moisture vegetables. Kapsalis et al. (1985) evaluated several oligoglycerols and polyglycerols and their esters for their potential as humectants. Although they were effective in lowering a_w, most resulted in unacceptable odors and taste that could, however, be a result of impurities in the synthesis method used; it is also possible that pure polyglycerols would have a better flavor.

Sinskey (1976) investigated the use of aliphatic diols as humectants. The 1,2-diols and 1,3-diols with chains containing three to nine carbon atoms exhibited lower humectancy than glycerol but showed significant antimicrobial action. The 1,2-diols were more effective than the 1,3-diols, and those with longer chains were more effective than those with shorter chains. Esterification generally increased antimicrobial activity.

Novel ingredients could also be successfully used as humectants. Several low-molecular compounds that have been proposed as low-impact nonnutritive sweeteners satisfy most of the aforementioned properties of an "ideal" humectant. Such a nonnutritive sweetener was described earlier as neosugar (Anonymous 1985). It is composed of glucose attached in a β-(2-1) linkage to two, three, or four fructose units and is accordingly designated GF2, GF3, or GF4. It is produced by the action of a fungal enzyme, fructosyltransferase, on sucrose. The sweetness of neosugar is 0.4 to 0.6 times that of sucrose. Thus, GF2, with about half the sweetness and 1.5 times the molecular weight of sucrose, would effectively accomplish lower a_w than an amount of sucrose of equal sweetening power. Its noncaloric property is an added benefit in a market of increased demand for lower-calorie foods. A number of other low-molecular carbohydrates have become available at costs that allow their widespread use as food ingredients. A lot of these ingredients have "generally accepted as safe" (GRAS) status and are promoted for their functional and/or prebiotic effect. Their use as effective humectants could offer alternative tools to the design of IMF products with high marketing appeal. Examples of such ingredients are trechalose, the nonreducing α,α glucose bimer (Portmann and Birch 1995, Takanobu Higashiyama 2002), oligofructoses (De Gennaro et al. 2000, Franck 2002), and oligogalactoses (e.g., tagatose) (Bertelsen et al. 1999). Oligofructose is produced using two different manufacturing techniques that deliver slightly different end products. Chicory oligofructose (e.g., Raftilose®) is obtained by partial enzymatic hydrolysis of inulin (using an endoinulinase), eventually followed by spray-drying, but it can also be synthesized from sucrose using fructosyl-transferase (Bornet 1994, De Leenheer 1996).

Complementary to the use of humectants is the addition of surface-active agents or semisolid fats to plasticize the texture of an IMF. Gelling or emulsion formation in the mixture of an IMF does not lower the a_w (Guilbert and Cheftel 1982) but substantially improves the texture of the product. The plasticizing effect of fats is fundamental in the confectionery industry, and the softening of texture by surfactants is widely practiced in the baking industry and has been applied in the production of intermediate-moisture pet foods.

Production of Intermediate-Moisture Food Products

Several manufacturing techniques can be applied for producing IMF products (Karel 1976, Guilbert 1984). They can be classified in four main categories:

1. *Partial drying* can be used in the production of IMFs only if the starting materials are naturally rich in humectants. This is the case with dried fruits (e.g., raisins, apricots, prunes, dates, apples, and figs) and syrups (e.g., maple syrup). The final a_w of these products is in the range of 0.6 to 0.8.

2. *Moist infusion, or osmotic dehydration,* involves soaking solid food pieces in a water–humectant solution of lower a_w. This technique has also been defined as dewatering-impregnation soaking (Torreggiani et al. 1988). The difference in osmolality forces water to diffuse out of the food into the solution. Simultaneously, the humectant diffuses into the food, usually more slowly than the water diffuses out. Salt or sugar solutions are usually employed. This is the method for the production of candied fruits. Also, novel meat and vegetable IMF products have been produced by infusion in solutions of salt, sugar, glycerol, or other humectants (Maltini et al. 1993, Torreggiani et al. 1995, Rastogi et al. 2002). Forni et al. (1997) studied the color stability of apricots subjected to osmotic dehydration, followed by air dehydration, and then freezing, which led to "frozen intermediate-moisture fruits" with low a_w (0.86).
3. *Dry infusion* consists of first dehydrating solid food pieces and then soaking them in a water–humectant solution of the desired a_w. This process is more energy intensive, but it results in high-quality products. It has been used extensively in the preparation of IMFs for the National Aeronautic and Space Administration (NASA) and the U.S. military. The latter is a major developer and user of IMF products.
4. The process of *direct formulation* involves weighing and direct mixing of food ingredients, humectants, and additives, followed by cooking, extrusion, or other treatment, resulting in a finished product with desired a_w. This method is fast and energy efficient and offers great flexibility in formulation. It is used for both traditional IMF (e.g., confections, preserves) and novel IMF (e.g., pet foods, snacks) products.

Microbial and Chemical Quality of Intermediate-Moisture Food Products

Lowering of a_w is the essential processing step and main microbial hurdle in IMF products. However, setting the a_w at a certain level cannot always by itself ensure shelf stability. Other factors and properties of food systems should be taken into account, and often additional measures must be taken to achieve the desired stability.

Microbial Stability

Microbial stability is the primary criterion for the viability of an IMF product under development. The microbiology of foods with reduced a_w was reviewed by Christian (2000). As mentioned above, inhibiting microbial growth on a given substrate is not achieved exclusively by lowering the a_w, but rather, it is a function of all contributing hurdles—i.e., a_w, pH, temperature, oxidation–reduction potential, preservatives, and existing microflora (Kannat et al. 2002, Leistner and Gould 2002). Numerous microorganisms of significance to spoilage have been shown to be able to grow at a_w in the range of 0.6 to 0.84 when other conditions are favorable (see Table 11.5). Thus, additional precautions, besides the adjustment of a_w, must be taken to inhibit or limit the proliferation of these microorganisms in IMFs.

The pathogenic microorganisms of major concern in foods are effectively inhibited by the reduction of a_w to the IMF zone (see Table 11.1). Thus, the growth of *Clostridia* is prevented by such reduced a_w, regardless of storage temperature and pH (Roberts and Smart 1976). However, because growth could conceivably occur during formulation and storage before the reduction of a_w, good hygienic and manufacturing practices are essential. *Bacillus* sp. require a minimum a_w of 0.89 to 0.90 for growth (Jakobsen 1985). At IMF water activities, *Salmonella* sp. cannot multiply, having a limit for growth of 0.95, but their resistance to heat is greatly increased, and they may persist in IMFs for long periods (Corry

Table 11.5 Microorganisms Growing in a_w Range of Intermediate-Moisture Foods.

a_w	Microorganisms Generally Inhibited by Lowest a_w in This Range	Examples of Traditional Foods With a_w in This Range
0.80–0.84	Most molds (mycotoxigenic penicillia), most *Saccharomyces* spp. (e.g., *S. bailii, Debaryomyces*)	Most fruit juice concentrates, sweetened condensed milk, chocolate syrup, maple and fruit syrups, flour, rice, pulses containing 15%–17% moisture, fruitcake, country-style ham, fondants, high-sugar cakes
0.75–0.80	Most halophilic bacteria, mycotoxigenic aspergilli	Jam, marmalade, marzipan, glacé fruits, some marshmallows
0.65–0.75	Xerophilic molds (*Aspergillus chevalieri, A. candidus, Wallemia sebi*), *Saccharomyces bisporus*	Rolled oats containing about 10% moisture, grained nougats, fudge, marshmallows, jelly, molasses, raw cane sugar, some dried fruits, nuts
0.60–0.65	Osmophilic yeasts (*Saccharomyces rouxii*), few molds (*Aspergillus echinulatus, Monascus bisporus*)	Dried fruits containing 15–20% moisture, some toffees and caramels, honey

Source: Adapted from Beuchat, L.R. 1981. Microbial stability as affected by water activity. *Cereal Foods World* 26:345–349.

1976). Pasteurization of the ingredients before formulation is generally necessary. *Listeria monocytogenes* can grow at considerably lower a_w, with the reported limit of growth at 0.92 (Cole et al. 1990, Tapia de Daza 1991). The only food pathogen able to grow at even lower a_w, and therefore be of significant concern for IMFs, could be *Staphylococcus aureus*. It has been shown to grow at a_w above 0.84 to 0.85, if the pH is favorable (Pawsey and Davies 1976). Enterotoxin A and D production was reported at all growth supporting a_w values, but enterotoxin B and C production was not reported below 0.93 (Notermans and Heuvelman 1983, Ewald and Notermans 1988). Formulation of IMF products at the highest possible moisture content, for improved texture and palatability, requires additional measures for the inhibition of *S. aureus*. The same is true for molds. The most often encountered ones, for example, the common *Aspergillus* and *Penicillium* spp., can grow at a_w above 0.77 to 0.85. The minimum a_w for mycotoxin production by these molds is usually higher (see Table 11.1). Several xerophilic and xerotolerant molds can grow at a_w down to 0.62 to 0.64. Water activity minima for growth and mycotoxin production were given by Richard-Molard et al. (1985), who reviewed the effect of a_w on molds. An extensive list of growth minima and relevant references was published by Christian (2000).

Yeast growth is another potential problem with IMFs. Compared with the most tolerant halophilic bacteria that can grow down to the a_w levels of saturated NaCl (i.e., a_w 0.75), osmophilic yeasts can grow at a_w down to 0.62 (e.g., *Zygosacharomyces rouxii*). An extensive list of yeasts tolerant to low a_w, occurrence, the methodology of detection, and control methods for IMFs was presented in a comprehensive review by Tilbury (1976). Good manufacturing practices, pasteurization of mixtures, and use of chemical preservatives, such as sulfites, benzoates, *para*-hydroxybenzoates, sorbates, and diethyl pyrocarbonate, are the usual control measures. General microbiological specifications for IMFs and methodology for the assessment of compliance were proposed by Mossel (1976).

Although the effects on microbial behavior observed in concentrated environments of humectants are usually correlated and explained as a direct function of the a_w value, in some instances, the response to a_w depends also on the type of humectant used. This "specific solute effect" has been termed and reviewed by Christian (1981, 2000). For example, glycerol exhibits a less inhibitory effect on several Gram-negative rods, such as *Salmonella* sp. and *Escherichia coli*, than in glucose or salt systems of the same a_w. The same is true for *Listeria monocytogenes*. In the case of these bacteria, the growth limit was 0.01 to 0.02 a_w unit lower with glycerol than with other humectants. On the other hand, Gram-positive cocci with the most prominent *S. aureus* are shown to be glycerol sensitive, i.e., glycerol is more inhibitory than other humectants and particularly NaCl on an a_w basis. Ballesteros et al. (1993), however, could find no correlation of this specific effect to *S. aureus* in a number of humectant solutions at the same a_w with other physicochemical properties, such as viscosity, dielectric constant, oxygen solubility, or diffusivity. Thus, the hypothesis of "specific solute effect" could possibly be replaced or coupled with a "specific cell response," the combinatory effect of which would unlikely be predicted by mere knowledge of solution properties. There is not yet a parameter that correlates better with the effects of concentrated humectant solutions on microbial growth than a_w.

Because *S. aureus* and molds are of primary concern in IMF products, a number of microbial studies and tests for *S. aureus* and molds in these foods have been used as indicators and challenge microorganisms. Haas et al. (1975) surveyed various microbes occurring in IMFs and chose *Aspergillus glaucus, Aspergillus niger*, and a *Staphylococcus* sp. as suitable challenge organisms.

Combinations of pH, a_w, and preservative concentrations that offer adequate protection can be established on the basis of tests with these microorganisms. Effective mold inhibitors that are often used are sorbates and propionates. Acott et al. (1976) evaluated the effectiveness of antimicrobial agents in IMF products using the microbes mentioned above. Very few inhibitor systems prevented the growth of all three organisms at pH above 5.4 and a_w 0.86 to 0.90. Only propylene glycol, a humectant with specific antimicrobial activity, achieved complete inhibition. At high a_w, acidification to pH 5.2, in conjunction with the mold inhibitors, was effective for inhibition of *S. aureus*. At pH 5.2 to 6, a combination of propylene glycol (4% to 6%) with either potassium sorbate or calcium propionate (0.1% to 0.3%) was required. However, the effectiveness of a microbial protection combination is very system dependent, and extrapolations to other systems are often not valid. Troller (1985), using *S. aureus* as a test organism, studied differences in the inhibitory effect of low pH resulting from different acids. Tests were conducted at various water activities. Organic acids were generally more effective at high a_w, but an inorganic acid, phosphoric acid, was the most effective at lower a_w, at the upper limit of the IMF range. Haas and Herman (1978) showed that besides staphylococci, bacteria such as *Streptococcus faecalis* and a *Lactobacillus* sp. can grow at the upper limit of a_w in IMFs. Both can grow at a_w above 0.87 to 0.88. Sorbate was less effective on these bacteria than on staphylococci. Propylene glycol had an equally inhibitory effect on staphylococci and *S. faecalis* but had no specific antimicrobial effect (besides the lowering of a_w) on *Lactobacillus*.

A novel approach to microbial stabilization of IMFs, with a minimum amount of chemical preservatives, is the optimum distribution of a preservative throughout the food. The more susceptible part of the food, the surface, should have a higher concentration of the preservative than the interior. Temperature changes during distribution and storage can result in local condensation of water on the surface, leading to microbial outgrowth on the surface. Two methods of improving surface stability by maintaining a high concentration of preservatives were demonstrated by Torres et al. (1985a, 1985b). The first involved zein, which is an impermeable, edible food coating. The second was based on the maintenance of a pH differential between the surface and the bulk of the food. The reduction of surface pH increases the surface availability of the most active form of sorbic acid and other lipophilic acids used as preservatives. A negatively charged macromolecule was immobilized in the form of a component of surface coating, whereas other molecules, particularly electrolytes, moved freely. A deionized mixture of λ-carrageenan and agarose resulted in a pH differential of up to 0.5 pH unit. Both methods were tested and increased the microbial stability of an IMF substantially with a_w 0.88 (Torres and Karel 1985).

Chemical Stability

As illustrated in Figure 11.1, the only region of the stability map in which reactions are simultaneously at a minimum is where the a_w is approximately 0.25 to 0.35, well below the intermediate-moisture range. At water activities in the IMF range, chemical reactions increase rapidly and reach a maximum. Because enzymatic activity is usually prevented by enzyme inactivation with an initial thermal treatment, lipid peroxidation and nonenzymatic browning are the major deterioration reactions in IMF products (Williams 1976). Water has a dual effect on the rate of lipid peroxidation (Labuza 1975). It can retard oxidation by hydrating or diluting heavy metal catalysts or even precipitating them as hydroxides. Water forms hydrogen bonds with hydroperoxides and slows down the steps of peroxide decomposition. By promoting radical recombination, it can terminate the chain reaction. On the other hand, water can speed up the reaction by lowering the viscosity, thereby increasing the mobility of reactants and bringing catalysts into solution. It also swells the solid matrix of the system, with the result that new surfaces are exposed for catalysis. These contrary effects, occurring simultaneously, result in a minimum oxidation rate at an a_w close to that corresponding to the monolayer moisture content. In the IMF zone, the promoting action predominates; therefore, the oxidation rate increases with a_w. In systems high in trace metal catalysts, a maximum rate is reached at a_w 0.75 to 0.80, followed by a decline at higher a_w, the dilution effect again predominating (Labuza and Chou 1974). The method used to achieve a given level of a_w must also be considered. Foods prepared by dry infusion (adsorption) oxidize much more rapidly than those prepared by blending (desorption), at the same a_w (Labuza et al. 1972, 1972a; Chou et al. 1973). Thus, the actual water content is important, as the systems prepared by adsorption have higher moisture contents than those prepared by desorption, because of sorption hysteresis. Lipid peroxidation is a serious problem in IMFs, leading to unacceptable rancid products when control measures are not taken. Oxidation can be prevented by the elimination of oxygen through vacuum-packing and oxygen-impermeable packaging materials, by antioxidants, or by oxygen scavenger sachets. Fat-soluble free radical scavengers, such as butylated hydroxytoluene (BHT) or butylated hydroxyanisole (BHA), or water-soluble metal chelators, such as ethylenediaminetetraacetic acid (EDTA) or citric acid, may be used as antioxidants. Chelators, although more effective in model systems of high a_w, proved less effective than BHA in actual IMF systems (Labuza et al. 1972a), probably because of binding to proteins.

The effect of water on nonenzymatic browning (Maillard reaction) was reviewed by Labuza and Saltmarch (1981). As shown in the stability map (see Figure 11.1), the maximum reaction rate for nonenzymatic browning occurs in the IMF a_w range, usually at a_w 0.65 to 0.70. The observed maximum rate of browning can be attributed to a balance of viscosity-controlled diffusion, dilution, and concentration effects. At low water activities, the slow diffusion of reactants limits the rate. At higher water activities, faster diffusion enables reactions to occur faster until dilution of the reactants again slows them down. Also, the higher concentration of water retards the reversible reaction steps that produce water, e.g., the initial condensation stage. Up to 3.5 moles of water is formed per mole of sugar consumed in the reaction. On the other hand, water may increase deamination reactions, such as the production of furfural or hydroxymethyl furfural in the browning reaction sequence. The maximum browning rate occurs at different a_w values depending on the humectant used to reduce the a_w. The overall effect of liquid humectants is to shift the maximum reaction rate to a lower a_w. In both liquid and solid model systems containing glycerol, Warmbier et al. (1976) found the maxima of nonenzymatic browning rates with a_w in the range of 0.41 to 0.55. Liquid humectants influence the rate of browning by acting as solvents and thus increasing reactant mobility at lower moisture contents. However, increasing the viscosity by the addition of viscosity agents, such as sorbitol, can dramatically decrease the reaction rate at all water activities (Labuza 1980).

Nonenzymatic browning, although sometimes desirable, as in the production of confectionery and bakery products, can have deleterious effects on the shelf-life of IMF products. It results in loss of protein quality and the undesirable production of off-flavors and dark pigments. Loss of protein quality refers mainly to the loss of the essential amino acid lysine via reactions involving its free ϵ-amino group. When an IMF formulation is considered, the reactivity of the ingredients with respect to nonenzymatic browning must also be considered. The use of reducing sugars (especially pentoses) and amino acids as humectants should be avoided when the Maillard reaction is a major concern. Methods of controlling nonenzymatic browning, besides the use of low-reactive ingredients, include lowering the pH, maintaining low storage temperatures, and adding sulfites or alternative antibrowning agents.

When considering chemical food stability, however, one should bear in mind that food materials are often nonequilibrium systems, exhibiting time-dependent changes (Roos 1995). The a_w concept should be used in conjunction with the glass transition approach to assess IMF stability and quality (Lievonen and Roos 2002). In low-moisture food systems, various diffusion-controlled chemical reactions are affected by molecular mobility (Karel 1985, Karmas et al. 1992, Roos et al. 1996). The temperature dependence of nonenzymatic browning rates of foods or food models with various water activities has been the subject of several publications in the past decade (Karmas et al. 1992, Roos and Himberg 1994, Bell 1996, Bell et al. 1998, Lievonen et al. 1998, Lievonen and Roos 2000). A common thesis was that browning rates below T_g are very low, increasing with increasing $(T - T_g)$. Other important chemical reactions such as sucrose inversion by invertase (Kouassi and Roos 2000), aspartame degradation (Bell and Hageman 1994), thiamin stability (Bell and White 2000), and glycine loss (Bell et al. 1998) were also studied considering the combined effect of molecular mobility and a_w. The general conclusion of these studies was that reactions in amorphous foods are controlled by several factors, including temperature and T_g, water content, crystallization, and other structural changes, making it difficult to model the temperature dependence of food deterioration via a single equation.

When focusing on the safety perspective of IMFs, the effect of the nonequilibrium state of many semimoist foods on the value of a_w has been thoroughly studied in a critical review covering the role of a_w in the microbial stability of IMF (Chirife and Buera 1999). The conclusion was that the amorphous state of many IMF products, and the possible glass transitions occurring, does not significantly affect the safety of IMFs. Nonequilibrium effects (e.g., inability of water to diffuse in a semimoist food) may be, in most cases, either too slow within the time-frame of food life cycles or too small to play an important role in the food microbial growth range. However, it has been suggested that phase transitions and molecular mobility phenomena in IMFs should also be considered (Slade and Levine 1991, Roos 1995) when addressing quality and safety issues of IMFs.

Commercial Development of Intermediate-Moisture Foods

The basic principle of reducing a_w to prolong the shelf-life of food has been practiced for ages, as already mentioned. The first techniques involved drying or partial drying of foods. Sun-drying and roasting of meats over a fire to reduce water content were practiced by prehistoric cave dwellers.

Numerous drying methods have been used throughout history (Labuza 1976). For example, the Incas of South America made a dried potato product called *chuño* using a process involving freezing, pressing, and sun-drying (Salaman 1940). Ancient Mediterranean civilizations used salt to preserve meats and fish, as revealed in Egyptian hieroglyphics. This is probably the first case of reducing a_w by solute addition rather than by water removal. The same principle was practiced in the use of sugars to preserve fruits as jams or jellies; however, in this case, water is also partially removed by boiling.

The Sioux of North America made a product called pemmican, composed of semidry buffalo meat with added fat, pounded dry nuts, and chokecherries, which are very acidic because they contain benzoic acid (Binkerd et al. 1976). It was their major winter storage food and was carried by hunters and warriors as travel food. Binkerd et al. (1976) described the role of Armour & Company in making pemmican for Arctic and Antarctic explorers and the U.S. naval submarine fleet during World War II. Beef jerky is another traditional American meat product with intermediate moisture.

Traditional IMF products based on meat are also made in Asia (*tsusou-gan, sou-gan,* and *njorsou-gan* in China; *dendeng* in Indonesia) and Africa (*biltong, khundi,* and *quanta*). In Europe, fermented sausages known as *Brühdauerwurst, Speckwurst* (blood sausage), and *Bundnerfleisch* (dried beef), when intensively dried, have a_w in the range of less than 0.90, and thus could be classified as IMFs.

Biltong is considered a delicacy in South Africa, manufactured from long lean strips of meat cured in salt and spices for several hours, briefly plunged into hot water and vinegar, and air-dried for 1 to 2 weeks. Its reported average a_w is 0.7. Leistner (1985) reported a very wide variation of a_w among 25 tested samples of *biltong*, and those with high a_w became spoiled by molds. He concluded that a combination of a_w less than 0.77 and pH less than 5.5 is required for stability.

Leistner (1985) also surveyed samples and production methods for *sou-gan*, a Chinese intermediate-moisture pork product. Lean meat strips are pickled in a mixture of sugar, salt, soy sauce, monosodium glutamate, and spices for 24 hours and then dried in a single layer at 50° to 60°C to 35% moisture. A few minutes of grilling at 130°C may follow. The final a_w range is 0.55 to 0.69, and all tested samples were stable. This kind of product is simple to prepare, nutritious, and safe, with an acceptable sweet taste, and it could be introduced to developing countries outside Asia.

The Indonesian *dendeng* is very similar to *sou-gan*, with the curing mixture consisting of coconut sugar, salt, and spices; it has a final a_w range of 0.55 to 0.60 (Purnomo et al. 1983). Pastes or jams made by heating cooked dried legumes (usually beans) with sugar are consumed in large quantities in Asia. Some of these products also contain optional ingredients, such as gums, fat, salt, artificial color, and flavorings. Chinese bean pastes may

Table 11.6 Examples of Traditional Intermediate-Moisture Foods.

Food	Water Activity (a_w)	Principal Humectants[a]
Dried foods		
Raisins	0.51–0.62	a, b (naturally occurring)
Prunes, figs	0.65–0.83	a, b (naturally occurring)
Apricots, peaches	0.73–0.81	a, b (naturally occurring)
Sugared fruits	0.57–0.79	a, b
Jams, jellies, marmalade	0.82–0.84	a, b
Fruit fillings of pastries	0.65–0.71	a, b
Sweet chestnut puree	0.90	a
Dry salami	0.82–0.85	c, d
Country ham	0.88–0.90	a, c, f
Dried or salted fish	0.74–0.82	c, e, f
Honey	0.58–0.68	a, b
Maple syrup	0.82–0.86	a, b
Chocolate	< 0.55	a
Chutney sauce	0.86	b, c
Parmesan	0.66–0.74	c, e

Sucrose, a; other sugars, b; NaCl, c; food acids, d; amino acids, e; various humectants, f.

also contain fat, e.g., lard. The general term for these products in Japan is *an,* also known as *ann* or *ahn*. They probably originated in China and were introduced into Japan about 600 AD. The present annual consumption of *an* in Japan is about half a million tons. More than 50 different types of pulses (edible seeds of certain pod bearing plants, such as peas and beans) are used for *an*-making in Japan alone. The Japanese generally prefer *an* made from the adzuki bean (*Vigna angularis*), also commonly called red bean. Just as many different types of cheeses are made from milk, many different types of *an* are made by varying the pulse type, sugar type and concentration, moisture content, content of optional ingredients, and manufacturing process. Sucrose is the major sugar type, but corn or rice syrups (glucose) and honeys are also used. Basically, four broad categories of *an* are made in Japan. *Tsubu-an*, also called *ogura-an,* is made by infusing cooked whole beans with sugar, whereas *tsubushi-an* is mashed *tsubu-an*. *Neri-an,* also called *koshi-an,* is produced by sieving cooked beans to remove the seed coats, rinsing the sieved product, pressing to remove water, and kneading and heating with added sugar; the unsugared press cake is called *nama-an*. *Yokan* is similar to *neri-an* but contains added agar-agar to give it a more gelatinous texture and a higher moisture content; the *yokan* products with the highest moisture are called *misa-yokan*. *Tsubushi-an, tsubu-an*, and *neri-an* are available in consumer-sized metal cans in most Asian food shops in the United States. They are commonly produced by a hot-fill, vacuum-seal process without further heat-sterilization.

Other foods classified as traditional IMFs include products dried without added humectants (raisins, prunes, apricots, dates, figs, etc.), products to which sugar has been added (jams, jellies, honey, candied fruits, marshmallows, soft candies, pie fillings, and syrups), products to which salts and sugars have been added (country ham), and bakery products such as fruitcakes and pie fillings. Table 11.6 shows some traditional IMFs and their usual ranges of a_w.

The Burgess patent for pet foods (Burgess and Mellentin 1965) was the first major step leading to the technological advance of formulated IMF products in the United States. The patent described the mixing of meat with dry water binders, such as soy flakes and wheat

Table 11.7 A Typical Intermediate-Moisture Dog Food Formula.

Ingredient	Quantity[a] (%)
Meat byproducts	32
Soy flakes	20
Wheat feed flour	12
Sucrose	12
Corn syrup solids	12
Salt	1.6
Propylene glycol	5
Bone meat	3
Animal fat	2
Vitamins, minerals, color	0.3
Potassium sorbate	0.3

[a] Total exceeds 100% because of rounding.

flour, and solutes such as glycerol, sugar, and salt. After being mixed into a dough, the mass was extruded under high temperature (60° to 150°C) and pressure (1,000 and 5,000 psi) into a hamburger-like textured product. Other ingredients, such as vitamins and minerals, were added to the dough so that the final product was formulated to meet the total nutrient requirements of the animal. A typical formulation is shown in Table 11.7. Since this patent in 1965, many companies have developed similar products for both dogs and cats, and a large number of patents now exist in this area. These IMF products are convenient to use and, unlike canned pet products, leave no mess or odor, and they are more palatable to pets than dry products.

The direct application of solute-addition technology to products for humans has not been easy, because of the incompatible tastes of these foods and the humectants (e.g., sweetness of sucrose, saltiness of sodium chloride, and sweet, metallic off-flavor of glycerol). Also, textural requirements must be taken into account. To satisfy the requirements of taste and texture, infusion techniques have been tried. The dry or moist food items (meat, fish, or vegetables) are soaked in a solution containing several solutes instead of being ground into a dough, thus preserving the original texture (Brockman 1970, Kaplow 1970).

A number of IMF products for human consumption have been developed commercially. Among the first were shelf-stable breakfast pastries and meal-replacer bars. These products are formulated with a jam filling having lowered a_w in a dough shell of similar a_w, which is then baked and sealed in a foil pouch. They can be eaten directly from the pouch or after being heated in a toaster. Meal-replacer bars are intended primarily for breakfast. Because most people accept sweet foods for breakfast, a generation of such products has evolved, using direct-blending IMF technology and extrusion or bakery technology. Several of such products can be found in the marketplace.

Overall, the IMF commercial application story includes a great deal of research and development, several patents, and a few significant commercial applications. With accumulated knowledge of the principles of IMF production and experience gained from the first products, new ideas for solving problems related to IMF technology are continuously evolving. New humectants and new ways to achieve microbiological and chemical stability are being introduced. Some of these have led to successful commercial products. Other potential applications are patented but not yet commercially developed.

Description and ingredient analysis of IMF products available in the 1980s were reported by Taoukis et al. (1986) and are listed in Table 11.8.

Shelf-stable flour or corn tortillas, plastic wrapped in a cardboard box, are a characteristic example of a modern IMF product produced by blending and formulation based on all the concepts covering humectants and microbial and chemical stability discussed in the preceding sections. The same technology can be applied to other dough products, such as crepes (French), phyllo dough (Greek and Middle Eastern), and won ton wrappers (Chinese), which are usually marketed in dry, frozen, or refrigerated forms.

Soft cookies marketed by several manufacturers have higher moisture content than traditional cookies, in order to have increased chewiness. Some have a drier outer crust for textural stability. They have high moisture content but are produced at the lower end of the IMF a_w zone (0.5 to 0.6 a_w) using IMF technology. Similarly, filled pasta products (ravioli, tortellini) contain a filling with lowered a_w (0.5 to 0.6), which makes them shelf stable at room temperature.

Table 11.8 Composition of Some Commercial Intermediate-Moisture Foods.

Product and Manufacturer	Measured or Estimated Water Activity (a_w)	Declared Ingredient List
Blueberry Pop-Tarts (Kellogg's)	0.6	Enriched wheat flour; blueberry preserves; dextrose[a]; vegetable shortening; corn syrup; sugar; whey[a]; cracker meal; hydrogenated vegetable oil; gelatinized wheat starch; slat: tricalcium phosphate[a]; baking powder; citric acid; baking soda[b]; natural flavoring; vitamins and minerals; BHA or BHT[b,c]
Cherry Toaster Pastries (First National)	0.6	Wheat flour; sugar; corn syrup[a]; shortening[a] (with BHA); invert syrup; dextrose[a]; corn flour[a]; whey solids; cherries; glycerine; gelatinized cornmeal[a]; gelatinized cornstarch; ground dried apples; salt; leavening[a]; citric acid; imitation flavor[b]; wheat starch; potassium sorbate; artificial color[b]; vitamins
Breakfast Bar (Carnation)	0.5	Sugar; partially hydrogenated vegetable oil[a]; peanuts; chocolate; soy protein isolate; high-fructose corn syrup; toasted oats[a]; flour; calcium caseinate; dried corn syrup; crisp rice[a]; nonfat milk; cocoa; glycerol; vitamins and minerals[a]
Ready-to-Spread Frosting (General Mills)	0.7	Sugar; animal or vegetable shortening or both[a]; water; corn syrup; wheat starch[a]; monoglycerides and diglycerides; nonfat milk; polysorbate 60; cocoa; citric acid; soy lecithin[b]; sodium acid pyrophosphate; dextrose; pectin[a]; potassium sorbate; BHA[d], BHT[b]
Pie Crust Sticks (General Mills)	0.65	Enriched bleached flour; animal or vegetable fat or both (with BHA or BHT); modified cornstarch; salt; dried corn syrup[a]; monoglycerides and diglycerides[a]; sodium caseinate; colors
Pizza Crust (Fairmont Foods/Keebler)		Bleached flour; water; sugar; salt[a]; ethanol[a,d]
Special Cuts (Quaker Oats)	0.85	Beef byproducts, sucrose; water[a]; sodium caseinate; cornstarch; propylene glycol[a]; soy protein concentrate[a,d]; pregelatinized wheat flour; animal fat (with BHA, BHT, and citric acid); potassium sorbate[d]; vitamins and minerals

Table 11.8 Composition of Some Commercial Intermediate-Moisture Foods (*continued*).

Product and Manufacturer	Measured or Estimated Water Activity (a_w)	Declared Ingredient List
Slim Jims (General Mills)	0.8	Beef; water; salt; corn syrup[a]; flavoring[a]; dextrose; hydrolyzed vegetable protein[a]; spice[a]; sodium nitrite; lactic acid starter culture[d]
Caramel-Nut Granola Dipp (Quaker Oats)	0.5	Milk chocolate; caramel; crisp rice, rolled oats; brown; peanuts[a]; corn syrup; partially hydrogenated vegetable oil[a]; invert sugar; rolled whole wheat[a]; corn syrup solids; peanut butter[a]; glycerol; dried unsweetened coconut[a]; nonfat dry milk; almonds; sorbitol; honey[a]; salt[a]; natural and artificial flavors[a]; BHA; citric acid[b]
Flour tortillas (Old El Paso)	0.85	Enriched wheat flour; water; corn syrup solids; partially hydrogenated soybean oil[a]; glycerol; salt[a]; potassium sorbate[a]; calcium propionate[d]; monoglycerides and diglycerides[d]; fumaric acid; sodium metabisulfite[e]
Strawberry Fruit Roll-Ups (General Mills)	0.55–0.6	Strawberries; pear puree concentrate; maltodextrin; sugar[a]; partially hydrogenated soybean oil[a]; citric acid; natural flavor[b]; guar gum; monoglycerides and diglycerides; ascorbic acid; xanthan gum; ethyl maltol; artificial color
Chocolate chip soft cookies (Procter & Gamble)	0.57	Bleached flour; semisweet chocolate chips; sugar; high-fructose corn syrup[a]; partially hydrogenated soybean oil[a]; water; modified food starch; molasses; baking soda[a]; salt; artificial flavors[a]
Fruit Chewy Cookie (Nabisco)	0.65	Sugar; enriched wheat flour[a]; evaporated apples; vegetable shortening; corn syrup; water[a]; dextrose; high-fructose corn syrup[a]; natural flavor[a]; modified corn starch; carob bean gum; malic and citric acids; whey[b]; salt; spices[a]; baking soda; sodium benzoate; lecithin[d]
Shelf-stable imitation cheese (Universal Foods)	0.89	Italian Natural Cheese; water; sodium caseinate and calcium caseinate; partially hydrogenated vegetable oil; glycerol; salt[a]; tricalcium phosphate[a]; adipic acid; propylene glycol; disodium phosphate[a,d]; sorbic acid; artificial color[d]

Source: Adapted from Taoukis, P.S., Breene, W.M., and Labuza, T.P. 1988. Intermediate-moisture foods. In: *Advances in Cereal Science and Technology, Vol. IX*, ed. Y. Pomeranz, pp. 91–128. St. Paul, MN: American Association of Cereal Chemists. [Data from ingredients listed on product packages.]

[a] Humectant; [b] antioxidant; [c] BHA, butylated hydroxyanisole; BHT, butylated hydroxytoluene; [d] antimicrobial agent; [e] browning inhibitor.

Military Development of Intermediate-Moisture Foods

According to Moody (2000), "There are unique military requirements that distinguish a soldier's chow from everyday cuisine; military rations need a much longer shelf-life for both perishable and semi-perishable food." Requirements for shelf-stable individual rations are 3 years at 27°C (80°F) or 6 months at 38°C (100°F). Current troops are highly mobile and have very little time to stop and prepare food and consume it. Because some IMF meals can directly feed a faster, lighter mobilized military, they are appropriate for military field feeding, as summarized below:

1. They have an extended shelf-life, i.e., they are microbiologically stable and do not require refrigeration.

2. They are ready to eat, i.e., they require very little or no preparation.
3. They are low in moisture, giving a reduced weight and volume.
4. They are pliable, allowing them to be molded for packaging, transportation, and storage.

However, according to Brockman (1970), traditional IMF products were either not suitable for consumption or had limited application in military field feeding because of high salt (intermediate-moisture meats) or sugar content (intermediate-moisture jellies and confectionaries). Because the adjustment of a food's a_w is an important approach to controlling spoilage by microorganisms, both the Department of Defense (DOD) and NASA have been exploiting IMF technology, in order to develop ration components that are more suitable and nutritious and provide acceptability, portability, and stability for soldiers, astronauts, and civilians.

Pilot studies conducted at General Foods Corporation, under a contract with the U.S. Army, developed IMF products (diced chicken, ground beef, diced carrots, beef stew, barbecue pork, and apple pie filling) with an a_w between 0.75 and 0.90 (Hollis et al. 1968). Two infusion methods were used to adjust a_w: (1) infusing dehydrated foods with a solution containing glycerol, propylene glycol, potassium sorbate, salt, sugar, and various seasonings for flavor and (2) soaking/cooking raw foods in a similar, but more concentrated solution. Except for the apple pie, both procedures produced satisfactory samples. The products were microbiologically stable after storage for 4 months at 38°C. The instability of the beef stew gravy emulsion and the browning of the apples were two problems encountered. Sensory evaluations of all other items showed products to be acceptable "as is" and closer to commercial products when rehydrated.

The instability of the beef stew gravy emulsion and the browning of the apples in the previous study were corrected through reformulation (Hollis et al. 1969). This study has also shown that casseroles (chicken a la king and ham in cream sauce at a_w 0.85), inoculated with *S. aureus,* were microbiologically stable after 4 months at 38°C. Similar observations were seen after inoculation with *E. coli*, *Salmonella,* and vegetative cells of *Clostridium perfringes.*

Other studies conducted by General Foods (contracted with Natick) prepared IMFs using wet infusion. Beef, pork, chicken, lamb, ham, and tuna, as well as a variety of nonmeat items such as peas, carrots, mushrooms, onions, potatoes, pineapple, celery, macaroni, and egg noodles, were soaked and/or cooked in an aqueous solution of glycerol, salt, and antimycotic agents (Brockmann 1970). Representative examples of various IMFs prepared by the equilibration method are shown in Table 11.9. The composition of the equilibration solutions are given in Table 11.10.

Under contract with the U.S. Air Force, Swift & Company produced several bite-sized IMF products by the dry infusion method (Pavey and Schack 1969). The products were freeze-dried and subsequently infused. An optimum infusion technique, with regard to the sequence and method of adding ingredients, was used for each product. The best physical binding properties were generally achieved when 5% to 10% glycerol, 5% gelatin, and about 3% sorbitol were used in the infusion solution and 7% to 12% fat was used in the dry product. The formula of one of these products is given in Table 11.11. The final moisture content and a_w of the 10 tested products are given in Table 11.12.

Under a contract with the U.S. Army, Swift & Company incorporated glycerol, salt, and potassium sorbate into 14 cooked items to produce an aw of 0.83 ± 0.02 after drying (Pavey 1972). The products included beef, ground beef, chicken (white meat), pork tenderloin, omelet, carrots, pineapple, turkey (dark meat), halibut filet, ham, bologna, and pan-

Table 11.9 Preparation of Representative Intermediate Moisture Foods by Equilibration.

Initial Material	% H_2O	Ratio[a]: Initial wt. / Solution wt.	Processing	Equilibrated Product % H_2O	a_w
Tuna, canned water-packed pieces 1 cm thick	60.0	0.59	Cold soak	38.8	0.81
Carrots diced, 0.9 cm cooked	88.2	0.48	Cook 95°–98°C refrig.	51.5	0.81
Macaroni, elbow cooked, drained	63.0	0.43	Cook 95°–98°C refrig.	46.1	0.83
Pork loin, raw, 1 cm thick	70.0	0.73	Cook 95°–98°C refrig.	43.0	0.81
Pineapple canned, chunks	73.0	0.46	Cold soak	43.0	0.85
Celery 0.6 cm cross-cut, blanched	94.7	0.52	Cold soak	39.6	0.83
Beef, rib eye, 1 cm thick	70.8	2.35	Cook 95°–98°C refrig.	—	0.86

Source: Adapted from Brockmann, M.C. 1970. Development of intermediate moisture foods for military use. *Food Technology (Chicago)* 24:896–900.

[a] For composition of equilibration solutions see Table 11.10.

Table 11.10 Composition of Equilibration Solution.

Components of Solution (%)	Tuna	Carrot	Macaroni	Pork	Pineapple	Celery	Beef
	Products Equilibrated						
Glycerol	53.6	59.2	42.7	45.6	55.0	68.4	87.9
Water	38.6	34.7	48.8	43.2	21.5	25.2	—
Sodium chloride	7.1	5.5	8.0	10.5	—	5.9	10.1
Sucrose	—	—	—	—	23.0	—	—
Potassium sorbate	0.7	0.6	0.5	0.7	0.5	0.5	—
Sodium benzoate	—	—	—	—	—	—	2.0

Source: Adapted from Brockmann, M.C. 1970. Development of intermediate moisture foods for military use. *Food Technology (Chicago)* 24:896–900.

Table 11.11 Formula of Ready-to-Eat Semimoist Roast Pork Bites (Ground and Compressed).

Ingredients	Quantity (%)
Pork, cooked, ground, freeze-dried	45.00
Water, distilled	10.000
Water as steam	7.7815
Glycerol	8.00
Pregelatinized starch	7.00
Gelatin (175 Bloom)	5.00
Nondairy creamer (Coffee Mate)	4.00
Sorbitol, dry	3.00
Hydrolyzed vegetable protein, beef	3.00
Applesauce, dehydrated	2.50
Sugar (sucrose)	2.00
Salt	2.00
Monosodium glutamate	0.40
Sorbic acid	0.20
Ascorbic acid	0.045
Garlic powder	0.020
White pepper	0.020
Ribotide	0.020
Onion powder	0.010
Citric acid	0.0035
Total	100.0000

Source: Pavey, R.L., and Schack, W.R. 1969. Formulation of intermediate moisture bite-size food cubes. Tech. Rep. Contract F4160967-C-0054. U.S. Air Force School of Aerospace Medicine, San Antonio, TX.

Table 11.12 Example Analysis of Some Intermediate-Moisture Foods.

Cubes	Water Content (%)	Average Salt Content (%)	pH	Water Activity (a_w)
Roast beef	22.2	3.0	5.75	0.79
Barbecue beef	16.2	2.7	5.05	0.66
Roast pork	22.4	3.6	5.70	0.74
Barbecue chicken	19.7	4.0	5.20	0.70
Chicken a la King	14.9	3.6	5.90	0.61
Beef stew	17.3	3.7	5.80	0.65
Corned beef	16.2	5.4	5.85	0.62
Chili with beans	13.9	2.6	5.65	0.79
Sausage	24.2	4.5	4.90	0.78
Ham	19.9	4.5	5.90	0.72

Source: Adapted from Pavey, R.L., and Schack, W.R. 1969. Formulation of intermediate moisture bite-size food cubes. Tech. Rep. Contract F4160967-C-0054. U.S. Air Force School of Aerospace Medicine, San Antonio, TX.

cakes, sweet potato, and peaches. The products were prepared by soaking the foods in four different equilibration solutions or by adding the equilibration solution directly into the product prior to cooking (e.g., omelet, bologna, and pancakes). Equilibration was followed by vacuum drying, canning, and storing products for 3 months at 38°C. All products were microbiologically stable after the 3-month storage period. Sensory results showed poor acceptance for all products equilibrated with high concentrations of glycerol. These products tended to have a sweet-bitter flavor. The products equilibrated in low glycerol concentrations tended to be dry, tough, hard, or rubbery.

NASA, and the U.S. space program, also supported IMF development. Several IMF products were used during space flights in the Gemini and Apollo programs. The first solid food consumed on the moon was an IMF product. It was a gelatin–fruit–sugar bar that was fit inside the pressure suit so that it could be consumed without manipulation by hand. Being an IMF, the bar provided a concentrated source of energy without increasing thirst. Strawberry, lemon, cherry, and apricot fruit bars were used. Other intermediate-moisture space foods were jellied fruit candy, pecans, peaches, pears, apricots, bacon bites, fruitcake, and nutritionally complete food sticks. The formula of the nutritionally complete fruitcake, consumed on the moon by Apollo 17 astronauts, is shown in Table 11.13. The food sticks were a blended candy product developed with the same technology used for intermediate-moisture pet food and confectionery. The Pillsbury Company later marketed the product to the public, first as Space Sticks and then as Food Sticks. The basic ingredients for the peanut butter–flavored product were syrup, corn syrup, peanut butter, vegetable oil, starch, soy, protein, glycerol, emulsifiers, salt, vitamins, and minerals. The a_w was about 0.6, preventing nutrient loss and chemical reactions. Fat was added to impart plasticity.

The effects of high-temperature storage on the stability and acceptability of four intermediate-moisture entrées were evaluated by Johnson et al. (1972). The products, having a_w between 0.81 and 0.86, were pork with barbecue sauce, pork with sweet and sour sauce, pork with Oriental sauce, and ham with sweet mustard sauce. The intermediate-moisture entrées were produced using wet infusion (similar to products produced under the General Foods Contract) and then packaged and placed into storage at 38°C for 6 months. Microbiological, nutritional, and sensory evaluations were conducted at 0, 3, and 6 months. Microbiological data indicated that all products were microbiologically stable

Table 11.13 Apollo 17 Fruitcake Ingredients.

Ingredient	**Quantity (% Weight)**
Flour, wheat, soft	7.3
Flour, soy	7.3
Sugar	19.0
Shortening	7.8
Eggs, whole, fresh	6.96
Salt	0.4
Baking powder	0.4
Water	2.2
Cherries, candied	10.4
Pineapple, candied	8.6
Pecans, shelled	13.8
Raisins, bleached	15.6
Clove powder	0.06
Nutmeg	0.06
Cinnamon	0.12
Total	100.00

Source: Developed by U.S. Army Natick Laboratories: Klicka, M.V., and Smith, M.C. Food for U.S. manned space flight. Technical Report Natick/TR-82/019, 1982. United States Army Natick Research & Development Laboratories, Natick MA 01760.

under nonsterile conditions after 6 months. There were only slight differences in color and appearance between the control and samples stored for 3 months. However, there was noticeable darkening of meats stored for 6 months. Sensory scores decreased for all four products during the 6-month storage period. The Oriental pork and ham with sweet mustard had no significant differences. However, differences in the barbecue pork and sweet and sour pork were highly significant ($p < 0.1$). Nutritional data showed that IMF products were comparable in nutrient content to existing meat components of combat rations. Vitamin retention during storage was similar to retention of foods processed using commercial methods. Johnson et al. concluded that IMF products were comparable in nutrient content to existing combat ration meat components. The four intermediate-moisture pork products—barbecue, sweet and sour, Oriental, and sweet mustard—had protein contents of 18.4%, 19.8%, 19.3%, and 16.6%, respectively. The protein content for canned pork slices in juice was 21.2%. In this same study, it was concluded that vitamin retention during storage was similar to retention of foods processed using commercial methods. Frozen pork has 40% thiamin loss after 6 months, canned pork has 48% loss after 43 weeks at 27° to 39°C and complete loss at 10 weeks for denatured pork. Some thiamin loss was evident in all intermediate-moisture products stored for 6 months at 38°C; ham with sweet mustard sauce only had a 10.5% loss, pork with Oriental sauce had a 68.8% loss, and barbecue pork had a 26.5% loss. All showed a decrease in pyridoxine while there was little or no loss of riboflavin during the 6-month storage period.

In another study, long-term storage of IMFs at different temperatures was evaluated (Secrist et al. 1977). Five entrées were prepared using the same wet infusion method as in the previous study. The products were stored at temperatures of 4°, 21°, and 38°C for 12 months. All five entrées were microbiologically stable at all three temperatures during the 12-month storage period. Nutrient losses were comparable to losses incurred by commercial products. Products stored at 4° and 21°C were acceptable after 12 months of storage. However, there was a decrease in acceptability in samples after 3 months at 38°C.

Table 11.14 Acceptability of Control, IMD fish, IMD Fish with Sauce, and Fried IMD Fish.

Product	Acceptance scores[a] (mean ± standard deviation)
Control (boiled fresh fish)	4.9 ± 2.1
Control w/sauce)	6.0 ± 1.6
IMD fish	4.7 ± 1.5
IMD fish w/sauce	6.2 ± 1.4
Control (boiled fresh fish)	3.4 ± 1.6
Fried control	6.8 ± 1.6
IMD fish	5.8 ± 1.4
Fried IMD fish	7.8 ± 0.9

Source: Adapted from Dymsza, H.A. and Silverman, G. 1979. Improving the acceptability of intermediate-moisture fish. *Food Technol.* 23:52en53.
a Based on a 9-point hedonic scale: 1 = dislike extremely; 9 =like extremely

Dymsza and Silverman (1979) developed a rapid method for producing intermediate-moisture fish using an infusion solution, as well as a method to improve product acceptability by desorptive removal of the infused solute. The desorption method involved boiling the intermediate-moisture fish in water (1:3) for 15 minutes. This process was used to remove the objectionable flavor imparted by the infusion solution, consisting of 43% distilled water, 40% glycerol, 10% sodium acetate, and 7% sodium chloride and resulted in "intermediate-moisture desorbed (IMD) fish" (Dymsza and Silverman 1979). The intermediate-moisture fish was unacceptable; it had a strong objectionable sugar/salt flavor and a mean acceptance rating of 2.7. However, acceptability was increased when the intermediate-moisture fish was desorbed; acceptability of the IMD fish was further increased when served with sauce or fried (see Table 11.14).

Commercial precooked bacon was examined to determine if it complied with military specifications for moisture-to-salt ratio (percent moisture divided by percent salt) of 9.0 or less (Powers et al. 1978, 1981; Walker et al. 1979). Microbiology and a_w were evaluated in addition to the moisture-to-salt ratio. It was determined that the bacon did not comply with the moisture-to-salt ratio, and coagulase-positive staphylococci were found in numbers as high as 1.7×10^5 per gram. Because of the possibility that bacon could have a moisture-to-salt ratio of 9.0 or lower and still have a_w above 0.86, it was suggested that a_w might provide greater assurance of microbiological safety than a moisture-to-salt ratio of 9.0.

In 1985, Washington State University was awarded a U.S. Army contract to produce an intermediate-moisture white bread. In research directed by H. Leung, various humectant combinations were tried. As a result, an organoleptically acceptable bread that was microbiologically stable for up to 20 weeks was produced. However, problems associated with changes in texture and development of off-flavors during storage were found, which must be solved for intermediate-moisture bread to meet all the requirements for military use (Leung 1986). The DOD Combat Feeding Program (CFP) initiated the development of the Meals Ready-to-Eat (MRE) pouched bread in the mid-1980s; it was produced in 1988 and successfully fielded during Operation Desert Storm. The pouched bread technology was patented in 1991 (patent number 5,059,432). Formulation, packaging, and storage environment are important aspects of bread making. Without proper formulation, freshly baked bread will undergo unwanted, rapid change in microbial load and texture. The pouched bread was formulated to produce both microbiological and textural stability through lowered water activities. The initial formulation, intended for storage not exceeding 70°C, in-

Table 11.15 Leavened Bread with Extended Shelf-Life.

Ingredient	% Weight
Flour, high gluten	51.03
Water	29.15
Shortening	8.55
Glycerol	6.00
Yeast, instant, dry	1.90
Salt	1.29
Sucrose ester, emulsifier	1.00
Polyvinylpyrrolidone	1.00
Potassium sorbate	0.05
Cream flavor	0.03

Source: Adapted from Berkowitz, D., and Oleksyk, L.E. 1991. Leavened bread with an extended shelf life. U.S. Patent 5,059,432, October 22, 1991.

cluded high gluten flour, water, shortening, glycerol, sugar, instant dry yeast, sucrose ester, polyvinylpyrrolidone, potassium sorbate, and cream flavor. Table 11.15 shows a formulation for bread intended for storage at a temperature not exceeding 38°C. Current formulations eliminate the polyvinylpyrrolidone and add xantham gum and gum Arabic. Additional quality protection is afforded by maintaining low oxygen levels in the pouch using oxygen-scavenging sachet and flexible pouches to prevent gas and moisture transmission.

The combination of formulation, oxygen-scavenging sachets and packaging material, allows the pouched bread to be stored for at least 6 months at 100°F or 3 years at 80°F while preserving quality flavor and nutrition. Studies have shown that the morale of soldiers is significantly increased when bread is provided. It also increases the acceptability of other ration components.

The development of shelf-stable bread led to an expanded variety of shelf-stable bakery items for both individual and group feeding. This was followed by the development of a shelf-stable cake with a_w as high as 0.890. The cake batter is filled into half-size steam table pans (Tray-Packs) that are baked with the lid on, cooled, and sealed. The Tray-Pack cake has been available to the troops since 1986, has a 3-year shelf-life at 27°C, and can be stored and served in the same pan. The aluminum tray can has recently been replaced with pouches, fabricated from high-barrier films, or polymeric trays with lids made from high-barrier films. The pouches contain intermediate-moisture bread, waffles, and bacon; the polymeric trays contain intermediate-moisture cakes, brownies, and cookies.

A family of intermediate-moisture pound cakes and brownies was developed with a_w greater than 0.85. The six flavors developed include vanilla, lemon, orange, pineapple, chocolate mint, and fudge brownie. These intermediate-moisture cakes replaced the lower-quality, highly dense thermoprocessed nut cakes in 1988. Other intermediate-moisture items developed include shelf-stable waffles (1994) and pancakes (1996) with a_w greater than 0.85.

Intermediate-moisture fermented meats are very popular items with military troops. They are not only a morale booster but provide good nutrition. The U.S. Army developed intermediate-moisture beef sticks (pepperoni and nacho cheese flavored) for use alone and as an ingredient in shelf-stable sandwiches. This was done by carefully manipulating a_w and pH and preventing lipid oxidation (Briggs et al. 1991, 1992; Richardson et al. 1995). The beef sticks were processed with 0%, 3%, 6%, and 9% glycerol to determine the optimum level that would give the lowest a_w without negatively affecting flavor or texture. The optimum level of glycerol was between 3% and 6%, giving a final a_w of 0.876 and 0.846, respectively. The texture of the product with no added glycerol became chewy and fibrous. At 6% and 9% glycerol levels, the products were grainy and less chewy and had an objectionable sweet note. As the concentration of glycerol increased, the amount of oil that leeched out of the beef stick also increased, thus decreasing palatability. Prevention of lipid oxidation was achieved using a synergistic antioxidant system consisting of BHA, ascorbyl palmitate, and vitamin E. The mechanism of synergism in the antioxidant system is apparent in that the ascorbyl palmitate was retained.

Conca (1995) evaluated a shellac-based coated material as an edible film for intermediate-moisture beef sticks to preserve quality and to reduce packaging volume, weight, and waste. Samples were coated with a two-phase, shellac-based coating and stored for 4 weeks at room temperature. After 4 weeks, the coated beef sticks lost significantly less weight than the noncoated beef sticks. Also, significant color, texture, and size changes were observed for the noncoated beef sticks. Barret et al. (1998) looked at texture and storage stability of processed beef sticks with 0%, 2%, and 4% glycerol levels. Results show that glycerol functioned as an effective textural plasticizer in the beefsticks. Yang (1997) prepared turkey slices using a combination of microwave-assisted freeze-drying and humectant infusion with 20% glycerol for 30 minutes, to reduce a_w while maintaining texture. Microwave-assisted freeze-drying accelerated the drying rate by 30% and produced a product with good texture and rehydratability. The water activities of the untreated and treated samples were 0.89 and 0.85, respectively. The glycerol-treated samples were softer.

The U.S. Army has further expanded IMF technology to develop a family of multicomponent shelf-stable products, e.g., bread/meat, pizza/burrito-type combinations (ABC News-World News Tonight 2002; BBC News 2002; CNN.com 2002; Discovery Channel Canada 2002; Tech TV 2002). Barbecue chicken, barbecue beef, Italian pocket (pepperoni and sausage in tomato sauce), pepperoni, nacho cheese–flavored beef, and peanut butter and jelly are some of the varieties currently available. This project embodies the hurdle approach for development of safe, stable, and highly acceptable food products. The hurdles used to develop the shelf-stable sandwich include (1) a_w, which is controlled by incorporating various humectants such as salt, glycerol, and rice syrup into the formulation; (2) pH, which is controlled by choosing foods that are naturally acidic or by incorporating food-grade acids; (3) preservatives, such as nitrites and antioxidants; (4) thermal pasteurization through baking; (5) high-barrier packaging films; and (6) oxygen-scavenging sachets. The sandwiches were further developed and commercialized under a Cooperative Research and Development Agreement (CRDA) with Good Mark Foods, Inc. The combination of meat with the bread at differing a_w and pH makes both the safety and acceptability of the product concerns. The shelf-stable sandwiches currently have water activities above 0.86 and pH values above 4.9. However, the FDA states that foods not thermally processed must have an a_w of less than 0.86 or a pH less than 4.5. Challenge studies were conducted on several types of shelf-stable sandwiches to ensure their safety and to provide guidelines to manufacturers, by determining a_w and pH factors that prevent or influence the growth of *S. aureus* (Powers et al. 1999). An inoculum cocktail of three strains of *S. aureus* was delivered at the interface between the bread and meat filling; the product were packaged in trilaminated pouches with oxygen-scavenging sachets and put into storage for 6 months at 35°C. The barbecue chicken sandwich was tested at two different a_w levels, 0.920 and 0.890. The pH was 4.8 for both. The counts of *S. aureus* declined immediately and were at nondetectable levels (<100/g) within 7 to 14 days (see Figure 11.2). The nacho cheese–flavored beef sandwich was also stable at three water activities, 0.90, 0.89, and 0.86, with pH at 4.3, 5.3, and 5.2, respectively (see Figure 11.3).

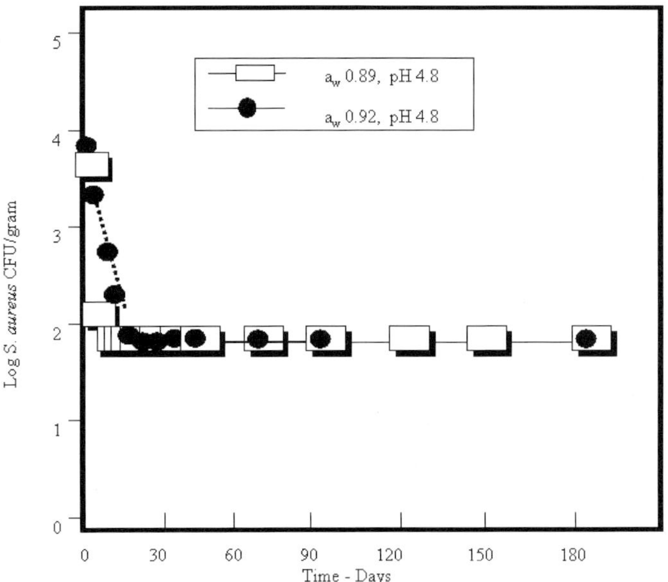

Figure 11.2 The effects of a_W on the growth and survival of a three-strain cocktail of *S. aureus* in barbecue chicken sandwiches stored at 35°C for 6 months.

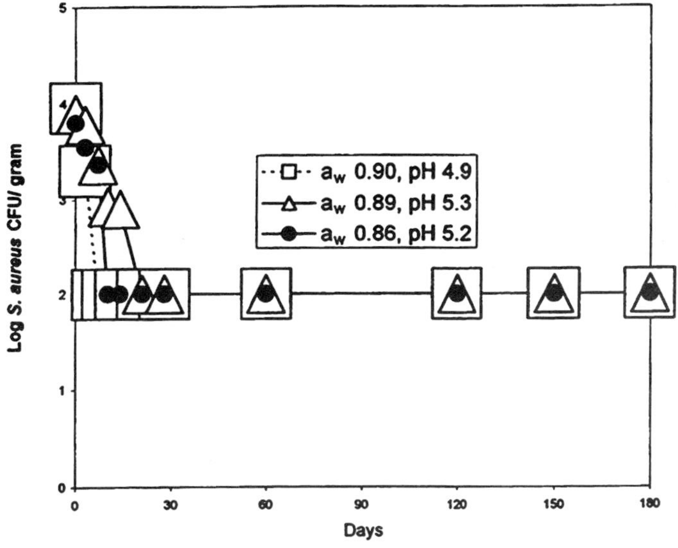

Figure 11.3 The effects of a_W on the growth and survival of a three-strain cocktail of *S. aureus* in nacho cheese–flavored beef sandwiches stored at 35°C for 6 months.

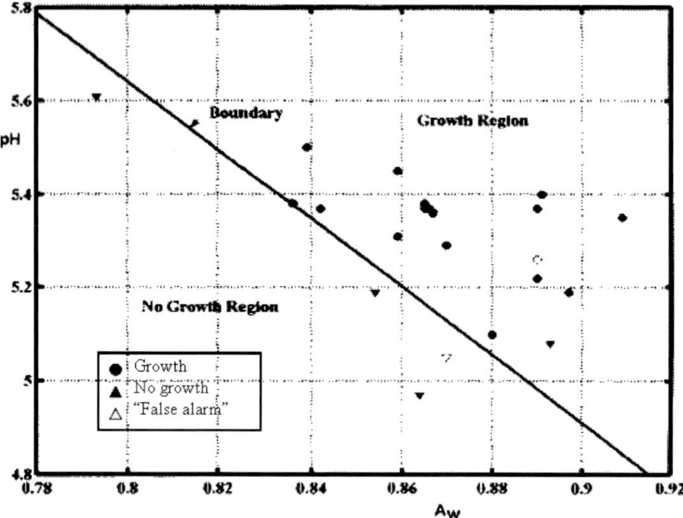

Figure 11.4 The growth–no-growth boundary region under pH and a_w conditions characterizing the domains of growth and no-growth for *S. aureus* in bread.

A quasi chemical kinetics model was developed for IMFs (Taub et al. 2000a, 2000b, 2003; Feherry et al. 2001, 2002b). This model fits the growth/death kinetics for *S. aureus* in ground bread crumbs with respect to variations in a_W, pH, and temperature. Figure 11.4 shows the growth–no growth boundary region for a_W and pH based on the statistics derived from the quasi chemical modeling results.

Current work in the Combat Feeding Program has focused on (1) increasing the variety, acceptability, and commercialization of military unique IMF products, (2) finalizing generic hazard analysis and critical control points (HAACP) for the commercial processing of shelf-stable sandwiches, and (3) applying the quasi chemical model to other IMF products. Intermediate-moisture technology will be further exploited along with intercomponent film and coating technologies to further develop a family of multicomponent breakfast items. Integrating the two technologies will allow the use of ingredients previously impossible due to moisture or fat migration. Proposed items include bacon and cheddar pocket, syrup-filled French toast, and breakfast pizza and/or burrito.

Future Intermediate-Moisture Foods

A systematic approach to the development of new IMF products comprises the following according to Brimelow (1985):

1. *Feasibility of the proposed product.* Is it feasible to make a proposed product by intermediate-moisture technology? Does IMF technology improve on alternative potential technologies? Is it cost-competitive?
2. *Water activity range of the product.* What type of product is being proposed, and what is its microbiological, chemical, and physical stability? Is reduced a_W alone sufficient to preserve the food? Would a combined approach be more advantageous?

3. *Proposed packaging methods.* What are the barrier requirements, handling resistance, consumer appeal, and cost?
4. *Proposed manufacturing methods.* Different methods must be evaluated, and quality must be optimized versus cost.
5. *Proposed quality assurance schemes.* The humectants–solution composition and the a_w of the ingredients and product must be monitored at all stages of processing. Other normal testing for quality assurance must be carried out.
6. *Selling the total project to decision makers.* A new IMF product must have demonstrated advantages in quality, novelty, and convenience compared to existing products.

IMF technology can provide easy to eat, easy to store and prepare, individually sized, and nutritionally dense foods, offering new marketing opportunities. It is up to product development scientists to innovatively use known principles and to come up with improved or new IMF products. Research in the IMF field involves (1) the development of new formulations containing combinations of new humectants of high organoleptic acceptability, (2) the development of new antimicrobial agents suitable for IMFs and new methods to limit chemical deterioration and loss of quality during storage, and (3) the development of improved processes for large-scale production.

Development for special applications of IMF products, such as for the military, and applications of IMF principles and technology in traditional and novel food products for the developing world are foreseen to increase. Commercial application could be boosted in combination with new technological and marketing trends for functional foods and nutraceutics.

References

ABC News World News Tonight. ABC News, Inc. May 22, 2002.

Acott, K., Sloan, A.E., and Labuza, T.P. 1976. Evaluation of antimicrobial agents in a microbial challenge study for an intermediate moisture dog food. *Journal of Food Science* 41:541–546.

Alzamora, S.M., Chirife, J., and Gerschenson, L.N. 1994. Determination and correlation of the water activity of propylene glycol solutions. *Food Research International* 27(1):65–67.

Anderson, C.B., and Witter, L.D. 1982. Water binding capacity of 22 L-amino acids from water activity 0.33 to 0.95. *Journal of Food Science* 47:1952–1954.

Anonymous. 1985. Neosugar: A fructooligosaccharide nonnutritive sweetener. *Nutrition Review* 43(5):155.

Ballesteros, S.A., Chirife, J., and Bozzini, J.P. 1993. Specific solute effects on *Staphylococcus aureus* cells subjected to reduced water activity. *International Journal of Food and Microbiology* 20:51–66.

Barret, A.H., Briggs, J., Richardson, M., and Reed, T. 1998. The texture and storage stability of processed beefsticks as affected by glycerol and moisture levels. *Journal of Food Science* 63:84–87.

BBC News. 2002 U.S. military unveils "super sandwich." BBC London, U.K.

Bell, L.N. 1996. Kinetics of non-enzymatic browning in amorphous solid systems: Distinguishing the effects of water activity and the glass transition. *Food Research International* 28:591–597.

Bell, L.N., Touma D.E., White, K.L., and Chen, Y.H. 1998. Glycine loss and Maillard browning as related to the glass transition in a model food system. *Journal of Food Science* 63: 625–628.

Bell, L.N., Hageman, M.J. 1994. Differentiating between the effects of water activity and glass transition dependent mobility on a solid state chemical reaction: Aspartame degradation. *Journal of Agricultural and Food Chemistry* 42:2398–2401.

Bell, L.N., and Labuza, T.P. 2000. *Moisture Sorption: Practical Aspects of Isotherm Measurement and Use*. 2nd Edition. St. Paul, MN: American Association of Cereal Chemists.

Bell, L.N., and White, K.L. 2000. Thiamin stability in solids as affected by the glass transition. *Journal of Food Science* 65:498–501.

Benmergui, E.A., Ferro Fontan, C., and Chirife, J. 1979. The prediction of water activity in aqueous solutions in connection with intermediate moisture foods. a_w prediction in single aqueous electrolyte solutions. *Journal of Food Technology* 14:625–637.

Berkowitz, D., and Oleksyk, L.E. 1991. Leavened Bread with an Extended Shelf Life. U.S. Patent 5,059,432, October 22, 1991.

Bertelsen, H., Jensen, B.B., and Buemann, B. 1999. D-Tagatose: A novel low calorie bulk sweetener with prebiotic properties. *World Review of Nutrition and Dietetics* 85:98–109.

Beuchat, L.R. 1981. Microbial stability as affected by water activity. *Cereal Foods World* 26:345–349.

Binkerd, E.F., Kolari, O.E., and Tracy, C. 1976. Pemmican American Meat Scientists Association Reciprocal Meat Conference, Provo, UT.

Bone, D.P. 1969. Water activity: Its chemistry and applications. *Food Product Development* 3(5):81–94.

Bone, D.P., Shannon, E.L., and Ross, K.D. 1975. The lowering of water activity by order of mixing in concentrated solutions. In: *Water Relations of Foods,* ed. R.B. Duckworth, pp. 613–626. New York: Academic Press.

Bornet, F.R.J. 1994. Undigestible sugars in food products. *American Journal of Clinical Nutrition* 59:763S–769S.

Briggs, J., Dunne, C., Richardson, M.J., Yang, T.C.S., and Taub, I.A. 1991. Synergistic antioxidant systems for extending the shelf life of intermediate moisture, high acid, fermented meats. Institute of Food Technologists Annual Meeting Technical Program Book of Abstracts, p. 442. Presented at the Institute of Food Technologists Annual Meeting.

Briggs, J., Richardson, M.J., Yang, T., Berkowitz, D., and Tartarini, K. 1992. Shelf stable sandwich for military rations. Institute of Food Technologists Annual Meeting Technical Program Book of Abstracts, p. 821.

Brimelow, C.J.B. 1985. A pragmatic approach to the development of new intermediate moisture foods. In: *Properties of Water in Foods*, eds. D. Simatos and J.L. Multon, pp. 405–419. NATO Advanced Science Institutes Series E, Applied Sciences, No. 90. Dordrecht, the Netherlands: Martinus Nijhoff.

Brockmann, M.C. 1970. Development of intermediate moisture foods for military use. *Food Technology (Chicago)* 24:896–900.

Burgess, H.M., and Mellentin, R.W. 1965. Animal food and method of making the same. U.S. patent 3,202,514.

Bussiere, G., and Serpelloni, M. 1985. Confectionery and water activity: Determination of a_w by calculation. In: *Properties of Water in Foods,* eds. D. Simatos and J.L. Multon, pp. 627–645. NATO Advanced Science Institutes Series E, Applied Sciences, No. 90. Dordrecht, the Netherlands: Martinus Nijhof.

Chirife, J., Ferro Fontan, C., and Scorza, O.C. 1980a. A study of the water activity lowering behavior of some amino acids. *Journal of Food Technology* 15:383–387.

Chirife, J., Ferro Fontan, C., and Bemergui, E.A. 1980b. The prediction of water activity in aqueous solutions in connection with intermediate moisture foods IV. a_w prediction in aqueous non electrolyte solutions. *Journal of Food Technology* 15:59–70.

Chirife, J., Resnik, S.L., and Ferro Fontan, C. 1985. Application of Ross' equation for prediction of water activity in intermediate moisture food systems containing a non-solute solid. *Journal of Food Technology* 20:773–779.

Chirife, J., and Buera, M.P. 1995. A critical review of some non-equilibrium situations and glass transitions on water activity values of foods in the microbiological growth range. *Journal of Food Engineering* 25:531–552.

Chou, H.E., Acott, K.M., and Labuza, T.P. 1973. Sorption hysteresis and chemical reactivity: Lipid oxidation. *Journal of Food Science* 38:316–319.

Christian, J.H.B. 1981. Specific solute effects on microbial water relations. In: *Water Activity: Influences on Food Quality,* eds. L.B. Rockland and G.F. Stewart, pp. 825–854. New York: Academic Press.

Christian, J.H.B. 2000. Drying and reduction of water activity. In: *The Microbiological Safety and Quality of food,* vol. 1, eds. B.M. Lund, T.C. Baird-Parker, and G.W. Gould, pp. 146–174. Gaithersburg, MD: Aspen Publishers.

Chuang, L., and Toledo, R.T. 1976. Predicting the water activity of multicomponent systems from water sorption isotherms of individual components. *Journal of Food Science* 41:922–927.

Cole, M.B., Jones, M.V., and Holyoak, C. 1990. The effect of pH, salt concentration, and temperature on the survival and growth of Listeria monocytogenes. *Journal of Applied Bacteriology* 69(1):63–72.

Conca, K.R., and Yang, T.C.S. 1993. Evaluation of a shellac-based coating material as an edible film for intermediate moisture foods. Institute of Food Technologists Annual Meeting Technical Program Book of Abstracts, p. 821.

Corry, J.E.L. 1976. The safety of intermediate moisture foods with respect to salmonella. In: *Intermediate Moisture Foods*, eds. R. Davies, G.G. Birch, and K.J. Parker, pp. 215–236. London: Applied Science.

De Gennaro, S., Birch, G.G., Parke, S.A., and Stancher, B. 2000. Studies on the physicochemical properties of inulin and inulin oligomers. *Food Chemistry* 68:179–183.

De Leenheer, L. 1996. Production and use of insulin: Industrial reality with a promising future. In: *Carbohydrates as Organic Raw Materials III*, eds. H. Van Bekkum, H. Roper, and A.G.J. Voragen, pp. 67–92. New York: VCH Publishers.

Duckworth, R., Allison, J., and Clapperton, H.A. 1976. The aqueous environment for chemical change in intermediate moisture foods. In: *Intermediate Moisture Foods*, eds. R. Davies, G.G. Birch, and K.J. Parker, pp. 89–99. London: Applied Science.

Dymsza, H.A., and Silverman, G. 1979. Improving the acceptability of intermediate-moisture fish. *Food Technology* 23:52–53.

Erickson, L.E. 1982. Recent developments in intermediate moisture foods. *Journal of Food Protection* 45:484–491.

Ewald, S., and Notermans, S. 1988. Effect of water activity on growth and enterotoxin D production of *Staphylococcus aureus*. *International Journal of Food Microbiology* 6(1):25–30.

Feeherry, F.E., Ross, E.W., and Taub, I.A. 2001. Modeling the growth and death of bacteria in intermediate moisture. *Acta Horticulturae (ISHS)* 566:123–128.

Feeherry, F.E., Ross, E.W., and Taub, I.A. 2002. Predictive modeling of pathogen growth and death in intermediate moisture (IM) rations. In: Proceedings of the 23rd Army Science Conference, December 3-7, 2002, Orlando, FL.

Ferron Fontan, C., Chirife, J., and Boquet, R. 1981. Water activity in multicomponent non-electrolyte solutions. *Journal of Food Technology* 18:553–559.

Flink, J.M. 1977. Intermediate moisture food products in the American marketplace. *Journal of Food Processing and Preservation* 1:324–339.

Forni, E., Sormani, A., Scalise, S., and Torreggiani, D. 1997. The influence of sugar composition on the colour stability of osmodehydrofrozen intermediate moisture apricots. *Food Research International* 30(2):87–94.

Franck, A. 2002. Technological functionality of inulin and oligofructose. *British Journal of Nutrition* 87(Suppl. 2):S287–S291.

Gee, M., Farkas, D., and Rahman, A.R. 1977. Some concepts for the development of intermediate moisture foods. *Food Technology* 32(4):58–63.

Glasstone, S. 1946. *Textbook of Physical Chemistry*. New York: Van Nostrand.

Gould, G.W. 1985. Present state of knowledge of a_w effects on microorganisms. In: *Properties of Water in Foods,* eds. D. Simatos and J.L. Multon, pp. 229–245. NATO Advanced Science Institutes Series E, Applied Sciences, No. 90. Dordrecht, the Netherlands: Martinus Nijhoff.

Gould, G.W. 1996. Methods for preservation and extension of shelf life. *International Journal of Food Microbiology* 33:51–64.

Grover, D.W. 1947. The keeping properties of confectionary as influenced by its water vapor pressure. *J. Soc. Chem. Ind. (Lond.)* 66:201.

Guilbert, S. 1984. Additifs et agents d' presseurs de lÕaw. In: *Additifs et Auxiliaires de Fabrication dans les Industries Agro-alimentaires*, ed. J.L. Multon, pp. 199–227. Aprix, Paris: Tec et Doc Lavoisier.

Guilbert, S., and Cheftel, J.C. 1982. Influence de la gelification et de l'émulsification sur l'activité de l'eau d'aliments modelés aa humidité intermédiaire. *Sciences Des Aliments Series II* 2:43–52.

Guilbert, S., Clement, O., and Cheftel, J.C. 1981. Relative efficiency of various a_w-lowering agents in aqueous solutions and in intermediate moisture foods. *Lebensittel-Wissenschaft und-Technologie* 14:245–251.

Haas, G.J., and Herman, E.B. 1978. Bacterial growth in intermediate moisture food systems. *Lebensittel-Wissenschaft und-Technologie* 11:74–78.

Haas, G.J., Bennett, D., Herman, E.B., and Collette, D. 1975. Microbial stability of intermediate moisture foods. *Food Product Development* 9(3):86.

Hagenmaier, R.D., Cater, C.M., and Mattil, K.F. 1975. Coconut skim milk as an intermediate moisture product. *Journal of Food Science* 40:717–720.

Hollis, F., Kaplow, M., Halik, J., and Nordstrom, H. 1969. Parameters for moisture content for stabilization of food products (Phase II). General Foods Corporation, NY. Contract No. DAAG-17-67-C-0098, Technical Report 70-12-FL, 1969 (AD 701 015). U.S. Army Natick Research and Development Command, Natick, MA 01760.

Hollis, F., Kaplow, M., Klose, R., and Halik, J. 1969. Parameters for moisture content for stabilization of food products. General Foods Corporation, NY. Contract No. DAAG-17-67-C-0098, Technical Report 69-26-FL, 1968 (AD 675473). U.S. Army Natick Research and Development Command, Natick, MA 01760.

Iglesias, H.A., and Chirife, J. 1982. *Handbook of Food Isotherms: Water Sorption Parameters for Food and Food Components*. Orlando, FL: Academic Press.

Jakobsen, M. 1985. Effect of a_w on growth and survival of Bacillaceae. In: *Properties of Water in Foods*, eds. D. Simatos and J.L. Multon, pp. 259–272. NATO Advanced Science Institutes Series E, Applied Sciences, No. 90. Dordrecht, the Netherlands: Martinus Nijhoff.

Johnson, R.G, Sullivan, D.B., Secrist, J.L., and Brockmann, M.C. 1997, 1972. The effect of high temperature storage on the acceptability and stability of intermediate moisture foods. Technical Report 72-76-FL, 1972 (AD 747 362). U.S. Army Natick Laboratories, Natick, MA 01760.

Kanatt, S.R., Chawla, S.P., Chander, R., and Bongirwar, D.R. 2002. Shelf-stable and safe intermediate-moisture meat products using hurdle technology. *Journal of Food Protection* 65:1628–1631.

Kaplow, M. 1970. Commercial development of intermediate moisture foods. *Food Technology (Chicago)* 24:889–893.

Kapsalis, J.G., Ball, D.H., Alabran, D.M., and Cardello, A.V. 1985. Polyglycerols and polyglycerol esters as potential water activity reducing agents. Chemistry and sensory analysis. In: *Properties of Water in Foods*, eds. D. Simatos and J.L. Multon, pp. 481–496. NATO Advanced Science Institutes Series E, Applied Sciences, No. 90. Dordrecht, the Netherlands: Martinus Nijhoff.

Karel, M. 1973. Recent research and development in the field of low-moisture and intermediate-moisture foods. *Critical Reviews in Food Technology* 3:329–373.

Karel, M. 1975. Physico-chemical modification of the state of water in foods. In: *Water Relations of Foods*, ed. R.B. Duckworth, pp. 639–657. New York: Academic Press.

Karel, M. 1976. Technology and application of new intermediate moisture foods. In: *Intermediate Moisture Foods,* eds. R. Davies, G.G. Birch, and K.J. Parker, pp. 4–28. London: Applied Science.

Karel, M. 1985. Effects of water activity and water content on mobility of food components, and their effect on phase transitions in food systems. In: *Properties of Water in Foods,* eds. D. Simatos and J.L. Multon, pp. 153–169. NATO Advanced Science Institutes Series E, Applied Sciences, No. 90. Dordrecht, the Netherlands: Martinus Nijhoff.

Karmas, E., and Chen, C.C. 1975. Relationship between water activity and water binding in high and intermediate moisture foods. *Journal of Food Science* 40:800–801.

Karmas, R., Buera, M.P., and Karel, M. 1992. Effect of glass transition on rates of nonenzymatic browning in food systems. *Journal of Agricultural and Food Chemistry* 40:873–879.

Kirk, J., Dennison, D., Kokoczka, P., Heldman, D., and Singh, R. 1977. Degradation of ascorbic acid in a dehydrated food system. *Journal of Food Science* 42(5):1274–1279.

Kouassi, K., and Roos, Y.H. 2001. Glass transition and water effects on sucrose inversion by invertase in noncrystalline carbohydrate food systems. *Food Research International* 34:895–901.

Kusik, C.L., and Meissner, H.P. 1973. Vapor pressures of water over aqueous solutions of strong electrolytes. *Industrial Engineering and Chemical Product Research Development* 12(1):112–115.

Labuza, T.P. 1970. Properties of water as related to the keeping quality of foods. In: *Proceedings of the Third Annual Meeting of the International Congress of Food Science and Technology*, pp. 618–635. Chicago: Institute of Food Technology.

Labuza, T.P. 1975. Oxidative changes in foods at low and intermediate moisture levels. In: *Water Relations of Foods*, ed. R.B. Duckworth, pp. 455–474. New York: Academic Press.

Labuza, T.P. 1976. Drying food: Technology improves on the sun. *Food Technology (Chicago)* 30(6):37–46.

Labuza, T.P. 1977. The properties of water in relationship to water binding in foods: A review. *Journal of Food Processing and Preservation* 1:167–190.

Labuza, T.P. 1980. The effect of water activity on reaction kinetics of food deterioration. *Food Technology* 34(4):36–41.

Labuza, T.P. 1984. *Moisture Sorption: Practical Aspects of Isotherm Measurement and Use*. St. Paul, MN: American Association of Cereal Chemists.

Labuza, T.P. 1985. Water binding of humectants. In: *Properties of Water in Foods*, eds. D. Simatos and J.L. Multon, pp. 421–445. NATO Advanced Science Institutes Series E, Applied Sciences, No. 90. Dordrecht, the Netherlands: Martinus Nijhoff.

Labuza, T.P., and Chou, H.E. 1974. Decrease of linoleate oxidation due to water at intermediate water activity. *Journal of Food Science* 39:112–113.

Labuza, T.P., and Saltmarch, M. 1981. The nonenzymatic browning reaction as affected by water in foods. In: *Water Activity: Influences on Food Quality*, eds. L. Rockland and G.F. Stewart, pp. 605–650. New York: Academic Press.

Labuza, T.P., and Sloan, A.E. 1981. Forces of change: From Osiris to open dating. *Food Technology* 35(7): 34–43.

Labuza, T.P., McNally, L., Gallagher, D., Hawkes, J., and Hurtado, F. 1972a. Stability of intermediate moisture foods. 1. Lipid oxidation. *Journal of Food Science* 37:154–159.

Labuza, T.P., Cassil, S., and Sinskey, A.J. 1972b. Stability of intermediate moisture foods. 2. Microbiology. *Journal of Food Science* 37:160–162.

Laing, B., Schlueter, D., and Labuza, T.P. 1978. Degradation kinetics of ascorbic acid at high temperature and water activity. *Journal of Food Science* 43:1440–1443.

Leiras, M.C., Alzamora, S.M., and Chirife, J. 1990. Water activity of galactose solutions. *Journal of Food Science* 55:1174–1176.

Leistner, L. 1985. Hurdle technology applied to meat products of the shelf stable product and intermediate moisture food types. In: *Properties of Water in Foods,* eds. D. Simatos and J.L. Multon, pp. 309–329. NATO Advanced Science Institutes Series E, Applied Sciences, No. 90. Dordrecht, the Netherlands: Martinus Nijhoff.

Leistner, L., and Gould, G. 2002. *Hurdle Technologies. Combination Treatments for Food Stability, Safety and Quality,* p. 194. New York: Kluewer Academic/Plenum Publishers.

Leistner, L., and Rodel, W. 1976. The stability of intermediate moisture foods with respect to microorganisms. In: *Intermediate Moisture Foods*, eds. R. Davies, G.G. Birch, and K.J. Parker, pp. 120–137. London: Applied Science.

Leung, H.K. 1986. Bread quality: Effects of water binding ingredients. Res. Dev. Assoc. Annual Meeting Report, U.S. Army Natick Laboratory, Natick, MA. pp. 43–48.

Leung, H., Moriss, H.A., Sloan, A.E., and Labuza, T.P. 1976. Development of an intermediate moisture processed cheese food product. *Food Technology (Chicago)* 23:42–44.

Lievonen, S.M., and Roos, Y.H. 2002. Water sorption of food models for studies of glass transition and reaction kinetics. *Journal of Food Science* 67:1758.

Lievonen, S.M., Laaksonen, T.J., and Roos, Y.H. 1998. Glass transition and reaction rates: Nonenzymatic browning in glassy and liquid systems. *Journal of Agricultural and Food Chemistry* 46:2778–2784.

Lievonen, S.M., and Roos, Y.H. 2002. Nonenzymatic browning in amorphous food models: Effects of glass transition and water. *Journal of Food Science* 67(6):2100–2106.

Linko, P., Kervinen, R., Karpinen, R., Rautalinna, E.K., and Vainionp$da$$da$, J. 1985. Extrusion cooking for cereal-based intermediate-moisture products. In: *Properties of Water in Foods*, eds. D. Simatos and J.L. Multon, pp. 465-479. NATO Advanced Science Institutes Series E, Applied Sciences, No. 90. Dordrecht, the Netherlands: Martinus Nijhoff.

Loncin, M. 1975. Basic principles of moisture equilibria. In: *Freeze Drying and Advanced Food Technology*, eds. S.A. Goldblith, L. Rey, and W.W. Rothmayer. London: Academic Press.

Maltini, E., Torreggiani, D., Brovetto, B.R., and Bertolo, G. 1993. Functional properties of reduced moisture fruits as ingredients in food systems. *Food Research International* 26:413–419.

Mizrahi, S., Labuza, T.P., and Karel, M. 1970. Computer aided predictions of extent of browning in dehydrated cabbage *Journal of Food Science* 35:799.

Moody, S.M. 2000. Chow time: Military feeding from Bunker Hill to Bosnia the history of the development and utilization of military rations in the United States Armed Forces. Master thesis, Kansas State University, Manhattan, KS.

Mossel, D.A.A. 1976. Microbiological specifications for intermediate moisture foods with special reference to methodology used for the assessment of compliance. In: *Intermediate Moisture Foods*, eds. R. Davies, G.G. Birch, and K.J. Parker, pp. 248–259. London: Applied Science.

Motoki, M., Torres, J., and Karel, M. 1982. Development and stability of intermediate moisture cheese analogs from isolated soybean proteins. *Journal of Food Processing and Preservation* 6:41–53.

Multon, J.L. 1981. L'Etat actuel des travaux de la Commission "Aliments aa humidité interm é diaireó du C.N.E.R.N.A." *Ind. Aliment. Agric.* 98:291–302.

Narashima Rao, D. 1997 Intermediate moisture foods based on meats: A review. *Food Review International* 13(4):519–551.

Norrish, R.S. 1966. An equation for the activity coefficients and equilibrium relative humidities of water in confectionery syrups. *Journal of Food Technology* 1:25–39.

Notermans, S., and Heuvelman, C.J. 1983 Combined effect of water activity, pH and sub-optimal temperature on growth and enterotoxin production of *Staphylococcus aureus*. *Journal of Food Science* 48, 1832–1840.

Obanu, Z.A. 1981. The applicability of intermediate moisture food technology for preservation of meat and fish in Nigeria. In: Proceedings of the National Conference on From Food Deficiency to Food Sufficiency, Rivers State University, Port Harcout, Nigeria, May 3-8.

Obanu, Z.A., Ledwand, D.A., and Lawrie, R.A. 1975. The protein of intermediate moisture meat stored at tropical temperature. II. Effect of protein changes on some aspects of meat quality. *Journal of Food Technology* 10:667–674.

Pavey, R.L., and Schack, W.R. 1969. Formulation of intermediate moisture bite-size food cubes. Technical Report Contract F4160967-C-0054. U.S. Air Force School of Aerospace Medicine, San Antonio, TX.

Pavey, R.L. 1972. Controlling the amount of internal aqueous solution in intermediate moisture foods. Contract No. DAAG-17-70-C-0077, Swift & Company Research and Development Center, Illinois. Technical Report 73-17-FL, 1972 (AD 757 766). U.S. Army Natick, Laboratories, Natick, MA 01760.

Pavey, R.L., and Schack, W.R. 1969. Formulation of intermediate moisture bite-size food cubes. Tech. Rep. Contract F4160967-C-0054. U.S. Air Force School of Aerospace Medicine, San Antonio, TX.

Pawsey, R., and Davies, R. 1976. The safety of intermediate moisture foods with respect to *Staphylococcus aureus*. In: *Intermediate Moisture Foods,* eds. R. Davies, G.G. Birch, and K.J. Parker, pp. 182–201. London: Applied Science.

Pelaez, J., and Karel, M. 1980. Development and stability of intermediate moisture tortillas. *Journal of Food Processing and Preservation* 4:51–65.

Pitzer, K.S., and Mayorga, G. 1973. Thermodynamics of electrolytes. II. Activity and osmotic coefficients for strong electrolytes with one or both ions univalent. *Journal of Physical Chemistry* 77:2300–2308.

Portmann, M.O., and Birch, G. 1995. Sweet taste and solution properties of α,α-trehalose. *Journal of Science and Food Agriculture* 69:275–281.

Powers, E.M., Berkowitz, and Walker, G.C. 1981. Bacteriology, water activity, and moisture/salt ratio of six brands of precooked canned bacon. *Journal of Food Protection* 44:447–449.

Powers, E.M., Briggs, J., DeFao, A., Lee, C., Racicot, K., Richardson, M., Senecal, A., and Wong, C. 1999. Effect of water activity on the microbiological stability of mobility-enhancing ration components. Technical Report Natick/TR-00/003. U.S. Army Natick Soldier Center, Natick, MA 01760.

Powers, E.M., Latt, T.G., Johnson, D.R., and Rowley, D.B. 1978. Occurrence of *Staphylococcus aureus* in and the moisture content of precooked canned bacon. *Journal of Food Protection* 41:708–711.

Prausnitz, J.M. 1969. *Molecular Thermodynamics of Fluid Phase Equilibria.* Englewood Cliffs, NJ: Prentice-Hall.

Purnomo, H., Buckle, K.A., and Edwards, R.A. 1983. A preliminary study on a traditional intermediate moisture beef product. *Journal of Food Science Technology* 20:177–178.

Raoult, F.M. 1888. *Z. Phys. Chem.* 2:353 (cited in Van Den Berg, C., and Bruin, S. 1981).

Rastogi, N.K., Raghavarao, K.S.M.S., Niranjan, K., and Knorr, D. 2002. Recent developments in osmotic dehydration: Methods to enhance mass transfer. *Trends in Food Science and Technology* 13(2):48–59.

Reid, D.S. 1976. Water activity concepts in intermediate moisture foods. In: *Intermediate Moisture Foods*, eds. R. Davies, G.G. Birch, and K.J. Parker, pp. 54–65. London: Applied Science.

Richard-Molard, D., Lesage, L., and Cahagnier, B. 1985. Effect of water activity on mold growth and mycotoxin production. In: *Properties of Water in Foods.* eds. D. Simatos and J.L. Multon, pp. 273–292. NATO Advanced Science Institutes Series E, Applied Sciences, No. 90. Dordrecht, the Netherlands: Martinus Nijhoff.

Richardson, M., Briggs, J. Senecal, A. Dunne, P., and Lee, C. 1995. Development of intermediate moisture meats for use in military shelf-stable sandwiches. Proceedings of the 41st Annual International Congress of Meat Science and Technology.

Roberts, T.A., and Smart, J.L. 1976. Control of *Clostridia* by water activity and related factors. In: *Intermediate Moisture Foods*, eds. R. Davies, G.G. Birch, and K.J. Parker, pp. 203–211. London: Applied Science.

Robinson, R.A., and Stokes, R.H. 1970. *Electrolyte Solutions*, 2nd Edition. London: Butterworths.

Rockland, L.B., and Nishi, S.K. 1980. Influence of water activity on food product quality and stability. *Food Technology* 34(4):42–52.

Ross, K.D. 1975. Estimation of water activity in intermediate moisture foods. *Food Technology* 29(3):26–34.

Roos, Y.H. 1995. *Phase Transitions in Foods*. San Diego: Academic Press.

Roos, Y.H. 1995. Glass transition-related physicochemical changes in foods. *Food Technology* 49(10):97–102.

Roos, Y.H., Karel, M., and Kokini, J.L. 1996. Glass transitions in low moisture and frozen foods: Effects on shelf-life and quality. *Food Technology* 50:95–108.

Salaman, R.N. 1940. The biology of the potato with reference to use as a wartime food. *Chem. Ind.* 59:735.

Scott, W.J. 1957. Water relations of food spoilage microorganisms. *Adv. Food Res.* 7:83–127.

Seow, C.C. 1988. *Food Preservation by Moisture Control*, p. 277. London/New York: Elsevier Applied Science.

Shapero, M., Nelson, D.A., and Labuza, T.P. 1978. Ethanol inhibition of *Staphylococcus aureus* at limited water activity. *Journal of Food Science* 43:1467–1469.

Short, P.M., Levesque, G.S., McAllister, D.R., Kanter, D.G., Rosado, J.E., and Popper, R.D. A systems analysis to define the types of rations needed to support amphibious operations. U.S. Army Natick Research Development and Engineering Center, Technical Report 86-071.

Singh, R.K., Lund, D.B., and Buelow, F.H. 1983. Storage stability of intermediate moisture apples: Kinetics of quality change. *Journal of Food Science* 48:939–944.

Singh, R.K., Lund, D.B., and Buelow, F.H. 1984. Computer simulation of storage stability in intermediate moisture apples. *Journal of Food Science* 49:759–764.

Sinskey, A.J. 1976. New developments in intermediate moisture foods: Humectants. In: *Intermediate Moisture Foods*, eds. R. Davies, G.G. Birch, and K.J. Parker, pp. 260–277. London: Applied Science.

Slade, L., and Levine, H. 1991. Beyond water activity: Recent advances based on an alternative approach to the assessment of food quality and safety. *Critical Reviews in Food Science and Nutrition* 30(2,3):115–360.

Sloan, A.E., and Labuza, T.P. 1975. Investigating alternative humectants for use in foods. *Food Product Development* 9(7):75–88.

Sloan, A.E., and Labuza, T.P. 1976. Prediction or water activity lowering ability of food humectants at high a_w. *Journal of Food Science* 41:532–535.

Sloan, A.E., Waletzko, P.T., and Labuza, T.P. 1976. Effect of order-of-mixing on a_w lowering ability of food humectants. *Journal of Food Science* 41:536–540.

Smith, M.C., Huber, C.S., and Heidelbaugh, N.D. 1971. Apollo 14 food system. *Aerospace Medicine* 42:1185.

Suzuki, S., Takei, H., Nemoto, Y., Matoba, K., Yachida, T., and Watanabe, N., eds. 1975. *The An Handbook*, p. 532. Tokyo: Korinshoin Publishing (in Japanese).

Takanobu, Higashiyama. 2002. Novel functions and applications of trehalose. *Pure Applied Chemistry* 74(7):1263–1269.

Taoukis, P.S., Breene, W.M., and Labuza, T.P. 1988. Intermediate-moisture foods. In: *Advances in Cereal Science and Technology, Vol. IX*, ed. Y. Pomeranz, pp. 91–128. St. Paul, MN: American Association of Cereal Chemists.

Taoukis, P., Labuza, T.P., and Saguy, I. 1997. Kinetics of food deterioration and shelf-life prediction. In: *The Handbook of Food Engineering Practice*, eds. K.J. Valentas, E. Rotstein, and R.P. Singh, pp. 361–403. New York: CRC Press.

Tapia de Daza, M., Villegas, Y., and Martinez, A. 1991. Minimal water activity for growth of Listeria monocytogenes as affected by solute and temperature. *International Journal of Food and Microbiology* 14:333–337.

Taub, I.A., Feeherry, F.E., Ross, E.W., Kustin, K., and Doona, C.J. 2003. A quasi-chemical kinetics model for the growth and death of *Staphylococcus aureus* in intermediate moisture bread. *Journal of Food Science* 68(8):2530–2537.

Taub, I.A., Ross, E.W., and Feeherry, F.E. 2000a. Model for predicting the growth and death of pathogenic organisms. In: Proceedings of the Third International Conference on Predictive Modeling in Foods, eds. J.F.M. Van Impe and K. Gernaerts, September 12-15, 2000. Leuven, Belgium: Katholieke University.

Taub, I.A., Ross, E.W., and Feeherry, F.E. 2000b. Model for predicting the growth and death of pathogenic bacteria. In: Proceedings of the 22nd Army Science Conference, December 12-15, 2000, Norfolk, VA.

Taylor, S.L., and Bush, R.K. 1986. Sulfites as food ingredients: A scientific status summary by the IFT Expert Panel on Food Safety and Nutrition. *Food Technology* 40(6):47–52.

Teng, T.T., and Seow, C.C. 1981. A comparative study of methods for prediction of water activity of multicomponent aqueous solutions. *Journal of Food Technology* 18:409–419.

Tilbury, R.H. 1976. The microbial stability of intermediate moisture foods with respect to yeasts. In: *Intermediate Moisture Foods*, eds. R. Davies, G.G. Birch, and K.J. Parker, pp. 138–165. London: Applied Science.

Torreggiani, D., Forni, E., Erba, M.L., and Longoni, F. 1995. Functional properties of pepper osmodehydrated in hydrolyzed cheese whey permeate with or without sorbitol. *Food Research International* 28(2):161–166.

Torreggiani, D. 1995. Technological aspects of osmotic dehydration in foods. In: *Food Preservation by Moisture Control*, eds. G.V. Barbosa-Caanovas and J. Welti-Chanes, pp. 281-304. New York: CRC Press.

Torres, J.A., and Karel, M. 1985. Microbial stabilization of intermediate moisture food surfaces. III. Effects of surface preservative concentration and surface pH control on microbial stability of an intermediate moisture cheese analog. *Journal of Food Processing and Preservation* 9:107–119.

Torres, J.A., Motoki, M., and Karel, M. 1985a. Microbial stabilization of intermediate moisture food surfaces. I. Control of surface preservative concentration. *Journal of Food Processing and Preservation* 9:75–92.

Torres, J.A., Bouzas, J.O., and Karel, M. 1985b. Microbial stabilization of intermediate moisture food surfaces. II. Control of surface pH. *Journal of Food Processing and Preservation* 9:93–106.

Troller, J.A. 1980. Influence of water activity on microorganisms in foods. *Food Technology* 34(5):76–80.

Troller, J.A. 1985. Effects of a_w and pH on growth and survival of *Staphylococcus aureus*. In: *Properties of Water in Foods,* eds. D. Simatos and J.L. Multon, pp. 247–257. NATO Advanced Science Institutes Series E, Applied Sciences, No. 90. Dordrecht, the Netherlands: Martinus Nijhoff.

Troller, J.A., and Christian, J.H.B. 1978. *Water Activity and Food.* New York.

Vaamonde, G., Chirife, J., and Scarmato, G. 1984 Inhibition of *Staphylococcus aureus* in laboratory media adjusted with polyethylene glycols. *Journal of Food Science* 49:296–297.

Van Den Berg, C., and Bruin, S. 1981. Water activity and its estimation in food systems: Theoretical aspects. In: *Water Activity: Influences on Food Quality,* eds. L. Rockland and G.F. Stewart, pp. 1–64. New York: Academic Press.

Walker, G.C., Tuomy, J.M., Stark, O.J., and Powers, E.M. 1979. Chemical and microbiological analysis of sliced precooked canned bacon. Technical Report, NATICK/TR-79/038. U.S. Army Natick Research and Development Command, Natick, MA 01760.

Warmbier, H.C., Schnickles, R.A., and Labuza, T.P. 1976. Effect of glycerol on non-enzymatic browning in a solid intermediate moisture model food system. *Journal of Food Science* 41:528–531.

Williams, J.C. 1976. Chemical and nonenzymic changes in intermediate moisture foods. In: *Intermediate Moisture Foods,* eds. R. Davies, G.G. Birch, and K.J. Parker, pp. 100–119. London: Applied Science.

Yang, A.P.P., and Taub, I.A. 1997. Preparation of intermediate moisture sandwich meats. Institute of Food Technologists Annual Meeting Technical Program Book of Abstracts, pp. 21–27.

12 Desorption Phenomena in Food Dehydration Processes

Gustavo V. Barbosa-Cánovas and Pablo Juliano

Some Definitions

A number of common food preservation processes lower water activity (a_w) to eliminate deterioration from microorganisms, enzyme activity, and potentially adverse chemical reactions that occur between food components. Reducing the amount of "free" or "unbound" water minimizes the undesirable chemical and structural changes that can take place in foods during storage. The processes used to reduce the amount of "free" water in foods include dehydration techniques such as concentrating and drying. *Dehydration* is defined as the removal of water due to simultaneous heat and mass transfer during heating under controlled conditions of temperature, humidity, and air flow (Okos et al. 1992, Barbosa-Cánovas and Vega-Mercado 1996).

The *water activity* in a substance is a measure of the amount of "free" or "unbound" water (i.e., active water) present in a food sample. Quantitatively, a_w is equal to the vapor pressure water in the material divided by the vapor pressure of pure water (at the same temperature and external pressure). A portion of the total water content available in foods is strongly bound to (or associated with) specific sites on chemicals that comprise the food. These sites may include hydroxyl groups of polysaccharides, carbonyl and amino groups of proteins, and other polar sites that may interact with water through hydrogen bonding, ion–dipole bonds, or other strong chemical bonds. In addition to strongly bound water molecules, other water molecules in foods are "bound" less firmly but are still not available as a solvent for various water-soluble food components. The different types of water are shown schematically in Figure 12.1.

Water activity describes the continuum of energy states of water in a system. Because water is present in varying degrees of "free" and "bound" states, analytical methods that attempt to measure total moisture content at different conditions use different structural pathways for water removal in the food. In contrast, a_w relates to "unbound" water, being a more descriptive parameter of "free" water inside the food. Changes in a_w affect microbial survival and reproduction, as well as enzymatic and chemical reactions that occur in "unbound" water. Microbes cannot readily access the "bound" water because they would have to compete with chemical structures in the food itself.

Some properties used to characterize air–water vapor mixtures, such as absolute humidity, equilibrium relative humidity, dry and wet bulb temperature, and sorption isotherms, are basic to understanding how the concept of a_w can be used during dehydration.

Humidity is the driving force behind drying operations and is a measure of water vapor in the air due to "free" water in the food. The *absolute humidity,* or moisture content in the air (W), can be expressed as the ratio of the mass of water vapor (m_{water}) in a volume of air to the mass of dry air (m_{air}) in the same volume.

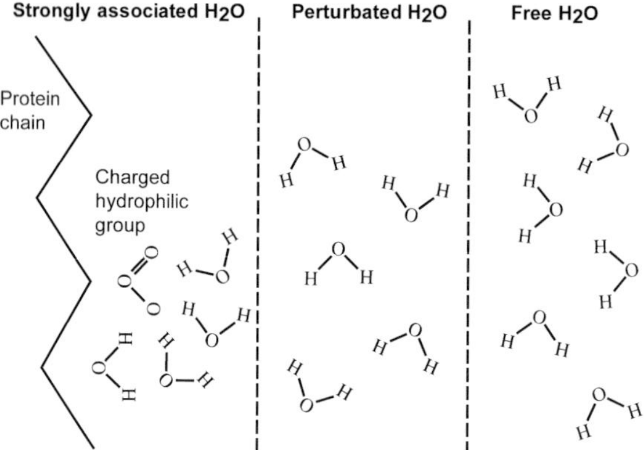

Figure 12.1 Strongly associated, perturbated, "free" water in a food matrix.

$$W = \frac{m_{water}}{m_{air}} = 0.622\frac{P_{water}}{P_{air}} \tag{12.1}$$

where P_{water} is the partial pressure of water vapor and P_{air} is the partial pressure of air. Multiplier 0.622 is derived from the application of the ideal gas law and the ratio of the molecular weight of water (18.01 g/mol) to air (28.96 g/mol). Absolute humidity is important in computing the amount of energy and water added or removed from a volume of air.

Relative humidity Φ (%) is a ratio of the mass of water vapor in a volume of air to the maximum mass of water vapor that a volume of air can hold at a given temperature and pressure.

$$\Phi = 100\frac{m_{water}}{m_{water-sat}} = 100\frac{P_{water}}{P_{water-sat}} \tag{12.2}$$

where m_{water} is the water mass in the mixture, $m_{water\text{-}sat}$ is the mass of water that saturates the air-water mixture at the same temperature, and P_{water} and $P_{water\text{-}sat}$ are respective partial pressures. Moisture in the air can be related to relative humidity in the following expression (Mujumdar and Menon 1995):

$$W = 0.622\frac{\Phi P_{water-sat}}{P_{water} + P_{air} - \Phi P_{water-sat}} \tag{12.3}$$

When a wet food is placed in a sealed enclosure and dry air is brought in contact with the food, equilibrium between the air and food material is eventually reached. After a suitable equilibration period, the air surrounding the surface of the food reaches a constant relative humidity, designated as equilibrium relative humidity (ERH). Equilibrium water vapor pressure at the surface of a food indicates how strongly the water is "bound." At a constant temperature, this equilibrium vapor pressure is represented by the a_w in the material in the following expression:

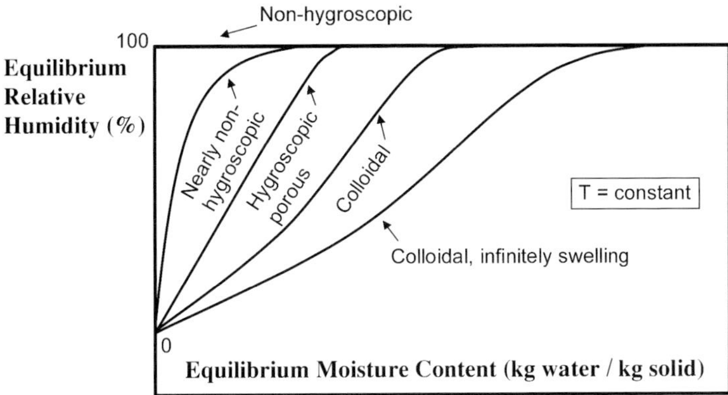

Figure 12.2 Equilibrium moisture curves from various types of solids (adapted from Mujumdar 1997).

$$ERH = 100\ a_w \tag{12.4}$$

Dry bulb and *wet bulb temperatures* are important in determining the amount of water removed from the food and absorbed by the air, as well as in evaluating energy exchanged in the form of heat during the drying process. Dry bulb temperature is directly measured with a dry temperature probe (thermometer) immersed in an air water mixture without any modification to the thermometer, whereas wet bulb temperature is measured with a wet temperature probe (e.g., a thermometer covered with a wet cloth). The wet bulb is at a lower temperature because latent energy (or enthalpy of vaporization) is being transferred from the surrounding air to water around the wet bulb, changing the liquid to vapor. The lower the initial moisture in the air, the longer it takes the wet probe to attain equilibrium, and the greater is the decrease in bulb temperature. Thus, the difference between dry and wet bulb temperature is related to the amount of moisture present in the air. Moreover, if a dry air current is applied to a moist food, the temperature at the surface of the food will equal the wet bulb temperature.

At completion of a dehydration process, the water vapor pressure in the food is in equilibrium with its surroundings (Heldman and Singh 1981) and reaches the corresponding *equilibrium moisture content*. As seen in Chapter 5, equilibrium relative humidity, or the a_W of different foods, can be related to the corresponding equilibrium moisture content in curves termed *sorption isotherms*. In these curves, the partial pressure of the water in the food is plotted against its water content at constant temperature. In order to facilitate the prediction of moisture content during desorption of different concentrated or dried foods, computerized methods are available (Labuza 2003). Figure 12.2 shows the different equilibrium moisture contents for different types of solids.

Mass and Heat Transfer in Food Dehydration

When a food is subjected to thermal drying, the surrounding environment transfers energy (mostly as heat) to the water in the food, generating its release as vapor through the surface. Inside the food, the water is transferred toward the food surface in the form of a slurry or vapor. External water transfer from the surface to the surrounding atmosphere occurs through a thin film of air in contact with the surface. The removal of water as vapor depends on the external conditions, such as air temperature, amount of vapor carried by the air, and air flow or amount of air passing over the food (Fellows 2000). Other factors such as area of exposed surface (microstructural shape of the solid) and pressure are also relevant (Mujumdar and Menon 1995). External drying conditions are especially important during the initial stages of drying when "free" moisture is being removed, provoking in certain cases considerable shrinkage. Excessive surface evaporation after initial "free" moisture has been removed sets up high-moisture gradients from the interior of the food to the surface, leading to varied energy and mass transfer mechanisms.

From an energy transfer point of view, there are two main classifications for drying processes: adiabatic and nonadiabatic (Heldman and Singh 1981). Adiabatic processes are those in which the heat of evaporation is supplied by sensible heat from drying gas, which carries the vapor away from the food. Some examples of adiabatic drying are particle drying (e.g., spray or fluidized bed drying) or bed drying (e.g., tray drying). Convective (or adiabatic) drying accounts for more than 90% of dehydrated food production, and most convective dryers work at temperatures below the boiling point of the feed mixture (Mujumdar 1997). Nonadiabatic drying occurs when radiation and conduction (provided through or by contact with a solid material) are the main forms of heat transfer. Vacuum drying (e.g., freeze-drying), purge drying (e.g., drum and pan drying), dielectric drying (e.g., microwave and radiofrequency drying), and infrared drying are among many subgroups of nonadiabatic processes. Most food engineering textbooks focus on adiabatic drying processes to describe heat transfer mechanisms.

Mechanisms of Mass and Heat Transfer

Dehydration of foods is accomplished by external and internal heat transfer, i.e., from the surroundings to food surface and from the surface to the interior. Simultaneously, temperature increase induces internal moisture transfer through the interior of the food to the surface. External mass transfer mainly occurs through surface water evaporation to the surroundings. Figure 12.3 shows a simplified diagram illustrating this complex phenomenon of heat and mass transfer during drying.

Heat transfer mechanisms

The heat transfer mechanisms, convection, conduction, and radiation participate to a different extent in the drying process depending on whether they occur internally or externally (Fito et al. 1998). External heat transfer can occur via any of the three transfer mechanisms, whether the process is adiabatic or nonadiabatic. In adiabatic drying, convection transfer occurs mainly when a hot air current flows externally over the food structure, whereas conduction transfer is more common inside the product due to an internal gradient of temperature (see Figure 12.3). When a phase change occurs in the food structure, convection transfer of heat can also occur in the material to a lesser extent, causing moisture migration to the surface. Turbulent moisture migration is accompanied by evaporation–condensation of "free" water and sorption–desorption of bonded water. In the case of microwave or radiofrequency drying, conduction occurs as a result of internal heat generation.

Air variables that control heat transfer mechanisms to different extents include velocity, absolute humidity, and temperature. Control of air velocity is crucial for heat transfer during adiabatic and convective processes (Geankopolis 1993) but less important for con-

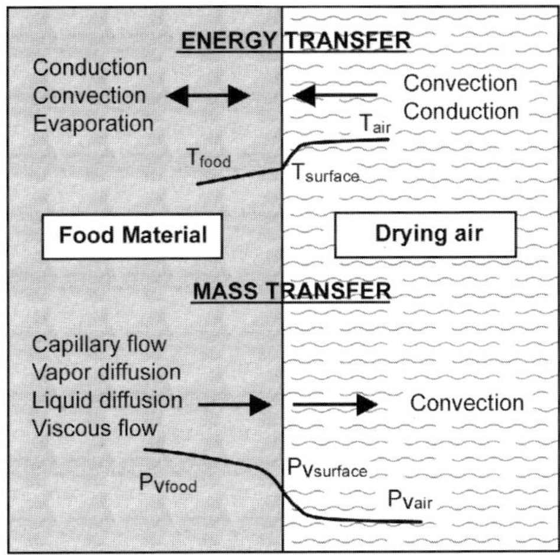

Figure 12.3 Schematic of food heat and mass transfer mechanism during adiabatic drying (adapted from Crapiste and Rotstein 1997). The drying air is at temperature T_{air} and vapor pressure P_{vair}; the food is at temperature T_{food} and vapor pressure P_{vfood}; the surface boundary region is at temperature $T_{surface}$ and vapor pressure $P_{vsurface}$.

duction and radiation. A higher air velocity causes greater bulk air movement and, hence, more efficient convective transfer. The absolute humidity of the air and temperature (relative to wet bulb temperature) determine the amount of heat transferred in a given timeframe.

Thickness of the solid being dried is another parameter to consider, because it can affect the drying time and the extent of heat penetration. Interface (or surface) heat and mass transfer are important during the early stages of drying, where external drying conditions (air velocity, temperature, and humidity) affect the drying rate (Saravacos and Kostaropoulos 2002, Ibarz and Barbosa-Cánovas 2003). When a piece of food is heated or cooled by a fluid, the extent of convective heat transfer at the surface of the food is given by the heat transfer coefficient h. The surface heat transfer coefficient h (W/m^2K) is determined by the amount of heat Q (W) transferred to a cross-sectional surface area A (m^2). The equation relating these terms is as follows:

$$Q = hA(T - T_w) \tag{12.5}$$

This equation includes the temperature difference between the heated medium T(K) and heated surface of the material T_w (K), which is close to the wet bulb temperature of the air.

Moisture transport mechanisms

Heat transfer from the surrounding environment causes internal water to be either transported to the surface of the product in liquid form and then evaporated, or evaporated internally at a liquid–vapor interface and then transported as vapor to the surface (Okos 1992). For instance, when a food encounters hot dry air, the moist air that forms a film surrounding the food surface is carried away by the moving air current. Water vapor removal establishes a pressure gradient from the interior of the food to the dry air, which provides the driving force behind water removal. The film is of variable thickness depending on the velocity of the air it receives. The thicker the film, the lower is the heat transfer coefficient and the lower the amount of water removed.

The transport mechanisms listed below and their combinations provide the moisture movement in food materials (Geankopolis 1993, Crapiste and Rotstein 1997):

- *Capillary flow*—due to gradients of capillary suction pressure, especially in powder beds and porous solids where material surface tension is involved
- *Liquid diffusion*—due to concentration gradients between depths in the food and at the surface; the most commonly considered mechanism for modeling drying processes
- *Vapor diffusion*—due to partial vapor–pressure gradients
- *Viscous flow*—due to total pressure gradients, caused by external pressure or high temperatures

Liquid steady capillary diffusion can be represented by the following mathematical expression:

$$J_W = -D \cdot \rho_s \cdot \frac{dx_w}{dZ} \tag{12.6}$$

where J_W (kg water/sec · m^2) is the mass transfer rate per unit mass and time, D (m^2/sec) is the mass diffusivity through capillaries, ρ_s is the apparent density of the dried food (including closed pores), Z (m) is the capillary distance in the direction of water flow, and x_W is the dry basis moisture content of the food (kg bulk water/kg dry solid). The minus sign is included because capillary diffusion occurs in the direction of lower moisture in the food. In convective drying, the evaporation rate X_m (kg bulk water/s), with respect to a cross-sectional surface area A (m^2), is proportional to the difference between the vapor pressure of water in the food, P_{wfood} (Pa), and vapor pressure of the water in air, P_{wair} (Pa), as indicated in the following simplified expression.

$$X_m = K_m \cdot A \cdot (P_{wfood} - P_{wair}) \tag{12.7}$$

The mass transfer coefficient K_m (kg/Pa · m^2·sec) depends on the air temperature, velocity, and viscosity. In addition to temperature, P_{wfood} depends on the moisture of the food at the surface and P_{wair} depends on the air relative humidity. The heat transfer coefficient and mass transfer coefficient can be interrelated through the density of the air ρ_{air} (kg/m^3).

The mass transfer flux J (kg/m^2 · sec) at the surface of the solid can be defined by using either the surface mass transfer coefficient h_m (kg/m^2 · sec) or the mass transfer coefficient k_c (m/sec) in the following two expressions:

$$J = h_m \cdot \Delta x = k_c \cdot \Delta C \tag{12.8}$$

where the expression containing h_m uses Δx, the difference in moisture content (kg water/kg air), and the expression with k_c is based on the driving force being the concentration difference, ΔC (kg/m^3). The two mass coefficients (see Equation 12.8) are interrelated by the density of the air (see Equation 12.9). For most air–moisture systems at normal dry-

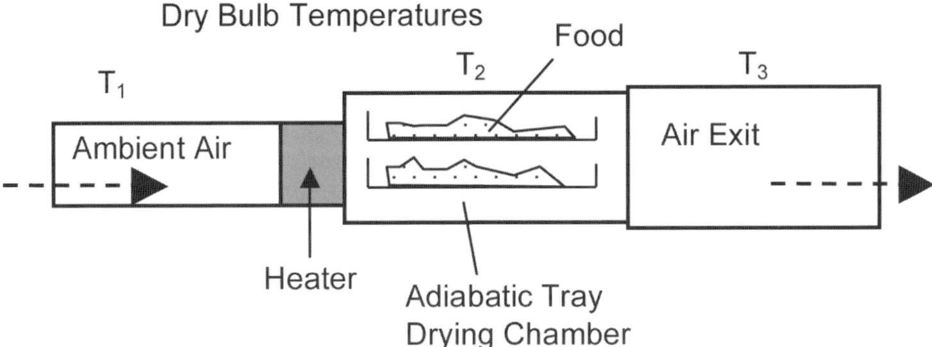

Figure 12.4 Tray dryer showing dry bulb temperatures T_1, T_2, and T_3 for air at chamber entrance, air in chamber, and air at chamber exit, respectively.

ing pressure (1 atm) and temperature (less than 100°C), the air density is approximately 1 kg/m^3 (Crapiste and Rotstein 1997).

$$h_m = k_c \cdot \rho_{air} \tag{12.9}$$

Other mechanisms like surface diffusion and flow due to shrinkage pressure or gravity forces may have a minor role in mass transfer (Crapiste and Rotstein 1997), for instance, when colloidal and fibrous materials such as vegetables and meats undergo shrinkage. During shrinkage, a hard layer develops at the surface, increasing resistance to the flow of liquid and vapor to the surface (Geankopolis 1993).

Psychrometrics

The properties of air–water vapor mixtures are important in calculating air-drying processes (convective and atmospheric drying), which can be calculated using equations from material and energy balances (ASAE 1982) or obtained from psychrometric charts or computer software, such as PsyCalc (Vega-Mercado et al. 2001). A psychrometric chart is a graphical representation of hygroscopic properties in air (Saravacos and Kostaropulos 2002). Air absolute humidity (W, kg water/kg dry air) is plotted along the vertical axis versus the dry bulb temperature at the horizontal axis. The curved upper axis represents the air–water saturation point and gives the wet bulb temperature, whereas the curves parallel to the saturation curve represent lines of constant relative humidity (% RH). The specific enthalpy (H, kJ/kg dry air) and the specific volume (V, m^3 mixture/kg dry air) can also be determined from the chart. In particular, the enthalpy represents the heat content of moist air and is a relative measure of external energy and the flow of work per unit mass (Okos et al. 1992).

Parameters of a typical convective adiabatic tray dryer containing a certain moist food (without air recirculation) are shown in Figure 12.4. Air inside the drying chamber is initially at dry bulb temperature T_1 and is mixed with penetrating hot air (at fixed temperature, relative humidity, and flow rate), increasing its temperature at constant moisture content (Y_1) up to dry bulb temperature T_2, and increasing in enthalpy by ΔH (kJ/kg dry air). At this point, moisture is adiabatically removed from the food (enthalpy H_2 is constant) at

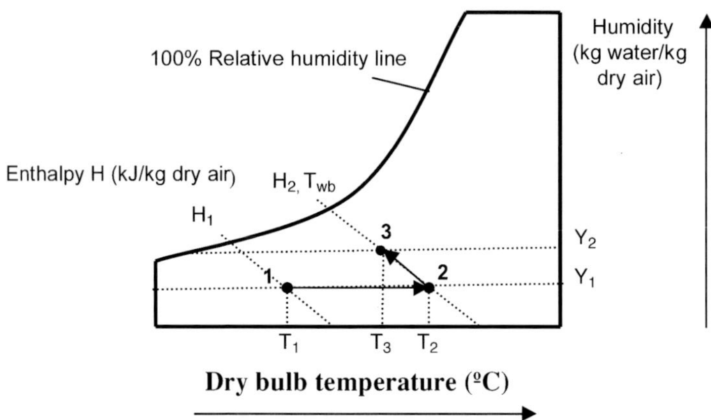

Figure 12.5 A two-step process showing the paths of air in an adiabatic dryer; step 1 to 2 represents heating at constant humidity Y_1; step 2 to 3 represents adiabatic cooling at constant enthalpy H_2 and wet bulb temperature T_{wb} (adapted from Saravacos and Kostaropoulos 2002).

a constant wet bulb temperature T_2, whereas moisture in the air is increased up to Y_2. After a moisture uptake of ΔY (kg water/kg dry air), the air exits the chamber at dry bulb temperature T_3, which is lower than T_2.

Figure 12.5 shows how the process is represented in a psychrometric chart, where relative humidities and enthalpies of various mixtures can be calculated with only two properties of air. Note that in the final stage of the drying process, water vapor pressure in the food achieves equilibrium with its surroundings and moisture content reaches a constant value. Thus, the air relative humidity in surface surroundings will be close to the product's equilibrium relative humidity (i.e., once mass transfer has stopped, a_w of the dried food, expressed as ERH, will match the relative humidity of the atmosphere at the surface).

Mass and Energy Balances

Overall, mass and energy balances usually give important information on dryer performance and several variables of dryer design. A simplified representation of a drying process where air is recirculated into the drying chamber is sketched in Figure 12.6. Food-related variables are mass flow rate F (kg dry food/hr) and moisture X (kg water/kg dry solid), while drying air variables are air flow rate G (kg dry air/hr) and moisture Y (kg vapor/kg dry air). Assuming F and G remain constant throughout the drying process, the total amount of water W transferred per hour from the food to the air, based on a steady water mass transfer of water, can be obtained from the following overall moisture balance equations.

Food $$W = F\,(X_{inlet} - X_{outlet}) \tag{12.10}$$

Air $$W = G\,(\mathrm{Y}_{\mathrm{outlet}} - \mathrm{X}_{\mathrm{inlet}}) \tag{12.11}$$

Note that the analysis above also applies to co-current systems and to the dryer's heater. The net amount of energy Q transferred from the air to the food can be determined from overall energy balances. Thermal properties are needed such as the specific heat of solids,

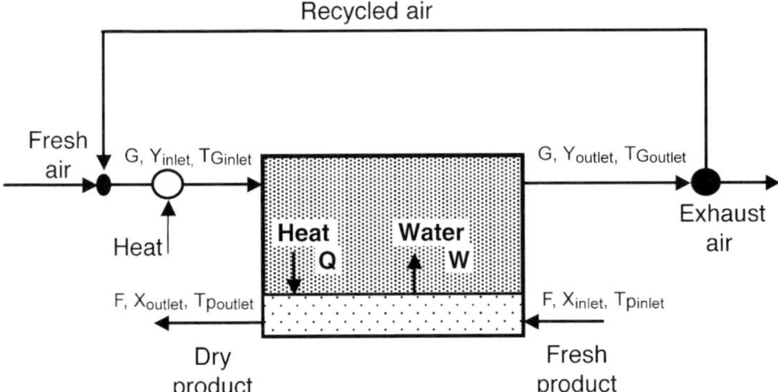

Figure 12.6 Simplified representation of a convective drying system, where F(kg dry food/h) and G (kg dry air/h) are the mass flow rate and air flow rate, respectively; X_{inlet} and X_{outlet} represent the moisture content (kg water/kg dry solid) in the solid, and Y_{inlet} and Y_{outlet} represent the moisture (kg vapor/kg dry air) in the air; TG_{inlet}, TG_{outlet}, Tp_{inlet}, and Tp_{outlet} are the temperature of gas (air) and the food product at the entrance and exit of the system, respectively.

liquid water, dry air, and water vapor (C_p, C_w, C_g, and C_v, respectively), as well as the heat of vaporization ΔH_v (approximate total heat of desorption) at reference temperature T_r (usually taken as 0°C).

Food $$Q = FC_{P_{outlet}}(T_{P_{outlet}} - T_{P_{inlet}}) + W[\Delta H_v - C_w(T_{P_{inlet}} - T_r)] \quad (12.12)$$

Air $$Q = G(C_g + C_v Y_{inlet})(T_{G_{inlet}} - T_{G_{outlet}}) - WC_v(T_{G_{outlet}} - T_r) \quad (12.13)$$

Heat of vaporization and specific heat values (25°C), $C_w = 4.197$ kJ/kg · K and $C_g = 1.007$ kJ/kg · K, are tabulated, and values $\Delta H_v = 2.443$ kJ/kg and $C_v = 1.876$ kJ/kg · K can be used. Heat of vaporization and specific heat values are tabulated, and values $\Delta H_v = 2.443$ kJ/kg (25°C), $C_w = 4.197$ kJ/kg · K, $C_g = 1.007$ kJ/kg · K, and $C_v = 1.876$ kJ/kg · K can be used (Barbosa-Cánovas and Vega-Mercado 1996). C_p depends on the composition of the solid. For industrial calculations, additional heat losses must be accounted for because the systems are not completely adiabatic. Heat and mass balances can help estimate, for example, the outlet gas temperature and moisture content gained by the air. They can also be used to measure the amount of energy involved in the process for a specific rate of mass being dried.

Dehydration Kinetics

The drying rate at a specific air temperature and humidity is useful for estimating the dryer size selection, as well as drying time. Because data for drying rates in food materials are scarce, drying rates must be obtained experimentally from the slope of the "free" moisture content curves as a function of drying time (Okos et al. 1992). Initially, data can be obtained as the weight of the product over time (see Figure 12.7), which can then be converted into drying rate.

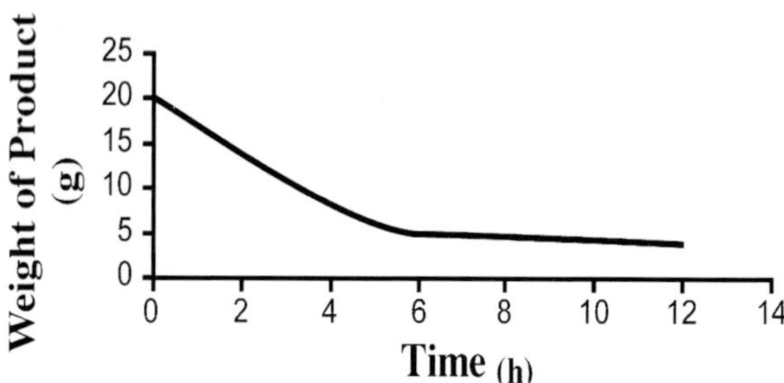

Figure 12.7 Weight variation of product during drying period.

Dehydration Curves

The dry-basis moisture content X_t of the product at time t is defined as the ratio between the amount of water in the food and the amount of dry solids (kg total water/kg dry solids). In order to express a characteristic drying curve, the food must be evaluated with respect to the equilibrium moisture content in a given drying system. The so-called free moisture content, X, is defined as the difference between the moisture content X_t at drying time t and the equilibrium moisture content X_{eq}.

$$X = X_t - X_{eq} \tag{12.14}$$

Drying rates can be obtained from a drying curve by plotting the free moisture content against drying time (see Figure 12.7). This plot is then converted to a drying rate curve by calculating the derivative of the free moisture over time. The experimental procedure for obtaining the plot involves placing a layer of food material in a laboratory scale dryer. The dryer is operated under controlled temperature, air velocity, and humidity, and the weight and temperature of sample are monitored by appropriate instrumentation as a function of time.

The drying rate, R, is proportional to change in free moisture with time dX/dt. Then, the value dX/dt for each point on the drying curve can be obtained (see Figure 12.8) from the value of the tangent to the curve at each point. Hence, the drying rate can be expressed as (Geankoplis 1993):

$$R = -\frac{L}{A}\frac{dX}{dt} \tag{12.15}$$

where L is the amount (kg) of dry solid used and A is the area of the drying surface. When plotting drying rate versus free moisture content at different times, the following curve is obtained (see Figure 12.9).

The typical drying cycle consists of three stages: heating the food to drying temperature, evaporation of moisture from product surface, and falling of drying rate. At time zero, the moisture content of a solid is given by point A if the solid is at a cold temperature and by A′ if hot. After the drying temperature is reached (point B), the drying curve is divided into two distinct portions: (1) *constant rate period* (line BC) where "unbound" water is removed and (2) *falling rate period* (line CD). Initially, temperature increase in the product

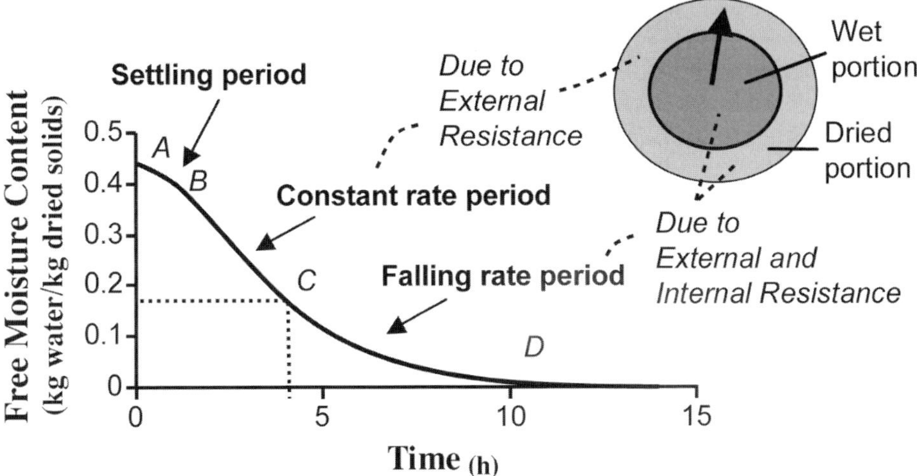

Figure 12.8 Free (dry-base) moisture content changes during drying. Constant rate (line BC) and falling rate (CD) periods are distinguished on the curve and associated with the external drying resistance of former period and with both the internal and external resistances of latter. Line AB represents a transition state before the stationary constant rate in which most "free" water is released.

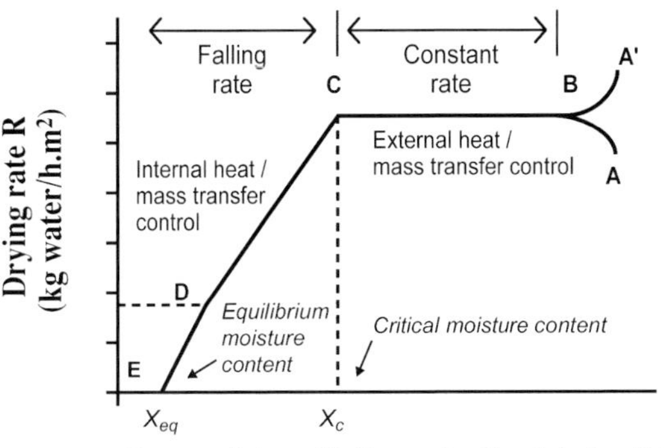

Figure 12.9 Drying rate curve versus free moisture content. X_c represents critical free moisture content, X_{eq} the equilibrium moisture content. Lines AB, BC, and CD represent the transition rate, the constant rate, and the falling rate, respectively (adapted from Geankopolis 1993; Ibarz and Barbosa-Cánovas 2003).

causes the surface of product to be very wet, with an a_W value close to one. Water evaporates at the surface as if no solid is present, and where the rate of evaporation is not dependent on the solid. In porous solids, the water removed from the surface is compensated by the flow of water from the interior of the solid. The constant rate period continues, while the evaporated water on the surface is compensated by the internal water. During the constant rate period, the temperature at the surface of the product corresponds approximately to wet bulb temperature (Okos 1992, Geankoplis 1993). Point B represents the condition of equilibrium temperature at the product surface, and point C represents the critical moisture content when the surface of the solid is no longer wet.

The falling rate period starts when the drying rate begins to decrease and the a_w at the surface becomes smaller than 1. From C to D, the wet areas on the surface become completely dry and the drying rate is governed by the internal flow of water and water vapor. When the surface is dry and the solid is warmer (point D), the evaporation sites will continue moving toward the center of the solid. The heat required to remove moisture is transferred through the solid to the evaporation surface. From D to E, water removed from the center of the solid moves to the surface as vapor. When the drying rate slowly approaches 0, it also approaches the equilibrium moisture content, i.e., the food achieves equilibrium with the drying air (Fellows 2000).

Different interpretations, other than the theory described above, have been made regarding drying curves. Some authors in the food engineering field maintain that no marked differences exist distinguishing between the first and second falling rate periods (Ibarz and Barbosa-Cánovas 2003). Moreover, Saravacos and Kostaropoulos (2002) reported that many food materials experience a very short (or negligible) constant rate period and can even dry completely in the falling rate period, during which moisture diffusion to the surface is the controlling mass transfer mechanism.

Although the amount of water removed in the falling rate period is relatively small, drying time can be considerably longer than during constant-rate period (Geankopolis 1993). In general, increased air velocity and air temperature can increase the drying rate, while increased humidity and solid thickness decrease it (Okos et al. 1992). The size of the food piece selected is relevant in both the constant and falling rate periods. In the constant rate period, smaller pieces provide larger surface areas for evaporation, whereas in the falling rate period, smaller pieces provide shorter distances for moisture transit through the food to the surface.

Constant Rate Period

During the constant rate period, the rate of moisture removal from the product is limited only by the rate at which mass evaporates at the surface of product, depending on external conditions (air flow, temperature, and humidity). The surface of the material remains saturated with water during the drying process. This drying rate will continue as long as the migration of moisture to the evaporative surface is more rapid than evaporation from this surface. The rate R at which evaporation occurs at the surface depends on both the heat and mass transfer coefficients; these are interrelated as shown in Equation 12.16 (Heldman and Singh 1981, Okos et al. 1992, Barbosa-Cánovas and Vega-Mercado 1996).

$$R = \frac{dX}{dt} = \frac{hA_q(T_{air} - T_{food})}{A_w \Delta H_v} = k_m(X_{food} - X_{air}) \tag{12.16}$$

where h is the heat transfer coefficient at the surface, k_m is the mass transfer coefficient for moisture transfer to the surrounding air, T_{air} is the air dry bulb temperature, T_{food} is the wet bulb temperature (or temperature at the surface of the food), X_{air} and X_{food} are the air absolute humidity and moisture content of food at the food surface, respectively, corresponding to dry and wet bulb temperatures, and ΔH_v is the latent heat of vaporization. A_q is the surface area through which heat is received and is not necessarily equal to the evaporation area A_w (Fito et al. 1998). Water evaporation rates may be limited by low convective heat transfer rates from the air to the surface or by low mass transfer rates of moisture moving from the water surface to the air. In some situations, heat transfer (e.g., in solar or infrared drying) and heat-transfer coefficient will be radiation dependent.

Saravacos and Kostaropoulos (2002) described an empirical first-order drying rate equation for thin-layer drying as a function of free moisture content X.

$$\frac{dX}{dt} = -KX \tag{12.17}$$

where K (sec^{-1}) is the drying constant. Variable separation and integration of this expression between X_0 and X_t and corresponding times 0 and t, and substitution by free moisture X using Equation 12.14 provides a semilog expression.

$$\log\left(\frac{X_o - X_t}{X_o - X_{eq}}\right) = Kt \tag{12.18}$$

Drying constant K depends on the food and air dry bulb temperature. However, K of several food materials is known to change during the drying process since significant changes in the physical structure of the materials will alter mass transfer mechanisms within the materials; this will be seen as changes in the slope of the semilog drying curve.

Falling Rate Periods

Following the period where drying occurs at a constant rate, drying rate R decreases with time when the moisture content of the food is lower than the critical moisture content X_C and the rate of internal mass transfer to the material surface typically controls the process (Okos 1992, Ibarz and Barbosa-Cánovas 2003). In this case, the surface vapor pressure of the solid decreases as moisture content drops. Movement of water in the solid can be explained by simultaneous occurrence of more than one mass transfer mechanisms (listed above), including diffusion of liquid due to concentration gradients, diffusion of water vapor due to partial vapor pressure, movement of liquid due to capillary forces, gravity forces, and/or surface diffusion (Barbosa-Cánovas and Vega-Mercado 1996).

The movement of water through the food depends on its porous structure, as well as on the different interactions of water within the food matrix; note that this differs from the external heat and mass transfer coefficients, which depend only on external flow conditions. Internal mass transfer resistance during drying is related to the resistance that the structure exerts on the water to flow outward to the surface, whereas mass transfer resistance is related to the transfer of water from the surface to the heating medium. The importance of internal versus external mass transfer resistance can be inferred from drying studies conducted on samples of different sizes (i.e., varying slab thickness or sphere and cylinder radii). Development of related theories can be found in the literature (Chen and Johnson 1969, Bruin and Luyben 1980, Fortes and Okos 1980, Geankoplis 1993, Barbosa-Cánovas and Vega-Mercado 1996).

The main mechanism in the drying of solids is water *diffusion*. For example, in solids of fine structure, the capillaries, pores, and smaller voids within fill with water vapor. Water in a vapor or liquid phase diffuses until reaching the surface. At the surface, evapo-

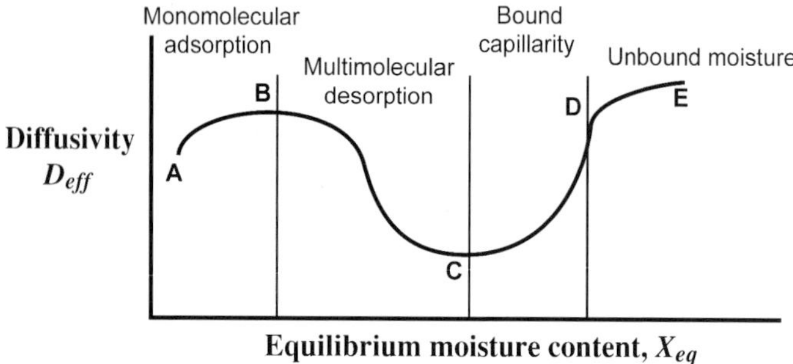

Figure 12.10 Dependence of diffusivity with equilibrium moisture content (adapted from Barbosa-Cánovas and Vega-Mercado 1996). The figure shows regions of adsorption, desorption, and capillarity according to the equilibrium moisture content.

ration is completed and vapor mixes with the air layer surrounding the food. Application of Fick's law to a unidimensional system can be expressed as:

$$\frac{\partial X}{\partial t} = D_{eff} \frac{\partial^2 X}{\partial x^2} \tag{12.19}$$

In this equation, X is the moisture content of the product, t is the time, x is the dimension in transfer direction, and D_{eff} (m^2/sec) is the diffusion coefficient. Depending on type of geometry considered (e.g., slab, cylinder, or sphere), the solution to Fick's differential equation can take different forms, some of which are used to determine D_{eff} experimentally.

A rough estimate of D_{eff} can be obtained from the drying constant K in Equation 12.18, after plotting $[(X_t - X_{eq})/(X_0 - X_{eq})]$ versus time on a semilog scale. If the drying slope K obtained is constant, assuming symmetrical geometries such as plate, slab, or sphere, K will directly depend on the diffusivity D_{eff} as well as on the thickness L of a food slab or plate, or eventually, on the radius r of a sphere (Saravacos and Maroulis 2001).

Plate or slab $\quad K - \pi^2 D / L^2 \quad$ (12.20)
Sphere $\quad K - \pi^2 D / r^2 \quad$ (12.21)

The relationship between diffusivity D_{eff} and moisture is illustrated in Figure 12.10. Region A–B represents the monomolecular adsorption at the surface of the solid and involves movement of water in vapor phase by diffusion. Region B–C involves multimolecular desorption, in which moisture begins to move in the liquid phase. Microcapillarity plays an important role in region C–D, where moisture easily migrates from water-filled pores. In region D–E, moisture exerts its maximum vapor pressure and the migration of moisture is due essentially to capillarity.

The dependence of D_{eff} on temperature is commonly represented in an Arrhenius-type equation (Barbosa-Cánovas and Vega-Mercado 1996, Crapiste and Rotstein 1997).

$$D_{eff} = D_0(X)\exp\left[\frac{\Delta E_d(X)}{RT}\right] \tag{12.22}$$

where $D_0(X)$ is a moisture-dependent diffusivity term, $\Delta E_d(X)$ (cal/mol) is the activation energy for the diffusion process, X is the fixed moisture content, *R* is the gas constant (1.987 cal/mol • K), and T(K) is the temperature of the food system. Extensive information on the diffusivity of different food products at different temperatures and moisture contents is available from research presented by Sablani et al. (2000).

Diffusion theory using Fick's law does not account for shrinkage and case hardening, nor does it consider other factors, such as pore space, tortuosity, composition of polymers, and transition state of foods (Barbosa-Cánovas and Vega-Mercado 1996). Non-Fickian behavior has been attributed to moisture profiles inside the food held at temperatures near the glass transition temperature. However, good prediction with Fick's law only occurs when the food is in the rubbery state or in the glassy region, that is, when the food is far from its glass transition temperature (Singh et al. 2003). Glass transition temperatures are useful for determining the rate at which a food matrix deforms or collapses during drying (Achanta and Okos 2000).

Drying Resistance

Apart from moisture diffusion theory, other mechanisms can supplement the description of internal and external resistance in water removal during drying. Among these mechanisms, capillary theory (Fortes and Okos 1980), evaporation–condensation theory (Bruin and Luyben 1980), and regular regimen theory (Kerhoff 1994) can be considered.

A liquid can flow through pores and over the food surface by way of *capillary movement* due to molecular attraction between the liquid and the solid. Capillary flow depends on the pressure difference between flowing water and air–water mixtures at the liquid–gas interface present in a capillary. Flow through capillaries also depends on the permeability of the material, which depends on the pore distribution as well as on the surface tension and liquid dynamic viscosity. These material properties directly influence the overall internal resistance to water release occurring during drying.

During capillary movement or diffusion vapor condensation within the product, water may condense near the surface. *Evaporation–condensation* theory assumes that the rate of condensation is equal to the rate of evaporation at the surface of the food, so that liquid does not accumulate in the pores near the surface. In this case, heat and mass diffuse simultaneously and continuously through the complex porous polymeric networks within the solid. Because there is condensation through the pores, internal resistance to liquid flow becomes significant. Hence, tortuosity of the diffusion path must be considered for heat and mass balances.

The *Phillip and De Vries theory* (1957) works with principles similar to those in the evaporation–condensation theory but considers condensation at the pore entrance and evaporation at the exit of the pore. By considering the chemical potential of water, vapor flux in a capillary has been described as a function of a_w or relative humidity within the product. Barbosa-Cánovas and Vega-Mercado (1996) presented a transport phenomena model for the gradient of water vapor density through a capillary as a function of temperature, moisture content, chemical potential, and a_w.

Another approach to characterizing drying mechanisms in terms of resistance is to use the Biot number, which is widely used to compare the resistance to mass transfer at the surface of a solid to the resistance inside the solid (Welti-Chanes et al. 2003). The Biot number (Bi) expression includes surface mass transfer coefficient k_c (m/sec) with the solid

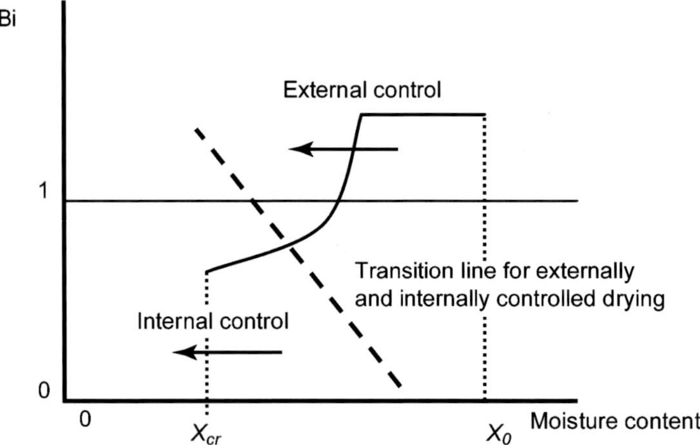

Figure 12.11 Biot numbers and moisture content, defining zones of drying internal and external resistance (adapted from Barbosa-Cánovas and Vega-Mercado 1996). The angled dotted line indicates the transition for externally and internally controlled drying.

thickness d (m), and the effective diffusion coefficient D_{eff} (m^2/sec) so that $Bi = k_c d/D_{eff}$. Alternatively, Kerhof (1994) proposed a modified Biot number to determine whether the drying process is controlled externally or internally. The proposed Biot number is:

$$Bi = \frac{k_{g,eff} L \rho^{s}_{wg}}{D_{eff} \rho_s x^{*}_{cr}} \qquad (12.23)$$

where $k_{g,eff}$ is the effective mass transfer coefficient (m/sec) for the change in moisture content of air at the food surface, L is the dimension of the product particle (m), ρ^{s}_{wg} is the saturation vapor concentration (kg/m^3), D_{eff} is the effective diffusion coefficient (m^2/sec), ρ_s is the concentration of solids (kg/m^3), and x_{cr}^{*} is the critical moisture content of product. Figure 12.11 diagrams controlling resistances for different Biot numbers and moisture content combinations.

Drying is controlled by external resistance when both the heat and mass transfer resistance are located on the air side at the surface of the product (Crapiste and Rotstein 1997). Generally, externally controlled drying occurs with high moisture contents in food products. During external control, the product is maintained around the wet bulb temperature and therefore the drying temperature can be increased. In this process, the surface water content drops off rapidly and the drying rate is determined by the internal rate of diffusion. When the Biot number is less than 0.1, the temperature and moisture profiles inside the product are flat. In foods like pasta, fruits, and vegetables where a hard crust is formed on the surface that retards evaporation, external control can be maintained by decreasing air velocity or increasing relative humidity of drying air. External control can avoid excessive temperature increase inside the product and subsequent product deterioration, especially at the surface of the food.

External control has been modeled in cellular material by Crapiste et al. (1988). They developed a model based on a_w in a diffusion-like equation that includes the shrinkage of the solid, which is indicated by gradients in the dimensionless coordinate z expressed with the differential operator ∇.

$$\frac{\partial X}{\partial t} = \nabla_z (D_{eff} \nabla_z X) \tag{12.24}$$

When drying is internally controlled, the a_w at the surface decreases according to the sorption isotherm (Barbosa-Cánovas and Vega-Mercado 1996) or the vapor pressure of the water within the food structure. In a drying process controlled by internal resistance, thermal conductivity limits the rate of heat transfer. In the inner parts of the food, far from the surface, the higher vapor pressure of liquid present leads to more vapor formation (Cook and DuMont 1991). Drying stops when the vapor pressure of water placed in the inner parts of the food equals the vapor pressure of water at outer regions inside the food. Vapor pressure of strongly "bound" water in the solid is lower than that of "free" liquid, and the equilibrium partial pressure of the water vapor is correspondingly lower. The following equation is an analytical solution for Equation 12.24, which includes the concept of internal and external resistance, and is valid for one-dimensional transport at constant diffusivities in a slab (Crapiste and Rotstein 1997).

$$\frac{X_t - X_{eq}}{X_0 - X_{eq}} = \sum_{i=0} \frac{2Bi^2}{\lambda_i^2 + B_i^2 + Bi} \exp\left[\frac{\lambda_i^2 D_{eff} t}{L_p^2}\right] \tag{12.25}$$

where L_p is the half-thickness of the slab and λ_i are solutions of $\tan(\lambda_i) = Bi/\lambda_i$.

Resistance can also occur when the outer zone of the material is drier, being moist in the vapor phase only (receding front model). However, at any given depth in the food, moisture can be in either liquid or mixed form (Crapiste and Rodstein 1997).

Sorption Isotherms and Drying Process Characterization

Drying processes, in general, involve applying heat to a product for water removal, which results in reduced moisture content and a_w within the product (Okos et al. 1992). When air is brought in contact with a wet food material, equilibrium between the air and food material must eventually be reached. The plot of moisture content at equilibrium versus relative humidity at a certain temperature, or sorption isotherm, is often used to illustrate the effect of temperature and the extent of dehydration, or to study water desorption thermodynamics. Equilibrium relative humidity, as already mentioned, is the a_w (in percent) corresponding to the partial vapor pressure at a space boundary surrounding the material. It is important to remember that sorption isotherms involve an equilibrium situation, which does not account for dynamic (time-dependent) drying processes in which convective, conductive, or radiant heat is eventually exchanged, or water mass is simultaneously removed.

A moisture sorption isotherm prepared by adsorption (starting from the dry state) will not necessarily be the same as an isotherm prepared by desorption (from the wet state). Desorption isotherms are useful in the investigation of drying processes, while adsorption isotherms are used for observation of hygroscopic products. A desorption isotherm usually lies above the adsorption isotherm in the midrange of a_w values. This phenomenon is called hysteresis (see Chapter 5). Moisture sorption isotherms are sigmoidal in shape for most foods, although foods that contain large amounts of sugar or small soluble molecules have a J-shaped isotherm curve.

Equilibrium relative humidity (or a_w) is influenced by hydrogen-bond formation and the presence of dissolved solutes. Differences between electrolytic and nonelectrolytic solutions, as well as the amount of positively and negatively charged ions, also play a role in desorption curves. Water desorption (rate and amount) may be influenced by surface area and porosity of food materials. The pore size distribution or the presence of thermolabile capillaries in the food matrix may influence the rate and extent of water unbinding.

Temperature Effect on Sorption

Moisture content at equilibrium usually decreases with increase in temperature. Moreover, for constant moisture contents, the amount of "bound" water is also more easily desorbed with rising temperatures, wherein mobility of the water molecules as well as dynamic equilibriums between the vapor and adsorbed phases are affected (Kaminski and Kudra 2000). If the amount of water within the product is constant (line A in Figure 12.12), an increase in temperature increases the a_w (arrow labeled T in Figure 12.12) due to the greater amount of energy available for vaporization, as predicted by the Clausius-Clapeyron relationship (Barbosa-Cánovas and Vega-Mercado 1996, Okos et al. 1992).

$$\left[\frac{\delta \ln a_w}{\partial(1/T)}\right]_X = -\frac{E_b}{R} \tag{12.26}$$

where E_b is the binding energy of water in cal/mol, R is the gas constant (1.987 cal/mol • K), and X is a fixed moisture content. The binding energy is defined as the difference between the heat of adsorption of water and its enthalpy of condensation. The Clausius-Clapeyron equation can be integrated at a constant moisture level to achieve:

$$\ln\frac{a_T}{a_r} = \frac{E_b}{R}\left(\frac{1}{T_r} - \frac{1}{T}\right) \tag{12.27}$$

where a_r and a_T are the water activities at reference temperature T_r and drying temperature T, respectively, at a fixed moisture content. As temperature increases, the isotherm shifts down. The downward shift graphically results in an increase in a_w values along a line of fixed moisture constant (line A) for increasing temperatures; the a_w values at X_{eq} are higher on the isotherms for higher temperatures T_b and T_c than at lower temperature T_a.

Considering a drying temperature T_c and an initial a_w a_{wi}, the ultimate moisture value in the product will be X_{eq} or the equilibrium moisture content. The X_{eq} is a function of both the temperature of the drying process and the a_w (see Figure 12.12).

$$X_{eq} = f\ (T,\ a_{wi}) \tag{12.28}$$

Temperature can also affect the shape of desorption isotherms, because decreased temperature decreases the equilibrium moisture content X_{eq} due to increased heat of vaporization, even as equilibrium relative humidity decreases during desorption. In effect, the hysteresis decreases for increased temperature. Given a certain temperature, a lower relative humidity will need a higher enthalpy of vaporization to remove water molecules sorbed by hydrophilic and charged sites in proteins and polysaccharides (Okos 1992).

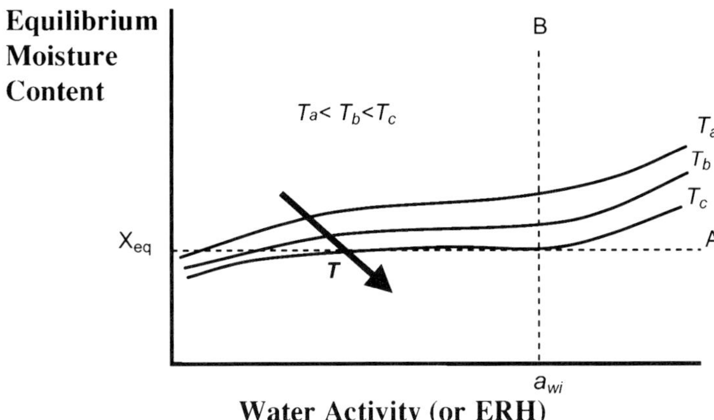

Figure 12.12 Effect of temperature on water sorption (adapted from Barbosa-Cánovas and Vega-Mercado 1996). The downward shift in equilibrium moisture content due to reduced temperature is represented across line A of constant equilibrium moisture content and on line B of constant water activity.

Temperature effects vary depending on the material tested. Foods high in protein, or starch, and high-molecular-weight polymers attain higher moisture contents at equilibrium than do those high in soluble solids, crystalline salts, and sugars. Thus, desorption isotherms vary drastically from one food product to another (see Figure 12.13).

In general, changes in desorption curves will depend on the initial state of the food (rubbery/amorphous versus crystalline) and the transitions taking place during desorption,

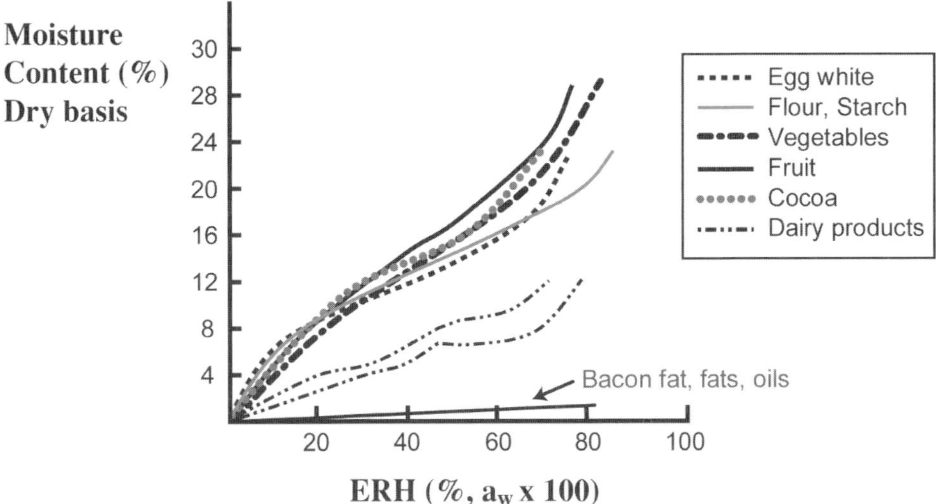

Figure 12.13 Typical equilibrium moisture contents of food materials (adapted from Mujumdar and Menon 1995).

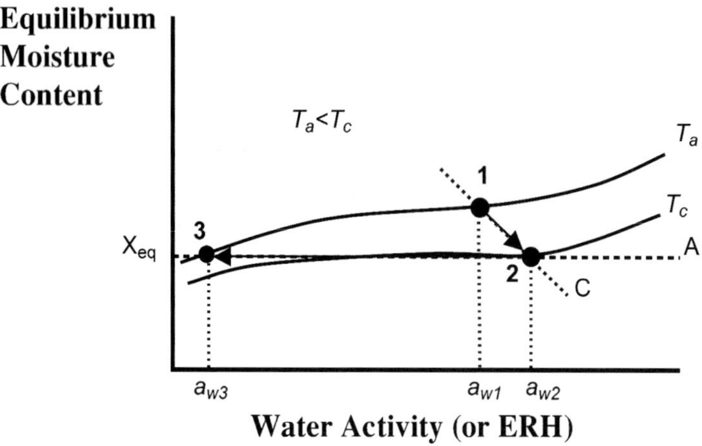

Figure 12.14 Drying process characterization with desorption isotherms (adapted from Barbosa-Cánovas and Vega-Mercado 1996). Line C represents the increase in water activity from a_{w1} to a_{w2} (or ERH) due to moisture release during drying, until constant equilibrium moisture content is reached (line A). A decrease in water activity to a_{w3} at constant equilibrium moisture content is represented by line A.

as well as the speed of desorption, which also depends on the temperature of the solid matrix during drying. Solubility of solutes can sometimes be a controlling factor, but control usually depends on the state of the matrix or food polymer to which the water is strongly "bound." Labuza (1984) showed that the effect of temperature on a_w is negligible in high moisture foods, but in intermediate- and low-moisture foods, a 10°C change in temperature can result in a small percentage change in a_w.

Drying Process Representation and Energy Changes in Sorption Isotherms

The drying process of a product with initial moisture content X_i and a_w a_{w1} at temperature T_a is represented by pathway 1-2-3 in Figure 12.14. The process has two main steps, which, depending on the type of product or drying mechanisms participating, may correspond to the constant rate and the falling rate period described earlier:

1. The moisture content will drop while drying and a_w will remain constant or increase toward a_{w2}, depending on how fast the water vapor is removed from the product.
2. The drying will continue until the equilibrium moisture content X_{eq} is reached at a_{w2}. Water activity and moisture content are defined by the air dry bulb temperature, or by the sorption isotherm corresponding to the drying temperature T_c. The cool-down stage results in a product with certain moisture content X_{eq}, and a reduced a_{w3}, based upon the Clausius-Clapeyron relationship, as represented by line A in Figure 12.14.

A desorption isotherm can be divided into three regions wherein bound/free water possesses enthalpies of vaporization of increasing magnitude as water activities become lower, depending on the binding energy. Water with considerably high enthalpy of vaporization is the monolayer water (see Figure 12.15, region A), where the first water molecules are

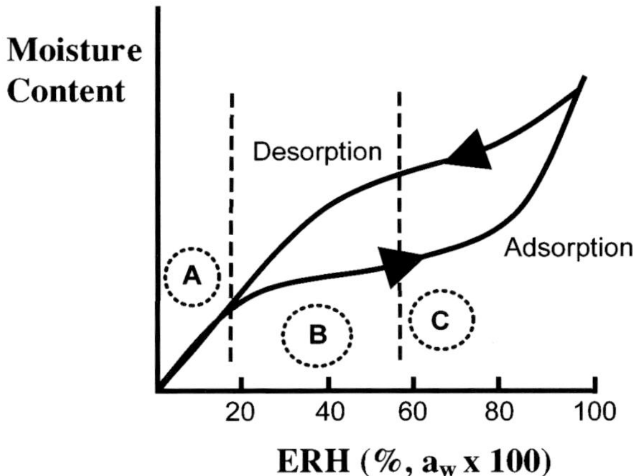

Figure 12.15 Labels A, B, and C represent regions in adsorption and desorption isotherms where water has different binding energy (from Fortes and Okos 1980).

sorbed to the hydrophilic, charged, and polar groups of food components, including H-bonded water and hydrophobic hydration water (Kinsella and Fox 1986). Based on the Brunauer-Emmett-Teller (BET) theory (Karel 1975), the heat of sorption should be constant up to monolayer coverage and then suddenly decrease.

When water molecules bind less firmly than in the first monolayer region (see Figure 12.15, region B), the vaporization enthalpy is lower due to transition from strongly "bound" water to capillary water. Generally, the heat of sorption for low moisture contents is higher than the BET theory indicates, and also falls more gradually, indicating the gradual change from monolayer to capillary water (see Figure 12.16). The main reason for increase in water content at high values of a_w is probably capillary condensation.

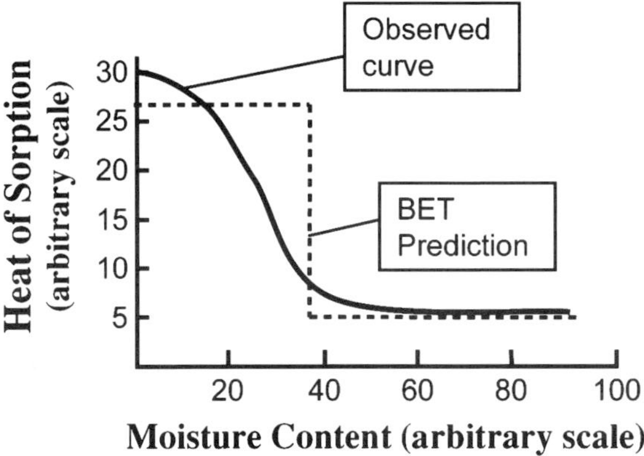

Figure 12.16 Comparison of experimental heat of sorption versus predicted BET theory as a function of moisture content (Labuza 1968).

The free energy change ΔF depends on the heat of water sorption in a food. It can be related to a_w and temperature T in the following expression, where *R* is the ideal gas constant.

$$\Delta F = RT \ln(a_w) \tag{12.29}$$

Rockland and Nishi (1980) and other authors (Barbosa-Cánovas and Vega-Mercado 1996) discussed the relationship of water content in food and the free energy change ΔF (or work done by the attractive forces) during the sorption, where *n* and *k* are empirical constants.

$$\log(-\Delta F) = nX_{eq} + k \tag{12.30}$$

Dehydration Methods for Water Activity Control

Water activity in a food can be decreased or controlled by several water removal methods, and each of these methods can be adapted to achieve certain conditions of temperature, product size, air relative humidity, and system pressure at which the food can maintain desirable and specified sensory and nutritional properties. The main reasons for drying are increased shelf-life, a desired physical form, color, flavor, or texture, and reduced volume and weight for transportation. The nature of certain products does not allow direct exposure to hot air, so heating is accomplished by means of heat exchangers that prevent direct contact between the product and the heating medium (Cook and DuMont 1991). Wet feedstock may be in the form of liquid (slurry, suspension, or solution), solid (particulate, sheet-like, pelletized and extruded forms), or paste. Furthermore, foods can be categorized into liquid solutions and gels, capillary-porous rigid materials, and capillary-porous colloidal materials.

Various applications for over 200 dryer types have been found in the food industry (Mujumdar 1997), but only 20 or so types are commonly used in practice. Figure 12.17 shows the basic types of dryers based on the mode of heat transfer input mentioned earlier. Dryers using hot air to heat solid surfaces are the most commonly found dryers in food processing, and these are classified under atmospheric drying operations. Among these, tray dryers, spray driers, and fluid beds bear mention as examples for continuous or batch processing. Freeze-drying permits a different approach for water vapor removal through the sublimation of ice.

Foods are dried commercially from two states: either from a natural state (e.g., vegetables, fruits, milk, spices, grains) or after processing (e.g., food powders like instant coffee, whey, soup mixes, nondairy creamers).

Conventional Air Drying

As mentioned, conventional air drying is based on exposing food products to a direct stream of hot air, thus heating the product to remove water vapor. Drying parameters include a 50° to 95°C air temperature, air relative humidity 10% to 40%, and air velocities 1 to 4 m/sec. Some examples of air drying systems include rotary dryers for solids in granular form, continuous tunnel and belt dryers, and steam dryers where superheated steam is used (Barbosa-Cánovas and Vega-Mercado 1996). Table 12.1 provides some examples of

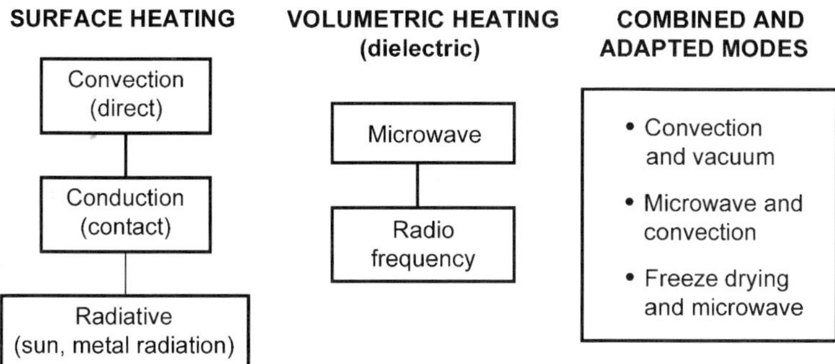

Figure 12.17 Basic dryer types classified according to the mode of heat transfer.

Table 12.1 Typical Drying Conditions and Dryers Found in the Fruit Processing Industry.

Product	Temperature	Dryer type
Apple slices, rings, cubes	60–93°C	Multiple-stage conveyor
Banana slices	60–82°C	Track, tunnel
Cherries	60–98°C	Multiple-stage conveyor, truck tunnel dryers
Coconut flakes	65–93°C	Two-stage conveyor
Cranberries	82–93°C	Multiple-stage conveyor
Nuts (peanuts, almonds, pistachios)	65–232°C	Single-stage multiple zone dryer
Pineapple rings, cubes	65–93°C	Truck and tunnel dryers

Source: Barbosa-Cánovas, G.V., and Vega-Mercado, H. 1996. *Dehydration of Foods*. New York: Chapman and Hall.

products, dryer types, and processing temperatures commonly found in the fruit processing industry.

Spray Drying

Spray drying is also a convective drying technique, but it deserves special attention due to its short residence time in the range of a few seconds (5 to 100). It involves both food particle formation and drying. The feed is transformed from a fluid state into droplets and then into dried particles by spraying the droplets continuously into a hot drying medium. The feed can be in a solution, suspension, or paste form, while the final product will be in the form of powders, agglomerates, or granules. The process can be summarized as consisting of three phases: spray formation, drying, and air-powder separation.

The typical drying process (see Figure 12.18) begins by pumping a liquid feed into an atomizer, which breaks up the feed, transforming it into a spray of fine droplets, and then ejecting it into the drying chamber. The spray is then in contact with and suspended by a heated drying medium (usually air), which allows the moisture to evaporate and the droplets to be transformed into dry particles that are almost identical in shape and size. Finally, the dried particles are separated from the drying medium and collected as the final product. The drying medium is cleaned by means of a cyclone or a scrubber and released to the environment or, in some cases, recirculated to the drying chamber.

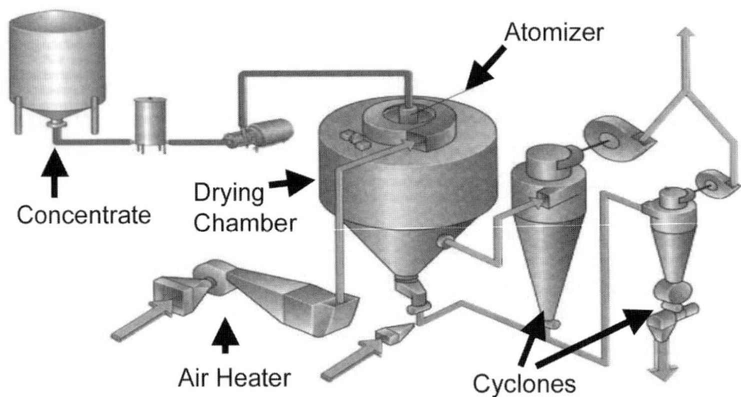

Figure 12.18 Typical co-current spray drying system (adapted from http://www.egr.msu.edu/~steffe/handbook/fig173.html).

Table 12.2 Typical Foods Produced Using Spray-Drying Process.

Dairy products	Spray-dried milk, whey, cheese, buttermilk, sodium caseinate, coffee creamer, butter, ice cream mixes, baby foods
Egg products	Spray-dried whole egg, egg yolk, and albumen
Protein sources	Dried meat purees, soy powders, isolated soy protein, whey proteins
Fruits and vegetables	Fruit and vegetable pulps, pastes and juices (e.g., tomato, banana, and citrus)
Extracts	Coffee and tea extracts
Condiments	Garlic and pimento

The shape and characteristics of spray-dried products depend on whether the drying air temperature is above or below the boiling point of the droplets (Vega Mercado et al. 2001). Table 12.2 shows some typical spray-dried products from different food sources.

Vacuum Dehydration and Freeze Drying

A vacuum drying system uses low pressure, which reduces the water boiling point and therefore prevents the use of high temperatures. The two drying variables usually controlled to monitor food quality are pressure (3 to 60 kPa) and air temperature (50° to 100°C).

Freeze drying (or lyophilization) is a more sophisticated method in which the system's pressure is kept so low that the water in the material remains frozen and sublimes away from the solids. Being a cold process and due to the absence of air, freeze-drying is especially useful for drying heat-sensitive foods, such as coffee and tea extracts, in which the volatile compounds responsible for aroma and flavor are easily lost during ordinary drying operations. Freezing of the product, sublimation of ice, and removal of water vapor are the three main steps in a freeze-drying operation. A schematic diagram of a common freeze-drying system is shown in Figure 12.19. The frozen material is loaded on top of heated plates, which provide the energy for ice sublimation and "bound" water desorption.

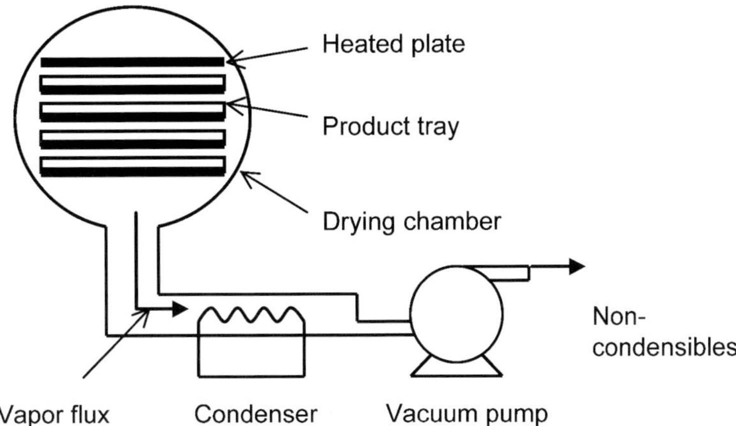

Figure 12.19. A common freeze-drying system (adapted from Liapis and Marchello 1984).

Heat transfer occurs primarily by conduction from the heated plates. Convection from the remaining air inside the drying chamber to the exposed surfaces and radiation (e.g., from microwaves), if used, will transfer heat to a lesser extent.

Initially, the drying rate is high due to little resistance to both the heat and mass flux. However, buildup of a resistive layer on the frozen material slows down the rate as drying proceeds (Vega-Mercado et al. 2001); the dry layer surrounding the product serves as insulation material, affecting heat conduction of the ice front. As thickness of the dry outer layer is increased, mass transfer is also reduced due to reduction in the diffusion process.

The advantage of freeze-drying is that the ice structure in the product during the drying process minimizes shrinkage of the product and thus promotes a rapid and nearly complete rehydration (Bruin and Luyben 1980). Therefore, freeze-drying is also very useful for drying certain food materials (e.g., vegetables, fruits, and meats) used for instant foods, such as soup mixes, where good reconstitution properties are highly desirable.

Osmotic Dehydration

The concentration of food products by means of product immersion in a hypertonic solution (i.e., sugar, salt, sorbitol, or glycerol) is known as osmotic dehydration (Lazarides 2001). Osmotic dehydration systems mainly consist of a storage tank, where an osmotic solution is prepared, and a pump to control flow rate at the processing tank. The product is placed at the processing tank where the osmotic solution is pumped at a constant rate (Vega-Mercado et al. 2003). In some systems, water removal can also be aided with the use of vacuum through cryostabilization technology (Martînez-Monzó et al. 2001).

Osmotic dehydration results in a product with an intermediate moisture content. This is due to low water flow from the product into the concentrated solution and solute transfer from the solution into the product. As a pretreatment technique, osmotic dehydration is claimed to achieve, simultaneously, dewatering and direct formulation of the product through impregnation plus leaching. The main drying parameters include solute concentration, process temperature, sample size, speed of agitation, and time of immersion.

Dielectric Drying Methods

Unlike convective-only drying, microwave drying is based on the application of an electromagnetic field to polarize, rotate, and provoke the lateral movement of water molecules in the system, which librates heat. Furthermore, some portions in proteins and carbohydrates can also line up in a microwave electric field (Giese 1992). Heat dissipated when exposed to an alternating electromagnetic field is a function of the frequency dielectric loss factor of the material ϵ'' as expressed in following.

$$E_G = 5.56 \cdot 10^{-4} \cdot f \varepsilon'' E^2 \tag{12.31}$$

where E_G is the conversion of microwave energy into thermal energy (W/cm^3), f is the frequency (GHz), and E is the electric field in (V/cm). Dielectric heating using microwaves has reduced the drying time for many agricultural products, and it is usually used as a pretreatment before conventional drying. Microwaves penetrate the core of a food, not just the surface, as in conventional air drying. Therefore, drying mechanisms function differently because heat randomly develops in different areas of the food, and evaporation, diffusion, and capillary movement of water are hard to predict.

There are two main frequencies available for microwave drying: 915 MHz and 2.45 GHz (for household ovens). Microwave energy is generally combined with hot air drying to process diced fruits (Giese 1992, Feng and Tang 1998, Feng et al. 1999) and with the use of vacuum (Vega-Mercado et al. 2003) to improve product quality and lower processing times. Microwave drying is currently used for freshly extruded pasta, potato chips, juice concentrates, and bread crumbs, among other foods (Giese 1992).

Radiofrequency is another type of dielectric method used for drying. The concept is similar to microwave drying, in that radiofrequency uses electromagnetic energy to heat the product, but the frequency spectrum is from 30 to 300 MHz, which is lower than that for microwaving. Microwave drying and radiofrequency technology affect materials differently and equipment varies. Radiofrequency energy mainly acts through the electrical conductivity of the material and ionic species in foods (e.g., dissolved salts), and provides deeper penetration than microwaves.

Final Remarks

Water activity can be used to characterize the dehydration process. Mass and heat transfer mechanisms for water removal are both process and product dependent. A desorption isotherm, as implied in its design protocol, will always use moisture content at equilibrium conditions, which should not be confused with the moisture content obtained at dynamic convective drying stages. During convective drying, either surface water evaporates into the air or liquid/vapor diffuses inside the food. As a result, heat and mass transfer properties of biological materials vary with concentration and temperature. Such limitations must be taken into account when modeling dehydration processes. Furthermore, food shrinkage during drying will affect drying rate determination due to changes in the surface area. Novel processes such as vacuum, osmotic, and dielectric drying offer very different water extraction pathways, where the porous structure of a food can be maintained avoiding shrinkage. Thus, products obtained using these processes will show different water binding sites with different sorption energies.

References

Achanta, S., and Okos, M.R. 2000. Quality changes during drying of food polymers. In: *Drying Technology in Agriculture and Food Sciences,* ed. A.S. Mujumdar, pp. 133–148. New York: Elsevier Science Publishers.

American Society of Agricultural Engineers (ASAE). 1982. *Agricultural Engineers Yearbook.* St. Joseph, MI: ASAE.

Barbosa-Cánovas, G.V., and Vega-Mercado, H. 1996. *Dehydration of Foods.* New York: Chapman & Hall.

Bruin, S., and Luyben, K. 1980. Drying of food materials. In: *Advances in Drying,* ed. A.S. Mujumdar. New York: Hemisphere Publishing.

Chen, C.S., and Johnson, W.H. 1969. Kinetics of moisture movement in hygroscopic materials. I. Theoretical consideration of drying phenomena. *Transactions of the ASAE* 12:109–113.

Cook, E.M., and DuMont, H.D. 1991. *Process Drying Practice,* p. 1. New York: McGraw-Hill.

Crapiste, G.H., and Rotstein, E. 1997. Design and performance evaluation of dryers. In: *Handbook of Food Engineering Practice,* eds. K.J. Valentas, E. Rotstein, and R.P. Singh, p. 125. Boca Raton, FL: CRC Press.

Crapiste, G.H., Whitaker, S., and Rotstein, E. 1988. Drying of cellular material. I. A mass transfer theory. II. Experimental and numerical results. *Chemical Engineering Science* 43(11):2919–2936.

Fellows, P.J. 2000. Dehydration. In: *Food Processing Technology: Principles and Practice,* pp. 311–340. Boca Raton, FL: CRC Press.

Feng, H., and Tang, J. 1998. Microwave finish drying of diced apples in a spouted bed. *Journal of Food Science* 63(4):679–683.

Feng, H., Tang, J., Mattinson, D.S., and Fellman, J.K. 1999. Microwave and spouted bed drying of frozen blueberries: The effect of drying and pretreatment methods on physical properties and retention of flavor volatiles. *Journal of Food Processing and Preservation* 23:463–479.

Fito, P., Andrés, A.M., Albors, A.M., and Barat, J.M. 1998. *Deshidratación de Productos Agrîcolas: Secado por Aire Caliente.* Valencia, Spain: Universidad Politécnica de Valencia.

Fortes, M., and Okos, M.R. 1980. Drying theories: Their bases and limitations as applied to foods. *Advances in Drying* 119–150.

Geankoplis, C.J. 1993. Drying of process materials. In: *Transport Processes and Unit Operations,* 3rd Edition, Chapter 9, pp. 520–583. Englewood Cliffs, NJ: Prentice-Hall.

Giese, J. 1992. Advances in microwave processing. *Food Technology,* Special Report 9:118–123.

Heldman, D.R., and Singh, R.P. 1981. Food dehydration. In: *Food Process Engineering,* p. 261. New York: Van Nostrand Reinhold.

Ibarz, A., and Barbosa-Cánovas, G.V. 2003. Dehydration. In: *Unit Operations in Food Engineering,* p. 573. Boca Raton, FL: CRC Press.

Kaminski, W., and Kudra, T. 2000. Equilibrium relations of foods and biomaterials. In: *Drying Technology in Agriculture and Food Sciences,* ed. A.S. Mujumdar, pp. 1–34. Enfield, NH: Science Publishers.

Karel, M. 1975. Water activity and food preservation. In: *Principle of Food Science. Part II: Physical Principles of Food Preservation,* eds. M. Karel, O.R. Fennema, and D.B. Lund. New York: Marcel Dekker.

Kerhof, P.J. 1994. The role of theoretical and mathematical modeling and scale-up. *Drying Technology* 12:1–46.

Kinsella, J.E., and Fox, P.F. 1986. Water sorption by proteins: Milk and whey proteins. *Critical Reviews in Food Science and Nutrition* 24(2):91–139.

Labuza, T.P. 1968. Sorption phenomena in foods. *Food Technology* 22(3):263–266.

Labuza, T.P. 1984. *Moisture Sorption: Practical Aspects of Isotherm Measurement and Use.* St. Paul, MN: American Association of Cereal Chemists.

Labuza, T.P. 2003. Water activity programs. Available at http://fscn.che.umn.edu/Ted_Labuza/Pages_Folder/aw.html.

Lazarides, H.N. 2001. Reasons and possibilities to control solids uptake during osmotic treatment of fruits and vegetables. In: *Osmotic Dehydration and Vacuum Impregnation. Applications in Food Industries,* eds. P. Fito, A. Chiralt, J.M. Barat, W.E.L. Spiess, and D. Behsnilian, pp. 33–41. Lancaster, PA: Technomic.

Liapis, A.I., and Marchello, J.M. 1984. Advances in the modeling and control of freeze-drying. In: *Advances in Drying,* ed. A.S. Mujumdar, pp. 217–244. Washington, DC: Hemisphere Publishing.

Martînez-Monzó, J., Martînez-Navarrete, N., Chiralt, A., and Fito, P. 2001. Combined vacuum impregnation-osmotic dehydration in fruit cryoprotection. In: *Osmotic Dehydration and Vacuum Impregnation. Applications in Food Industries,* eds. P. Fito, A. Chiralt, J.M. Barat, W.E.L. Spiess, and D. Behsnilian, pp. 61–78. Lancaster, PA: Technomic.

Mujumdar, A.S. 1997. Drying fundamentals. In: *Industrial Drying of Foods,* ed. C.G.J. Baker, pp. 7–30. New York: Chapman and Hall.

Mujumdar, A.S., and Menon, A. 1995. Drying of solids: Principles, classification, and selection of dryers. In: *Handbook of Industrial Drying,* ed. A.S. Mujumdar, pp. 1–40. New York: Marcel Dekker.

Okos, M.R., Narsimhan, G., Singh, R.K., and Weitnauer, A.C. 1992. Food dehydration. In: *Handbook of Food Engineering,* eds. D.R. Heldman and D.B. Lund, pp. 437–562. New York: Marcel Dekker.

Phillip, J.R., and De Vries, D.A. 1957. Moisture movement in porous materials under temperature gradient. *Trans. Am. Geophys. Union* 38:222–232.

Rockland, L.B., and Nishi, S.K. 1980. Influence of water activity on food product quality and stability. *Food Technology* 4:42–50.

Sablani, S., Rahman, S., and Al-Habsi, N. 2000. Moisture diffusivity in foods: An overview. In: *Drying Technology in Agriculture and Food Sciences,* ed. A.S. Mujumdar, pp. 35–60. Enfield, NH: Science Publishers.

Saravacos, G.D., and Kostaropoulos, A.E. 2002. Food dehydration equipment. In: *Handbook of Food Processing Equipment*, p. 331. New York: Kluwer Academic/Plenum Publishers.

Saravacos, G.D., and Maroulis, Z.B. 2001. *Transport Properties of Foods.* New York: Marcel Dekker.

Singh, P.S., Cushman, J.H., and Maier, E.D. 2003. Multiscale fluid transport theory for swelling biopolymers. *Chemical Engineering Science* 58:2409–2419.

Vega-Mercado, H., Góngora-Nieto, M.M., and Barbosa-Cánovas, G.V. 2001. Advances in dehydration of foods. *Journal of Food Engineering* 49:271–289.

Welti-Chanes, J., Mújica-Paz, H., Valdez-Fragoso, A., and Leon-Cruz, J. 2003. Fundamentals of mass transport. In: *Transport Phenomena in Food Processing,* eds. J. Welti-Chanes, J.F. Vélez-Ruiz, and G.V. Barbosa-Cánovas, pp. 3–24. Boca Raton, FL: CRC Press.

13 Applications of Water Activity Management in the Food Industry

Jorge Welti-Chanes, Emmy Pérez, José Angel Guerrero-Beltrán, Stella M. Alzamora, and Fidel Vergara-Balderas

The influence of a food's water content on perishability has been known since ancient times. During the period between 15,000 and 10,000 BC, our ancestors began to preserve foods by drying. They dried excess fish and meat in the wind and sun, and later did the same with excess fruits (Ray 1992). Around 8,000 BC, many innovations in preservation techniques were introduced along with a more settled pattern of life. To establish a steady supply of food, grains and fruits were stabilized naturally by drying, and excess meat and fish were preserved by smoking and dry-salting. In 1795, Masson and Challet applied an artificial drying technique to vegetables in a hot air room; later in the twentieth century, innovations included artificial drying of liquids by drum or spray drying, and freeze-drying.

These methods of preserving foods by drying and salting were developed over several millennia and were considered empirical and more an art than actual science. Considerable interest in the influence of water activity (a_w) on food product quality and stability began in the early 1950s (Scott 1953, 1957). The relationship between total moisture content and product stability was promoted through empirical and inconsistent observations. On the contrary, measured values of a_w generally correlate well with the potential for growth and metabolic activity; therefore a_w has been considered a good indicator of water availability for microbial activity (Lenovich 1987, Chirife 1995, Christian 2000).

As a result, in recent years the a_w principle has been included in various government regulations, suggesting that a_w (rather than total moisture content) controls the growth, death, survival, sporulation, and toxin production of diverse microorganisms. Of all the factors affecting microbial growth, death, and survival in foods, the influence of a_w on vegetative microorganisms and spores has been the most intensively studied by food microbiologists (Alzamora et al. 2003). On the other hand, in the 1980s and 1990s, in addition to the concepts of moisture and a_w, the glass transition concept and its quantification through the term "glass transition temperature" (T_g) was introduced to consider water mobility in foods and its relation to stability (Roos 1995).

At present, it clearly appears that the concepts of moisture, a_w, and T_g must be considered in the design of new food products and in transformation or preservation processes. The understanding of these concepts and their relationships to foods and preservation processes plays a critical role in achieving quality, stability, nutritional, and economical objectives. Therefore, the use of the a_w concept in the food industry can help in managing ingredients and intermediate and final food products, to increase stability and quality.

Water Activity and Industrial Importance

Water plays a key role in the stability and preservation of foods. It has been observed that several foods with the same water content differ significantly in perishability. Evaluation and control of a_w help to establish certain food characteristics: nutritional, textural, microbiological, and sensorial. In addition, a_w is a critical parameter addressed in food regulations such as the U.S. Food and Drug Administration (21 CFR and Potentially Hazardous Foods definition), Hazard Analysis and Critical Control Points (HACCP) systems, and American National Standards Institute (ANSI)/NSF75 norm.

Annual medical expenses, productivity losses, and cost of premature deaths of people related to pathogenic microorganisms in foods are estimated at around US$6.9 billion (Crutchfield and Roberts 2000). Many foods, cosmetics, and drug products are returned due to microbial contamination resulting from trades between countries. A quality assurance program including a_w determination, as well as other types of analyses, could be useful to ensure trade of safe, good-quality foods.

It is important to recognize that fresh, perishable foods have high moisture contents, as well as high a_w values, usually higher than 0.98, whereas low-moisture foods are more stable with a_w values less than 0.6, and in between, intermediate-moisture foods have diverse degrees of stability.

There is a close relationship between product stability and a_w, as several deteriorative reactions are a_w dependent. The a_w concept has been of paramount importance in making maps to predict what kinds of reactions will occur according to food composition. The general map was developed by Labuza (1968), who shows the relationship between a_w and reaction rates in several deteriorative situations, including microbial growth (see Chapter 5).

It would be ideal to have such a map for every food system; such a map would be very useful for practical applications in the food industry. It is well known, as shown in the general map, that most chemical reactions and microbial growth rates decrease with reduction in a_w. However, oxidation of unsaturated lipids behaves differently with an increased reaction rate when a_w is reduced, due to the catalytic activity of metal ions when water is removed from the neighborhood of these ions. Also, the oxidation of unsaturated lipids activity increases at high a_w values due to mobility of reactants.

Specific Water Activity Applications in the Food Industry

Knowledge of interaction of water in foods led to the rediscovery and optimization of traditional preservation techniques and to a renewed interest in foods that are shelf-stable using a_w control. This applies to dehydrated, intermediate, and high-moisture traditional foods with inherent hurdles and also to novel products, especially high-a_w foods for which the hurdles are intelligently selected and intentionally applied. Traditional low- and intermediate-moisture foods can be regarded as the oldest foods preserved by humans. However, in the quest for quality, the importance of considering the combined action of decreased a_w with other preservation factors as a way to develop new improved foodstuffs advanced almost simultaneously with modern food processing, such that, currently, consumers seek fresh-like characteristics in many processed products. The food industry responded to these demands by producing so-called minimally processed foods, which has resulted in a widespread industry that is receiving a good deal of attention.

Controlling a_w in food design is being used in many ways to cover quality and safety needs:

- In various stages of food distribution chain, during storage, processing, and/or packaging as a "back-up" hurdle in existing minimally processed products with short shelf-life

to diminish microbial pathogenic risk and/or increase shelf-life (i.e., by a slight reduction in a_w in addition to refrigeration)
- Traditionally, for obtaining long–shelf-life products (low- and intermediate-moisture foods); actual trends in these applications are to obtain very high sensory quality products by using the advances in knowledge of water sorption phenomena, a_w prediction, deleterious physicochemical reactions, and polymer science, as well as more controlled and/or sophisticated drying techniques
- As a preservative factor (together with other emerging and/or traditional preservative factors) to obtain high moisture novel foods by hurdle techniques (Alzamora et al. 2003)

In industrialized countries, with readily available energy and infrastructure and widespread use of refrigeration, the control of a_w has been mainly applied to develop a great variety of mild thermally processed, chill- and frozen-distributed foods. Applications include fermented meats (sausages, raw hams) and shelf-stable mild heated meats (ready-to-eat, fresh-like); *sous vide* and cook-chill dishes, healthy foods (low-fat and/or low-salt and functional foods); and foods processed by emerging techniques (e.g., hydrostatic high pressure) (Leistner and Gould 2002).

On the contrary, in many developing countries, refrigeration is expensive and not always available. Thus, the emphasis is on an a_w reduction approach for the development of ambient-stable foods, which have minimal energy, machinery, and infrastructure requirements for processing, storage, and distribution (Leistner and Gould 2002). Common applications demand foods with reduced a_w (achieved by partial drying or addition of salt or sugar), usually combined with acidification (i.e., a reduction of pH), addition of preservatives (e.g., sorbic, benzoic or propionic acid), fermentation, and addition of antioxidants (e.g., ascorbic acid).

Most of the traditional foods that remain stable, safe, and tasty during long-term storage (without refrigeration) in developing countries like Africa, Asia, and Latin America are intermediate-moisture foods, in which lowering of a_w is one of the main preservative factors or hurdles (Leistner and Gould 2002). Many of the manufacturing processes for intermediate-moisture foods were empirically developed, but today the hurdles and their specific roles are better understood and can be rationally selected to design or optimize the preservation system.

Dried Foods

Dried foods are those with a_w in the range 0.1 to 0.4. Food drying is the oldest and most used means of preservation. The method is based on water removal to attain stability and to reduce transportation and storage costs. Some criteria used in the design process and improved stability of dried foods are presented next.

Monolayer value

It has been demonstrated in several studies that for most dry foods, there is a moisture content below which the rates of quality loss are negligible (Iglesias et al. 1976, 1984; Peleg and Mannheim 1977; Katz and Labuza 1981). This moisture content is designated as the monolayer value (M_o) and is generally around 0.1 to 0.4 a_w. Because drying is a process that needs energy for water removal, the monolayer moisture value must be the final water content of a dried food to obtain the most stable product. According to the energy required

Table 13.1 BET Monolayer Moisture Value (g water/100 g dry solids) Predicted at Two Temperatures.

Product	20°C	35°C
A. Starchy foods		
Corn	2.01	1.88
Potato	1.83	1.76
Wheat flour	1.87	1.78
B. Protein foods		
Chicken, raw	1.91	1.73
Eggs, dried	1.30	1.27
Gelatin	2.60	2.30
C. Fruits		
Banana	1.51	1.24
Peach	2.39	2.05
Pineapple	3.11	2.68
D. Vegetable and spices		
Celery	1.69	1.62
Cinnamon	1.87	1.67
Ginger	1.92	1.71
Onion	1.69	1.41
Horseradish	1.86	1.68
Thyme	1.57	1.40

Source: Iglesias, H.A., and Chirife, J. 1984. Technical note: Correlation of BET monolayer moisture content in food with temperature. *Journal of Food Technology* 19:503–507.

to bind water to the solid structure, water removal below the monolayer value is an expensive process. This is why the final moisture during drying is close to the monolayer value but not lower. Consequently, it is important to take into account the economy of the process and product stability.

The monolayer value can be determined from the BET (Brunauer-Emmett-Teller) isotherm equation as applied to the experimental moisture sorption data. The BET isotherm model is a two-parameter equation and fits the sorption isotherm in the a_w range of 0.1 to 0.5. Another model to obtain the monolayer value is the GAB (Guggenheim-Anderson-de Boer) equation (see Chapter 5). This model is a three-parameter equation and can be applied in the a_w range of 0.1 to 0.9 (Timmermann et al. 2001). An example of predicted monolayer moisture BET values, for different kinds of foods, is showen in Table 13.1. Most of these values are in the moisture (d.b.) range of 1.30 to 3.11 at 20°C and 1.24 to 2.68 at 35°C, with the increase of temperature as the principal variable reducing the monolayer values. The application of monolayer moisture is generally adequate to define the final moisture level in dehydrated foods, in terms of economy and efficiency of the process, and to achieve maximum product stability. However, for foods with high sugar content, lower values than those of the monolayer moisture content are recommended to improve stability.

Critical water activity value and mechanical properties

Mechanical properties of dehydrated foods are related to water vapor sorption properties. Mechanical properties of low-moisture products change drastically when the moisture content related to the critical a_w value is reached. At this value, there is an increase in water mobility associated with glassy to rubbery transition, which, in turn, affects mechanical properties.

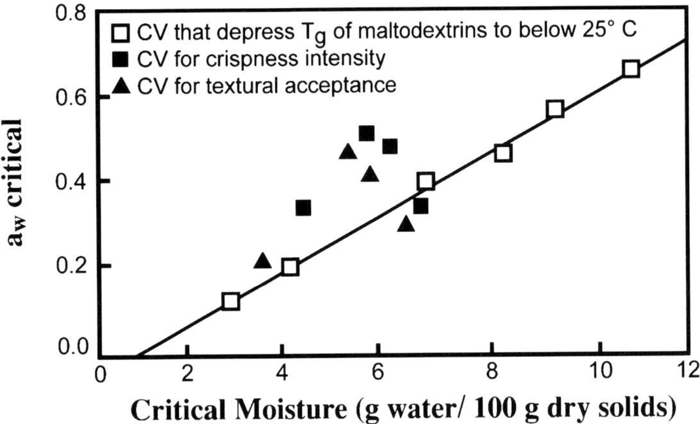

Figure 13.1 Critical values (CV) for a_w and moisture affecting mechanical properties (adapted from Roos, Y. 1995. Water activity and glass transition temperature: How do they complement and how do they differ. In: *Food Preservation by Moisture Control (Fundamentals and Applications)*, eds. G. Barbosa-Cánovas and J. Welti-Chanes. Lancaster, PA: Technomic).

Low-a_w products that are crisp when a force is applied, and soft at higher a_w values, are designated as crispy foods. On the other hand, intermediate- or high-a_w moist products that fold when a force is applied, and are hard at low a_w values, are designated as soft foods. For hardness of several crispy foods, the critical a_w is around 0.5 to 0.6. Phase transitions occur when the product gains water at a_w values below the critical a_w, around 0.4 (Katz and Labuza 1981). The critical a_w value can be used to calculate the maximum gain or loss of water by a product during formulation, packaging, and storage. To prevent changes in the mechanical properties of foods, the a_w value must be kept below the critical limits to maintain a suitable texture and to prevent water migration in a multicomponent food; otherwise, a packaging barrier must be used (Katz and Labuza 1981).

Most dry foods have a fragile texture in crispness; this texture is an essential characteristic of dry cereal and snack food products. Katz and Labuza (1981) showed that the crispness loss is due to water plasticization and softening. These types of foods show a crispness loss at a_w values exceeding the critical a_w value. In addition, they pointed out that dry cereals have a crisp texture in the glassy state, but an increase in water content or temperature may change the material to the rubbery state; if ambient temperature becomes higher than T_g, it causes sogginess. Roos (1993) estimated critical moisture content and a_w values for maltodextrins with various molecular weights. The results suggested that both the critical water content and corresponding a_w that depressed the T_g to below 25°C increased with increasing molecular weight. It was also noticed that a plot with critical a_w values against corresponding water content gave a straight line. This plot is shown in Figure 13.1 with experimentally determined critical water content and a_w values for crispness intensity and texture acceptance of snack foods (Katz and Labuza 1981). It may be concluded that critical values for depression of T_g to below 25°C increase with increasing molecular weight. Above critical a_w values, crispness of low moisture foods is lost and they

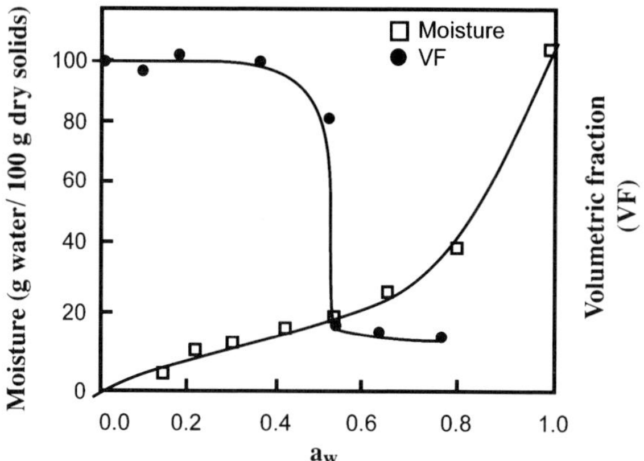

Figure 13.2 Moisture sorption isotherm (19°C) and collapse (volumetric fraction [VF]) of freeze-dried fish hydrolyzate (adapted from Aguilera, J.M., Levi, G., and Karel, M. 1993. Effect of water content on the glass transition and caking of fish protein hydrolizates. *Biotechnology Progress* 9:651–654).

become soggy. The loss of crispness can also be determined using mechanical measurements probably due to decreasing viscosity. In sensory testing, loss of crispness and textural acceptance is observed (Roos 1995).

Water activity, flow properties of powders, caking, and agglomeration

Caking is a process affected by a_w, temperature, and time. Free-flowing powders are transformed into lumps and eventually into an agglomerated solid resulting in a loss of functionality. This problem is common in foods and pharmaceutical industries. Stages in caking phenomena involve humidification, bridging, agglomeration, compaction, and liquefaction (Peleg and Manheim 1977). To maintain the flow properties of a powder and to avoid caking, maintain a low a_w value, subject the powder to a low-moisture atmosphere, use moisture-barrier packaging materials, store at low temperatures, or use anticaking agents (Peleg and Mannheim 1977).

Collapse refers to loss of structure that may occur in freeze-drying due to plasticization of the freeze-concentrated solids or loss of structure resulting from the plasticization of low-moisture foods (Roos 1995).

Figure 13.2 shows water content and volumetric fraction as a function of a_w for a freeze-dried fish protein hydrolyzate. In this plot, at a_w 0.52, the protein hydrolyzate has collapsed (Aguilera et al. 1993). Collapse of food powders has been shown to be a time-dependent phenomenon occurring at a temperature determined by water content. Collapse results from viscous flow above T_g, probably at a rate defined by $T - T_g$ (Roos and Karel 1991). Collapse is a time-dependent phenomenon that occurs in foods during storage above a critical a_w value (Roos 1995) (see Figure 13.2).

During drying and storage of foods, collapse and stickiness must be avoided; however, in some foods, water plasticization is useful to control stickiness in agglomeration of food powders. Stickiness of low-moisture foods results in agglomeration and caking of food

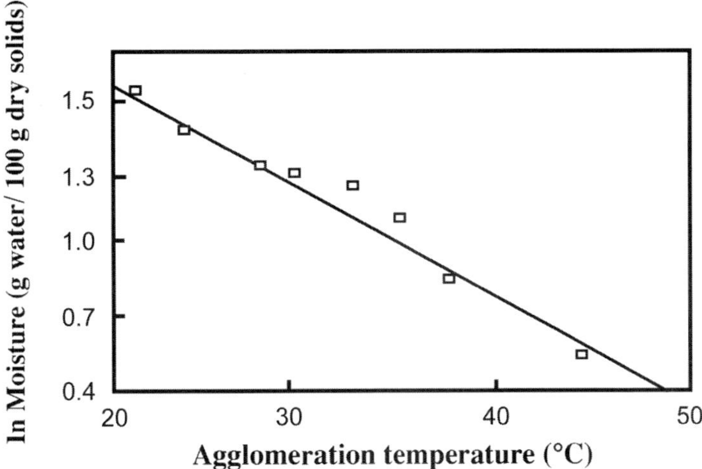

Figure 13.3 Agglomeration temperature as a moisture function of dried orange comminuted.

materials. Stickiness of food powders during spray-drying is often caused by a product's high water content and temperature. In spray-drying, stickiness is detrimental due to caking of a viscous material on surfaces inside the drying chamber. To increase the particle size of food powders, increasing water content and temperature to achieve stickiness is used in agglomeration, followed by dehydration. The sticky point temperature decreases with increasing water content. Stickiness may also occur during storage of food materials and results in caking under such conditions, which causes plasticization and decreases T_g to below ambient temperature (Roos 1995). Figure 13.3 shows the agglomeration temperature of dried orange comminuted as a function of water content.

Formulation and product mixing

Several food products involve the partitioning of two or more ingredients that have different a_w values and thus different chemical potentials. During storage, moisture can migrate because of the chemical potential difference between ingredients, until a final equilibrium a_w is reached. This process can cause problems, such as changes in texture and sensory properties, as well as product safety.

With careful use of the a_w concept, it is possible to design safe products with the highest quality and increased shelf-life. Water activity data can be used to predict potential risks in new products. Some simple mathematical models can be used to predict final a_w at equilibrium for a mixture of two or more ingredients, as well as the time needed to reach that value when ingredients are mixed. In addition, methods can be developed whereby the important physical parameters used to predict equilibrium and rate of change of mixture's a_w can be easily obtained in the laboratory using moisture sorption isotherm information. This will speed product development and quickly determine whether a potential problem exists. The procedure used to solve this problem is based on the assumption that the total initial moisture in the system equals the total moisture at equilibrium and that the moisture

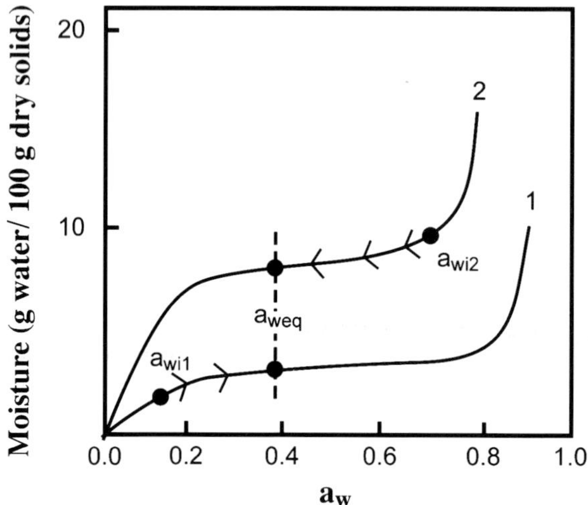

Figure 13.4 Moisture exchange in ingredients 1 and 2 of a mixture that takes place in order to reach equilibrium.

isotherm of each ingredient follows a straight line in this range of moisture. The mathematical model for the isotherm in this moisture range is in the form:

$$m = ba_w + c \tag{13.1}$$

where b is slope, c is intercept, and m is moisture content.

With this idea, Salwin and Slawson (1959) pointed out that the equilibrium a_w can be predicted using the equation:

$$a_{weq} = \frac{\sum a_{wi} b_i w_i}{\sum b_i w_i} \tag{13.2}$$

where a_{wi} is initial a_w of each ingredient i, b_i is absolute value of slope of moisture isotherm of each ingredient i, and w_i is weight of dry solids of each ingredient i. The concepts expressed in Equation 13.2 are illustrated in Figure 13.4, and its understanding is a great tool for developing mixtures of food ingredients.

To predict a_w at equilibrium, the moisture isotherms of each component at the temperature which the product will be stored are needed. The equation can also be used to design the food system; therefore, the percent of each ingredient is such that no ingredient crosses the critical a_w, making it unacceptable. To do this, the a_w at equilibrium is substituted for the critical a_w in Equation 13.1. If this cannot be done, then ingredient modification is a possible solution to obtaining a more stable or better-quality product (Labuza 1984).

Shelf-life and packaging of dry foods

Moisture sorption isotherms are used to predict the shelf-life of packaged foods because packages differ in water vapor permeabilities. They are useful for evaluating the critical

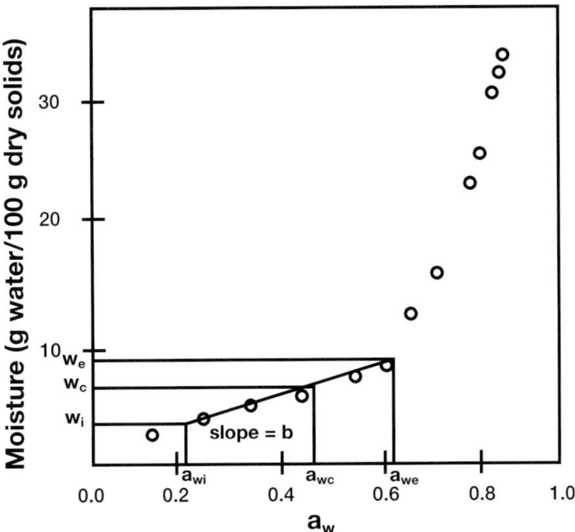

Figure 13.5 Moisture and a_w values required to apply mathematical model for gain or loss moisture.

moisture content at which any characteristic (texture, color, nutritional value, microbial growth) of a dehydrated food can be modified generating an inadequate product.

The permeability of the package will define the air properties surrounding the food and, hence, the maximum storage time before the critical moisture is reached. Equation 13.3, based on Fick's law, can be used to evaluate the maximum storage time (t_{st}) to reach the critical moisture content (w_c). It has been shown that it is useful for selection of packaging materials for dry foods, when both the moisture isotherms and the critical moisture values are known (Martinez et al. 1998):

$$t_{st} = \frac{M}{A}\frac{\Delta\left(p/p^o\right)_t}{q}bp^o\left(2.3031\log\frac{w_e - w_i}{w_e - w_c}\right) \tag{13.3}$$

where M is dry mass of food (kg), A is surface area of package (m^2), $\Delta\left(p/p^o\right)$ is change of a_w in a specified time, q is water vapor permeability of package (g water/day • m^2 • mm Hg), b is isotherm slope (g H_2O/g solids at each a_w), p^o is vapor pressure of pure water at temperature T (mm Hg), w_e is equilibrium moisture content (g H_2O/g solids), w_i is initial moisture content (g H_2O/g solids), and w_c is critical moisture content (g H_2O/g solids).

To use this model, it is necessary to know the moisture sorption isotherm, package permeability, environmental conditions during storage, as well as packing conditions. It is assumed that the following aspects exist: constant relative humidity outside the package; a linear isotherm with a straight line form (see Equation 13.1) (see Figure 13.5); an equilibrium moisture content (w_e) on the linear isotherm; and, finally, the critical moisture content (w_c) is lower than the equilibrium moisture content (w_e).

Thermodynamic properties evaluation and application to design and stability of foods

An analysis of the food–water vapor system is possible when moisture sorption data are known at several temperatures. The Clausius-Clapeyron equation originally was used for a vapor–liquid equilibrium. This equation applying the temperature–vapor pressure data of a food can be used to evaluate the enthalpy change associated with a sorption process at several moisture contents (isosteric heat of sorption):

$$\ln a_w = -\frac{Q_S}{R} \bullet \frac{1}{T} \quad (13.4)$$

where Q_S is isosteric heat of sorption, R is gas constant, and T is absolute temperature.

Thermodynamic properties that describe the relationship between water and food are helpful in evaluating the energy requirements in concentration and drying processes and in predicting optimal storage conditions for maximum stability of dry foods. In addition, the evaluation of several thermodynamics properties (enthalpy, entropy, Gibbs free-energy, etc.) is important in the design and optimization of dryers. Several thermodynamic properties are moisture content dependent and can be useful in describing a sorption process (Beristain et al. 1996).

The isosteric heat of sorption (Q_S) can be used to evaluate the energy requirements of dehydration processes. The moisture content in a product, in which the net isosteric heat of sorption is close to the latent heat of vaporization of water, can give an idea of the quantity of "tied water" (or monolayer water) in the food. Around this a_W, isosteric heat of sorption increases when moisture content decreases below this critical limit of "tied water" (Kiranoudis et al. 1993).

Enthalpy changes are related to the energy changes occurring when water molecules are mixed with solute during an adsorption process. Entropy changes (ΔS) can be related to attraction–repulsive forces in the system. Also, entropic changes are related to spatial arrangements in the water interphase during adsorption and are related to a disorder level in a water–solute system. This concept is useful in dissolution and crystallization processes. Gibbs free-energy changes (ΔG) indicate the affinity of solute for water and give the criterion for process spontaneity (Beristain et al. 1996). Gibbs free-energy can be calculated with the following equation (Iglesias et al. 1976):

$$\Delta G = RT \ln a_w \quad (13.5)$$

In addition, adsorption enthalpy and entropy data can be used to predict moisture adsorption isotherms at different temperatures with a sound theoretical base. For example, Figure 13.6 shows a comparison between the experimental adsorption isotherm and the predicted isotherm using enthalpy–entropy adsorption data at 45°C for freeze-dried pulp of *Aloe vera*. This thermodynamic and mathematical approach is a good tool for prediction and useful for analysis and process design.

Intermediate- and High-Moisture Foods

There are actually two categories of foods with reduced a_W for which stability is based on a combination of factors: intermediate- and high-moisture foods.

Intermediate-moisture foods range generally from 0.60 to 0.90 a_W and 10% to 50% water

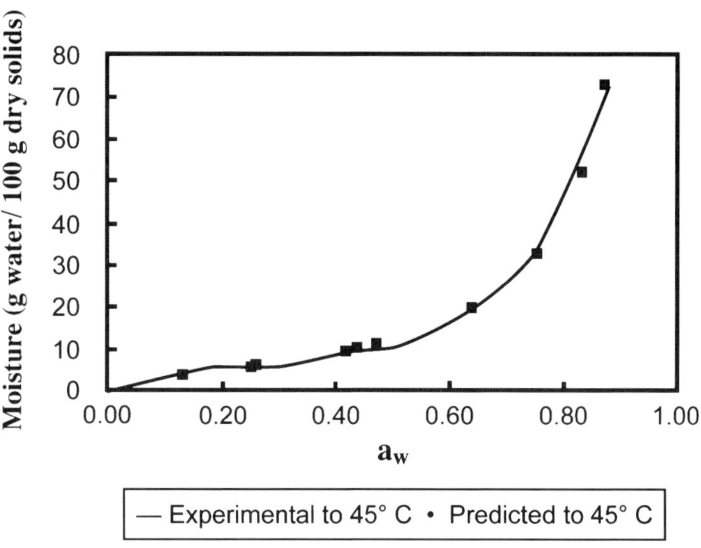

Figure 13.6 Adsorption moisture isotherm of freeze-dried pulp of *Aloe vera* at 45°C.

by weight (Jayaraman 1995). Additional hurdles can provide the limit of safety against spoilage by microorganisms resistant to a_w (mainly molds and yeasts, which can grow at a_w as low as 0.60), and also against some bacterial species that are likely to grow when the a_w value of the intermediate-moisture foods is near the upper limit of water activities (i.e., 0.90 a_w). With these objectives, the lowering of a_w is often combined with chemical preservatives (i.e., nitrites, sorbates, sulphites, benzoates, antimicrobials of natural origin, smoke components); reduction of pH (which usually inhibits or decreases bacterial growth, accentuates the action of preservatives, and increases the minimum a_w values for bacterial growth); and sometimes with the addition of competitive microorganisms. Other intermediate-moisture products receive a thermal treatment during the manufacturing process that inactivates heat-sensitive microorganisms; in some cases, the subsequent hot filling in containers further improves the microbial stability (Leistner and Gould 2002).

Most intermediate-moisture foods are designed to be storable for several months at room temperature, even in tropical climates, and eaten without any rehydration. They are wet enough to be ready-to-eat without giving a sensation of dryness when eaten, but dried enough to be shelf-stable (Karel 1973, 1976; Jayaraman 1995). Many intermediate-moisture products, due to the addition of very high amounts of solutes (e.g., sugar or salt) in order to reduce a_w to the desired level, are too sweet or salty. They become undesirable from the nutritional and sensory point of view. Therefore, this category of products has been subjected in the past decade to continuous revision and discussion regarding formulations.

On the other hand, high-moisture foods have a_w values above 0.90. Thus, in this category, the reduction of a_w is a hurdle of less relative significance because most microorganisms are able to proliferate (Leistner and Gould 2002). Stability at room temperature is reached by applying intentional and carefully designed hurdle technology. High-moisture fresh-like fruits and cooked meat products are preserved by the interaction of a_w with mild heat treatment and pH with preservatives. This represents a rational application of the combined approach making the products storable without refrigeration (Alzamora et al. 1995, 2000; Leistner and Gould 2002).

In 1994, a survey was conducted in 11 countries through the Science and Technology for Development (CYTED) Program, under a project named "Development of Intermediate-Moisture Foods from Iberoamerica." Information on 260 traditional intermediate- and high-moisture foods was collected (Tapia et al. 1994, Welti et al. 1994). Many of these products, which are also common in different parts of the world, are safe and storable without refrigeration and require inexpensive packaging. Most shelf-stable foods do not rely solely on a_w for microbial control but also on other preservation factors. The binary combination of a_w and pH acts as relevant hurdles in many of these products preventing proliferation of pathogenic microorganisms, while other hurdles (antimicrobials, thermal treatment, etc.) play a secondary role, mainly against spoilage flora (Tapia et al. 1994).

Different approaches have been explored for obtaining shelf-stability and fresh-like characteristics in fruit products. Commercial, minimally processed fruits are fresh with high moisture content and are prepared in a fresh-like state for convenient consumption and distribution to the consumer. Minimal processing includes minimum preparation procedures like washing, peeling and/or cutting, packing, etc. The fruit product is usually placed in refrigerated storage where its stability varies depending on the type of product, processing, and storage conditions. However, product stability without refrigeration is an important issue not only in developing countries but also in industrialized countries. The principle used by Leistner for shelf-stable high-moisture meats ($a_w > 0.90$), where only mild heat treatment is used and the product still exhibits a long shelf-life without refrigeration, can be applied to other foodstuffs. Fruits would be a good choice. Leistner states that for industrialized countries, production of shelf-stable products is more attractive than intermediate-moisture foods because the required a_w for shelf-stable products is not so low and less humectants and/or less drying of the product is necessary (Leistner 2000).

If fresh-like fruit is the goal, dehydration should not be used in processing. Reduction of a_w by addition of humectants should be used at a minimum level to maintain the product in a high-moisture state. To compensate the high moisture left in the product (in terms of stability), a controlled blanching can be applied without affecting sensory and nutritional properties; pH reductions can be made that will not impair flavor; and preservatives can be added to reduce the risk of potential spoilage microflora. In conjunction with the above mentioned factors, a slight thermal treatment, pH reduction, slight a_w reduction, and the addition of antimicrobials (sorbic or benzoic acid, sulfite), all placed in context with the hurdle technology principles applied to fruits, can be an interesting alternative to intermediate-moisture preservation of fruits, as well as to commercial minimally processed refrigerated fruits (Alzamora et al. 1989, 1993, 1995; Guerrero et al. 1994; Alzamora 1997; Cerrutti et al. 1997).

Pulsed vacuum osmotic dehydration is a new technique used to develop minimal processing of foods. This new method of osmotic dehydration takes advantage of the porous microstructure of tissues, and uses vacuum impregnation to reduce process time and improve incorporation of additives. Vacuum impregnation shows faster water loss kinetics in short time treatments compared with time-consuming atmospheric "pseudo-diffusional" processes. This is due to the occurrence of a specific mass transfer phenomenon, the hydrodynamic mechanism, and the result produced in the solid–liquid interface area. Many fruits and vegetables have a great number of pores and offer the possibility of being im-

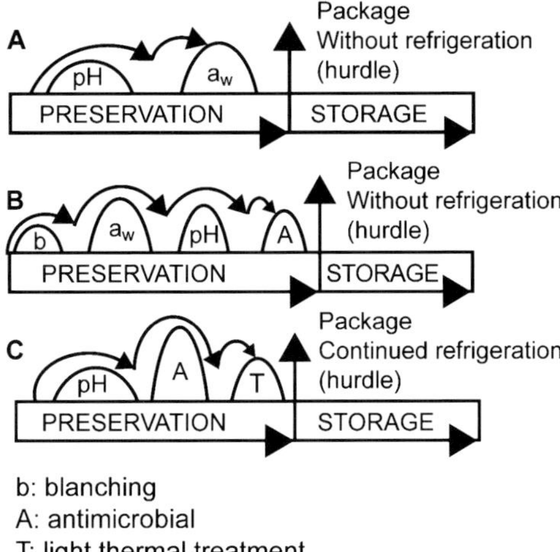

Figure 13.7 Role of a_w and the hurdles applied to three preservation systems for fruits: **(A)** intermediate-moisture fruit, **(B)** high-moisture fruit, and **(C)** minimally processed fruit.

pregnated by a determined solution of solute and additives. Thus, the product composition as well as its physical and chemical properties may be changed to improve its stability. An important advantage of using low pressures (approximately 50 mbar) for minimal preservation of fruit is that equilibration times are shorter than at atmospheric pressure (e.g., 15 minutes under vacuum versus a few hours in forced convection at atmospheric conditions, or a few days in media without agitation for reducing a_w to 0.97) (Alzamora et al. 2003).

High-moisture foods have a_w values higher than 0.90. Thus, in these products, a_w reduction is a less important preservation factor and several microorganisms can proliferate (Leistner and Gould 2002). The CYTED Program defined minimally processed food as a partially processed food using hurdle technology, resulting in a product that is shelf-stable without refrigeration and having fresh-like properties of texture, flavor, and appearance (CYTED 1991). Figure 13.7 shows the role of a_w as being a hurdle in intermediate- and high-moisture foods. It shows that the intelligent combination of hurdles results in stable, sensorial, and economic food products, with or without refrigeration.

Water activity prediction

A main tool to evaluate and predict the stability of intermediate- and high-moisture foods is a_w evaluation. Thus, quality assurance laboratories and other industrial activities exist that use several mathematical models to predict the most stable a_w value. This information can be used to reformulate a food product for quality and stability improvement. Studies by van den Berg and Bruin (1981), Leung (1986), Rizvi (1986), Chirife et al. (1987), Toledo (1991), and Welti-Chanes and Vergara-Balderas (1997) show several reviews about procedures used to evaluate or predict a_w. In each case, theoretical and empirical models, as well as several examples, are presented. Also, studies by Benmergui et al. (1979), Ferro-Fontán et al. (1979, 1980), Chirife et al. (1980), Ferro-Fontán and Chirife (1981a, 1981b), and Favetto and Chirife (1985) present important information about a_w prediction in aqueous solutions and foods.

Water activity can be influenced in at least three ways during the preparation of intermediate- and high-moisture foods:

1. Water can be removed by dehydration, evaporation, or concentration process.
2. Additional solute can be incorporated. The impregnation of solute can be performed by moist or dry infusion. Moist infusion consists of soaking the food pieces in a water–solute solution of lower a_w, while dry infusion involves direct mixing of food pieces and solutes in adequate proportions. This process is called "osmotic dehydration" and allows incorporating not only the solute used to control a_w but also the desired quantities of antimicrobial and antibrowning agents, or any solute for improving sensory and nutritional quality. By controlling these above complex exchanges, it is possible to conceive different combinations of water loss and solid gain, from use of a simple dewatering process (with substantial water removal and only marginal sugar pickup) to a candying or salting process (in which solute penetration is favored and water removal limited) (Torregiani and Bertolo 2002). For porous foods, moist infusion can be also performed under vacuum, as previously mentioned. The internal gas or liquid occluded in the open pores is exchanged for an external liquid phase (of controlled composition) due to pressure changes.
3. Through a combination of methods 1 and 2—in this case, the food pieces are infused with the solutes and additives and then partially dried. The advantages of this combination, compared with drying only, are an increase in stability of the pigments responsible for color, an enhancement of natural flavor, a better texture, and a greater load capacity of the dryer (Alzamora et al. 2003).

Whatever the procedure used to reduce a_w, it is necessary to know the a_w–moisture content relationship in the food. Important contributions have been made in the field of a_w prediction over the past 50 years, and comprehensive analysis of the procedures traditionally used to calculate a_w have been performed by Vigo et al. (1980), van den Berg and Bruin (1981), Muñozcano et al. (1987), Aguilera et al. (1991) Chirife (1995), and Welti-Chanes and Vergara-Balderas (1997). In each case, the applicability of various theoretical and empirical equations was analyzed and some descriptive examples were presented.

There is no model in existence with a simple mathematical structure representing the sorption or a_w lowering characteristics of foods or their components in the whole a_w range, because the depression of a_w in foods is due to a combination of mechanisms, each capable of predominating in a given range of a_w.

In high- and intermediate-moisture foods, a_w is mainly determined by the nature and concentration of soluble substances (i.e., sugars, NaCl, polyols, amino acids, organic molecules, other salts) in the aqueous phase of the food (Chirife 1995). A number of equations based on the thermodynamic properties of binary and multicomponent electrolyte and nonelectrolyte solutions have been studied theoretically and experimentally for calculating or predicting the a_w of these foods. Figure 13.8 summarizes several theoretical and empirical models suggested for the calculation of a_w in semimoist and moist foods (van den Berg and Bruin 1981, van den Berg 1986, Chirife 1995).

Final Remarks

Water activity is a parameter of great importance in developing and improving food stability, as the role of water is critical in terms of microbial, chemical, and sensory quality.

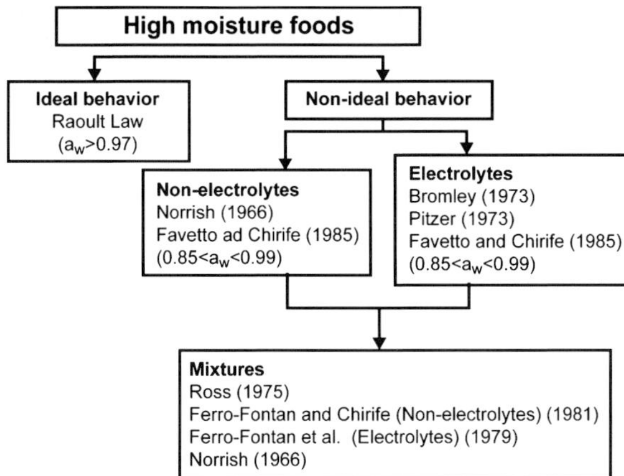

Figure 13.8 van den Berg diagram for a_w prediction.

Evaluation and quantification of a_w are the main elements used by food scientists to understand the more important changes occurring in processing, formulating, or reformulating of foods. Therefore, industrial understanding of the basic principles of a_w is fundamental today in food product development.

References

Aguilera, J.M., Chirife, J., Tapia de Daza, M.S., and Welti, J. 1991. *Inventario de Alimentos de Humedad Intermedia Tradicionales de Iberoamérica.* México: Instituto Politécnico Nacional.

Aguilera, J.M., Levi, G., and Karel, M. 1993. Effect of water content on the glass transition and caking of fish protein hydrolizates. *Biotechnology Progress* 9:651-–654.

Alzamora, S.M. 1997. Preservación I. Alimentos conservados por factores combinados. In: *Temas de Tecnología de Alimentos*, vol. 1, ed. J.M. Aguilera. México: Programa Iberomericano de Ciencias y Tecnología (CYTED)/Instituto Politécnico Nacional (IPN).

Alzamora, S.M., Cerrutti, P., and López-Malo, A. 1995. Minimally processed fruits by combined methods. In: *Food Preservation by Moisture Control (Fundamentals and Applications)*, eds. G. Barbosa-Cánovas and J. Welti-Chanes, pp. 463–492. Lancaster, PA: Technomic.

Alzamora, S.M., Fito, P., López-Malo, A., Tapia, M.S., and Parada-Arias, E. 2000. Minimally processed fruit using vacuum impregnation, natural antimicrobial addition, and/or high hydrostatic pressure techniques In: *Minimally Processed Fruits and Vegetables. Fundamental Aspects and Applications,* eds. S.M. Alzamora, M.S. Tapia, and A. López-Malo, pp. 293–315. Gaithersburg, MD: Aspen Publishers.

Alzamora, S.M., Gerschenson, L.N., Cerrutti, P., and Rojas, A.M. 1989. Shelf stable pineapple for long-term non-refrigerated storage. *Lebensittel-Wissenschaft und-Technologie* 22:233–236.

Alzamora, S.M., Tapia, M.S., Argaiz, A., and Welti, J. 1993. Application of combined methods technology in minimally processed fruits. *Food Research International* 26:125–130.

Alzamora S.M., Tapia, M.S., López-Malo, A., and Welti-Chanes, J. 2003. The control of water activity. In: *Food Preservation Techniques*, eds. P. Zeuthen and L. Bogh-Sorensen, pp. 126–153. Cambridge, UK: Woodhead Publishing Limited.

Benmergui, E.A., Ferro Fontán, C., and Chirife, J. 1979. The prediction of water activity in aqueous solutions in connection with intermediate moisture foods. I. a_w prediction in single aqueous electrolyte solutions. *Journal of Food Technology* 14:625–637.

Beristain, C., García, H., and Azuara, E. 1996. Enthalpy-entropy compensation in food vapor adsorption. *Journal of Food Engineering* 30:405–415.

Cerrutti, P., Alzamora, S.M., and Vidales, S.L. 1997. Vanillin as antimicrobial for producing shelf-stable strawberry purée. *Journal of Food Science* 62:608–610.

Chirife, J. 1995. An update on water activity measurements and prediction in intermediate and high moisture foods: The role of some non-equilibrium situations. In: *Food Preservation by Moisture Control (Fundamentals and Applications),* eds. G. Barbosa-Cánovas and J. Welti-Chanes, pp. 169–189. Lancaster, PA: Technomic.

Chirife, J., Argaíz, A., and Welti, J. 1987. *Predicción de la Actividad de Agua en Alimentos.* México: Programa Iberomericano de Ciencias y Tecnología (CYTED)/Instituto Politécnico Nacional (IPN), Información técnica No. 1.

Chirife, J., Ferro-Fontán, C., and Benmergui, E.A. 1980. The prediction of water activity in aqueous solutions in connection with intermediate moisture foods. IV. a_w prediction in aqueous non-electrolyte solutions. *Journal of Food Technology* 15:59–70.

Christian, J.H.B. 2000. Drying and reduction of water activity. In: *The Microbiological Safety and Quality of Foods*, eds. B.M. Lund, T.C. Baird-Parker, and G.W. Gould, pp. 146–174. Gaithersburg, MD: Aspen Publishers.

Crutchfield, C.R., and Roberts, T. 2000. *Food Review* 23(3):44–49.

CYTED-D Program. 1991. *Desarrollo de Alimentos de Humedad Intermedia Importantes para Iberoamérica. Subproyecto: Frutas, Hortalizas*, ed. J. Welti-Chanes. Puebla, México: Universidad de las Américas.

Favetto, G.J., and Chirife, J. 1985. Simplified method for the prediction of water activity in binary solutions. *Journal of Food Technology* 20:631–636.

Ferro-Fontán, C., and Chirife, J. 1981a. A refinement of Ross equation for predicting the water activity of non-electrolyte mixtures. *Journal of Food Technology* 16:219–221.

Ferro-Fontán, C., and Chirife, J. 1981b. The evaluation of water activity in aqueous solutions from freezing point depression. *Journal of Food Technology* 16(1):21–30.

Ferro-Fontán, C., Chirife, J., and Benmergui, E.A. 1979. The prediction of water activity in aqueous solutions in connection with intermediate moisture foods. II. On the choice of the best a_w lowering single strong electrolyte. *Journal of Food Technology* 14:639–646.

Ferro-Fontán, C., Chirife, J., and Benmergui, E.A. 1980. The prediction of water activity in aqueous solutions in connection with intermediate moisture foods. III. Prediction in multicomponent strong electrolyte aqueous solutions. *Journal of Food Technology* 15:47–58.

Guerrero, S., Alzamora, S.M., and Gerschenson, L.N. 1994. Development of a shelf-stable banana purée by combined factors: Microbial stability. *Journal of Food Protection* 57:902–907.

Iglesias, H.A., and Chirife, J. 1984. Technical note: Correlation of BET monolayer moisture content in food with temperature. *Journal of Food Technology* 19:503–507.

Iglesias, H.A., Chirife, J., and Viollaz, P. 1976. Thermodynamics of water vapor sorption by sugar beet root. *Journal of Food Technology* 11:91–101.

Jayaraman, K.S. 1995. Critical review on intermediate moisture fruits and vegetable. In: *Food Preservation by Moisture Control (Fundamentals and Applications)*, eds. G.V. Barbosa-Cánovas and J. Welti-Chanes, pp. 411–442. Lancaster, PA: Technomic.

Karel, M. 1973. Recent research and development in the field of low moisture and intermediate moisture foods. *CRC Critical Reviews in Food Technology* 3:329–373.

Karel, M. 1976. Technology and application of new intermediate moisture foods. In: *Intermediate Moisture Foods*, eds. R. Davies, G.G. Birch, and K.J. Parker, pp. 4–31. London: Applied Science Publishers.

Katz, E.E., and Labuza, T.P. 1981. Effect of a_w on the sensory crispness and mechanical deformation of snack food products. *Journal of Food Science* 46:403–409.

Kiranoudis, C.T., Maroulis, Z.B., Tsami, E., and Marinos-Kouris, D. 1993. Equilibrium moisture content and heat of desorption of some vegetables. *Journal of Food Engineering* 20:55–74.

Labuza, T.P. 1984. *Moisture Sorption: Practical Aspects of Isotherm Measurement and Use.* St. Paul, MN: American Association of Cereal Chemists.

Labuza, T.P. 1968. Sorption phenomena in food. *Food Technology* 22:263–272.

Leistner, L. 2000. Hurdle technology in the design of minimally processed foods. In: *Minimally Processed Fruits and Vegetables. Fundamental Aspects and Applications*, eds. S.M. Alzamora, M.S. Tapia, and A. Lopez-Malo, pp. 13–27. Gaithersburg, MD: Aspen Publishers.

Leistner, L., and Gould, G.W. 2002. Hurdle technologies. *Combination Treatments for Food Stability. Safety and Quality*. New York: Kluwer Academic/Plenum Publishers.

Lenovich, L.M. 1987. Survival and death of microorganisms as influenced by water activity. In: *Water Activity: Theory and Applications to Food*, eds. L.B. Rockland and L. Beuchat, pp. 119–36. New York: Marcel Dekker.

Leung, H.K. 1986. Water activity and other colligative properties of foods. In: *Physical and Chemical Properties of Foods*, ed. M.R. Okos, pp. 138–185. St. Joseph, MI: American Society of Agricultural Engineers.

Martínez, N.N., Grau, A.A., Chiralt, B.A., and Fito, M.P. 1998. *Termodinámica y Cinética de Sistemas Alimento Entorno*. Valencia, Spain: Universidad Politécnica de Valencia, Servicio de Publicaciones.

Muñozcano, A., Wesche, H., Arana, R., García, H., Argaíz, A., Vergara, F., and Welti, J. 1987. Determinación y predicción de la actividad acuosa en alimentos mexicanos de humedad media y alta. *Tec. Alim. (Mex.)* 22:10.

Peleg, M., and Mannheim C.H. 1977. The mechanism of caking of powdered onion. *Journal of Food Processing and Preservation* 1:3–11.

Ray, B. 1992. The need for food biopreservation. In: *Food Biopreservatives of Microbial Origin*, eds. B. Ray and M. Daeschel, pp. 1–24. Boca Raton, FL: CRC Press.

Rizvi, S.S.H. 1986. Thermodynamic properties of foods in dehydration. In: *Engineering Properties of Foods*, eds. M.A. Rao and S.S.H. Rizvi, pp. 133. New York: Marcel Dekker.

Roos, Y. 1993. Water activity and physical state diagrams effects on amorphous food stability. *Journal of Food Processing and Preservation* 16:433–447.

Roos, Y. 1995. Water activity and glass transition temperature: How do they complement and how do they differ? In: *Food Preservation by Moisture Control (Fundamentals and Applications)*, eds. G. Barbosa-Cánovas and J. Welti-Chanes, pp. 133–153. Lancaster, PA: Technomic.

Roos, Y., and Karel, M. 1991. Plasticizing effect of water thermal behavior and crystallization of amorphous food models. *Journal of Food Science* 56:38–43.

Salvin, H., and Slawson, V. 1959. Moisture transfer in combination of dehydrated foods. *Food Technology* 13:815–818. In: *Termodinámica y Cinética de Sistemas Alimento-entorno* (1998), eds. N.N. Martínez, A.A. Grau, B.A. Chiralt, and M.P. Fito. Valencia, Spain: Universidad Politecnica de Valencia, Servicio de Publicaciones.

Scott, W.J. 1953. Water relations of *Staphylococcus aureus* at 30°C. *Australian Journal of Biological Science* 6:549–564.

Scott, W.J. 1957. Water relations of food spoilage microorganisms. *Advances in Food Research* 7:83–127.

Tapia, M.S., Aguilera, J.M., Chirife, J., Parada, E., and Welti, J. 1994. Identification of microbial stability factors in traditional foods from Iberoamerica. *Rev. Española Ciencia Tecnol. Alim.* 34:145–163.

Timmermann, E.O., Chirife, J., and Iglesias, H.A. 2001. Water Sorption isotherms of food and foodstuffs: BET or GAB parameters? *Journal of Food Engineering* 48:19–31.

Toledo, R.T. 1991. *Fundamentals of Food Process Engineering*, 2nd Edition. New York: Van Nostrand Reinhold.

Torregiani, D., and Bertolo, G. 2002. The role of an osmotic step: combined processes to improve quality and control functional properties in fruit and vegetables. In: *Engineering and Food for the 21st Century*, eds. J. Welti-Chanes, G.V. Barbosa-Cánovas, and J.M. Aguilera, pp. 651–670. Boca Raton, FL: CRC Press.

van Den Berg, C. 1986. Water activity. In: *Concentration and Drying of Foods,* ed. D. MacCarthy, pp. 11–61. London: Elsevier Applied Science.

van Den Berg, C., and Bruin, S. 1981. Water activity and its estimation in food systems: Theoretical aspects. In: *Water Activity: Influences on Food Quality*, eds. L.B Rockland and G.F. Stewart, pp. 1–64. New York: Academic Press.

Vigo, S., Chirife, J., Scorza, O.C., Bertoni, M.H., and Sarrailh, P. 1980. Estudio sobre alimentos tradicionales de humedad intermedia elaborados en Argentina. Determinación de la a_w, pH, humedad y sólidos solubles. *Rev. Agroq. Tecnol. Alimentos* 21:91.

Welti, J., Tapia, M.S., Aguilera, J., Chirife, J., Parada, E., López-Malo, A., López, L.C., and Corte, P. 1994. Classification of intermediate moisture foods consumed in Ibero-America. *Rev. Española Ciencia Tecnol. Alim.* 34:53–63.

Welti-Chanes, J., and Vergara-Balderas, F. 1997. Actividad de agua: Concepto y aplicación en alimentos con alto contenido de humedad. In: *Temas de Tecnología de Alimentos*, vol. 1, ed. J.M. Aguilera. México: Programa Iberomericano de Ciencias y Tecnología (CYTED)/Instituto Politécnico Nacional (IPN).

14 Applications of Water Activity in Nonfood Systems

Anthony J. Fontana, Jr., and Gaylon S. Campbell

In 1958, Sterling Taylor published a paper advocating the use of water activity (a_w) to describe the energy status of water in soil (Taylor 1958). The idea of describing or measuring the energy status of soil water was not new to soil physics; it had been extensively used for more than half a century by that time. However, expressing the energy status in terms of a_w was new. The idea was never implemented, but it establishes a link between soil physics and food physics that can provide some useful insights in both directions.

In this chapter, we consider applications of a_w to a number of different areas, including soil physics, plant water relations, seed physiology and technology, medicine, and pharmaceuticals. Each area tends to have its own terminology and measurement techniques, so practitioners of these sciences are often unaware of the link among methods. We will begin by presenting units and methods of measurement and show how these relate to each other. We then give some examples of a_w applications in each of the areas mentioned.

Water Potential

The energy status of soil and plant water is typically described using the term "water potential." Other terms used in some fields are "suction," "tension," "head," and "capillary pressure." *Water potential* is defined as the energy required per unit mass or volume of water to transport, reversibly and isothermally, an infinitesimal quantity of water from the system under test to a pool of pure, free water (Taylor 1972). Units are J/kg or $J \cdot m^{-3}$, indicating energy per unit mass or volume of water. The unit $J \cdot m^{-3}$ is equivalent to $N \cdot m^{-2}$, which is a unit of pressure, the Pascal (abbreviated Pa). Potential energy per unit volume can therefore be expressed as a pressure.

There are several important things to note about this definition. The first is that it is a measure of the energy of the water in the system, or its ability to do work. The reference is pure, free water, which is defined to have a potential of 0. Requiring the measurement to apply to an infinitesimal quantity of water makes clear the fact that water potential is a differential property. Water potential does not apply to the total quantity of water held by the system but just to the infinitesimal quantity of water held least tightly. As a porous system is dried, each infinitesimal quantity of water removed requires more energy than the previous quantity. In other words, dry materials hold water more tightly than the same material when wet.

The energy of water in a porous system can be changed, relative to that of pure, free water in several ways. One can change the hydrostatic or pneumatic pressure on the water; one can change the position of the water in a gravitational field; one can adsorb the water

Table 14.1 Equivalent Values for Water Potential, Water Activity, Freezing Point Depression, and Osmolality.

	Water Potential (J/kg)	Water Potential (MPa)	Water Activity (a_w)	Freezing Point (°C)	Osmolality (osmol/kg)
	−1	−0.001	0.999993	−0.00081	0.0004
	−10	−0.01	0.999926	−0.00813	0.0041
Field capacity	−33	−0.033	0.999756	−0.02683	0.0135
	−100	−0.1	0.999261	−0.08131	0.0410
	−1000	−1.0	0.992638	−0.81497	0.4105
Permanent wilt	−1500	−1.5	0.988977	−1.22399	0.6157
	−10,000	−10	0.928772	−8.33341	4.1049
Air dry	−100,000	−100	0.477632		41.049
Oven dry	−1,000,000	−1000	0.000618		410.49

on a solid surface; or one can dilute the water with solute. The total water potential (ψ) is often assumed to be the sum of all of these component potentials (although there are cases where this is not true). Thus,

$$\Psi = \Psi_p + \Psi_g \Psi_m \Psi_o \tag{14.1}$$

where the subscripts p, g, m, and o are the pressure, gravitational, matric, and osmotic components of the water potential. Of these, only the matric and osmotic components are significant in intermediate-moisture foods. The matric potential is the portion of the water potential that is attributed to the attraction of the matrix for water, and water held in pores due to capillary action. The water molecules immediately adjacent to the matrix surfaces are held in a tightly adhering (absorbed) layer, but farther away, the bonds become weaker until the water becomes more bulk-like. The osmotic or solute potential is the portion of the water potential that is attributed to the interactions of solutes with water.

Water activity and water potential are related by the well known equation

$$\Psi = \frac{RT}{M_w} \ln a_w \tag{14.2}$$

where *R* is the gas constant (8.3143 J • $\text{mol}^{-1}\ \text{K}^{-1}$), T is Kelvin temperature, and M_w is the molecular mass of water (0.018 kg/mol). Table 14.1 shows a_w values for several water potentials. While both water potential and a_w are measures of the energy status of water in a system, the range of a_w is 0 to 1, while the range of water potential is $-\infty$ to 0 (water potential is a negative number for activities below 1 because one must do work on the system to remove water). Because of its much larger range and the expansion of the scale near zero (pure water), water potential tends to emphasize the biologically significant water energy range. It is easy to think there is not much difference between an a_w of 1.00 and 0.99, but this is the entire range of available water for a plant from water fully available to permanent wilting. Water potential appears to be much more significant than a_w in the range 0 to -1500 J/kg.

As with a_w, water potential is a measure of the driving force for transport. Darcy's law states that the flux density of water in a porous system is directly proportional to the gradient in water potential in the system (Taylor 1972). The condition for equilibrium is that the water potential is the same everywhere within the system.

Energy Status of Water

A lowering of water potential reduces vapor pressure, lowers freezing point, and increases boiling point. These changes are sometimes called the colligative properties of water, and can be used to determine water potential or a_w. For example, one can measure the vapor pressure of water in a sample by sealing it in a chamber and measuring the dew point temperature of the air in the chamber, which is in equilibrium with the sample. The vapor pressure can be computed from the dew point temperature using (Buck 1981)

$$e = a\exp\left(\frac{bT}{c+T}\right) \tag{14.3}$$

where a = 0.611 kPa, b = 17.502 C^{-1}, and c = 240.97 C. The temperature T is the dew point temperature in degrees Celsius. The a_w is defined as the ratio of the vapor pressure of the air in equilibrium with a sample to the saturation vapor pressure at sample temperature and external pressure. The a_w is therefore computed from

$$a_w = \frac{e_a}{e_s} = \exp\left(\frac{bT_d}{c+T_d} - \frac{bT_s}{c+T_s}\right) \tag{14.4}$$

Equation 14.2 is used to convert this value to water potential as

$$\Psi = \frac{RT}{M_w}\left(\frac{bT_d}{c+T_d} - \frac{bT_s}{c+T_s}\right) \tag{14.5}$$

Water potential is therefore roughly proportional to the dew point temperature depression of the air in equilibrium with the sample.

Equations 14.3 and 14.4 can also be used to compute the water potential of a system when an ice phase is present, or to compute the freezing point depression of a system. The a_w when an ice phase is present is

$$a_w = \frac{e_i}{e_w} = \exp\left(\frac{b_i T}{c_i + T} - \frac{bT}{c+T}\right) \tag{14.6}$$

The constants for vapor pressure over ice are $b_i = 21.87$ and $c_i = 265.5$. When an ice phase is present, the a_w of the sample is completely determined by its temperature. The freezing point temperature for any a_w value can be determined by solving Equation 14.6 for temperature. Table 14.1 shows freezing point values for a range of water activities and water potentials.

One other measure of a_w or water potential is used mainly in the medical field. It is the concentration of an ideal solute that gives the same water potential as the test sample. The water potential of an ideal solute is given by the van't Hoff equation:

$$\Psi = -CRT \tag{14.7}$$

where C is the solute concentration in mol/kg. The concentration of an ideal solute with water potential of ψ is $C = \psi/RT$. The concentration is expressed as osmol/kg or mosmol/kg. Table 14.1 shows this measurement along with water potential, a_w, and freezing point.

Table 14.1 clearly shows the equivalence of the various methods of computing a_w or water potential. In general, the relationships are not linearly related, so there are no simple conversions and the equations given or the table must be used to convert between them. The point to be made regarding this widely disparate set of units is that all of these measures, from many different fields, describe the energy status of water.

Measurement Methods

Measurement methods were covered in Chapter 6. The various instruments may read out in any of the units in Table 14.1, but it should be clear that they could read out in other units as well. In other words, an instrument that measures freezing point depression can read out directly in the freezing point temperature, but it could also read out in water potential or a_w units. An osmometer that reports osmolality could just as easily report a_w or freezing point depression.

Soil–Plant Applications

Liquid water moves from the soil to and through the roots, through the xylem of plants, to the leaves, and eventually evaporates in the substomatal cavities of the leaf. The driving force for this flow is a water potential gradient. In order for water to flow the leaf water potential must be more negative than that of the soil. The entire system is sometimes thought of as being similar to a resistor network in an electronic circuit where water and current flow are analogous, and where the potential differences are like voltage differences in the circuit. Ohm's law is then used to describe the flow of water in the soil–plant system.

The transpiration stream flows through and around individual cells, so a local water potential equilibrium is established between the cell and its environment. The total water potential of the cell is the sum of its components, as discussed earlier, so the sum of osmotic, matric, and pressure components must add to the total water potential at that location in the transpiration stream. Within the cell, the main components of the water potential are the pressure and osmotic potentials. The cytoplasm is so highly hydrated that there is little contribution from the matric potential. In the cell wall, the main component is a matric potential. Solute concentration is low in the cell wall, so the osmotic contribution is small.

Figure 14.1 shows a moisture-release curve, or "isotherm," for leaf tissue (Campbell et al. 1979). Two regions are clear, one where pressure within the cells is positive (positive turgor), and another where the cells are flaccid. When the cell is turgid, small changes in water content will result in large changes in total potential and, therefore, turgor pressure, with minimal change in osmotic potential. The turgor pressure is 0 when the tissue is flaccid, so the water potential change is much less per unit change in water content, and the change in total potential is equal to the change in osmotic potential. Plants growing in environments to which they were adapted typically remain turgid at all evaporative demands. The consequence is that the osmotic environment inside the cell remains almost constant in the face of wildly fluctuating values of total potential (diurnal variation due to evaporative demand). The osmotic potential, rather than the total potential, determines the biochemical suitability of the cell environment. Maintaining a relatively constant osmotic po-

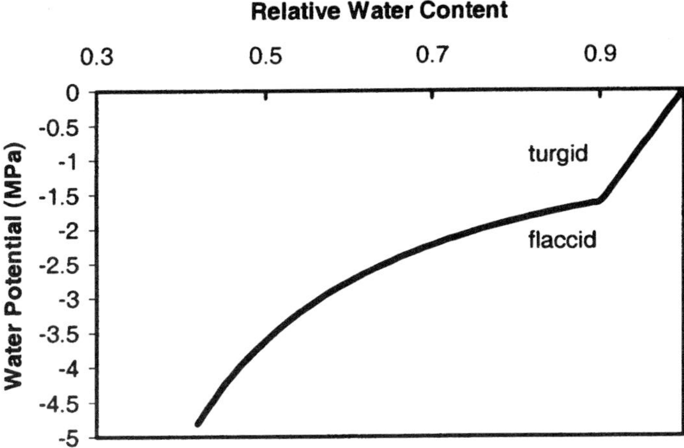

Figure 14.1 Moisture-release curve for a wheat leaf.

Table 14.2 Hypothetical but Typical Water Potentials in a Soil–Plant System Under Well-Watered and Dry Conditions.

	Wet Soil, Low Demand (MPa)	Dry Soil, High Demand (MPa)
Soil	–0.03	–1.0
Root surface	–0.05	–1.1
Root xylem	–0.6	1.7
Leaf xylem	–0.9	–1.9
Leaf mesophyll	–1.0	–2.0

tential assures a stable cell environment suitable for photosynthesis, respiration, meiosis, and other biochemical processes that take place in a living cell. Cells with rigid walls are responsible for this relatively constant osmotic environment within the cell.

Table 14.2 gives typical water potentials and corresponding water activities for the soil, xylem, and leaf at low evaporative demand, high soil moisture conditions, and high demand/low soil moisture conditions. Two water potential values that have significance in plant water relations correspond to "field capacity" and "permanent wilting point." Field capacity is the water potential of a soil profile, which has been thoroughly wetted, and then allowed to drain without water use by evaporation or plant uptake for 2 to 3 days. Many factors influence the actual value attained, but in a deep, uniform, medium texture soil, the field capacity water potential is likely to be around −33 J/kg. This corresponds to an a_w of 0.99976. Virtually all water movement in soil therefore occurs when a_w is above 0.9998.

If a dwarf sunflower is grown in a small pot and allowed to use water from the soil in the pot until the sunflower wilts and cannot recover turgor overnight in a humid atmosphere, the soil is said to be at the permanent wilting point. If the water potential of the soil is measured at permanent wilting, it will be around −1500 J/kg, which corresponds to an a_w of 0.989. While plants other than dwarf sunflowers, in rooting conditions other

Table 14.3 Water Activity and Corresponding Water Potential Values, and the Wet Basis Water Contents of Rape (*Brassica napus*), an Oily Seed, and Wheat (*Triticum vulgare*), a Non-oily Seed.

Water Activity	Water Potential (MPa)	Rape Water Content[a] (g/g)	Wheat Water Content[a] (g/g)
0.10	–314	0.031	0.060
0.20	–219	0.039	0.080
0.30	–164	0.045	0.093
0.40	–125	0.052	0.106
0.50	–94.4	0.060	0.120
0.60	–69.6	0.069	0.132
0.70	–48.6	0.080	0.147
0.80	–30.4	0.093	0.163
0.90	–14.3	0.121	0.215

[a] Water content data are from Roberts, E.H. 1972. *Viability of Seeds*. London: Chapman and Hall.

than small pots, can withdraw water to potentials lower than −1500 J/kg, the permanent wilting point still provides a useful benchmark. Two important points can be made here. One is that in plants, and soils that support plant growth, the a_w must always be very near 1.0, much higher than is typical of food systems. The other is that, by far, the largest a_w gradient in the soil–plant–atmosphere system is between the leaf and the atmosphere, because, within the leaf, the a_w is likely above 0.99, and outside the leaf, it may be 0.50 or below.

Water Activity and Seeds

Seed Viability

For historical reasons, a_w is not commonly used in seed storage but has several distinct advantages for specifying conditions related to seed longevity. Three environmental factors are of fundamental importance in determining the longevity of stored seeds: water, temperature, and oxygen. With respect to their response to water, seeds are classified into two groups: orthodox and recalcitrant (Roberts 1973). Orthodox seeds can be dried to low water contents without damage. Recalcitrant seeds are similar to other plant tissue in that they experience desiccation death if they are allowed to dry below some critical moisture level. Our discussion applies mainly to orthodox seeds, although the principles are similar for both.

As the water content of seeds changes, so does their a_w. For a particular seed sample, with a given wetting history, there exists a unique relationship between water content and a_w called the moisture sorption isotherm (Roberts and Ellis 1989). If one has the isotherm for a particular sample, it does not matter whether water content or a_w is measured, because the isotherm allows the other to be inferred. If no isotherm is available, then one should determine which of the two variables best represents the process of interest and measure that variable, because the isotherm for each species is unique and there is no general relationship that allows conversion from one to the other.

Table 14.3 shows data for rape (an oily seed) and barley (a nonoily seed). As can be seen, rape has much lower water contents than barley at all of the water activities shown. The table also shows the corresponding water potentials. Clearly, isotherms differ substantially from species to species. Isotherms may also differ from cultivar to cultivar, and can depend on the environment in which the seed was produced. Extensive tables relating a_w (equilibrium relative humidity) and water content of seeds are given in Roberts (1972).

Water content has long been used to describe the effect of moisture on seed viability, and recommendations for seed storage conditions for maximum longevity are often given in terms of water content. The relationships differ for each species, however, and must be determined for each. Roberts and Ellis (1989) have shown that a logarithmic relationship exists between longevity and seed water content, but the relationship differs for each species.

On the other hand, Roberts and Ellis (1989) show that the relationship between a_w and longevity is linear and similar from species to species. This is to be expected because the a_w measures the availability of the water to participate in chemical and physical processes. The use of a_w to describe seed water status therefore has an enormous advantage over the use of water content. It eliminates specific testing of each seed lot and provides a general and simple (linear) relationship between viability and seed water status.

Roberts and Ellis (1989) state that the rate of loss of viability in orthodox seeds increases with increasing water potential between water potentials of −350 and −14 MPa. These values correspond roughly to 0.1 and 0.9 a_w (10% and 90% RH). Below −350 MPa (0.077 a_w), there is little change in longevity with decreasing moisture. This lower limit corresponds to seed water contents ranging from 2% to 6%. Clearly, a simple and quick a_w measurement can supply the information needed to know whether seeds are at optimum moisture for storage, while substantial research is required to know what water content would be appropriate.

Seed Coating

Coating is a process designed to create uniform layers of material over the seed. Coating helps to protect, smooth, and round the seed surface, which improves flowability and machine planting. The coating process also allows application of nutrients, fungicides, insecticides, colors, and other additives to create a nutritious environment in the immediate vicinity of the germinating seed. This provides a "boost" for the seedling in its critical early stages of development.

Water activity is the driving force for moisture migration; thus, when seeds are coated, it is the equilibrium a_w that best determines longevity and viability. Water content becomes even less useful as a measure of seed moisture when coatings are applied to seeds. The seed and coatings can have the same a_w but vastly different water contents. The mass of the coating is often several times the mass of the enclosed seed, and the isotherm for the coating material is entirely different from that of the seed (Taylor et al. 1997). At an a_w of 0.10, the water content of the seed might be 0.06 g/g, while that of the coating might be 0.02 g/g. Because the coating mass is so much larger than the seed mass, the overall water content would be close to 0.02 g/g. If the specification for safe seed storage calls for a seed water content of below 0.06 g/g, the water content measurement on the coated seed is of almost no value in determining whether or not a given seed drying operation meets that specification.

The a_w, on the other hand, is the same for the seed and the coating and can be easily measured on both with no pretreatment (Taylor et al. 1997). If the specification calls for an a_w below 0.1 for safe storage, seeds and their coatings are simply dried to this a_w and stored.

Seed Priming

Seeds germinate when a_w reaches a critical physiological level in the seed (Wilson and Harris 1966, Bradford 1995). This varies within and between plant species, but in general occurs when the seed environment is between 0.985 and 1.0 a_w (-2 and 0 MPa). Exceptions occur when seeds have impenetrable seed coats or contain dormancy resulting in chemicals that must be removed before germination occurs. Seeds having permeable seed coats usually go through three recognizable phases of germination: (1) imbibition—a_w of the seed environment is higher than that inside the seed, causing water molecules to flow through the seed epidermis into the embryo, leading to (2) the activation phase, in which stored seed hormones and enzymes stimulate physiological development, leading to (3) growth of the radical, ending the germination phase.

Dormant (dry) seeds are usually at very low a_w, in the range of 0.08 to 0.70 a_w. Some metabolism occurs even at these low water activities. Water movement into dry seeds during the imbibition phase is at first rapid but slows as the a_w of the seeds approaches the a_w of the environment. If imbibition is too rapid (from an environment where $a_w = 1$), damage to hydrating cells often occurs. Seed technologists control the imbibition of seeds through priming to improve seed performance.

Priming describes an osmotic seed treatment to enhance seed processes for rapid and uniform germination (Heydecker 1977, Heydecker and Coolbear 1977, Kahn 1977, Bradford 1986). It is only during the activation phase that seed priming can be successfully accomplished. Seed priming involves controlled hydration without germination by suspension of the seeds in a liquid or solid imbibing medium that is conditioned to an a_w or water potential near but less than germination level. Seeds are hydrated until the a_w or water potential of the seeds and the imbibing medium reach equilibrium. A theoretical example of germination at a series of high water potential levels is given in Figure 14.2, showing the effect of near-germination water potential levels. Germination at high water potential, as shown in the 0-MPa curve, occurs quickly, without allowing priming opportunity. By allowing the priming medium and seeds to come to equilibrium with each other at lower water potential levels than the dotted line, time in phase II is lengthened, allowing priming activities to proceed. At low water potential (-1.5 MPa) or a_w (0.989 a_w), water is not sufficiently available for radical emergence to occur.

The purpose of priming is to reduce the germination time in the field, make germination occur over a short period, improve stand establishment, and improve percentage germination. Priming materials include substances for either matric or osmotic priming, although some seed priming carriers include both osmotic and matric processes (Kahn 1992). Some are available commercially, while others are held for private competitive business use.

Medical Applications

The osmotic potential of human blood is around -0.7 MPa. The osmotic potential of fresh sweat in a normal subject is between -0.3 and -0.4 MPa. That of urine is -2.0 to -3.0 MPa (Campbell and Norman 1998). It is clear that the body uses metabolic energy to concentrate solutes in urine and reduce salt loss in sweat. In patients with cystic fibrosis, the ability to reduce salt loss in sweat is impaired, so sweat in these patients has an osmotic potential near that of blood. The sweat water potential, or osmolality, can be measured in infants as an early screening procedure to detect cystic fibrosis (Webster and Lochlin 1977,

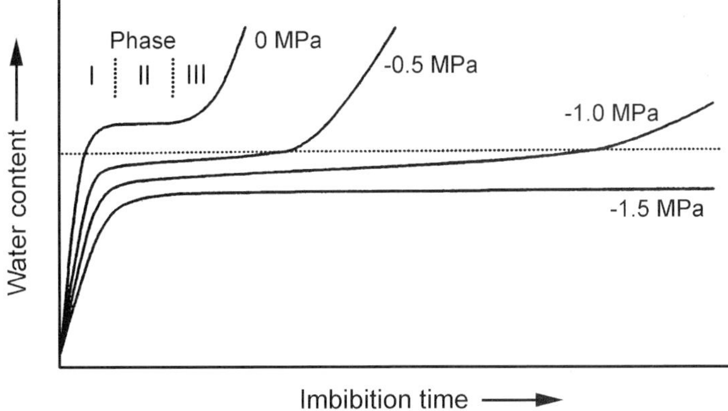

Figure 14.2 Time courses of seed water potential during three phases of germination. Imbibition at reduced water potential lowers seed water content, extends the length of phase II, and delays entry into phase III. Radical emergence and growth will occur if the water content exceeds a critical level indicated by the horizontal dotted line (reprinted from Bradford, K.J. 1995. Water Relations in Seed Germination. In: *Seed Development and Germination*, eds. J. Kigel and G. Galili, Chapter 13, pp. 351–396. New York: Marcel Dekker, with permission from Taylor and Francis Group, LLC).

Webster and Barlow 1981, Barben et al. 2005). Other applications of these measurements relate to dehydration.

Sugar, honey, and sugar paste have been used to treat wounds and burns for thousands of years. Majno (1975) and Selwyn and Durodie (1985) give a historical review of this therapy, starting from ancient Egypt to the Roman physician Galen in 170 AD and through scientific studies by physicians in the eighteenth, nineteenth, and twentieth centuries. Despite its use as a traditional remedy for burns and wounds, the method is not well recognized in mainstream medical care (Lusby and Coombes 2002). Sugar, when used for wound management, has rapid antimicrobial action, enhanced granulation formation, epithelialization, accelerated wound healing, and minimal cost and is readily available. Sugar paste was reported as being used successfully in 605 patients with wounds, burns, and ulcers. Its use has been associated with lower requirements for skin grafting, antibiotics, and lower hospital costs (Knutson et al. 1981). As well as having antimicrobial properties, honey and sugar paste are associated with scarless healing in some cavity wounds (Topham 2000, 2002).

Sugar therapy has been successfully used for injuries such as infected surgical wounds, necrotizing fasciitis, skin ulcers, crush injuries, deep tissue infections, malodorous wounds, or other skin defects that need a healthy granulation bed. The sugar therapy process consists of the wound being cleansed and dried with a sterile towel. Sugar is then poured onto the wound to a depth of about 1 cm. Some sort of dressing is placed over the wound and the dressing is changed once or twice per day. Normally, the wound is debrided, and healthy tissue is formed in 5 to 9 days. After the wound is decontaminated and healthy granulation tissue is restored, the wound may be closed by conventional means. Repeated applications over 3 to 6 weeks are generally required for complete healing.

An important function of sugar in the treatment of infected wounds is to create an environment of low a_w, which inhibits or stresses bacterial growth (Chirife et al. 1982, 1983). Therefore, sugar, honey, and sugar paste rapidly become bactericidal. Electron microscopy results are in keeping with this viewpoint (Selwyn and Durodie 1985). When samples of the sugar paste were inoculated with 105 cfu/g of *Staphylococcus aureus, Streptococcus faecalis, Escherichia coli,* or *Candida albicans*, less than 10 cfu/g was detectable after 1 hour at 25°C (Gordon et al. 1985). Sugar pastes diluted with serum have a reduced bactericidal effect; −75% paste in serum gave an 80% reduction in viable numbers of *S. aureus* within 2 hours and a 99% reduction in viable numbers of *P. mirabilis* within 1 hour (Ambrose 1986). Although sugar has a dramatic effect on microbial cells, it is harmless to human tissues. The removal of water from tissue surface in the presence of hypertonic sucrose solutions is balanced by the continuous influx of water from deeper layers of the tissue (Chirife et al. 1983).

Pharmaceutical Applications

Knowledge of the a_w of pharmaceuticals (proteins, drugs, and excipients) is essential in obtaining a dosage form with optimal chemical, physical, microbial and shelf-life properties. Moisture content routinely found in pharmaceutical specifications is only a qualitative parameter when the composition is constant. Pharmaceuticals have varying compositions, and thus a_w, not moisture content, allows evaluation of chemical stability, microbial stability, flow properties, compaction, hardness, and dissolution rate of dosage forms of pharmaceuticals, proteins, biopharmaceuticals, nutraceuticals, and phytochemicals.

The a_w concept has served the microbiologist and food technologist for decades and is the most commonly used criterion for safety and quality. Water activity is a better index for microbial growth than water content, as described in Chapter 10. Knowledge of the behavior of microorganisms in pharmaceutical products at different a_w levels is important in meeting federal food, drug, and cosmetic laws. Many pharmaceutical systems and consumer products can be formulated to be self-preserving or naturally preserving due to a_w (Curry 1985, Enigl and Sorrells 1997; Enigl 1999). Even when using a_w for microbial control, it is essential to follow strict good manufacturing practices to ensure the lowest possible load of microorganisms.

The concept of using a_w for microbial limits testing is gaining interest in the pharmaceutical industry (Friedel and Cundell 1998, Friedel 1999). The USP (United States Pharmacopeia) Method <1112> Microbiological Attributes of Nonsterile Pharmaceutical Products—Application of Water Activity Determination provides guidance on the influence of a_w as it pertains to the susceptibility of a product formulation to microbial contamination. USP <1112> discusses the potential for improving product preservation by maintaining low a_w. The determination of a_w of nonsterile pharmaceutical dosage forms aids in the following:

- Optimizing product formulations to improve antimicrobial effectiveness of preservative systems
- Reducing the degradation of active pharmaceutical ingredients within product formulations susceptible to chemical hydrolysis
- Reducing the susceptibility of formulations (especially liquids, ointments, lotions, and creams) to microbial contamination

- Providing a tool for the rationale for reducing the frequency of microbial limit testing and screening for objectionable microorganisms for product release and stability testing using methods contained in the general test chapter Microbial Limit Tests <61> (USP-NF 2006)

Similar to microbial control, the reaction rates of many chemical reactions that impact drug formulation stability are dependent on a_w. Water may influence chemical reactivity by acting as a solvent or reactant, or by changing the mobility of the reactants by affecting the viscosity of the pharmaceutical dosage form, as discussed in Chapter 7. Hydrolytic reactions, lipid peroxidation, Maillard reactions, and enzymatic activity are examples of chemical reactions that show a dependence on a_w (Labuza and Saltmarch 1981, Drapron 1985, Karel 1986, Leung 1987, Carstensen 1988, Bell and Labuza 1994).

Protein, enzyme, and biopharmaceutical stabilities are influenced significantly by a_w due to their relatively fragile nature. Great care must be taken to prevent aggregation under pharmaceutically relevant conditions. Most proteins, enzymes, and biopharmaceuticals also must maintain integrity to remain active. Maintaining critical a_w levels to prevent dissolution, aggregation, and conformational changes from occurring is important to deliver the correct dosage (Costantino et al. 1994, Hageman 1988).

Because a_w describes the thermodynamic energy status of the water within a system, there is a close relationship between a_w and the physical stability of pharmaceutical forms. The difference in a_w levels between components or a component and the environmental humidity is a driving force for moisture migration. Knowledge of whether water will absorb or desorb from a particular component is essential to prevent degradation, especially if one of the substances is moisture sensitive. For example, if equal amounts of component A at 2% and component B at 10% moisture content are to be blended together, will there be moisture exchange between the components? The final moisture content of the blended material would be 6%, but was there any moisture exchange between components A and B? The answer depends on the water activities of the two components. If the water activities of the two components are the same, then no moisture will exchange between the two components.

Likewise, two ingredients at the same moisture content may not be compatible when mixed together. If two materials of differing water activities but the same water content are mixed together, the water will adjust between the materials until an equilibrium a_w is obtained. Thus, for gelatin capsule design, one should match the a_w of the capsule to the internal drug dosage formulation. If the capsule material is at a higher a_w than the dosage formulation, water will migrate into the dosage formulation and may increase the rate of chemical degradation or cause the capsules to shrink or crack and leak. In contrast, if the drug dosage formulation is at a higher a_w than the capsule, water will migrate to the capsule and may cause the capsules to swell or become sticky.

The importance of a_w as opposed to total water content is shown in preformulation compatibility studies involving moisture-sensitive drugs (Heidemann and Jarosz 1991). Using a_w, hygroscopic excipients (starch, cellulose, and magaldrate) have successfully been formulated for use with moisture sensitive drugs. The excipients preferentially bind moisture and make the dosage form less susceptible to changes in relative humidity during manufacture, shipment, storage, or patient use, thus extending shelf-life. This is also applicable to other polymer systems of pharmaceutical interest, such as proteins (e.g., gelatin and keratin) and various synthetic hydrogels.

Additionally, a_w may be used in studying the response of solid dosage forms to changing environments. More robust package designs and formulations resistant to customer abuse can be developed using a_w information. Aging and package testing can be easily accomplished by measuring a_w changes over time. When tablets are in the process of equilibrating to a higher or lower value or have a coating, the tablets should be crushed to obtain an accurate a_w for the entire tablet.

Water activity of powders affects the flow, caking, compaction, and strength properties of solid dosage forms (Dawoodbhai and Rhodes 1989, Ahlneck and Zografi 1990). Many pharmaceutical materials are manufactured as wet granulations and then dried down to a specified moisture content. However, due to hystersis within the moisture sorption isotherm, the a_w is different when a formulation is dried down (desorption), as opposed to wetted up (adsorption) to the same moisture content. What looks like the same material with the same moisture content will have very different water activities. This difference will dramatically affect further processing, preservation, and shelf-life of the material.

Water activity also has use in the design and development of coating and capsule technology. The confectionery industry has long used a_w for coating candies and gums. Several pharmaceutical companies are beginning to investigate these technologies for new drug delivery systems.

This chapter lists only a few of the many applications that a_w measurements could offer the pharmaceutical industry. Water activity should therefore become an official pharmacopoeial method, as the importance of measuring a_w has long been recognized by the food industry and would provide a similar benefit to the pharmaceutical industry.

References

Ahlneck, C., and Zografi, G. 1990. The molecular basis of moisture effects on the physical and chemical stability of drugs in the solid state. *International Journal of Pharmaceutics* 62:87–95.

Ambrose, U. 1986. An investigation into the mode of action of Northwick Park Hospital sugar pastes. Applied biology thesis. Hartfield Polytechnic.

Barben, J., Ammann, R.A., Metlagel, A., and Schoeni, M.H. 2005. Conductivity determined by a new sweat analyzer compared with chloride concentrations for the diagnosis of cystic fibrosis. *Journal of Pediatrics* 146:183–188.

Bell, L.N., and Labuza, T.P. 1994. Influence of the low-moisture state on pH and its implication for reaction kinetics. *Journal of Food Engineering* 22:291–312.

Bradford, K.J. 1986. Manipulation of seed water relations via osmotic priming to improve germination under stress conditions. *HortScience* 21(5):1105–1112.

Bradford, K.J. 1995. Water relations in seed germination. In: *Seed Development and Germination*, eds. J. Kigel and G. Galili, Chapter 13, pp. 351–396. New York: Marcel Dekker.

Buck, A.L. 1981. New equations for computing vapor pressure and enhancement factor. *Journal of Applied Meteorology* 20:1527–1532

Campbell, G.S., and Norman, J.M. 1998. *An Introduction to Environmental Biophysics,* 2nd Edition. New York: Springer Verlag.

Campbell, G.S., Papendick, R.I., Rabie, E., and Shayo-Ngowi, A.J. 1979. A comparison of osmotic potential, elastic modulus, and apoplastic water in leaves of dryland winter wheat. *Agronomics Journal* 71:31–36.

Carstensen, J.T. 1988. Effect of moisture on the stability of solid dosage forms. *Drug Development and Industrial Pharmacy* 14(14):1927–1969.

Chirife, J., Herszage, L., Joseph, A., and Kohn, E.S. 1983. In vitro study of bacterial growth inhibition in concentrated sugar solutions: Microbiological basis for the use of sugar in treating infected wounds. *Antimicrobial Agents and Chemotherapy* 23(5):766–773.

Chirife, J., Scarmato, G., and Herszage, L. 1982. Scientific basis for use of granulated sugar in treatment of infected wounds. *The Lancet* 1(8271):560–561.

Costantino, H.R., Langer, R., and Klibanov, A.M. 1994. Solid-phase aggregation of proteins under pharmaceutically relevant conditions. *Journal of Pharmacy Science* 83:1662–1669.

Curry, J. 1985. Water activity and preservation. *Cosmetics and Toiletries* 100:53–55.

Dawoodbhai, S., and Rhodes, C.T. 1989. The effect of moisture on powder flow and on compaction and physical stability of tablets. *Drug Development and Industrial Pharmacy* 15:1577–1600.

Drapron, R. 1985. Enzyme activity as a function of water activity. In: *Properties of Water in Foods*, eds. D. Simato and J.L. Multon, pp. 171–190. Dordrecht, the Netherlands: Martinus Nijhoff.

Enigl, D. 1999. Creating "natural" preservative systems by controlling water activity. *Pharmaceutical Formulation and Quality* 3(5):29–34.

Enigl, D., and Sorrells, K. 1997. Water activity and self-preserving formulas. In: *Preservative-Free and Self-Preserving Cosmetics and Drugs*, eds. J. Kabara and D. Orth, pp. 45–73. New York: Marcel Dekker.

Friedel, R.R. 1999. The application of water activity measurement to microbiological attributes testing of raw materials used in the manufacture of nonsterile pharmaceutical products. *Pharmacopeial Forum* 25:8974–8981.

Friedel, R.R., and Cundell, A.M. 1998. The application of water activity measurement to the microbiological attributes testing of nonsterile over-the-counter drug products. *Pharmacopeial Forum* 24:6087–6090.

Gordon, H., Middleton, K., Seal, D., and Sullens, K. 1985. Sugar and wound healing. *The Lancet* 2(8456):663–664.

Hageman, M.J. 1988. The role of moisture in protein stability. *Drug Development and Industrial Pharmacy* 14:2047–70.

Heidemann, D.R., and Jarosz, P.J. 1991. Performulation studies involving moisture uptake in solid dosage forms. *Pharmaceutical Research* 8:292–297.

Heydecker, W. 1977. Stress and seed germination. In: *The Physiology and Biochemistry of Seed Dormancy and Germination*, ed. A.A. Khan, pp. 240–282. Amsterdam: Elsevier.

Heydecker, W., and Coolbear, P. 1977. Seed treatments for improved performance—Survey and attempted prognosis. *Seed Science and Technology* 5:353–425.

Karel, M. 1986. Control of lipid oxidation in dried foods. In: *Concentration and Drying of Foods,* ed. D. MacCarthy, pp. 37–51. London: Elsevier Applied Science.

Khan, A.A. 1977. Preconditioning, germination and performance of seeds. In: *The Physiology and Biochemistry of Seed Dormancy and Germination*, ed. A.A. Khan, pp. 283–316. Amsterdam: Elsevier.

Khan, A.A. 1992. Preplant physiological seed conditioning. In: *Horticultrual Reviews*, Vol. 13, ed. J. Janick, pp. 131–181. New York: John Wiley.

Knutson, R.A., Merbitz, L.A., Creekmmore, M.A., and Snipes, H.G. 1981. Use of sugar and povidone-iodine to enhance wound healing: Five years' experience. *South Medical Journal* 74:1329–1335.

Labuza, T.P., and Saltmarch, M. 1981. The nonenzymatic browning reaction as affected by water in foods. In: *Water Activity: Influences on Food Quality*, eds. L.B. Rockland and G.F. Stewart, pp. 605–650. New York: Academic Press.

Leung, H.K. 1987. Influence of water activity on chemical reactivity. In: *Water Activity: Theory and Applications to Food,* eds. L.B. Rockland and L.R. Beuchat, pp. 27–54. New York: Marcel Dekker.

Lusby, P.E., and Coombes, A. 2002. Honey: A potent agent of wound healing? *Journal of Wound, Ostomy and Continence Nursing* 29(6):295–300.

Majno, G. 1975. *The Healing Hand: Man and Wound in the Ancient World.* Cambridge, MA: Harvard University Press.

Roberts, E.H. 1972. *Viability of Seeds*. London: Chapman and Hall.

Roberts, E.H. 1973. Predicting the storage life of seeds. *Seed Science and Technology* 1:499–514

Roberts, E.H., and Ellis, R.H. 1989. Water and seed survival. *Annals of Botany* 63:39–52.

Selwyn, S., and Durodie, J. 1985. The antimicrobial activity of sugar against pathogens of wounds and other infections of man. In: *Properties of Water in Foods,* eds. D. Simatos and J.L. Multon, pp. 293–308. Dordrecht, the Netherlands: Martinus Nijhoff.

Taylor, A.G., Grabe, D.F., and Paine, D.H. 1997. Moisture content and water activity determination of pelleted and film-coated seeds. *Seed Technology* 19(1):24–32.

Taylor, S.A. 1958. The activity of water in soils. *Soil Science* 86:83-90.

Taylor, S.A. 1972. *Physical Edaphology: The Physics of Irrigated and Nonirrigated Soils.* San Francisco: Freeman.

Topham, J. 2000. Sugar for wounds. *Journal of Tissue Viability* 10(3):86–89.

Topham, J. 2002. Why do some cavity wounds treated with honey or sugar paste heal without scarring? *Journal of Wound Care* 11(2):53–55.

USP-NF. 2006. *Method <1112> Application of water activity determination to nonsterile pharmaceutical products.* United States Pharmacopeia-National Formulary, pp. 3802–3803.

Webster, H.L., and Barlow, W.K. 1981. New approach to cystic fibrosis diagnosis by use of an improved sweat-induction/collection system and osmometry. *Clinical Chemistry* 27:385.

Webster, H.L., and Lochlin, H. 1977. Cystic fibrosis screening by sweat analysis. A critical review of techniques. *Medical Journal of Australia* 1:923.

Wilson, A.M., and Harris, G.A. 1966. Hexose, inositol, and nucleoside phosphate esters in germinating seed of crested wheatgrass. *Plant Physiology* 41:1416–1419.

15 The Future of Water Activity in Food Processing and Preservation

Cynthia M. Stewart, Ken A. Buckle, and Martin B. Cole

Although the control of water activity (a_w) is perhaps the earliest means of preserving food, a number of recent trends and developments in food science are likely to have a significant impact on the way it is used in the future. The development of predictive microbial models and quantitative risk assessment is leading to a paradigm shift in the way that food safety risks are managed internationally. This will lead to the use of glass transition temperature (T_g) as a well-defined control measure within a risk management framework. Another recent development in food preservation is the use of nonthermal processes for pasteurizing and sterilizing foods; the safe application of these new food processes will require that the effect of T_g is well understood. Finally, developments in whole cell genomics give the potential to examine which genes or gene products are switched on in response to microbial stresses. As with other preservative factors, it is hoped that these techniques will allow a more detailed mechanistic understanding of the effects of reduced a_w on microbial cells to be developed.

Developments in Risk Management

New developments in risk management are likely to lead to the formalization of how different preservation hurdles, such as reduced a_w, can be used in combination to control particular microbial hazards. Predictive modeling and risk assessment now offer the potential of linking exposure to a microbial hazard to the likely number of cases of illness in the population, and they are driving new risk management approaches based on the concept of "food safety objectives" (FSOs). Although quantitative aspects of the scheme are still being advanced, the framework is helping to facilitate transparent communication of the food safety responsibilities of the different stakeholders across the food chain. The appropriate level of protection (ALOP), as derived from a microbiological risk assessment, is typically expressed in terms relevant to public health, e.g., as the number of cases in a population of 100,000. While this serves a purpose when informing the public, especially when communicating a desired reduction in disease, the ALOP is not a useful measure in the further implementation of food safety measures, for example, at the level of food control/inspection or food production.

The FSO concept aims to translate public health risk into a definable goal: "the maximum frequency and/or concentration of a hazard in a food at the time of consumption, that provides or contributes to the appropriate level of protection (ALOP)" (Codex Committee on Food Hygiene [CCFH] report 2004). The approach enables the food industry to meet a specific FSO through the application of the principles of Good Hygienic Practice, Hazard Analysis and Critical Control Points (HACCP) systems, performance criteria, process/

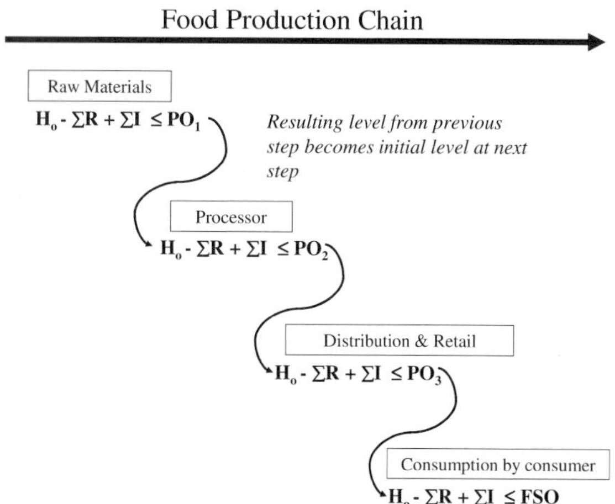

Figure 15.1 Schematic demonstrating the use of a performance objective at each step in the food chain to achieve the food safety objective at the point of consumption. H_0 = initial level of the hazard, ΣR = the total (cumulative) reduction of the hazard, ΣI = the total (cumulative) increase of the hazard, PO = performance objective, FSO = food safety objective (from Stewart, C.M., Cole, M.B., and Schaffner, D.W. 2003. Managing the risk of staphylococcal food poisoning from cream-filled baked goods using food safety objectives. *Journal of Food Protection* 66:1310–1325. Reprinted with permission from the *Journal of Food Protection*. Copyright held by the International Association for Food Protection, Des Moines, Iowa, U.S.A.).

product criteria, and/or acceptance criteria. Using the FSO concept provides a scientific basis that allows industry to select and implement measures of hazard control for a specific food or food operation. It enables regulators to better develop and implement inspection procedures to assess the adequacy of the control measures implemented by industry and to quantify the equivalence of inspection procedures in different countries. Thus, the practical value of using the FSO concept is that it offers flexibility of operation; it does not prescribe how an operation achieves compliance—it defines the goal.

Performance Objectives

In recognition of the need to express the level of a hazard at different points in the value chain, the term *performance objective* (PO) has been defined as "the maximum frequency and/or concentration of a hazard in a food at a specified step in the food chain before the time of consumption that provides or contributes to an FSO or ALOP, as applicable" (CCFH 2004) (see Figure 15.1). From the information provided in an FSO, regulatory authorities and food operators can select appropriate control measures to achieve the intended safe levels of pathogens. A control measure is "any action and activity that can be used to prevent or eliminate a food safety hazard or reduce it to an acceptable level" (CCFH 2004). One or more control measures may be necessary at each stage along the food chain to ensure a food is safe when consumed. In the design of control measures, it is necessary to establish *performance criteria,* and how it will be achieved; *process criteria;* and *product criteria.*

Performance, Process, and Product Criteria

When designing and controlling food operations, it is necessary to consider pathogen contamination, destruction, survival, growth, and possible recontamination. Consideration must be given to the subsequent conditions to which the food is likely to be exposed, including further processing and potential abuse (time, temperature, cross-contamination) during storage, distribution, and preparation. The ability of those in control of foods at each stage in the food chain to prevent, eliminate, or reduce food safety hazards varies with the type of food processed and the effectiveness of available technology.

A *performance criterion* is defined as "the effect in frequency and/or concentration of a hazard in a food that must be achieved by the application of one or more control measures to provide or contribute to a PO or FSO" (CCFH 2004). When establishing performance criteria, account must be taken of the initial levels of the hazard and changes of the hazard during production, processing, distribution, storage, preparation, and use. An example of a performance criterion is a 6-$\log_{10}$ reduction of salmonellae when cooking ground beef.

Process criteria are the control parameters (e.g., time, temperature, pH, a_w) at a step, or combination of steps, that can be applied to achieve a performance criterion. For example, the control parameters for milk pasteurization in the United States are 71.7°C for 15 seconds. This combination of temperature and time will ensure the destruction of *Coxiella burnetii*, as well as other non–spore-forming pathogens that are known to occur in raw milk.

Product criteria consist of parameters that are used to prevent unacceptable multiplication of microorganisms in foods. Microbial growth is dependent on the composition and environment of the food. Consequently, pH, a_w, temperature, gas atmosphere, etc., can influence the safety of particular foods. For example, it may be necessary for a food to have a certain pH (e.g., 4.6 or below) or a_w (e.g., 0.86 or below) to ensure that it meets an FSO for a particular pathogen, for which growth in the product must be limited (e.g., *Clostridium botulinum* or *Staphylococcus aureus*).

A performance criterion is preferably less but at least equal to the FSO and can be expressed by the following equation:

$$H_0 - \Sigma R + \Sigma I \leq FSO \qquad (15.1)$$

where H_0 is initial level of the hazard, ΣR is total (cumulative) reduction of the hazard, and ΣI is total (cumulative) increase of the hazard (FSO, H_0, R, and I are expressed in $\log_{10}$ units).

It should be recognized that the parameters used in the above equation are point estimates, whereas in practice, they will have a distribution of values associated with them. If data exist for the variance associated with the different parameters, then the underlying probability distributions may be established using an approach similar to that in risk assessment.

In the following example, Stewart et al. (2003) illustrated the application of International Commission on Microbiological Specifications for Food (ICMSF) principles in setting an FSO and the use of performance, process, and product criteria to demonstrate how an FSO could be met for *S. aureus* in cream-filled baked goods. Although there may be a cooking step in the production of the cream filling, which could adequately control *S. aureus* in the product, the potential for recontamination exists while the filling is being cooled or, more likely, during handling of the filling while the pastry is being filled.

Table 15.1 Considerations for H_0, ΣI, and ΣR in Each Step of the Food Chain for *Staphylococcus aureus* in Cream-Filled Baked Goods.

Processor:		
Preventing growth		
Growth (ΣI)	+	Product not formulated to prevent growth
Growth ($\Sigma I = 0$)	–	Product formulated to prevent growth
Cooking of cream filling		
Pathogen present (H_0)	+	Present, low numbers
Pathogen reduced by process (ΣR)	+	Eliminated
Preventing recontamination		
Recontamination (ΣI)	+	Minimize through use of aseptic/hygienic filling and packaging
In-pack pasteurization		
Pathogen reduced by process (ΣR)	+	Eliminated
Distribution and retail		
Recontamination (ΣI)	±	Unlikely if appropriately packaged
Growth (ΣI)	±	Unlikely if chilled distribution OR formulated to prevent growth
Consumer:		
Recontamination (ΣI)	±	Possible through product handling
Growth (ΣI)	±	Possible depending on time-temperature storage

Source: Stewart, C.M., Cole, M.B., and Schaffner, D.W. 2003. Managing the risk of staphylococcal food poisoning from cream-filled baked goods using food safety objectives. *Journal of Food Protection* 66:1310–1325. Reprinted with permission from the *Journal of Food Protection*. Copyright held by the International Association for Food Protection, Des Moines, Iowa, U.S.A.

Therefore, a combination of measures to control *S. aureus* in the product may be required in order to meet the following illustrative FSO: the concentration of *S. aureus* shall not exceed 10^4 cfu/g at any time in the filling before the consumption of cream-filled baked goods. These measures include controlling initial levels in raw materials, reducing the levels during cooking of the cream or custard filling, minimizing the recontamination of the filling between the cooking step and the filling of pastry shells or cakes, and preventing an unacceptable increase in levels by formulation (reduced a_w, reduced pH and/or preservatives) or by good temperature control throughout the distribution chain (see Table 15.1). For example, if the increase in concentration due to recontamination of cooked cream filling and subsequent growth due to improper holding before filling the baked good is assumed to be as high as 100 cfu/g and the FSO of 10^4 cfu/g or less at the time of consumption is still to be met, there can be no greater than a 100-fold increase in the level of *S. aureus* during storage and distribution of the finished cream-filled product before the food is consumed, as given below:

$$
\begin{aligned}
H_0 - \Sigma R + \Sigma I &\leq FSO \\
2 - 0 + \Sigma I &\leq 4 \\
\Sigma I &\leq 2 \log_{10} \text{cfu/g}
\end{aligned}
$$

If recontamination and subsequent growth could occur, then an increase of 100-fold or more could be controlled by appropriate formulation (e.g., through appropriate a_w and pH and/or the addition of preservatives that inhibit growth). Growth boundary models can provide an estimate of the likelihood of the pathogen's ability to grow under given conditions.

Stewart et al. (2002) developed growth boundary models for *S. aureus* based on relative humidity (RH), commonly referred to as a_W (adjusted with various humectants) and pH, both with and without the addition of potassium sorbate (see Figure 15.2). Shelf-life determinations based on various RH, pH, and/or potassium sorbate levels are given in Stewart et al. (2003). Appropriate product criteria could be developed using these models, which would then need to be validated by performing challenge studies of cream fillings. This would allow for assessment of the likely increase (ΣI) of *S. aureus* to 10^4 cfu/g or greater before the recommended use-by-date and thus evaluate the role of the product criteria in achieving the FSO.

Acceptance procedures, which include finished product specifications (chemical, physical, organoleptic, and microbiological), should be established. Additionally, the control measures must be validated. Validation is the obtaining of evidence that the control measures are effective and that the FSO and/or PO will be met. Validation can involve reports from expert panels, scientific literature, laboratory research (e.g., developing data through challenge studies), collecting data during normal processing in food operations (commercial experience with process and products), or comparison with similar processes or products (ICMSF 2002). Each method has its strengths and weaknesses, and in certain cases, two or more methods may be desirable. Additionally, the variability that occurs in a food operation must be considered when establishing the critical limits associated with control measures. For example, factors that can influence variability of a process include equipment performance and reliability, integrity of container seals, processing times and temperatures, pH, humidity, flow rates, and turbulence. Data collected during production (e.g., measurements of a_W, time–temperature, pH, and microbiological data at initial, in-process, and final product points) will allow for the determination of process variability.

After effective control measures have been established, it is necessary to establish procedures to monitor each CCP in HACCP plans and to verify that the control measures are being implemented as planned. Monitoring and verification can consist of a variety of measurements, such as (ICMSF 2002):

- Sensory assessments based on visual inspection, aroma, taste, touch, and sound
- Chemical measurements such as for sodium chloride, water content, and acetic acid
- Physical measurements such as a_W, pH, and temperature–time measurements
- Packaging such as integrity of container closure (e.g., hermetic seal)
- Records for incoming raw materials
- Microbial tests
- Environmental sampling

FSOs and POs define the expected level of control and provide the basis for auditing or inspecting food operations with regard to their GHP, HACCP plan, process or product criteria, validation data, verification records, and process control records. FSOs define the tolerable level of hazards in foods when consumed; POs define the acceptability of processes; microbiological criteria define the acceptability of one or more lots of food.

Food safety management systems based on FSOs and POs provide greater flexibility in how food operators can control hazards. Confidence in the safety of a food depends on the ability of the food industry to control variability in the initial numbers of the hazard (H_0), in processes that reduce the hazard (ΣR) and in controlling the increase of the hazard (ΣI) throughout the food chain. Variability must be considered during process validation to en-

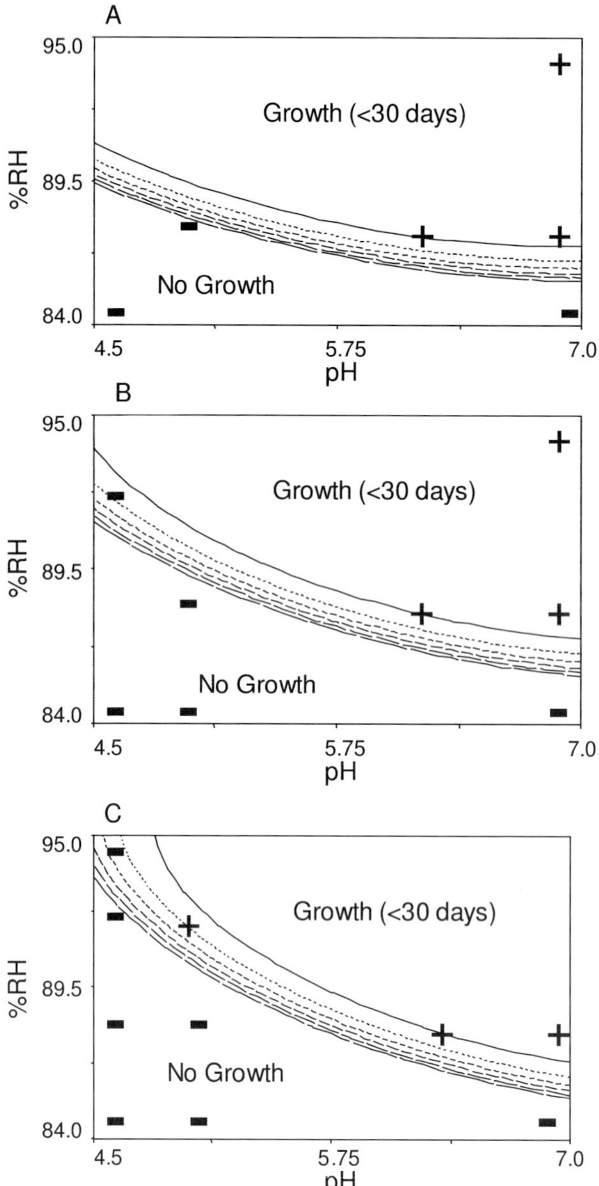

Figure 15.2 Predicted time for growth (TTG) of *Staphylococcus aureus* based on the combined effects of relative humidity (%RH), pH, and potassium sorbate. The contours are scaled in intervals of 30 to 180 days. Moving down from the growth in <30-day region, the first contour encountered is the 30-day line. Moving up from the no-growth region, the first contour encountered is the 180-day line. The region above the 180-day contour line is not a plateau, but a series of ever-increasing contours or longer TTG. (**A**) 0 ppm potassium sorbate, (**B**) 500 ppm potassium sorbate, (**C**) 1000 ppm potassium sorbate. + = positive for enterotoxin, – = negative for enterotoxin (from Stewart, C.M., Cole, M., Legan, J.D., Slade, L., Vandeven, M.H., and Schaffner, D. 2001. Modeling the growth boundaries of *Staphylococcus aureus* for risk assessment purposes. *Journal of Food Protection* 64:51–57. Reprinted with permission from the *Journal of Food Protection*. Copyright held by the International Association for Food Protection, Des Moines, Iowa, U.S.A.).

sure safety but also to avoid overprocessing. In the future, FSOs and POs will become increasingly important in achieving national public health goals and in determining equivalency of safety for foods in international trade.

Nonthermal Processing Technologies and Impact of Water Activity

With few exceptions, the quality of food decreases over time following harvest, slaughter, or manufacture and the nature of this loss is dependent on the type of food, its composition, formulation, processing, and the conditions of storage (Gould 2002). The most important targets for preservation are those most important for maintaining the safety and quality of the product and include microbiological, enzymic, chemical, and physical reactions. When food preservation systems fail, the consequences can be severe (microbial food safety compromised) and economically important (with loss of product due to spoilage), while in some cases, there may be a relatively minimal loss of quality (Gould 2002).

There are a limited number of methods available for food production that combat pathogenic and spoilage microorganisms and delay reactions resulting in quality losses. Most antimicrobial techniques act by delaying or preventing microbial outgrowth (e.g., through reduced a_w, acidification, chilled storage), whereas there are few food processing technologies that primarily inactivate microorganisms in foods (e.g., heat treatment and irradiation) (Gould 2002). However, most of the new and emerging food processing technologies coming into commercial use (or being developed) are focused on microbial inactivation.

While there may be a limited number of food processing and preservation options, consumers are increasingly demanding fresh-tasting food products that are therefore generally less heavily preserved (e.g., with lower added NaCl or sugars) and/or processed, less reliant on food preservatives, and more nutritious and convenient. At the same time, consumer concerns about food safety have steadily increased as the incidence of reported food poisoning outbreaks has continued to rise. These trends have fueled interest in nonthermal processing technologies, such as high-pressure processing (HPP) and pulsed electric fields (PEF). Compared with traditional thermal processing methods, which are often harsh and cause detrimental changes in the texture, flavor, color, and/or nutrient value of foods, the new preservation technologies have a limited detrimental effect on food quality and can be used to minimize or eliminate extensive thermal processing and/or use of chemical preservatives. Preservation of freshness and protection of flavor, appearance, and nutritional value results in a high-quality food product, often with extended shelf-life, which is in line with key market drivers. For these reasons, nonthermal processing technologies offer the ability to produce foods with improved quality, increased consumer appeal, and a value-added premium price. Although commercialization of these technologies has been slow to date, these trends plus improvements in efficiency and reductions in cost mean that the rate of adoption of nonthermal processes is likely to increase significantly.

High-Pressure Processing

HPP has advanced further than the other alternative physical food processing technologies, with the exception of irradiation, which is to some extent due to the advancement of the engineering of equipment such that commercially economic processes have become viable within the last decade or so. Today, high-pressure "pasteurization" has become a commercial reality with several fruit- and vegetable-based refrigerated food products currently on the international market, including a range of juices and fruit smoothies, jams, apple sauce, fruit blends, guacamole and other avocado products, tomato-based salsas and fajita meal kits containing acidified sliced bell peppers and onions, and heat-and-serve beef or chicken slices (acidified and precooked). Additionally, ready-to-eat meat products and seafood, including oysters, are on the market in the United States and Europe (Smelt 1998, Stewart and Cole 2001).

HPP uses pressure (300 to 700 MPa), for a few seconds or minutes, to destroy vegetative microorganisms and can be thought of as a "cold pasteurization" process. One of the major processing advantages is that pressure is transmitted uniformly and instantaneously throughout the food product; therefore, there is no gradient of effectiveness from outside to inside as there is with thermal processing. The chemical effects of HPP are quite different from the effects of heat. The fundamental difference is that HPP causes disruption of hydrophobic and ionic bonds in molecules, while heat affects covalent bonds (Heremans 1992). For example, proteins are affected differently by HPP compared with thermal processing. Structural changes may occur that can lead to unique texture changes in high-protein foods. Enzyme response also varies; some can be partially or fully inactivated by HPP, while others are actually stimulated. Effects on the activity of enzymes may be due to a pressure-induced conformational change in the protein, in which the effect of pressure is on the reaction itself or the disassociation of the enzyme into subunits (Heremans 1992). It is difficult to predict the effect of high pressure on enzyme activity because the new shape that the enzyme takes after pressurization changes its functionality. Additionally, no correlation can be made between the heat sensitivity of an enzyme and sensitivity of the enzyme to pressure.

Microorganisms are variable with regard to their sensitivity to HPP. Results of experiments conducted by Shigehisa et al. (1991) suggest that the order of sensitivity to HPP is Gram-negative bacteria > yeast > Gram-positive bacteria > bacterial spores. Typically, a 10-minute exposure to HPP in the range of 250 to 300 MPa or a 30- to 60-second exposure to HPP in the range of 545 to 600 MPa results in what could be called "cold pasteurization" (Hoover 1997). However, bacterial spores are highly resistant to HPP, even to pressures up to 1000 MPa (Timson and Short 1965, Sale et al. 1970, Cheftel 1992), and hence a combined treatment of parameters such as pressure, mild heat, and low pH is typically required for inactivation. Further information on food preservation by HPP can be found in several review papers, including those by Hoover et al. (1989) and Smelt (1998).

Although reduced a_w can preserve foods, it may also preserve microorganisms within these foods; in other words, microorganisms may survive longer in foods with lower a_w. Additionally, the resistance of microbial vegetative cells to heat (Corry 1974, 1976; Mattick et al. 2001) and bacterial spores (Murrell and Scott 1966) may be greatly enhanced at low a_w. There is a large amount of data available in the literature demonstrating the protective effects of low a_w (for proteins and microorganisms) against heat and numerous reports showing a protective effect of low a_w against pressure (Smelt et al. 2003).

The effect of medium or food composition on pressure resistance of microorganisms was first studied in-depth by Timson and Short (1965). They reported the protective effect of solutes (including glucose and NaCl) on microorganisms pressurized in milk, in addition to the synergistic effects of extreme pH values with pressure treatment (Gould 2002). Ogawa et al. (1992), Oxen and Knorr (1993), and Hashizume et al. (1995) found that pressure resistance of fungi increases as sugar (sucrose, fructose, glucose) concentration in the media increases. For example, *Rhodotorula rubra* was pressure treated at 400 MPa, 25°C, 15 minutes, and showed a 7-$\log_{10}$ reductions, 2-$\log_{10}$ reduction, and no reduction in sucrose solutions with a_w of 0.96, 0.94, and 0.91, respectively (Oxen and Knorr 1993). Knorr (1993, 1995) and Palou et al. (1997) also reported the baroprotective effect of reduced a_w (or increasing soluble solid concentration) on microorganisms undergoing HPP. All of this work demonstrates that yeasts and molds may be protected from inactivation when pressurized at ambient temperature.

Most recently, Goh et al. (2006) reported the osmoprotection effect on six species of fungi *(Saccharomyces cerevisiae, Kloeckera apiculata, Pichia anomala, Penicillium expansum, Rhizopus stolonifer,* and *Fusarium oxysporum*) in sucrose solutions from 50° to 60°Brix after HPP treatment at 600 MPa for 60 seconds. An increase in surviving cells was observed when cells were pressurized with increasing concentrations of sucrose for all the species investigated. For some species (e.g., *K. apiculata* and *P. anomala*), there was a gradual increase in osmoprotection against HPP as the solute concentration increased. However, for other species (e.g., *R. stolonifer* and *F. oxysporum*), there was a sharp increase in resistance once the °Brix of the sucrose solution reached a specific level.

Additionally, Goh et al. (2006) studied how different solutes would affect the baroprotection of the fungi; glycerol and NaCl with equivalent a_w to the sucrose solutions used (0.925, 0.903, and 0.866 a_w) were used. Sucrose gave the most consistent protective effect for *S. cerevisiae*, while NaCl was more protective for the molds studied. Glycerol provided little baroprotection, even with 0.86 a_w. This work shows that apart from the general osmotic effect on the cells and spores, solutes have a specific effect, which is in agreement with Smelt et al. (2003), who reported that NaCl confers less protection than carbohydrates against either heat or pressure damage to microbial cells. Smelt et al. (1998) reported that carbohydrates protect bacterial cells against pressure in the decreasing order of trehalose > sucrose > glucose > fructose > glycerol and that the membrane protective effect of these carbohydrates is also in this order. For example, there was no difference in the level of inactivation (greater than 5-$\log_{10}$ cfu/mL) of *Lactobacillus plantarum* when glycerol was used to achieve various water activities (0.93, 0.95, or 0.97 a_w) and pressure treatment was carried out at 400 MPa, 25°C initial temperature, for 15 minutes. However, little or no inactivation was observed when sucrose was utilized to reduce the a_w to 0.95 or 0.93, respectively, while at 0.97 a_w, slightly less inactivation was reported compared with results with the glycerol medium.

There have been similar a_w-related effects reported with regards to the barotolerance of enzymes. For example, Seyderhelm et al. (1996) reported that the deactivation of pectinmethylesterase was strongly affected in media with different sucrose contents. The higher sucrose concentrations led to increased barostability of the enzyme. The impact on the effectiveness of pressure treatments on enzyme activity can be significantly affected by the medium, thus necessitating a case-by-case approach when examining pressure effects in food systems (Knorr 1998).

Although low a_w protects cells against environmental stress during pressure treatments, surviving microorganisms that have been injured by pressure treatment (or heat) are more difficult to recover on media with a_w that is suboptimal for growth (Patterson et al. 1995). Therefore, the net effect of lower a_w may not always be easy to predict, particularly when food products are treated with pressure in their final retail package.

Pulsed Electric Fields

PEF technology uses high-strength electric fields, typically up to 50 kV/cm, in extremely short time pulses ranging from a few microseconds to milliseconds. Other parameters of

PEF processing include the number of pulses given, typically less than 100, as well as the pulse shape, including exponential decay, square, wave, or oscillatory pulses. The pulses may also be monopolar or bipolar. With PEF processing, food is treated for a short period of time and the energy lost due to heating of food is minimal; therefore, the efficiency of the process is high (Barbosa-Cánovas et al. 1998).

A processing system using high-intensity PEF consists of a power source, capacitor bank, switch, treatment chamber, voltage, current and temperature probes, and aseptic packaging equipment (Barbosa-Cánovas et al. 1998). The power source charges the capacitor bank while the switch is used to discharge energy from the capacitor bank across the food in the treatment chamber. In the treatment chamber, the food is pumped between a negative and positive electrode, where it is subjected to PEF. After treatment, the food is filled into containers aseptically or under ultraclean conditions, depending on the type of product being processed. Unlike HPP, food products cannot be packaged first and then treated by PEF; therefore, postprocessing recontamination is a potential risk.

Based on the dielectric rupture theory, PEF processing creates an electric potential difference across the membrane of microbial cells, which is called the transmembrane potential. When the transmembrane potential reaches a critical or threshold value, pores form in the cell membrane which increases membrane permeability. If the external electric field strength greatly exceeds the critical value, the increase in membrane permeability will be irreversible (Sale and Hamilton 1967, Zimmerman et al. 1974). When the cell membrane permeability increases, the intracellular contents leak as a result, leading to cell death. The efficacy of PEF is dependent on many variables including the electric field strength, the number of pulses, the shape of the pulse, the ionic concentration of the medium/food, conductivity of the suspension medium/food, temperature, and the stage of microbial growth. An excellent review on microbial inactivation by PEF is given by Wouters and Smelt (1998). Because the efficacy of this process is dependent on so many variables, defining the dose delivered by PEF is a complex issue.

The influence of a_w on microbial inactivation has not been widely studied. However, Min et al. (2002) reported that lower a_w, in either a chocolate liquor system or a model system of peptone water and glycerol, protected *Enterobacter cloacae* from PEF inactivation. For example, a PEF treatment of 24.5 kV/cm, for 320 µsec, gave a 1.3-$\log_{10}$ reduction in chocolate liquor with a_w 0.89, but only a 0.5-$\log_{10}$ reduction was achieved with the same treatment when a_w was reduced to 0.57. Similar results were reported for the model glycerol system. Aronsson and Rönner (2001) reported the influence of a_w on the inactivation of *E. coli* and *S. cerevisiae* by PEF. Generally, a decrease in the a_w (achieved by addition of glycerol) protected both *E. coli* and *S. cerevisiae* cells from PEF inactivation, with the effect being more pronounced at lower pHs for *E. coli*. The influence of a_w on inactivation of microorganisms certainly requires more study, and due to the complex nature of food systems, its effect should be studied on a case-by-case basis.

Irradiation

Irradiation was the first modern nonthermal food preservation technology to be commercialized. The World Health Organization has approved radiation dosages for foods of up to 10 kilo-Grays (kGy) as "unconditionally safe for human consumption." Irradiation as a food processing method is approved in 43 countries, including the United States, United Kingdom, Belgium, France, and the Netherlands. There is an extensive amount of information available in the literature about food irradiation, including reviews by Radomyski et al. (1994), Monk et al. (1995), and Farkas (1998) and a book by Barbosa-Cánovas et al. (1998).

There are currently two main methods of producing the ionizing radiation sources for food applications: (1) gamma-irradiation, the traditional radiation method in which the radioisotope used in most commercial facilities is cobalt-60, and (2) electron beam. Cobalt-60 is cost competitive with other processing methods, but it requires special handling during loading and the construction of proper storage for the source, as it will continue to irradiate and decay even when not in use. Massive shielding is required during processing to ensure safety of operators, and it is a process in which the food product is treated after packaging.

Electron beam technology is also cost-competitive, but unlike a radioactive source, it uses a beam of high-energy electrons produced by an accelerator. The source can be turned off when not in use, so no shielding or source storage is necessary, and portable electron beam accelerators exist. Additionally, it makes installation of irradiation facilities at the food production site possible, because electron beam technology uses in-line processing in a continuous system. One disadvantage is that electron-beam radiation does not penetrate the product very deeply (3.3- to 8-cm penetration); however, it can be modified to create x-rays for deeper penetration. X-ray technology, as an adjunct to electron beam technology, is currently in the process of being commercialized. The x-rays are generated via bombardment of a metal target by electrons, i.e., the same technology used in hospitals. X-ray technology also has the dual advantage of high penetration power and switch-off capability (Kilcast 1995, Barbosa-Cánovas et al. 1998).

The mechanisms for microbial inactivation by irradiation are well known. Irradiation irreversibly damages microbial DNA, leading to the inability of the organism to reproduce. There are over 45 years of extensive studies in the literature on the chemical and microbial effects of irradiation on food. Decimal reduction times (D-values) for most food pathogens and food spoilage microorganisms are widely published in literature. Irradiation resistance generally follows the order: Gram negative < Gram positive = molds < spores = yeasts < viruses. Bacterial spores are about 5 to 15 times more resistant to irradiation than are their corresponding vegetative cells. Generally, molds have equivalent resistance to vegetative bacteria and yeasts are generally as resistant as the more resistant bacteria (Barbosa-Cánovas et al. 1998). The efficacy of treatment is dependent on many parameters, including environmental conditions such as pH and temperature, type and species of target microorganism, chemical composition of the food, presence or absence of oxygen in the system, physical state of the food product, and stage of growth of the microorganism. A comprehensive review by Monk et al. (1995) presents a summary of irradiation D-values for most foodborne microorganisms.

Irradiation can cause oxidation of fats in foods leading to off-flavors and rancidity. However, this can be controlled by either vacuum packing or freezing the food before treatment. Irradiation can break down cellulose, and this can lead to softening of some fruits and vegetables. It does not inactivate food enzymes; therefore, they must be controlled by other means. Foods that are good candidates for radiation decontamination are poultry meat/carcasses, egg products, red meats, fishery products, and spices and other highly contaminated dry food ingredients (Farkas 1998).

While the inactivation of various foodborne pathogenic vegetative cells, spoilage microorganisms, and bacterial spores has been well studied in high-moisture products or dry ingredients (e.g., spices) and has been found to be effective in high- and low-a_w foods such as these, little has been published on the effects of reduced a_w on microbial inactivation by irradiation (e.g., for intermediate-moisture foods). Thayer et al. (1995) investigated the effect of NaCl, sucrose, and water content on the inactivation of *S. typhimurium* in irradiated pork and chicken. The addition of up to 8% NaCl (w/w) in mechanically deboned chicken meat and ground pork increased the survival of *S. typhimurium* in the irradiated meats. In contrast, when up to 8% (w/w) sucrose was added to ground pork, there was no significant increase in the survival of *S. typhimurium* in the irradiated pork product. The failure of sucrose to provide the same protection in the meat against irradiation is an argument against reduced a_w being the primary mechanism for protection. The authors suggested that it may be the chloride ion serving as an OH^- radical scavenger, providing protection to *S. typhimurium* when NaCl was added to the meat. Similar results were observed in a companion study on the retention of thiamin in pork (Fox et al. 1994).

Physiological Actions of Osmotic Stress

Over the past 20 years, our understanding of the homeostatic mechanisms by which bacteria resist preservative factors, such as reduction of a_w, has advanced enormously through the use of molecular biological techniques. Developments in whole cell functional genomics are likely to accelerate the development of this understanding and allow the molecular basis of sensing and responding to microbial stresses to be elucidated (Brul et al. 2002). Our current understanding of the mechanisms by which foodborne microorganisms respond to osmotic stress has been excellently reviewed by O'Bryne and Booth (2002). Although the specific responses of different bacteria vary, there are a number of common mechanisms. When exposed to hyperosmotic stress, such as that encountered in foods with reduced a_w, the main stress on the bacterial cell is the efflux of water across the plasma membrane, until the osmotic activity of the cytoplasm and the environment are balanced. The controlled accumulation of solutes allows the cells to restore turgor (Booth et al. 1988). However, even when the cells have restored turgor, the composition of the cytoplasm may not be optimal for enzymatic activity. Cells may exhibit limited growth potential as well as incur further stress through alterations in metabolism (Dodd and Stewart 1997). If the cells are transferred to a dilute medium/environment, water rapidly enters the cell and this must be compensated by the release of solutes into the environment via mechanosensitive channels (O'Byrne and Booth 2002; Stokes et al. 2003; Edwards et al. 2004, 2005). Through the collaboration of functional genomics, it is likely that there will soon be a molecular basis for understanding how mechanosensitive channels in bacteria allow the cell to sense osmotic stress.

Osmoregulation

The growth of bacterial cells, defined as the balanced increase in mass of major polymers (DNA, RNA, proteins, and lipids) until the cell achieves critical mass and divides, requires the control of water flow (O'Byrne and Booth 2002). Osmoregulation, which typically involves transport of compatible solutes into the cytoplasm, is all about the control over the influx and efflux of solutes from the cell, with water movement being essentially passive (O'Byrne and Booth 2002). Osmoregulation not only requires modulating the activity of transport and enzyme systems but also involves complex patterns of regulation of gene expression. The major compatible solutes used by foodborne microorganisms include quaternary amines (i.e., betaine and carnitine), amino acids (i.e., proline), amino acid derivatives (i.e., proline betaine), sugars (i.e., trehalose, mannitol), and peptides (i.e., prolyl-hydroxyproline).

Some of the mechanisms that Gram-positive bacteria, such as *S. aureus,* use to survive under osmotic stress are similar to those of Gram-negative enterics such as *E. coli* and *S. typhimurium*, both of which accumulate compatible solutes, including proline and glycine betaine, via transport. There are several major differences, as enterics generally can only grow in NaCl concentrations of up to 0.8 mol/L, whereas *S. aureus* has been reported to grow at NaCl concentrations as high as 4 mol/L (Miller et al. 1991). Additionally, enterics adapt to osmotic stress in a two-phase process. The process first allows the rapid accumulation of K^+/potassium glutamate, followed by the activation of transport systems leading to the accumulation of compatible solutes (Booth et al. 1988). In contrast, when *S. aureus* is under osmotic stress, it increases turgor by initially increasing the compatible solute pools in the cytoplasm, precluding the transport of K^+/potassium glutamate for regulating turgor (Pourkomailian and Booth 1992).

It is quite clear that food constitutes a rich source of compatible solutes and their precursors; thus, higher osmolarities may be required to achieve inactivation of pathogenic and spoilage microorganisms than would be necessary if compatible solutes were absent from foods (Gutierrez et al. 1995, Booth 1998, O'Byrne and Booth 2002). Because removing compatible solutes from foods would be virtually impossible, alternative strategies could include preventing the addition of inhibitors of transport systems for compatible solutes. This also is a difficult strategy to implement. Therefore, an understanding about osmoregulation, including the effects of solute-specific characteristics of various humectants used to depress the RH of foods, is needed. Further, information on regulation of the synthesis and activity of various transport systems, information on patterns of regulation of gene expression involved in osmoregulation, and understanding the mechanisms of cell death due to osmotic stress may generate ideas for optimally using osmotic stress as a food preservation method.

Solute-Specific Effects

The term *water activity* has been widely used for several decades in food science literature to predict microbial growth as well as the relationship between many common food deterioration reactions and a_w. Although a_w or RH is a better indicator for food stability and safety than the water content of a system, it is not always a reliable predictor. In recent years, increasing evidence based on glass transition theory has shown that molecular mobility may be an attribute that deserves further attention, as it is related to many important diffusion-limiting properties of foods (van den Berg 1986; Slade and Levine 1991, 1995; Champion et al. 2000). Molecular mobility is governed by glass transition temperature (T_g), a solute-specific property that is inversely linearly related to the measured RH of a system. Although the use of RH has served the food industry well for the past several decades since William James Scott first introduced the concept of a_w in 1953 (Scott 1953), further study to understand the solute-specific effects on microbial response to osmotic stress should continue. The use of molecular mobility in conjunction with RH to understand microbial response to osmotic stress is discussed in further detail by Stewart et al. (2002).

Conclusions

Since the concept was originally introduced in the 1950s, a_w has become a cornerstone of modern food preservation. In combination with other traditional preservation methods it is used as the basis for chemical and microbial stability in a wide range of products such as protein, cereal, fruit, and confectionary products.

Over the years, great progress has been made in understanding the mechanism of preservative action and how a_w can be used in combination with other preservative factors. It is likely that reduced a_w will continue to be an important consideration in the design of new food products. Developments in other areas of food science are likely to accelerate our understanding and use of a_w in food preservation.

Developments in predictive microbial modeling and quantitative risk assessment and management will mean that the effectiveness of preservative factors, such as a_w, either solely or in combination with other preservative factors, such as pH or preservatives, will be expressed in terms of a control measure with required performance in terms of managing food safety risk. Developments in novel nonthermal preservation technologies will drive the need to understand how a_w can be safely used in combination with treatments such as high-pressure and pulsed electric fields, etc. Finally, advances in functional genomics will greatly accelerate our ability to understand the molecular basis of how microbes respond to stresses such as reduced a_w, which may lead to new classes of preservatives or combinations of preservatives based on this understanding. These developments will be required if we are to meet the needs in the marketplace for foods that are healthier, more natural, fresher, and, at the same time, safe and convenient to eat.

References

Anonymous. 2004. Alinorm 04/27/13. Report of the 36th Session of the Codex Committee on Food Hygiene (CCFH). Available at http://www.codexalimentarius.net/reports.asp. Accessed April 1, 2007.

Aronsson, K., and Rönner, U. 2001. Influence of pH, water activity and temperature on the inactivation of *Escherichia coli* and *Saccharomyces cerevisiae* by pulsed electric fields. *Innovative Food Science and Emerging Technology* 2(2):105–112.

Barbosa-Cánovas, G.V., Pothakamury, U.R., Palou, E., and Swanson, B.G., eds. 1998. Food irradiation. *In: Nonthermal Preservation of Foods*, pp. 161–213. New York: Marcel Dekker.

Booth, I.R. 1998. Bacterial responses to osmotic stress: Diverse mechanisms to achieve a common goal. In: *The Properties of Water in Foods: ISOPOW 6*, ed. D.S. Reid, pp. 456–485. London: Blackie A & P (Chapman Hall).

Booth, I.R., Cairney, J., Sutherland, L., and Higgins, C.F. 1988. Enteric bacteria and osmotic stress: An integrated homeostatic system. *Journal of Applied Bacteriology Symposium Supplement* 17:35S–49S.

Brul, S., Coote, P., Oomes, S., Mensonides, F., Hellingwerf, K., and Kliss, F. 2002. Physiological actions for preservative agents: Prospective of use of modern microbiological techniques in assessing microbial behaviour in food preservation. *International Journal of Food Microbiology* 79(1): 55–64.

Centers for Disease Control and Prevention (CDC) FoodNet. 2000. 2000 Annual Report. Available at http://www.cdc.gov/foodnet/annuals.htm. Accessed April 1, 2007.

Centers for Disease Control and Prevention (CDC) FoodNet. 2002. 2002 Annual Report. Available at http://www.cdc.gov/foodnet/annuals.htm. Accessed April 1, 2007.

Champion, D., Le Meste, M., and Simatos, D. 2000. Towards an improved understanding of glass transitions and relaxations in foods: Molecular mobility in the glass transition range. *Trends in Food Science and Technology* 11(2):41–55.

Cheftel, J.C. 1992. Effects of high hydrostatic pressure on food constituents: An overview. In: *High Pressure and Biotechnology*, eds. C. Balny, R. Hayashi, K. Heremans, and P. Masson, Vol. 224, pp. 195–209. London: Colloque INSERM/John Libbey Eurotext Ltd.

Codex Alimentarius. 1999. Recommended international codes of practices: General principles of food hygiene, annexes to CAC/RCP 1-1969, Revised 3-1997, Amended 1999. Available at ftp://ftp.fao.org/codex/standard/en/CXP_001e.pdf. Accessed April 1, 2007.

Codex Alimentarius. 2001. Alinorm 03/13. Report of the 34th Session of the Codex Committee on Food Hygiene. Available at www.codexalimentarius.net/reports.asp. Accessed April 1, 2007.

Codex Alimentarius. 2004. Alinorm 04/27/13. Report of the 36th Session of the Codex Committee on Food Hygiene. Available at http://www.codexalimentarius.net/reports.asp. Accessed April 1, 2007.

Cole, M.B. 2004. Food safety objectives—Concept and current status. *Mitt. Lebensm. Hyg.* 95(11):13–20.

Corry, J.E.L. 1974. The effects of sugars and polyols on the heat resistance of salmonellae. *Journal of Applied Bacteriology* 37:31–43.

Corry, J.E.L. 1976. The effect of sugars and polyols on the heat resistance and morphology of osmophilic yeasts. *Journal of Applied Bacteriology* 40(3):269–276.

Dodd, C.E.R., and Stewart, G.A.S.B. 1997. Inimical process: Bacterial self-destruction and sub-lethal injury. *Trends in Food Science and Technology* 8:238–241.

Edwards, M.D., Booth, I.R., and Miller, S. 2004. Gating the bacterial mechanosensitive channels (MscS): A new paradigm. *Current Opinions in Microbiology* 7:163–167.

Edwards, M.D., Li, Y., Kim, S., Miller, S., Bartlett, W., Black, S., Dennison, S., Iscla, I., Blount, P., Bowie, J.U., and Booth, I.R. 2005. Pivotal role of the glycine-rich TM3 helix in gating the MscS mechanosensitive channel. *Nature Structural and Molecular Biology* 12:113–119.

FAO/WHO. 2004. Risk assessment of *Listeria monocytogenes* in ready-to-eat foods. Available at http://www.fao.org/es/esn/food/risk_mra_riskassessment_listeria_en.stm. Accessed April 1, 2007.

FDA Center for Food Safety and Applied Nutrition, USDA Food Safety and Inspection Service and Centers for Disease Control and Prevention. 2003. Quantitative assessment of relative risk to public health from foodborne *Listeria monocytogenes* among selected categories of ready-to-eat foods. Available at http://www.foodsafety.gov/~dms/lmr2-toc.html. Accessed April 1, 2007.

Farkas, J. 1998. Irradiation as a method for decontaminating food: A review. *International Journal of Food Microbiology* 44(3):189–204.

Fox, J.B., Jr., Lakritz, L., Kohout, K.M., and Thayer, D.W. 1994. Water concentration/activity and loss of vitamins B_1 and E in pork due to gamma irradiation. *Journal of Food Science* 59(6):1291–1295.

Goh, E.L.C., Hocking, A.D., Stewart, C.M., Buckle, K.A., and Fleet, G.H. 2006. Baroprotective effect of increased solute concentrations on yeast and moulds during high pressure processing. *Innovative Food Science and Emerging Technology,* in press.

Gould, G.W. 2002. The evolution of high pressure processing of foods. In: *Ultra High Pressure Treatments of Foods*, eds. M.E.G. Hendrickx and D. Knorr, pp. 3–21. New York: Kluwer Academic/Plenum Publishers.

Gutierrez, C., Abee, T., and Booth, I.R. 1995. Physiology of the osmotic stress response in microorganisms. *International Journal of Food Microbiology* 28(2):233–244.

Hashizume, C., Kimura, K., and Hayashi, R. 1995. Kinetic analysis of yeast inactivation by high pressure treatment at low temperatures. *Bioscience Biotechnology Biochemistry* 59(8):1455–1459.

Heremans, K. 1992. From living systems to biomolecules. In: *High Pressure and Biotechnology*, eds. C. Balny, R. Hayashi, K. Heremans, and P. Masson, Vol. 224, pp. 37–44. Montrouge: Colloque INSERM/John Libbey Eurotext Ltd.

Hoover, D.G. 1997. Minimally processed fruits and vegetables: Reducing microbial load by nonthermal physical treatments. *Food Technology* 51(6):66–71.

Hoover, D.G., Metrick, C., Papineau, A.M, Farkas, D.F., and Knorr, D. 1989. Biological effects of high hydrostatic pressure on food microorganisms. *Food Technology* 43:99–107.

Institute of Food Technologists (IFT). 2001. Processing parameters needed to control pathogens in cold-smoked fish. *Journal of Food Science* 66(Suppl.):S1055–S1132.

International Commission on Microbiological Specifications for Foods (ICMSF). 2002. Microorganisms in Foods 7: Microbiological Testing in Food Safety Management. New York: Kluwer Academic/Plenum Publishers.

Kilcast, D. 1995. Food irradiation: Current problems and future potential. *International Biodeterioration Biodegradation, Vol. 36,* pp. 279–296. Oxford, UK: Elsevier Science Limited.

Knorr, D. 1993. Effects of high-hydrostatic pressure processes on food safety and quality. *Food Technology* 47(6):156–161.

Knorr, D. 1995. Hydrostatic pressure treatment of food: Microbiology. In: *New Methods of Food Preservation,* ed. G.W. Gould, pp. 159–175. London: Blackie Academic and Professional.

Knorr, D. 1998. Advantages, possibilities and challenges of high pressure applications in food processing. In: *The Properties of Water in Foods,* ed. D.S. Reid, pp. 419–437. ISOPOW 6. London: Blackie Academic & Professional.

Mattick, K.L., Jørgensen, F., Wang, P., Pound, J., Vandeven, M.H., Ward, L.R., Legan, J.D., Lappin-Scott, H.M., and Humphrey, T.J. 2001. Effect of challenge temperature and solute type on heat tolerance of *Salmonella* serovars at low water activity. *Applied Environmental Microbiology* 67(9):4128–4136.

Miller, K.J., Zelt, S.C., and Bae, J. 1991. Glycine betaine and proline are the principal compatible solutes of *Staphylococcus aureus*. *Current Microbiology* 23(3):131–137.

Min, S., Reina, L., and Zhang, Q.H. 2002. Water activity and the inactivation of *Enterobacter cloacae* inoculated in chocolate liquor and a model system by pulsed electric field treatment. *Journal of Food Processing and Preservation* 26(5):323–337.

Monk, J.D., Beuchat, L.R., and Doyle, M.P. 1995. Irradiation inactivation of food-borne microorganisms. *Journal of Food Protection* 58(2):197–208.

Murrell, W.G., and Scott, W.J. 1966. Heat resistance of bacterial spores at various water activities. *Journal of General Microbiology* 43(3):411–425.

O'Byrne, C.P., and Booth, I.R. 2002. Osmoregulation and its importance to food-borne microorganisms. *International Journal of Food Microbiology* 74(3):203–216.

Ogawa, H., Fukuhisa, K., and Fukumoto, H. 1992. Effect of hydrostatic pressure on sterilization and preservation of citric juice. In: *High Pressure and Biotechnology,* eds. C. Balny, R. Hayashi, K. Heremans, and P. Masson, pp. 269–278. London: Colloque INSERM/John Libbey Eurotex Ltd.

Oxen, P., and Knorr, D. 1993. Baroprotective effects of high solute concentration against inactivation of *Rhodotorula rubra*. *Lebensmittel-Wissenschaft und Technologie* 26(3):220–223.

Palou, E., López-Malo, A., Barbosa-Cánovas, G.V., Welti-Chanes, J., and Swanson, B.G. 1997. Effect of water activity on high hydrostatic pressure inhibition of *Zygosaccharomyces bailii*. *Letters in Applied Microbiology* 24(5):417–420.

Patterson, M.F., Quinn, M., Simpson, R., Doyle, M.P., Beuchat, L.R., Montville, T.J., and Gilmour, A. 1995. The sensitivity of vegetative pathogens to high hydrostatic pressure treatment in phosphate buffered saline and foods. *Journal of Food Protection* 58(5):524–529.

Pourkomailian, B., and Booth, I. 1992. Glycine betaine transport by *Staphylococcus aureus*: Evidence for two transport systems and for their possible roles in osmoregulation. *Journal of General Microbiology* 138(12):2515–2518.

Radomyski, T., Murano, E.A., Olsen, D.G., and Murano, P.S. 1994. Elimination of pathogens of significance in food by low-dose irradiation: A review. *Journal of Food Protection* 57(1):73–86.

Sale, A.J.H., Gould, G.W., and Hamilton, W.A. 1970. Inactivation of bacterial spores by hydrostatic pressure. *Journal of General Microbiology* 60(3):323–334.

Sale, A.J.H., and Hamilton, W.A. 1967. Effects of high electric fields on microorganisms. II. Mechanisms of action of the lethal effect. *Biochimical Biophysical Acta* 148:789–800.

Scott, W.J. 1953. Water relations of *Staphylococcus aureus* at 30°C. *Australian Journal of Biological Science* 6(4):549–564.

Seyderhelm, I., Boguslawski, S., Michaelis, G., and Knorr, D. 1996. Pressure induced inactivation of selected food enzymes. *Journal of Food Science* 61(2):308–310.

Shigehisa, T., Ohmori, T., Saito, A., Taji, S., and Hayashi, R. 1991. Effects of high pressure on characteristics of pork slurries and inactivation of microorganisms associated with meat and meat products. *International Journal of Food Microbiology* 12(2-3):207–216.

Slade, L., and Levine, H. 1991. Beyond water activity: Recent advances based on an alternative approach to the assessment of food quality and safety. *Critical Reviews in Food Science and Nutrition* 30(2-3):115–360.

Slade, L., and Levine, H. 1995. Glass transitions and water-food structure interactions. In: *Advances in Food and Nutrition Research*, eds. S.L. Taylor and J.E. Kinsella, pp. 103–269. San Diego, CA: Academic Press.

Smelt, J.P., Hellemons, J.C., and Patterson, M. 2003. Effects of high pressure processing on vegetative microorganisms. In: *Ultra High Pressure Treatments of Foods*, eds. M.E.G. Hendrickx and D. Knorr, pp. 55–76. New York: Kluwer Academic/Plenum Publishers.

Smelt, J.P.P., Wouters, P.C., and Rijke, A.G.F. 1998. Inactivation of microorganisms by high pressure. In: *The Properties of Water in Foods*, ed. D.S. Reid, pp. 389–417. ISOPOW 6. London: Blackie Academic & Professional.

Smelt, J.P.P.M. 1998. Recent advances in the microbiology of high pressure processing. *Trends in Food Science and Technology* 9(4):152–158.

Stewart, C.M., and Cole, M.B. 2001. Preservation by the application of nonthermal processing. In: *Spoilage of Processed Foods: Causes and diagnosis*, eds. C.J. Moir, C. Andrew-Kabilafkas, G. Arnold, B.M. Cox, A.D. Hocking, and I. Jenson, pp. 53–61. Waterloo DC, Australia: Australian Institute of Food Science and Technology.

Stewart, C.M., Cole, M., Legan, J.D., Slade, L., Vandeven, M.H., and Schaffner, D. 2001. Modeling the growth boundaries of *Staphylococcus aureus* for risk assessment purposes. *Journal of Food Protection* 64(1):51–57.

Stewart, C.M., Cole, M., Legan, J.D., Slade, L., Vandeven, M.H., and Schaffner, D.W. 2002. *Staphylococcus aureus* growth boundaries: Moving towards more mechanistic predictive models based on solute specific effects. *Applied Environmental Microbiology* 68(4):1864–1871.

Stewart, C.M., Cole, M.B., and Schaffner, D.W. 2003. Managing the risk of staphylococcal food poisoning from cream-filled baked goods using food safety objectives. *Journal of Food Protection* 66(7):1310–1325.

Stokes, N.R., Murray, H.D., Subramaniam, C., Gourse, R.L., Louis, P., Bartlett, W., Miller, S., and Booth, I.R. 2003. A role for mechanosensitive channels in survival of stationary phase: Regulation of channel expression by RpoS. *Proceedings of the National Academy of Science* 100(26):15959–15964.

Thayer, D.W., Boyd, G., Fox, J.B., Jr., and Lakritz, L. 1995. Effects of NaCl, sucrose, and water content on the survival of *Salmonella typhimurium* on irradiated pork and chicken. *Journal of Food Protection* 58(5):490–496.

Timson, W.J., and Short, A.J. 1965. Resistance of microorganisms to hydrostatic pressure. *Biotechnology and Bioengineering* 7(1):139–159.

van den Berg, C. 1986. Water activity. In: *Concentration and Drying of Foods,* ed. D. MacCarthy, pp. 11–38. London: Elsevier Applied Science.

Wouters, P.C., and Smelt, J.P.P.M. 1998. Inactivation of microorganisms with pulsed electric fields: Potential for food preservation. *Food Biotechnology* 11:193–229.

Zimmerman, U., Pilwat, G., and Riemann, F. 1974. Dielectric breakdown on cell membranes. *Biophysical Journal* 14:881–899.

A Water Activity of Saturated Salt Solutions

Anthony J. Fontana, Jr.

°C	Cesium Fluoride	Lithium Bromide	Zinc Bromide	Water Activity Potassium Hydroxide	Sodium Hydroxide	Lithium Chloride	Calcium Bromide
10	0.049 ± 0.016	0.071 ± 0.007	0.085 ± 0.007	0.123 ± 0.014	—	0.113 ± 0.004	0.216 ± 0.005
15	0.043 ± 0.014	0.069 ± 0.006	0.082 ± 0.006	0.107 ± 0.011	0.096 ± 0.028	0.113 ± 0.004	0.202 ± 0.005
20	0.038 ± 0.011	0.066 ± 0.006	0.079 ± 0.005	0.093 ± 0.009	0.089 ± 0.024	0.113 ± 0.003	0.185 ± 0.005
25	0.034 ± 0.009	0.064 ± 0.005	0.078 ± 0.004	0.082 ± 0.007	0.082 ± 0.021	0.113 ± 0.003	0.165 ± 0.002
30	0.030 ± 0.008	0.062 ± 0.005	0.076 ± 0.003	0.074 ± 0.006	0.076 ± 0.017	0.113 ± 0.002	—
35	0.027 ± 0.006	0.060 ± 0.004	0.075 ± 0.003	0.067 ± 0.004	0.069 ± 0.015	0.113 ± 0.002	—
40	0.024 ± 0.005	0.058 ± 0.004	0.075 ± 0.002	0.063 ± 0.004	0.063 ± 0.012	0.112 ± 0.002	—
	Lithium Iodide	**Potassium Acetate**	**Potassium Fluoride**	**Magnesium Chloride**	**Sodium Iodide**	**Potassium Carbonate**	**Magnesium Nitrate**
10	0.206 ± 0.003	0.234 ± 0.005	—	0.335 ± 0.002	0.418 ± 0.008	0.431 ± 0.004	0.574 ± 0.003
15	0.196 ± 0.002	0.234 ± 0.003	—	0.333 ± 0.002	0.409 ± 0.007	0.432 ± 0.003	0.559 ± 0.003
20	0.186 ± 0.002	0.231 ± 0.003	—	0.331 ± 0.002	0.397 ± 0.006	0.432 ± 0.003	0.544 ± 0.002
25	0.176 ± 0.001	0.225 ± 0.003	0.308 ± 0.013	0.328 ± 0.002	0.382 ± 0.005	0.432 ± 0.004	0.529 ± 0.002
30	0.166 ± 0.001	0.216 ± 0.005	0.273 ± 0.011	0.324 ± 0.001	0.362 ± 0.004	0.432 ± 0.005	0.514 ± 0.002
35	0.156 ± 0.001	—	0.246 ± 0.009	0.321 ± 0.001	0.347 ± 0.004	—	0.499 ± 0.003
40	0.146 ± 0.001	—	0.227 ± 0.008	0.316 ± 0.001	0.329 ± 0.004	—	0.484 ± 0.004
	Sodium Bromide	**Cobalt Chloride**	**Potassium Iodide**	**Strontium Chloride**	**Sodium Nitrate**	**Sodium Chloride**	**Ammonium Chloride**
10	0.622 ± 0.006	—	0.721 ± 0.003	0.757 ± 0.001	0.775 ± 0.005	0.757 ± 0.002	0.806 ± 0.010
15	0.607 ± 0.005	—	0.710 ± 0.003	0.741 ± 0.001	0.765 ± 0.004	0.756 ± 0.002	0.799 ± 0.006
20	0.591 ± 0.004	—	0.699 ± 0.003	0.725 ± 0.001	0.754 ± 0.004	0.755 ± 0.001	0.792 ± 0.004
25	0.576 ± 0.004	0.649 ± 0.035	0.689 ± 0.002	0.709 ± 0.001	0.743 ± 0.003	0.753 ± 0.001	0.786 ± 0.004
30	0.560 ± 0.004	0.618 ± 0.028	0.679 ± 0.002	0.691 ± 0.001	0.731 ± 0.003	0.751 ± 0.001	0.779 ± 0.006
35	0.546 ± 0.004	0.586 ± 0.022	0.670 ± 0.002	—	0.721 ± 0.003	0.749 ± 0.001	—
40	0.532 ± 0.004	0.555 ± 0.018	0.661 ± 0.002	—	0.710 ± 0.003	0.747 ± 0.001	—

	Potassium Bromide	Ammonium Sulfate	Potassium Chloride	Strontium Nitrate	Potassium Nitrate	Potassium Sulfate	Potassium Chromate
10	0.838 ± 0.002	0.821 ± 0.005	0.868 ± 0.004	0.906 ± 0.004	0.960 ± 0.014	0.982 ± 0.008	—
15	0.826 ± 0.002	0.817 ± 0.004	0.859 ± 0.003	0.887 ± 0.003	0.954 ± 0.010	0.979 ± 0.006	—
20	0.817 ± 0.002	0.813 ± 0.003	0.851 ± 0.003	0.869 ± 0.003	0.946 ± 0.007	0.976 ± 0.005	—
25	0.809 ± 0.002	0.810 ± 0.003	0.843 ± 0.003	0.851 ± 0.004	0.936 ± 0.006	0.973 ± 0.005	0.979 ± 0.005
30	0.803 ± 0.002	0.806 ± 0.003	0.836 ± 0.003	—	0.923 ± 0.006	0.970 ± 0.004	0.971 ± 0.004
35	0.798 ± 0.002	0.803 ± 0.004	0.830 ± 0.003	—	0.908 ± 0.008	0.967 ± 0.004	0.964 ± 0.004
40	0.794 ± 0.002	0.799 ± 0.005	0.823 ± 0.003	—	0.890 ± 0.012	0.964 ± 0.004	0.959 ± 0.004

Source: Adapted from Greenspan, L. 1977. Humidity fixed points of binary saturated aqueous solutions. *Journal of Research of the National Bureau of Standards: A Phys. Chem.* 81A:89–96.

B Water Activity of Unsaturated Salt Solutions at 25°C

Anthony J. Fontana, Jr.

Molal Conc. Units	Lithium Bromide	Potassium Hydroxide	Sodium Hydroxide	Lithium Chloride	Calcium Bromide	Lithium Iodide	Potassium Acetate	Potassium Fluoride
	Water Activity at 25°C							
0.1	0.997	0.997	0.997	0.997	0.995	0.997	0.997	0.997
0.5	0.983	0.983	0.983	0.983	0.974	0.982	0.983	0.984
1.0	0.963	0.965	0.966	0.964	0.941	0.962	0.964	0.967
1.5	0.942	0.944	0.948	0.943	0.898	0.939	0.944	0.950
2.0	0.917	0.922	0.930	0.921	0.846	0.914	0.922	0.932
2.5	0.891	0.899	0.910	0.897	0.785	0.886	0.899	0.913
3.0	0.863	0.874	0.889	0.870	0.718	0.853	0.876	0.893
3.5	0.831	0.847	0.866	0.842	0.648		0.852	0.872
4.0	0.797	0.819	0.842	0.812	0.572			0.851
4.5	0.761	0.789	0.816	0.780	0.493			
5.0	0.724	0.760	0.789	0.747	0.417			
5.5	0.688	0.729	0.762	0.713	0.347			
6.0	0.651	0.699	0.734	0.679	0.284			
7.0	0.574	0.634	0.674	0.609				
8.0	0.496	0.569	0.612	0.539				
9.0	0.423	0.508	0.549	0.473				
10.0	0.352	0.450	0.488	0.412				
11.0	0.287	0.393	0.430	0.356				
12.0	0.235	0.340	0.376	0.307				
13.0	0.187	0.296	0.328	0.266				
14.0	0.149	0.262	0.285	0.230				
15.0	0.121	0.225	0.249	0.200				
16.0	0.098	0.191	0.218	0.175				
17.0	0.081		0.192	0.155				
18.0	0.067		0.170	0.138				
19.0	0.056		0.152	0.123				
20.0	0.048		0.136	0.110				

	Water Activity							
Molal Conc.	Magnesium Chloride	Sodium Iodide	Magnesium Nitrate	Sodium Bromide	Cobalt Chloride	Potassium Iodide	Strontium Chloride	Sodium Nitrate
0.1	0.995	0.997	0.995	0.997	0.995	0.997	0.995	0.997
0.5	0.975	0.983	0.975	0.983	0.975	0.984	0.976	0.984
1.0	0.942	0.965	0.944	0.966	0.944	0.967	0.947	0.970
1.5	0.900	0.946	0.906	0.948	0.904	0.951	0.911	0.956
2.0	0.848	0.925	0.862	0.929	0.859	0.933	0.870	0.942
2.5	0.788	0.903	0.813	0.908	0.810	0.916	0.822	0.929
3.0	0.722	0.880	0.758	0.887	0.758	0.899	0.768	0.916
3.5	0.652	0.855	0.701	0.865	0.709	0.881	0.711	0.904
4.0	0.580		0.643	0.841	0.664	0.863	0.654	0.892
4.5	0.509		0.585			0.846		0.880
5.0	0.439		0.526					0.868
5.5								0.856
6.0								0.843
	Sodium Chloride	Ammonium Chloride	Potassium Bromide	Ammonium Sulfate	Potassium Chloride	Strontium Nitrate	Potassium Nitrate	Potassium Sulfate
0.1	0.997	0.997	0.997	0.996	0.997	0.996	0.997	0.996
0.5	0.984	0.984	0.984	0.982	0.984	0.979	0.985	0.982
1.0	0.967	0.968	0.968	0.966	0.968	0.960	0.973	
1.5	0.950	0.952	0.952	0.951	0.952	0.941	0.962	
2.0	0.932	0.937	0.936	0.935	0.936	0.921	0.953	
2.5	0.913	0.921	0.920	0.919	0.920	0.902	0.920	
3.0	0.893	0.905	0.903	0.902	0.904	0.881	0.937	
3.5	0.873	0.889	0.887	0.885	0.887	0.860	0.930	
4.0	0.852	0.873	0.870	0.867	0.870	0.838		
4.5	0.830	0.857	0.853	0.849	0.853			
5.0	0.807	0.842	0.835	0.831				
5.5	0.784	0.826	0.818	0.813				
6.0	0.760	0.811	0.801					

C Water Activity and Isotherm Equations

Anthony J. Fontana, Jr.

WATER ACTIVITY EQUATIONS

Nonelectrolyte Solutions

1. Raoult's law:

$$a_w = \frac{N_{H_2O}}{N_{H_2O} + N_{Solute}}$$

where N_{H_2O} = moles of water and N_{solute} = moles of dissolved solute.
For dilute ideal solutions only (Raoult 1882).

2. Norrish equation:

$$a_w = X_w \left[e^{(KX_s^2)} \right]$$

where X_w = mole fraction of water, X_s = mole fraction of solute, and K is the empirical constant for the solute (Norrish 1966).
Some common K values:

Compound	K	Compound	K
DE 43	−5.31	Mannose	−2.28 ± 0.22
Galactose	−2.24 ± 0.07	Xylose	−1.54 ± 0.04
Glucose	−2.25 ± 0.02	Sucrose	−6.47 ± 0.06
Glucose	−2.11 ± 0.11	PEG 400	−26.6 ± 0.8
Fructose	−2.15 ± 0.08	PEG 600	−56 ± 2
Glycerol	−1.16 ± 0.01	Citric acid	−6.17 ± 0.49
Mannitol	−0.91 ± 0.27	Tartaric acid	−4.68 ± 0.5
Propylene glycol	−1	Malic acid	−1.82 ± 0.13
Alanine	−2.52 ± 0.37	Glycine	+0.87 ± 0.11

3. Grover model:

$$a_w = 1.04 - 0.1(E°) + 0.0045(E°)^2$$

where $E^o = \sum \left(\frac{E_i}{W_i} \right)$ and E_i is a constant for a solute and W_i is the total moisture content in grams of water per gram of ingredient *i* (Grover 1947).

Some common E_i values:

Component	E_i	Component	E_i
Fat	0	Invert sugar	1.3
Starch (DE < 50)	0.8	Protein	1.3
Gums	0.8	Acids	2.5
Pectin	0.8	Glycerol	4.0
Sucrose	1.0	Sodium chloride	9.0
Lactose	1.0		

4. Caurie model:

$$a_w = a_{w1}a_{w2}a_{w3} - \left[\frac{n(\omega_1\omega_2 + \omega_1\omega_3 + \omega_2\omega_3)}{55.5^2} + \frac{(n+1)\omega_1\omega_2\omega_3}{55.5^3}\right]$$

where ω is the molality of solute and n is the number of components in the mixture. The example shown is for three components, but the equation can be used for any number of components (Caurie 1986).

5. Money and Born equation:

$$a_w = \frac{1}{1+0.27n}$$

where n represents the moles of humectant/100 g H_2O. This equation is empirical and primarily used to determine the a_w of humectant-containing solutions (Money and Born 1951).

Electrolyte Solutions

1. Bromley equation: $a_w = e^{-0.018\Phi u m_s}$

where Φ is the osmotic coefficient, u is the number of ions per species, and m_s is the molality of the solute. This equation is for electrolyte solutions. In addition, the following equation is offered to calculate Φ:

$$\Phi = 1 + 2.303[T_1 + (0.06 + 0.6B)T_2 + 0.5BI]$$

$$T_1 = 0.511|z_m z_x|\frac{[1 + 2(1+\sqrt{I})\ln(1+\sqrt{I}) - (1+\sqrt{I})^2]}{(I(1+\sqrt{I}))}$$

$$T_2 = \frac{(1+2aI)}{(a(1+aI)^2)} - \frac{\ln(1+aI)}{(a^2 I)} \qquad a = \frac{1.5}{|z_m z_x|}$$

$$B = B_m + B_x + \delta_m\delta_z$$

where z_m and z_x are the charges of m and x ions, and v_m and v_x are the respective number of ions. I is the ionic strength. B, B_m, B_x, δ_m, δ_x are the Bromley coefficients (Bromley 1973).

2. Pitzer equation: $a_w = e^{-0.018\Phi\sum_i m_i}$

The Pitzer equation is similar to the Bromley equation where Φ is the osmotic coefficient and m is the molality of the solute. This equation also is for electrolyte solutions. The Pitzer equation differs from the Bromley equation in the way that Φ is calculated:

$$\Phi - 1 = |z_m z_x| F + 2m\left(\frac{v_m v_x}{v}\right) B_{mx} + 2m^2\left(\sqrt{\frac{v_m v_x}{v}}\right)^3 C_{mx}$$

$$F = \frac{-0.3921\sqrt{I}}{(1+1.2\sqrt{I}} \qquad I = 0.5\sum_i m_i z_i^2$$

$$|z_m z_x| = \frac{\sum_i m_i z_i^2}{\sum_i m_i} \qquad B_{mx} = B_{mx}(0) + B_{mx}(1)e^{-2\sqrt{I}}$$

where z_m and z_x are the charges of m and x ions, v_m and v_x are the respective number of ions, I is the ionic strength, and B(0), B(1), B_{mx}, and C_{mx} are the Pitzer coefficients (Money and Born 1951).

The Pitzer equation $a_w = e^{-0.018\Phi\sum_i m_i}$ also can be calculated using published values for Φ, many of which can be found in Robinson and Stokes (1965).

Multicomponent Mixtures

1. Ross equation:

$$a_{w\,final} = a_{w\,initial} \times a_{w1} \times a_{w2} \times \ldots a_{wi}$$

where $a_{w\,final}$ is the final water activity, $a_{w\,initial}$ is the initial a_w before adding solute i, and a_{wi} is the a_w that the solute would have if dissolved in all the water (Ross 1975). This equation requires determination of a_w for each component separately using another a_w prediction equation or using the component's isotherm.

2. Ferro Fontan-Benmergui-Chirifie equation:

$$a_w = \prod_i (a_{wi}(I))^{\frac{I_i}{I}}$$

where $I_i = 0.5 v_i m_i |Z_m Z_x|$ and $a_{wi}(I)$ is the water activity of a binary solution of I at the same total ionic strength (I) as the multicomponent solution. I_i is the ionic strength of the component i in the mixture, v_i is the total ions in the solute i in the solution, m_i is the molality of the solute, and z_m and z_x are the charges of m and x ions (Fontan et al. 1980).

3. Lang-Steinberg equation:

$$Log(1 - a_w) = \frac{(MW - \sum a_i w_i)}{\sum b_i w_i}$$

where a_i and b_i are the Smith equation constants for each component i in the mixture. M is the moisture content of the mixture, W is the total dry material in the mixture, w_i is the dry material of each component i in the mixture (Lang and Steinberg 1981). This equation can predict the water activity of multicomponent mixtures over the range 0.30 to 0.95.

4. Salwin-Slawson equation:

$$a_w = \frac{\sum_i a_i s_i w_i}{\sum_i s_i w_i}$$

where a_i is the initial water activity of component i used in the mixture, s_i is the sorption isotherm slope of component i at the mixture temperature, and w_i is the dry weight of component I (Salwin and Slawson 1959).

5. Ferro Fontan-Chirife-Boquet equation:

$$(a_w)_m = X_1 e^{(-K_m X_2^2)}$$

$$K_m = \sum_{s=1}^{n} K_s C_s \left[\frac{M_t}{M_s}\right] \quad M_t = \sqrt{\sum_{s=1}^{n} \left(\frac{C_s}{M_s}\right)}$$

where X_1 and X_2 are the molar fractions of the solvent and solutes, K_s is the Norrish K value for each solute in the mixture, C_s is the weight ration of each solute to the total of solids in the mixture, and M_s is the molecular weight of each component (Fontan et al. 1981).

ISOTHERM EQUATIONS

1. BET isotherm:

$$\frac{a_w}{(1-a_w)m} = \frac{1}{m_o c} + \left(\frac{c-1}{m_o c}\right) a_w$$

where m is the moisture in g/100 solids or g/g solids at water activity a_w and m_o is the monolayer value in same units. The constant c is calculated by:

$$c = e^{\frac{Qs}{RT}}$$

where Qs is the surface interaction energy cal/mole. The equation takes the form of a straight line and when $a_w/(1 - a_w)m$ is plotted against a_w, the intercept I and slope S can be determined (Brunauer et al. 1938). Then, the monolayer moisture content can be calculated as:

$$m_0 = \frac{1}{I+S}$$

2. GAB isotherm:

$$m = \frac{m_o k_b c a_w}{[(1 - k_b a_w)(1 - k_b a_w + c k_b a_w)]}$$

where k_b is a constant in the range of 0.70 to 1 and c is a constant in the range of 1 to 20. In addition, m_o is the monolayer water content in g water per gram of solid and a_w is the water activity at moisture m (van den Berg and Bruin 1981). This is a three-parameter equation and must be calculated using a nonlinear solution. This is done by calculating the moisture content (m) at a minimum of five water activity values (a_w) (ideally the values will be spread across the water activity range of 0.1 to 1). One point can be m = 0 and a_w = 0. Nonlinear regression can then be used to determine the values of k_b, c, and m_o. Once the constant values are known for the isotherm, any water activity value can be inserted in the GAB equation to determine the corresponding moisture content.

References

Bromley, L. 1973. Thermodynamic properties of strong electrolytes in aqueous solutions. *American Institute of Chemical Engineers Journal* 19:313.

Brunauer, S., Emmett, P.H., and Teller, E. 1938. Adsorption of gases in multimolecular layers. *Journal of American Chemical Society* 60:309.

Caurie, M. 1986. A general method for predicting the water activity of simple and multi-component mixtures of solutes and non-solutes. *Journal of Food Technology* 21:221–228.

Fontan, C.F., Benmergui, E.A., and Chirife, J. 1980. The prediction of water activity of aqueous solutions in connection with intermediate moisture foods. III. a_w prediction in multicomponent strong electrolyte aqueous solutions. *Journal of Food Technology* 15:47–58.

Fontan, C.F., Chirife, J., and Boquet, R. 1981. Water activity in multicomponent non-electrolyte solutions *Journal of Food Technology* 16:553–559.

Grover, D.N. 1947. The keeping properties of confectionary as influenced by its water vapor pressure. *J. Soc. Chem. Ind.* 66:201.

Lang, K.W., and Steinberg, M.P. 1981. Predicting water activity from 0.3 to 0.95 of multicomponent food formulation. *Journal of Food Science* 46:670.

Money, R.W., and Born, R. 1951. Equilibrium humidity of sugar solutions. *Journal of Science Food Agric.* 2:180.

Norrish, R.S. 1966. An equation for the activity coefficients and equilibrium relative humidities of water in confectionary syrups. *Journal of Food Technology* 1:25.

Raoult, F.M. 1882. Loi de congelation des solutions aqueuses de substances organiques. *CR Acad. Sci. Paris* 94:1517–1519.

Robinson, R.A., and Stokes, R.H. 1965. *Electrolyte Solutions; The Measurement and Interpretation of Conductance, Chemical Potential, and Diffusion in Solutions of Simple Electrolytes*. London: Butterworth.

Ross, K.D. 1975. Estimation of water activity in intermediate moisture foods. *Food Technology* 29:26–34.

Salwin, H., and Slawson, V. 1959. Moisture transfer in combinations of dehydrated foods. *Food Technology* 8:58–61.

van den Berg, C., and Bruin, S. 1981. Water activity and its estimation in food systems: Theoretical aspects. In: *Water Activity: Influences on Food Quality*, eds. L.B. Rockland and G.F. Stewart, pp. 1–61. New York: Academic Press.

D Minimum Water Activity Limits for Growth of Microorganisms

Anthony J. Fontana, Jr.

Water Activity	Microorganism		
	Bacteria	Molds	Yeast
0.97	*Clostridium botulinum* E *Pseudomonas fluorescens*	—	—
0.95	*Escherichia coli* *Clostridium perfringens* *Salmonella* spp. *Vibrio cholerae*	—	—
0.94	*Clostridium botulinum* A, B *Vibrio parahaemolyticus*	*Stachybotrys atra*	—
0.93	*Bacillus cereus*	*Rhizopus nigricans*	
0.92	*Listeria monocytogenes*		
0.91	*Bacillus subtilis*		
0.90	*Staphylococcus aureus* (anaerobic)	*Trichothecium roseum*	*Saccharomyces cerevisiae*
0.88			*Candida*
0.86	*Staphylococcus aureus* (aerobic)		
0.85		*Aspergillus clavatus*	
0.84		*Byssochlamys nivea*	
0.83		*Penicillium expansum* *Penicillum islandicum* *Penicillum viridicatum*	*Debarymoces hansenii*
0.82		*Aspergillus fumigatus* *Aspergillus parasiticus*	
0.81		*Penicillum cyclopium* *Penicillium patulum*	
0.80		*Penicillium citrinum*	*Saccharomyces bailii*
0.79		*Penicillum martensii*	
0.78		*Aspergillus flavus*	
0.77		*Aspergillus niger* *Aspergillus ochraceous*	
0.75		*Aspergillus restrictus* *Aspergillus candidus*	
0.71		*Eurotium chevalieri*	
0.70		*Eurotium amstelodami*	
0.62			*Saccharomyces rouxii*
0.61		*Monascus bisporus*	
<0.60	No microbial proliferation		

Sources: Adapted from Beuchat, L.R. 1981. Microbial stability as affected by water activity. *Cereal Foods World* 26:345–349; Beuchat, L.R. 1983. Influence of water activity on growth, metabolic activities, and survival of yeasts and molds. *Journal of Food Protection* 46:135–141, 150; Corry, J.E.L. 1978. Relationships of water activity to fungal growth. In: *Food and Beverage Mycology,* ed. L.R. Beuchat, pp. 45–82. Westport, CT: Avi Publishing; Russell, N.J., and Gould, G.W. 1991. Factors affecting growth and survival. In: *Food Preservatives,* eds. N.J. Russell and G.W. Gould, pp. 13–21. Glasgow: Blackie & Son; and Tilbury, R.H. 1976. The stability of intermediate moisture foods with respect to yeasts. In: *Intermediate Moisture Foods*, eds. R. Davies, G.G. Birch, and K.J. Parker, pp. 138–165. London: Applied Science Laboratories.

E Water Activity Values of Select Food Ingredients and Products

Shelly J. Schmidt and Anthony J. Fontana, Jr.

This appendix contains water activity values or ranges from various sources for select human and pet food products. The individual ingredients and products (called items) listed in the table have been organized alphabetically into main categories and then into alphabetized subcategories. The moisture content [% wet basis − (g water/g sample) × 100%] of the material and/or temperature of the a_w measurement are also given as available. If the a_w of an item was measured at more than one temperature, the a_w and temperature values are listed in corresponding order (i.e., the first a_w listed corresponds to the first temperature listed, and so on). The a_w and moisture content methods used (when known) are included either in the table or in the list of sources. Dr. Schmidt would like to acknowledge the contribution of several U.S. Army Research and Engineering Apprenticeship Program High School students who assisted in the collection of a_w and moisture content data in her lab, as well as several University of Illinois undergraduate and graduate students. Special thanks to Laura Wardwell for assistance in assembling the composite table below.

MAIN CATEGORY, Subcategory, Item	**a_w Value or Range**	**M.C. (% w.b.)**	**Temperature (°C)**	**Source**
BAKED GOODS				
Bagels				
Lender's Plain Bagels—Crumb	0.957	38.5	25	2
Lender's Plain Bagels—Crust	0.926	31.1	25	2
Batter and Dough				
Batter, Pancake (Hygrometer)	0.883		35	14
Batter, Pancake (Manometric Technique)	0.99		35	14
Batter, Pancake (Microcrystalline Cellulose Method)	0.960		35	14
Dough, Refrigerated Biscuit (Equilibrium Moisture Adsorption)	0.942–0.940		25	8
Biscuits				
Biscuit	0.630–0.605		25	19
Breads				
Bread	0.939		25	1
Bread	0.960	37		9
Bread (Hygrometer)	0.950		35	14
Bread (Manometric Technique)	0.93		35	14
Bread (Microcrystalline Cellulose Method)	0.949		35	14
Bread, White (Equilibrium Moisture Adsorption)	0.950–0.942		25	8
Bread, White (Sinascope)	0.97–0.94		25	8
Bread, White (Chilled Mirror)	0.964		25	1
Butternut Enriched Bread—Crumb	0.944; 0.955	34.0	22; 30	2
Butternut Enriched Bread—Crust	0.926; 0.919	29.5	22; 31	2
Butternut Roman Meal—Crumb	0.938; 0.947	35.2	21; 30	2

(continues)

MAIN CATEGORY, Subcategory, Item	a_w Value or Range	M.C. (% w.b.)	Temperature (°C)	Source
BAKED GOODS *(continued)*				
Breads *(continued)*				
Butternut Roman Meal—Crust	0.932; 0.908	29.0	21; 31	2
County Market Split Top Wheat Bread—Crumb	0.912; 0.929	34.6	22; 30	2
County Market Split Top Wheat Bread—Crust	0.910; 0.914	26.4	22; 30	2
County Market White Bread Enriched—Crumb	0.938; 0.941	38.1	22; 30	2
County Market White Bread Enriched—Crust	0.929; 0.899	28.7	22; 30	2
Indiana Spud Potato Bread Low Fat, Saturated Fat Free—Crumb	0.935; 0.945	36.1	21; 30	2
Indiana Spud Potato Bread Low Fat, Saturated Fat Free—Crust	0.926; 0.905	30.3	21; 31	2
Pepperidge Farm 100% Whole Wheat Thin Sliced Bread—Crumb	0.952; 0.955	40.6	21; 30	2
Pepperidge Farm 100% Whole Wheat Thin Sliced Bread—Crust	0.949; 0.944	32.1	22; 31	2
Pepperidge Farm Light Style Wheat Bread Low Fat—Crumb	0.942; 0.937	35.6	22; 31	2
Pepperidge Farm Light Style Wheat Bread Low Fat—Crust	0.929; 0.919	30.5	22; 31	2
Pepperidge Farm Original White Sandwiches and Toast—Crumb	0.944; 0.941	34.2	22; 31	2
Pepperidge Farm Original White Sandwiches and Toast—Crust	0.936; 0.938	29.4	22; 31	2
Pepperidge Farm Swirl Bread Cinnamon—Crumb	0.896; 0.900	29.3	22; 31	2
Pepperidge Farm Swirl Bread Cinnamon—Crust	0.887; 0.884	23.9	22; 31	2
Pillsbury Enriched Buttermilk Bread—Crumb	0.930; 0.932	35.5	22; 31	2
Pillsbury Enriched Buttermilk Bread—Crust	0.919; 0.904	28.0	22; 31	2
Pillsbury Enriched Wheat Bread—Crumb	0.939; 0.934	33.8	21; 31	2
Pillsbury Enriched Wheat Bread—Crust	0.915; 0.907	30.1	22; 31	2
Purity Sunbeam Enriched Bread—Crumb	0.932; 0.936	38.8	21; 30	2
Purity Sunbeam Enriched Bread—Crust	0.922; 0.910	29.2	22; 30	2
Purity Sunbeam Honey Grain Wheat Bread—Crumb	0.946; 0.945	39.7	22; 30	2
Purity Sunbeam Honey Grain Wheat Bread—Crust	0.935; 0.921	28.6	22; 30	2
Buns				
Sara Lee Heart Healthy Wheat Bakery Bun—Crumb	0.923	34.3	25	2
Sara Lee Heart Healthy Wheat Bakery Bun—Crust	0.903	26.1	25	2
Sunbeam Enriched Hamburger Bun—Crumb	0.944	35.3	25	2
Sunbeam Enriched Hamburger Bun—Crust	0.916	26.8	25	2
Cake				
Baked Cake (Equilibrium Moisture Adsorption)	0.944–0.918		25	8
Baked Cake (Sinascope)	0.94–0.90		25	8
Cake	0.750–0.720		25	19
Fruit Cake Vanilla	0.777–0.772		25	1
Twinkie, Hostess	0.795	18.9	22	2
Twinkie, Low Fat, Hostess	0.786	21.0	22	2
Cookies				
Archway Homestyle Cookies—Oatmeal	0.517; 0.553	7.1	21; 31	2
Archway Homestyle Cookies—Sugar	0.617; 0.630	8.4	21; 31	2
Delicious Old Fashioned Goodness Lemon Sandwich Cookies—Cookie	0.245; 0.319	3.0	22; 31	2
Delicious Old Fashioned Goodness Lemon Sandwich Cookies—Filling	0.318; 0.339	2.5	22; 30	2
Delicious Skippy Peanut Butter Sandwich Cookies—Cookie	0.184; 0.245	2.3	22; 31	2

MAIN CATEGORY, Subcategory, Item	**a_w Value or Range**	**M.C. (% w.b.)**	**Temperature (°C)**	**Source**
BAKED GOODS *(continued)*				
Cookies *(continued)*				
Delicious Skippy Peanut Butter Sandwich Cookies—Filling	0.261; 0.270	1.7	22; 31	2
Keebler Soft 'n Chewy Chips Deluxe	0.567; 0.575	5.9	21; 31	2
Keebler Summer E.L. Fudge Butter Sandwich Cookies w/ Fudge Créme—Cookie	0.200; 0.381	2.5	22; 31	2
Keebler Summer E.L. Fudge Butter Sandwich Cookies w/ Fudge Créme—Filling	0.314; 0.410	1.5	22; 31	2
Nabisco Chewy Chips Ahoy!	0.647; 0.633	8.6	21; 31	2
Nabisco Nutter Butter Sandwich Cookies—Cookie	0.248; 0.256	2.1	22; 31	2
Nabisco Nutter Butter Sandwich Cookies—Filling	0.265; 0.285	0.65	22; 31	2
Nabisco Nutter Butter Soft Cookies	0.649; 0.637	11.7	21; 31	2
Nabisco Oreo Chocolate Sandwich Cookies—Cookie	0.309; 0.201	1.4	22; 31	2
Nabisco Oreo Chocolate Sandwich Cookies—Filling	0.315; 0.252	0.06	22; 31	2
Snackwell's Reduced Fat Créme Sandwich Cookies—Cookie	0.310; 0.292	4.2	22; 31	2
Snackwell's Reduced Fat Créme Sandwich Cookies—Filling	0.311; 0.295	8.2	22; 31	2
Crackers				
Animal Crackers, Nabisco Barnum's	0.303; 0.346	4.6	22; 31	2
Cheese Nips Baked Snack Crackers, Nabisco	0.173; 0.203	1.8	22; 31	2
Club Crackers—Original, Keebler	0.225; 0.275	1.8	22; 31	2
Crackers	0.100	9		9
Goldfish Tiny Crackers—Cheddar Cheese, Pepperidge Farm	0.153; 0.184	1.3	23; 31	2
Graham Crackers, Honey Maid, Nabisco	0.251	3.0	29	2
Ritz Crackers, Nabisco	0.112; 0.122	1.1	23; 31	2
Saltine Crackers, Nabisco Unsalted Tops Premium	0.464; 0.536	5.4	22; 31	2
Triscuit Baked Whole Wheat Wafers—Original, Nabisco	0.132; 0.153	2.1	22; 31	2
Wheat Thins Baked Snack Crackers, Nabisco	0.164; 0.176	1.7	22; 31	2
Wheat Thins Baked Snack Crackers—Reduced Fat, Nabisco	0.147; 0.158	1.5	22; 31	2
Muffins				
Blueberry Mini—Muffins, Hostess	0.844	21.4	24	2
Master English Muffin Toasting Bread—Crumb	0.952	40.5	22	2
Master English Muffin Toasting Bread—Crust	0.947	31.9	22	2
Master Wheat English Muffin Toasting Bread—Crumb	0.945	36.7	22	2
Master Wheat English Muffin Toasting Bread—Crust	0.935	31.5	22	2
Rolls				
Wheat Dinner Roll	0.890	27.8	25	2
Tortillas				
Flour Tortilla	0.922	30.8	25	2
BEVERAGES				
Carbonated Soda				
7-Up, Regular	0.977		10	2
7-Up, Diet	0.996		10	2
Coke, Regular	0.978		10	2
C2 (Mid-calorie Coke)	0.983		10	2
Coke, Diet	0.996		10	2
Pepsi, Regular	0.982		10	2
Pepsi Edge (Mid-calorie Pepsi)	0.985		10	2
Pepsi, Diet	0.999		10	2

(continues)

MAIN CATEGORY, **Subcategory**, Item	a_w Value or Range	M.C. (% w.b.)	Temperature (°C)	Source
BEVERAGES *(continued)*				
Carbonated Soda *(continued)*				
RC, Regular	0.979		10	2
RC, Diet	0.996		10	2
Shasta, Regular	0.991		25	2
Shasta, Diet	0.996		25	2
Sierra Mist, Regular	0.977		10	2
Sierra Mist, Diet	0.994		10	2
Sprite, Regular	0.979		10	2
Sprite, Diet	0.999		10	2
Coffee				
Green Beans	0.5–0.6			22
Instant Coffee Crystals, Folgers	0.168	2.6	25	2
Roasted Beans	0.10–0.30			22
Soluble Powders	0.10–0.30			22
Drink Mixes				
Lemonade, Kool-Aid (Dry)	0.309	0.37	25	2
Margarita Mix—Strawberry (Liquid)	0.972		24.0	1, 2
Tea				
Tea, Instant (Beckman-Sina)	0.192		30	4
Tea, Instant (Kaymont-Rotronics)	0.174		30	4
Tea, Instant (Protimeter)	0.267		30	4
Tea, Instant (Vapor Pressure Manometric Technique)	0.130		30	4
BREAKFAST CEREALS				
Hot Cereals				
Oatmeal, Old Fashion, Quaker Oats (Dry)	0.340	8.8	25	2
Oatmeal, Quick, Quaker Oats (Dry)	0.368	9.2	25	2
Ready-to-Eat				
Flavorite Apple Dapples	0.297; 0.312	2.0	22; 31	2
Flavorite Bite Size Frosted Shredded Wheat	0.255; 0.300	4.1	22; 31	2
Flavorite Cocoa Crispy Rice	0.295; 0.271	1.7	22; 31	2
Flavorite Corn Flakes	0.357; 0.271	2.8	22; 31	2
Flavorite Crispy Hexagons	0.228; 0.237	2.1	22; 31	2
Flavorite Crispy Rice	0.266; 0.251	2.5	22; 31	2
Flavorite Frosted Flakes	0.281; 0.242	1.1	22; 31	2
Flavorite Frosted Fruit O's	0.298; 0.379	2.6	21; 31	2
Flavorite Golden Corn Nuggets	0.184; 0.223	1.0	22; 31	2
Flavorite Raisin Bran—Flakes	0.455; 0.445	6.4	22; 31	2
Flavorite Raisin Bran—Raisins	0.465; 0.439	7.2	21; 31	2
General Mills Cheerios	0.157	3.2	24	2
General Mills Cheerios, Frosted	0.186	2.2	25	2
General Mills Cheerios, Honey Nut	0.162	2.6	26	2
General Mills Cinnamon Toast Crunch	0.242	2.0	29	2
General Mills Cocoa Puffs	0.183	1.7	25	2
General Mills Golden Grahams	0.181	1.8	25	2
General Mills Lucky Charms	0.227	3.6	30	2
General Mills Trix	0.175	3.1	28	2
Kellogg's Apple Jacks	0.384; 0.385	3.0	22; 31	2
Kellogg's Cocoa Krispies	0.335; 0.364	3.0	22; 31	2
Kellogg's Corn Flakes	0.352; 0.382	3.5	22; 31	2
Kellogg's Corn Pops	0.295; 0.323	1.8	22; 31	2
Kellogg's Crispix—Corn on one side, Rice on the other	0.287; 0.322	3.0	22; 31	2
Kellogg's Fruit Loops	0.360; 0.386	2.3	22; 31	2
Kellogg's Frosted Flakes of Corn	0.425; 0.375	2.6	22; 31	2

MAIN CATEGORY, Subcategory, Item	a_w Value or Range	M.C. (% w.b.)	Temperature (°C)	Source
BREAKFAST CEREALS *(continued)*				
Ready-to-Eat *(continued)*				
Kellogg's Frosted Mini-Wheats	0.362; 0.389	5.4	22; 31	2
Kellogg's Honey Crunch Corn Flakes	0.306; 0.341	2.6	22; 31	2
Kellogg's Raisin Bran—Flakes	0.458; 0.411	6.6	22; 31	2
Kellogg's Raisin Bran—Raisins	0.472; 0.411	8.1	22; 31	2
Kellogg's Rice Krispies	0.279; 0.342	2.8	22; 31	2
Kellogg's Smacks	0.242; 0.288	1.7	22; 31	2
Malt O Meal Golden Puffs	0.200; 0.177	0.48	22; 31	2
Quaker Captain Crunch	0.422; 0.447	3.1	22; 31	2
Quaker New! Honey Crisp Corn Flakes	0.254; 0.261	1.7	22; 31	2
CEREAL GRAINS, LEGUMES, AND PRODUCTS				
Beans				
Great Northern Beans	0.475	10.1	25	2
Lupini Beans in Brine	0.945		25	20
Flour				
All-Purpose Flour, Gold Medal	0.453	13.5	20	2
Flour	0.352–0.350		25	1
Flour, Wheat	0.523		25	1
Grains				
Wheat Grains, Whole	0.700–0.675		25	19
Wheat Grains, Ground	0.696–0.675		25	19
Nuts and Seeds				
Almonds, Sliced, Diamond of California	0.476	3.1	20	2
Cashews	0.75			17
Peanuts, Dry Roasted	0.147	0.56	29	2
Peanuts, Honey Roasted, Planters	0.249; 0.323	0.84	22; 31	2
Peanuts, Salted, Planters	0.260; 0.508	1.1	22; 31	2
Peanut Butter, Reese's Creamy	0.263; 0.268	0.37	22; 31	2
Peanut Butter, Reese's Crunchy	0.230; 0.205	0.42	22; 31	2
Popcorn Seeds, Black	0.742	10.3	24	2
Sunflower Seeds, Flavorite Shelled	0.308; 0.313	0.84	22; 31	2
Sunflower Seeds	0.75			17
Walnuts, Chopped	0.58–0.57			13
Walnut, Dried Meats	0.427		25	20
Pasta				
Ditalini, Barilla 100% Selected Durum Wheat Semolina, Dry	0.334	5.4	20	2
Fresh Filled Pasta	0.973–0.916			6
Gnocchi	0.983–0.936			6
Lasagna, Dry	0.373	8.3	30	2
Noodle, Dry	0.570		25	1
Penne Rigate, Dry	0.386	7.9	30	2
Ravioli in Tomato Sauce	0.988		20	21
Spaghetti, Dry	0.390	6.9	30	2
Rice				
Rice	0.531		25	1
Rice	0.591		25	20
Rice, Uncle Ben's Original Enriched Parboiled Long Grain	0.420	7.1	20	2
Rice, Whole Grain Brown	0.491	7.0	25	2
Soy				
Soy-based Energy Bar	0.461	5.27		18

(continues)

MAIN CATEGORY, **Subcategory**, Item	a_w Value or Range	M.C. (% w.b.)	Temperature (°C)	Source
CEREAL GRAINS, LEGUMES, AND PRODUCTS (*continued*)				
Starch				
Corn Starch, Argo 100%	0.287	8.7	20	2
Wheat Starch	0.56		25	19
CONDIMENTS				
Ketchup				
Ketchup	0.933		25	1
Mustard				
Mustard, Cognac Type	0.938		25	1
Salad Dressings				
French Dressing	0.924		25	1
Ranch Dressing, Fat Free, Hidden Valley	0.977	74.7	24	2
Ranch Dressing, Light Hidden Valley	0.972	61.1	24	2
Ranch Dressing, Regular, Hidden Valley	0.965	44.9	24	2
Salsa				
Salsa, Medium, Tostitos	0.985	86.8	25	2
Sauces				
Cocktail Sauce	0.947		20	21
Fudge Sauce	0.795		25	12
Fudge Sauce	0.85			17
Fudge Sauce	0.834		25	20
Molho Sauce (Hot Sauce)	0.956		25	1
Soy Sauce	0.810		25	20
Soya Sauce	0.98			17
Soybean Sauce	0.917		25	1
Worcestershire Sauce	0.967		25	1
CONFECTIONERY				
Candies				
Jolly Rancher—Watermelon	0.393	4.7	21	2
Marshmallows	0.629; 0.635	15.6	21; 30	2
Starburst Original Fruits Fruit Chews—Cherry	0.597; 0.485	7.0	21; 31	2
Starburst Original Fruits Fruit Chews—Lemon	0.588; 0.482	6.7	21; 31	2
Starlight Mints	0.413	4.7	21	2
Twizzlers Strawberry Twists	0.657; 0.677	9.6	21; 31	2
Chocolate				
Milk Chocolate Bar	0.60			17
Milk Chocolate Bar, Hershey's	0.557; 0.428	1.2	21; 30	2
Gum (Chewing)				
Wrigley's Doublemint Chewing Gum	0.498; 0.529	4.4	22; 31	2
Wrigley's Coated Pellet Gum	0.462		23	1, 2
Wrigley's Juicy Fruit Gum	0.300		25	1, 2
Wrigley's Spearmint Chewing Gum	0.486; 0.514	4.4	22; 31	2
Wrigley's Winter Fresh Chewing Gum	0.492; 0.542	4.5	22; 31	2
Icing and Frosting				
Icing, Cake (Equilibrium Moisture Adsorption)	0.841–0.805		25	8
Icing, Cake (Sinascope)	0.79–0.76		25	8
Icing, Vanilla	0.795	15	20	2
Frosting, Chocolate	0.816		25	12
Syrups				
Chocolate Syrup	0.862		25	1
Glucose/Fructose Syrup	0.729		25	1
High Fructose Corn Syrup, IsoSweet. A.E. Staley	0.738	28.2	25	2
Maize Syrup Solids	0.60			17

MAIN CATEGORY, Subcategory, Item	**a_w Value or Range**	**M.C. (% w.b.)**	**Temperature (°C)**	**Source**
CONFECTIONERY *(continued)*				
Syrups *(continued)*				
Maple Syrup	0.956		25	1
Maple Syrup	0.90	29		9
Sweeteners				
Corn Sweet	0.994			15
Honey	0.552		25	1
Honey, Argentine	0.638–0.530	21–15	25	10
Honey, German Floral	0.608–0.482	21.5–14.0		11
Honey, German Honeydew	0.602–0.477	18.4–12.6		11
Honey, SueBee Clover	0.517	16.7	20	2
Sucrose, Crystalline	0.227	0.04	25	2
DAIRY				
Butter				
Butter, Salted	0.894	15.9	24	2
Butter, Salted	0.91		25	2
Butter, Salted	0.952–0.827	17.0–11.7		16
Butter, Salted	0.949		20	21
Butter, Unsalted	0.961		24	2
Butter, Unsalted	0.98		25	2
Butter, Unsalted	0.976		20	21
Cheese				
Cheese	0.96	40		9
Cheese, Spread	0.965		25	12
Cheese, Spread	0.946		25	20
Colby Jack Cheese, Sargento	0.967; 0.966	44.4	20; 30	2
Cheddar Cheese	0.95			17
Cheddar Cheese, Mild, Sargento	0.953; 0.957	43.8	20; 30	2
Cheddar Cheese, Sharp, Kraft	0.975	43.4	22	2
Cream Cheese, Philadelphia Original	0.991	53.4	24	2
Parmesan Cheese, Frigo	0.881	33.9	25	2
Parmesan Cheese (Beckman-Sina)	0.713		30	4
Parmesan Cheese (Kaymont-Rotronics)	0.725		30	4
Parmesan Cheese (Protimeter)	0.721		30	4
Parmesan Cheese (Vapor Pressure Manometric)	0.693		30	4
Processed Cheese	0.969–0.930			1
Processed Cheese	0.93–0.91			13
Processed Cheese Spread	0.96			17
Swiss Cheese	0.96			17
Swiss Cheese, Kraft Singles	0.946; 0.946	41.9	20; 30	2
Cream				
Cream, 25% Fat, Canned	0.973		25	3
Cream, 40 % Fat	0.979			15
Daisy Light Sour Cream	0.996; 0.991	82.2	21; 29	2
Daisy Sour Cream	0.994; 0.990	75.0	21; 30	2
Prairie Farms Lite Sour Cream	0.993; 0.990	78.4	21; 30	2
Prairie Farms No Fat Sour Cream	0.990; 0.986	82.8	21; 28	2
Margarine				
Margarine, Liquid	0.88			13
Margarine, Salted	0.897–0.885	17.1–16.6		16
Margarine, Unsalted	0.951–0.936	17.5–16.9		16
Mayonnaises				
Mayonnaises	0.960–0.930	29.2–14.4		16

(continues)

MAIN CATEGORY, Subcategory, Item	a_w Value or Range	M.C. (% w.b.)	Temperature (°C)	Source
DAIRY *(continued)*				
Milk				
Milk, 1.5%	0.995			15
Milk, 2%	0.988	89.2	22	2
Milk, Evaporated	0.983		20	21
Milk, Fat Free	0.990	90.9	21	2
Milk, Jam (Dulce de Leche)	0.789		25	1
Milk, Jam	0.842		25	7
Milk, Nonfat Dry (Beckman-Sina)	0.272		30	4
Milk, Nonfat Dry (Katmont-Rotronics)	0.203		30	4
Milk, Nonfat Dry (Protimeter)	0.277		30	4
Milk, Nonfat Dry (Vapor Pressure Manometric)	0.137		30	4
Milk, Skim	0.996		20	21
Milk, Skim, Dried (Dewpoint)	0.75			17
Milk, Sweetened Condensed	0.833		25	7
Milk, Whole	0.988	88.1	22	2
Milk, Whole	0.995–0.994			15
Milk Proteins				
Whey Concentrate (Dewpoint)	0.83			17
Whey Concentrate (Hygrometer)	0.815		35	14
Whey Concentrate (Manometric Technique	0.88		35	14
Whey Concentrate (Microcrystalline Cellulose Method)	0.820		35	14
Yogurt				
Breyers 99% Fat Free Strawberry	0.980; 0.977	74.7	21; 30	2
Breyers Light Fat Free Raspberries ‘n Cream	0.988; 0.978	85.9	21; 30	2
Breyers Smooth and Creamy Classic 99% Fat Free Strawberry	0.984; 0.977	73.9	21; 30	2
Dannon 99% Fat Free Raspberry	0.974; 0.973	74.5	20; 30	2
Dannon Light Fat Free Raspberry	0.998; 0.993	88.5	20; 30	2
Flavorite Lowfat Strawberry	0.983; 0.979	78.7	20; 30	2
Yoplait Light Fat Free Strawberry	0.992; 0.987	83.1	20; 30	2
Yoplait Original 99% Fat Free Strawberry	0.985; 0.980	74.8	21; 30	2
DEHYDRATED/DRY PRODUCTS				
Flavors				
Grill Flavor, Spray Dried	0.173		23.0	1, 2
Orange Flavor, Drum Dried	0.262		25.4	1, 2
Orange Flavor, Spray Dried	0.206		24.4	1, 2
Smoke Flavor, Spray Dried	0.170		24.8	1, 2
Powders				
Chocolate Powder	0.258–0.251		25	1
Malted Milk	0.194	2.0	25	2
Spices/Seasonings				
Black Pepper, Pure Ground	0.715	10.3	25	2
Cinnamon, Whole Ground	0.587	10.9	25	2
FRUITS, VEGETABLES, AND PRODUCTS				
Fruits (Fresh and Processed)				
Apples	0.988–0.975			15
Apples	0.988		20	21
Apple, Gala	0.985	84.6	25	2
Apple, Red Delicious	0.984	85.3	20	2
Apples, Puree with Sugar, Canned	0.974		25	3
Apple Sauce, Musselmans	0.976	81.8	22	2
Apple Sauce, Natural, Musselmans	0.983	87.8	21	2

MAIN CATEGORY, Subcategory, Item	a_w Value or Range	M.C. (% w.b.)	Temperature (°C)	Source
FRUITS, VEGETABLES, AND PRODUCTS *(continued)*				
Fruits (Fresh and Processed) *(continued)*				
Apple, Slices in Light Syrup	0.981		25	3
Apricots	0.985–0.977			15
Apricots, Halves in Light Syrup	0.992		25	3
Baby Food, Mixed Fruit	0.974		25	3
Bananas	0.987–0.964			15
Bananas	0.979		20	21
Bilberries	0.989			15
Blackberries	0.989–0.986			15
Blueberries	0.982			15
Cherries	0.986–0.959			15
Cherries, Sour	0.983–0.971			15
Cherries, Sweet	0.975			15
Cherries, Maraschino, Canned	0.905		25	3
Cherries, in Syrup, Canned	0.973		25	3
Cranberries	0.989			15
Coconut, Flavorite Flake	0.834; 0.851	12.3	22; 31	2
Coconut, Flavorite Shred	0.839; 0.850	12.2	22; 31	2
Currants	0.990			15
Dates	0.974			15
Dewberries	0.985			15
Figs	0.974			15
Fruit Cocktail, Canned	0.988–0.982			3
Fruits	0.97	94		9
Fruits, Dried	0.80–0.72	20		9
Gooseberries	0.989			15
Grapefruit	0.985–0.980			15
Grapes	0.986–0.963			15
Lemon, Fresh	0.998	90.8	20	2
Lemons	0.989–0.982			15
Lime, Fresh	0.998	91.5	20	2
Limes	0.980			15
Mangoes	0.986			15
Melon	0.991–0.970			15
Nectarines	0.984			15
Oranges	0.987–0.979			15
Papaya	0.990			15
Peach, Fresh	0.998	84.6	20	2
Peaches	0.989–0.979			15
Peach, Yellow Cling in Syrup, Canned	0.986–0.980		25	3
Pears	0.989–0.979			15
Pear, Halves in Syrup, Canned	0.992–0.986		25	3
Persimmons	0.976			15
Pineapple	0.988–0.985			15
Pineapple, Slices in Syrup	0.976–0.965		25	3
Plums	0.982–0.969			15
Quinces	0.981–0.972			15
Raisins	0.54–0.51			13
Raisins, Sun Maid California, Whole	0.526; 0.506	10.3	22; 31	2
Raisins, Sun Maid California, Cut	0.529; 0.535	12.6	22; 31	2
Raspberries	0.994–0.984			15

(continues)

MAIN CATEGORY, Subcategory, Item	a_w Value or Range	M.C. (% w.b.)	Temperature (°C)	Source
FRUITS, VEGETABLES, AND PRODUCTS *(continued)*				
Fruits (Fresh and Processed) *(continued)*				
Strawberries	0.997–0.986			15
Strawberry, with Sugar and Citric Acid	0.955–0.950		25	3
Tangerines	0.987			15
Watermelon	0.992			15
Fruit Juices				
Juices	0.97	87		9
Apple Juice	0.986			15
Cherry Juice	0.986			15
Grape Juice	0.983			15
Orange Juice	0.988			15
Orange Juice Concentrate	0.80			17
Pineapple Juice, Canned	0.990–0.988		25	3
Strawberry Juice	0.991			15
Tomato Juice, Canned	0.993		25	3
Jams/Jelly/Preserves/Spreads				
Jams and Jellies	0.94–0.82	32		9
Jam, Orange	0.839		25	1
Jam, Peach	0.833		25	1
Jam, Strawberry	0.839		25	1
Jam Strawberry, Seedless Smucker's	0.834; 0.833	29.4	21; 30	2
Jelly, Grape (Beckman-Sina)	0.818		30	4
Jelly, Grape (Kaymont-Rotronics)	0.852		30	4
Jelly, Grape (Protimeter)	0.820		30	4
Jelly, Grape (Vapor Pressure Manometric)	0.802		30	4
Jelly, Strawberry Smucker's	0.839; 0.844	34.4	21; 30	2
Jelly, Strawberry Smucker's				2
Preserves, Raspberry	0.835		25	12
Preserves, Strawberry Smucker's	0.834; 0.836	32.0	21; 30	2
Spreadable Fruit, Strawberry Smucker's Simply 100% Fruit	0.879; 0.881	46.9	21; 30	2
Vegetables (Fresh and Processed)				
Artichokes	0.987–0.976			15
Asparagus	0.994–0.992			15
Avocado	0.989			15
Beets	0.988–0.979			15
Broccoli, Sprouting	0.991			15
Brussel Sprouts	0.990			15
Cabbage	0.992–0.990			15
Carrots	0.993–0.983			15
Cauliflower	0.990–0.984			15
Celeriac	0.990			15
Celery	0.994–0.987			15
Celery Leaves	0.997–0.992			15
Corn, Sweet	0.994			15
Cucumbers	0.998–0.992			15
Eggplant	0.993–0.987			15
Endive	0.995			15
Green Beans	0.987	90.9	20	2
Green Beans	0.996–0.990			15
Green Pepper	0.998	93.3	20	2
Green Onions	0.996–0.992			15
Lima Beans	0.994			15

MAIN CATEGORY, Subcategory, Item	a_w Value or Range	M.C. (% w.b.)	Temperature (°C)	Source
FRUITS, VEGETABLES, AND PRODUCTS *(continued)*				
Vegetables (Fresh and Processed) *(continued)*				
Leeks	0.991–0.976			15
Lettuce	0.996			15
Mushrooms	0.995–0.989			15
Onions	0.990–0.974			15
Olives, Salted	0.957		25	1
Palm Heart, Canned	0.984		25	3
Parsnips	0.988			15
Peas, Green	0.990–0.980			15
Peppers	0.997–0.982			15
Potato, Russet, Baking	0.993	79.1	20	2
Potatoes	0.997–0.988			15
Potatoes, Sweet	0.985			15
Pumpkins	0.992–0.984			15
Radishes	0.990–0.980			15
Radishes, Small	0.996–0.994			15
Rhubarb	0.989			15
Rutabagas	0.988			15
Salsify	0.987			15
Spinach	0.998–0.994			15
Squash	0.994–0.996			15
Tomatoes	0.998	94.9	20	2
Tomato, Whole Peeled, Canned	0.993		25	3
Tomatoes	0.998–0.991			15
Tomatoes with Salt, Onion, Spices, Sweet Pepper, (Portuguese Sauce) Canned	0.979		25	3
Tomato Paste (16.1% solids)	0.975		25	3
Tomato Paste (28.1% solids)	0.967		25	3
Tomato Paste (Triple Concentration)	0.934		25	7
Tomato Pulp	0.993			15
Tomato Puree, Canned	0.992–0.987		25	3
Turnips	0.988			15
Vegetables	0.97	90		9
MEAT				
Fresh and Deli Meats				
Bacon	0.968		20	21
Beef	0.992–0.980			15
Carl Buddig, Beef	0.962; 0.962	66.5	20; 30	2
Carl Buddig Chicken	0.971; 0.971	66.8	20; 30	2
Carl Buddig Ham	0.967; 0.967	68.6	20; 30	2
Carl Buddig Smoked Turkey	0.976; 0.973	67.9	20; 30	2
Corned Beef. Canned	0.979–0.972		25	3
Ground Beef, 8% Fat, Lean	0.992	71.2	20	2
Ham, Cooked Canned	0.971		25	3
Ham, Deviled Canned	0.977–0.970		25	3
Lamb	0.990			15
Liverwurst	0.972		25	1
Liver Paste, Canned	0.980–0.971		25	3
Meat	0.97	67		9
Meat Paste, Canned	0.984–0.974		25	3
Pork	0.990			15
Pork Loin Roast, Center Cut, Extra Lean, Farmland	0.978	73.2	20	2

(continues)

MAIN CATEGORY, Subcategory, Item	a_w Value or Range	M.C. (% w.b.)	Temperature (°C)	Source
MEAT *(continued)*				
Fresh and Deli Meats *(continued)*				
Roast Beef, Canned	0.982		25	3
Jerky				
Beef Jerky, Russler's	0.691	16.6	20	2
Beef Jerky, various	0.65–0.92		20–25	1
Pepperoni/Salami/Sausages				
Pepperoni (Italian Style), Armour Premium	0.878	31.6	21	2
Pepperoni, Hormel	0.860; 0.857	31.2	21; 30	2
Salami, Dry	0.875		25	1
Salami (Dewpoint)	0.96			17
Salami (Hygrometer)	0.968		35	14
Salami (Manometric Technique)	0.99		35	14
Salami (Microcrystalline Cellulose Method)	0.969		35	14
Sausage	0.740–0.667		25	19
Sausage	0.975		20	21
Sausage in Lard, Canned	0.896		25	3
Sausage, Pork (Equilibrium Moisture Adsorption)	0.973		25	8
Sausage, Pork (Sinascope)	0.99–0.97		25	8
Sausage, Snack, Fully Dry	0.87–0.66	29.0–10.6		5
Sausage, Snack, Medium Dry	0.83	38.7		5
Sausage, Snack, Semi Dry	0.88	49.1		5
PET FOOD				
Canned				
Alpo Gourmet Dinner Prime Cuts in Gravy	0.994	79.6	19	2
Cat Food, Moist	0.881		25	1
Dry				
Cat Chow, Purina	0.236	4.4	25	2
Intermediate Moisture				
Dog Food, IMF (Beckman-Sina)	0.749		30	4
Dog Food, IMF (Kaymont-Rotronics)	0.800		30	4
Dog Food, IMF (Protimeter)	0.815		30	4
Dog Food, IMF (Vapor Pressure Manometric)	0.814		30	4
Pet Food Burgers	0.842–0.869		25	1
Pet Food, Soft Moist	0.83	24		9
Pet Food Special Cuts	0.835–0.855		25	1
POULTRY AND PRODUCTS				
Poultry				
Chicken, Boned, Canned	0.982		25	3
Chicken Breast, Boneless Skinless With Rib Meat, Tyson	0.979	74.6	20	2
Pheasant Puree, Canned	0.983		25	3
Eggs				
Eggs	0.97	75		9
SEAFOOD				
Anchovies	0.938		20	21
Anchovies, Salted in Oil	0.875		25	1
Fish, Cod	0.994–0.990			15
Fish, Various Species	0.989			15
Mackerel, Canned	0.974		20	21
Salmon Fillet, Fresh Chilean Raised	0.985	68.3	20	2
Salmon Paste, Canned	0.970		25	3
Salmon, Smoked	0.965		20	21
Salmon Pate, Smoked	0.993		20	21
Sardines, Canned	0.969		20	21

MAIN CATEGORY, Subcategory, Item	a_w Value or Range	M.C. (% w.b.)	Temperature (°C)	Source
SEAFOOD (*continued*)				
Trout, Smoked	0.948		20	21
Tuna, Canned	0.968		20	21
Tuna Pate	0.951		20	21
SNACK FOODS				
Fruit Snacks and Leathers				
Betty Crocker Fruit Roll-Ups—Electric Yellow	0.568; 0.548	10.0	22; 31	2
Betty Crocker Fruit Roll-Ups—Sizzling Red	0.519; 0.540	9.3	22; 31	2
Betty Crocker Fruit String Thing Chewy Snack—Cherry	0.594; 0.576	8.6	22; 31	2
Brach's Supersour Dinosaur Snacks—Green	0.596; 0.566	13.9	22; 30	2
Brach's Supersour Dinosaur Snacks—Red	0.607; 0.567	12.5	22; 31	2
Farley's Troll Fruit Snacks—Red/Yellow	0.643; 0.602	16.6	22; 31	2
Grist Mill Variety Pack Fruit Snacks—Grape	0.586; 0.574	4.2	22; 31	2
Grist Mill Variety Pack Fruit Snacks—Strawberry	0.576; 0.575	7.0	22; 31	2
Gelatin				
Jell-O Gelatin Snacks—Strawberry	0.981	79.9	20	2
Jell-O Gelatin Dessert—Peach (Dry)	0.281	1.7	25	2
Granola Bars				
Nature Valley Crunchy Granola Bar	0.214	1.5	25	2
Popcorn				
Popcorn, White, Jays	0.071	0.28	25	2
Potato Chips and Other Crunchy Snacks				
Cheetos Crunchy	0.091; 0.114	0.32	22; 31	2
Doritos Nacho Cheesier!	0.118; 0.092	0.95	22; 31	2
Lay's Classic Potato Chips	0.165; 0.267	1.1	23; 31	2
Ruffles Potato Chip	0.182; 0.178	1.3	22; 31	2
Shoestring Potatoes	0.080	1.1	25	2
Pretzels				
Baked Pretzel Stix	0.049	1.8	25	2
Puddings				
Jell-O Pudding Snacks—Chocolate	0.979	73.1	25	2
My-T-Fine Pudding and Pie Filling—Chocolate—Dry Mix	0.345	1.6	27	2
Pudding, Vanilla (Equilibrium Moisture Adsorption)	0.991		25	8
Pudding, Vanilla (Sinascope)	0.97		25	8
Rice Cakes				
Quaker Fat Free Caramel Corn Mini Rice Cakes	0.220; 0.277	2.9	22; 31	2
Quaker Fat Free Honey Nut Mini Rice Cakes	0.269; 0.362	3.9	22; 31	2
SOUPS				
Seafood Soup	0.978		20	21

References

1. Decagon Devices, Inc. (http://www.decagon.com/). In-house Testing, Pullman, Washington. a_w values were obtained using Decagon Aqualab Water Activity Meters, Series 3TE (chilled mirror).
2. Schmidt, S.J. Food Chemistry Laboratory, Department of Food Science and Human Nutrition, University of Illinois at Urbana-Champaign, IL, a_w values were obtained using Decagon Aqualab Water Activity Meters, Series 3TE (chilled mirror) and moisture content values obtained using vacuum oven drying (60° 29 in Hg for 24 hours), except for crystalline sucrose where the mositure content was obtained by Karl Fischer titration.
3. Alzamora, S.M., and Chirife, J. 1983. The water activity of canned foods. *Journal of Food Science* 48:1385-1387.
4. Asbi, B.A., and Baianu, I.C. 1986. An equation for fitting moisture sorption isotherms of food proteins. *Journal of Agricultural and Food Chemistry* 34:494-496.

5. Palumbo, S.A., Kissinger, J.C., Miller, A.J., Smith, J.L., and Zaika, L.L. 1979. Microbiology and composition of snack sausages. *Journal of Food Protection* 42(3):211-213.
6. Schebor, C., and Chirife, J. 2000. A survey of water activity and pH values in fresh pasta packed under modified atmosphere manufactured in Argentina and Uruguay. *Journal of Food Protection* 63(7):965-969.
7. Favetto, G., Resnik, S., Chirife, J., and Ferro-Fontan, C. 1983. Statistical evaluation of water activity measurement obtained with the Vaisala Humicap humidity meter. *Journal of Food Science* 48:534-538.
8. Fett, H.M. 1973. Water activity determination in foods in the range 0.80 to 0.99. *Journal of Food Science* 38:1097-1098.
9. Kaplow, M. 1970. Commercial development of intermediate moisture foods. *Food Technology* 24:53-57.
10. Chirife, J., Zamora, M.C., and Motto, A. 2006. The correlation between water activity and % moisture in honey: Fundamental aspects and applications to Argentine honeys. *Journal of Food Engineering* 72:287-292.
11. Schroeder, A., Horn, H., and Pieper, H.J. 2005. The correlation between moisture content and water activity (aw) in honey. *Deutsche-Lebensmittel-Rundschau* 101(4):139-142.
12. Scott, V.N., and Bernard, D.T. 1983. Influence of temperature on the measurement of water activity of food and salt systems. *Journal of Food Science* 48:552-554.
13. Troller, J.A. 1983. Water activity measurements with a capacitance manometer. *Journal of Food Science* 48:739-741.
14. Vos, P.T., and Labuza, T.P. 1974. Technique for measurements of water activity in the high a_w range. *Journal of Agricultural and Food Chemistry* 22:326-327.
15. Chirife, J., and Fontan, C.F. 1982. Water activity of fresh foods. *Journal of Food Science* 47:661-663.
16. Gomez, R., and Fernandez-Salguero, J. 1992. Water activity and chemical composition of some food emulsions. *Food Chemistry* 45:91-93.
17. Kidambi, R.N. Wiebe, H.H., Ernstrom, C.A., and Richardson, G.H. 1979. An economical dewpoint instrument for rapid measurement of water activity in foods. *Journal of Dairy Science* 62(Suppl. 1):40-41.
18. Brisske, L.K., Lee, S-Y, Klein, B.P., and Cadwallader, K.R. 2004. Development of a prototype high-energy, nutrient-dense food product for emergency relief. *Journal of Food Science* 69(9):S361-367.
19. Multon, J.L., Savet, B., and Bizot, H. 1980. A fast method for measuring the activity of water in foods. *Lebensmittel-Wissenschaft-und-Technologie* 13(5):271-273.
20. Stoloff, L. 1978. Calibration of water activity measuring instruments and devices: Collaborative study. *Journal of Association of Official Analytical Chemists* 61(5):1166-1178.
21. Fernandez-Salguero, J., Gomez, R., and Carmona, M.A. 1993. Water activity in selected high-moisture foods. *Journal of Food Composition and Analysis* 6:364-369.
22. Clifford, M.N. 1986. Coffee technology outlook—Physical properties of the coffee bean. *Tea and Coffee Trade Journal* April:14-16.

F Water Activity Values of Select Consumer and Pharmaceutical Products

Anthony J. Fontana, Jr., and Shelly J. Schmidt

This appendix contains water activity values or ranges from various sources for select consumer and pharmaceutical products. The individual ingredients and products (called items) listed in the table have been organized alphabetically into main categories and then into alphabetized subcategories. All tests were done at 25°C.

MAIN CATEGORY, **Subcategory**, Item	a_w Value or Range	Source
CONSUMER PRODUCTS		
Hair Products		
Hair gel	0.982	1
Shampoo	0.982–0.987	1
Lotions		
Body cream	0.972–0.983	1
Deodorant gel bar	0.984	1
Lip balm, topical/oral	0.360	3
Sun blocker	0.940–0.981	1
Soap		
Soap, creamed	0.567	1
Soap, regular	0.740–0.757	1
Soap, with glycerin	0.659–0.759	1
Soap, with glycerin and lanolin	0.856	1
Toothpaste		
Toothpaste	0.585–0.984	1
PHARMACEUTICALS		
Analgesic	0.401	1
Analgesic (gelatin capsules) liquid	0.530	1
Analgesic (gelatin capsules) gelatin	0.533	1
Anti-allergic	0.443	1
Antibiotic pills (cefacilin)	0.441	1
Anti-migraine, pills	0.386	1
Anti-inflammatory cream	0.852	1
Anti-inflammatory ointment	0.975	1
Anti-inflammatory Suspension	0.870	2
Antimicotic cream	0.950	1
Antimicotic powder	0.537	1
Aspirin	0.440	1
Citrobioflavonoide and vitamin C syrup	0.801	1
Cough drop, liquid center	0.400	3
Cough suppressant	0.890	3
Cough syrup	0.912–0.965	1
Decongestant/antihistamine (liquid-filled capsule)	0.450	3
Epileptic syrup	0.835	1
Lactulose syrup (laxative)	0.823	1

(continues)

MAIN CATEGORY, Subcategory, Item	a_w Value or Range	Source
PHARMACEUTICALS (*continued*)		
Laxative	0.927	1
Menstrual pain, pills	0.459	1
Mucolitic elixir	0.904	1
Neurotonic syrup	0.935	1
Potassium gluconate, elixir	0.926	1
Rectal suppositories	0.290	3
Tonic syrup	0.950	1
Vaginal Suppositories	0.300	3
Vitamin C tablets	0.330	1
Vitamin, multivitamin tablet	0.300	3
Ointments/Creams		
Bactericidal cream	0.841	1
Canker sore gel, oral	0.860	3
Cicatrizant cream	0.978	1
Cream for dermatitus	0.951–0.952	1
Gel anti-inflammatory (topical use)	0.942	1
Rectal cream	0.970	3
Rectal ointment	0.260	3
Vaginal antimicotic cream	0.982	1

References

1. Decagon Devices in-house testing, Pullman, Washington. a_w values were obtained using Decagon Aqualab Water Activity Meters series 3TE (chilled mirror).
2. Labuza, T.P. 1993. Water activity: Theory, management, and applications. AACC Water Activity Course. February 16-19, 1993, St Paul, MN.
3. Friedel, R.R., and Cundell, A.M. 1998. The application of water activity measurement to the microbiological attributes testing of nonsterile over-the-counter drug products. *Pharmacopeial Forum* 24:6087-6090.

Index